Neuroaesthetics

Exploring Beauty Within and Around Us

Exploring Complexity

ISSN: 2382-5901

For four centuries our sciences have progressed by looking at its objects of study in a reductionist manner. In contrast complexity science, that has been evolving during the last 30–40 years, seeks to look at its objects of study from the bottom up, seeing them as systems of interacting elements that form, change, and evolve over time. Complexity therefore is not so much a subject of research as a way of looking at systems. It is inherently interdisciplinary, meaning that it gets its problems from the real non-disciplinary world and its energy and ideas from all fields of science, at the same time affecting each of these fields.

The purpose of this series on complexity science is to provide insights in the development of the science and its applications, the contexts within which it evolved and evolves, the main players in the field and the influence it has on other sciences.

Volume 12: *A Crude Look at the Whole*
A Decade Later: Insights into Real-World Complexity that Still Resonate
edited by Jan Wouter Vasbinder & Helena Hong Gao

Volume 11: *Neuroaesthetics: Exploring Beauty Within and Around Us*
edited by Yan Bao, Richu Wang, Ernst Pöppel & Semir Zeki

Volume 10: *Balanced Sustainability in a Changing World*
edited by Ernst Pöppel & Maria Reinisch

Volume 9: *Fit for Purpose?: The Futures of Universities*
edited by Jan Wouter Vasbinder & Jonathan Y. H. Sim

Volume 8: *Buying Time for Climate Action: Exploring Ways around Stumbling Blocks*
edited by Jan Wouter Vasbinder & Jonathan Y. H. Sim

For the complete list of volumes in this series, please visit www.worldscientific.com/series/ec

Exploring Complexity – Volume 11

Neuroaesthetics

Exploring Beauty Within and Around Us

Editors

Yan Bao
Peking University, China

Ernst Pöppel
Ludwig Maximilian University Munich, Germany

Richu Wang
Chinese Academy of Sciences, China

Semir Zeki
University College London, UK

World Scientific

NEW JERSEY • LONDON • SINGAPORE • BEIJING • SHANGHAI • HONG KONG • TAIPEI • CHENNAI • TOKYO

Published by

World Scientific Publishing Co. Pte. Ltd.
5 Toh Tuck Link, Singapore 596224
USA office: 27 Warren Street, Suite 401-402, Hackensack, NJ 07601
UK office: 57 Shelton Street, Covent Garden, London WC2H 9HE

Library of Congress Control Number: 2024947869

British Library Cataloguing-in-Publication Data
A catalogue record for this book is available from the British Library.

Exploring Complexity — Vol. 11
NEUROAESTHETICS
Exploring Beauty Within and Around Us

ISBN 978-981-12-9862-2 (hardcover)
ISBN 978-981-12-9863-9 (ebook for institutions)
ISBN 978-981-12-9864-6 (ebook for individuals)

For any available supplementary material, please visit
https://www.worldscientific.com/worldscibooks/10.1142/13998#t=suppl

Desk Editor: Muhammad Ihsan Putra

Typeset by Stallion Press
Email: enquiries@stallionpress.com

https://doi.org/10.1142/9789811298639_fmatter

Foreword

In 1979, one of the editors of this book (Ernst Pöppel) got a phone call from a representative of a foundation in Germany (the Werner Reimers Stiftung in Bad Homburg) with a surprising question. He was informed that the foundation still has money for an interdisciplinary project but that they need a suggestion immediately, better the same day, because of an upcoming business meeting. This unexpected request did not allow long meditation, or an extended search in the literature, or a discussion with others. The answer had to be given within an hour. Without too much reasoning, a risk was taken. The suggestion was to initiate a study group on "the biological aspects of aesthetics", or with even more ambition, "the biological basis of the arts". The proposal was surprisingly accepted, and then for several years, a selected group of scientists and artists came regularly together to discuss topics in the visual arts, music, or poetry, also related to theoretical issues.

Results of this study group were published in 1988 in the book *Beauty and the Brain. Biological Aspects of Aesthetics* by the editorial team Ingo Rentschler (a physicist and medical psychologist), Barbara Herzberger (a historian and medical doctor), and David Epstein (a musicologist, composer, and conductor); this editorial team reflects already interdisciplinarity of "neuroaesthetics", a term introduced at that time. During the work of the study group and motivated by the discussions, some experimental results on the visual arts (Stoerig *et al.*, 1983) or on poetry (Turner and Pöppel, 1983) were reported. Other publications followed like on music (Epstein, 1995; Pöppel, 1989), and then the seminal book on *Inner Vision* by Semir Zeki (1999) appeared. Neuroaesthetics as a scientific field has blossomed since then, and many laboratories worldwide dedicate their research to "art and science".

It has to be noticed, however, that there is already a long history linking the arts and the sciences. The founder of "psychophysics" Gustav Theodor Fechner published in 1876 *Vorschule der Aesthetik* (*Pre-School of Aesthetics*) in which a more psychological approach was taken. Strangely enough, in 1806, a book worthwhile to read was published by the German romantic writer Jean Paul with the same title *Vorschule der Aesthetik*; using the same title by Fechner was hardly a coincidence. But this is still not the beginning to relate conceptually art and science. The beginning of modern times can be seen in the book *Aesthetica* by Alexander Gottlieb Baumgarten in 1750. Here aesthetics is seen both as a science and an art. In fact, Immanuel Kant in his *Critique of Judgment* (1790) refers although critically to Baumgarten; Kant gives with his concept of "expediency" (in German "Zweckmäßigkeit") a new frame for experiencing pleasure in art and also in nature. But the basis of art and science goes far more back in history and has shaped our reasoning until today. In several dialogues of Plato like in Symposion, Sophistes, Philebos, or Phaidros, we find reference to our aesthetic sense as it is reflected in the experience of beauty. In Symposion, we read in free translation: "If there is something for which it is worth to live it is looking at beauty".

With this rich background and with the new experimental possibilities, it was decided to start a series of contributions on the general topic "neuroaesthetics" in the *PsyCh Journal.* The editors of this series Yan Bao, Ernst Pöppel, and Semir Zeki together with the managing editor Richu Wang collected contributions for the journal in the years 2020, 2021, and 2022. The publications in these years reflect the scientific paradigms of horizontal and vertical interdisciplinarity. With the term "horizontal," neuroaesthetics in visual art, music, or poetry is indicated; with the term "vertical," underlying neural processes in the appreciation of art being music, poetry, or visual art are considered. This complementarity of horizontal and vertical is a typical feature of neuroaesthetics.

From the contributions to the *PsyCh Journal*, we selected some 30 publications (adding three new manuscripts) which are included in this book. These publications are grouped into four sections: In the first section, "General Ideas" are presented on the concept of interdisciplinarity, on different knowledge systems, on the relationship of art and science, on the link of aesthetics to morality, or on the concept of mathematical beauty. In the second section, "Art in Space", for instance evolutionary aspects are outlined, the importance of the Golden Ratio is emphasized, and observations on abstract art, eye movements looking at artwork, or the use of color is reported; on a theoretical level, "predictive coding" in neural systems is considered. Of great importance and in fact fascination are "Faces in Art"; they are brought together in the third section. The neural basis of face recognition is discussed, the memorability of faces is analyzed, or the recognition of original faces in artwork is described. Finally, in the fourth section, we are dealing with "Art in Time". The relevance of a pre-semantic time window of some three seconds in poetry or in films are described. In the last contribution of this book, the neural activity of an eminent composer is reported while he was composing a new piece of music lying in the brain scanner. The editors believe that with this collection of recent publications, a useful overview is provided about research activities on an international level in neuroaesthetics.

References

Baumgarten, A. G. (1750). *Aesthetica. Traiecti cis Viadrum (Frankfurt an der Oder)*. Johannis Christian Kleyb.

Epstein, D. (1995). *Shaping Time: Music, the Brain, and Performance*. New York, NY: Simon & Schuster (Macmillan).

Fechner, G. T. (1876). *Vorschule der Aesthetik (Pre-School of Aesthetics)*. Leipzig, Germany: Breitkopf & Härtel.

Paul, J. (1804). *Vorschule der Aesthetik (Pre-School of Aesthetics)*. Hamburg, Germany: Meiner.

Kant, I. (1790). *Critik der Urtheilskraft (Critique of Judgment)*. Leipzig, Germany: Reclam.

Pöppel, E. (1989). The measurement of music and the cerebral clock: A new theory. *Leonardo*, 22, 83–89.

Rentschler, I., Herzberger, B. & Epstein D. (eds.) (1988). *Beauty and the Brain. Biological Aspects of Aesthetics*. Basel, Switzerland: Birkhäuser Verlag.

Stoerig, P., Sütterlin, C. & Pöppel. E. (1983). Rechts und links in Bildwerken: Ein neuropsychologischer Beitrag zum Kunstverständnis (Right and left in artwork: A neuropsychological contribution for the appreciation of art). *Umschau in Wissenschaft und Technik*, 83, 427–428.

Turner, F. & Pöppel, E. (1983). The neural lyre: Poetic meter, the brain and time. *Poetry*, August, 277–309.

Zeki, S. (1999). *Inner Vision — An Exploration of Art and Brain*. Oxford: Oxford University Press.

https://doi.org/10.1142/9789811298639_fmatter

Contents

Part I
General Ideas

https://doi.org/10.1142/9789811298639_0001

Chapter 1

Neuroaesthetics: The Art, Science, and Brain Triptychon*

Semir Zeki[†], Yan Bao[‡,§,¶] and Ernst Pöppel[§,¶]

[†]*University College London, London, UK*
[‡]*School of Psychological and Cognitive Sciences, and Beijing Key Laboratory of Behavior and Mental Health, Peking University, Beijing, China*
[§]*Institute of Medical Psychology, Ludwig Maximilian University Munich, Munich, Germany*
[¶]*Parmenides Centre for Art and Science, Pöcking, Germany*

Abstract

In this and subsequent issues in the *PsyCh Journal*, we are publishing a set of papers that can be grouped together broadly under the title of "neuroaesthetics". This is a relatively new field, or rather a subfield, which aims to understand the brain mechanisms that are engaged during aesthetic and allied experiences in the widest sense. That, of course, encompasses a vast area: It amounts to trying to understand what neural mechanisms underpin inherited aesthetic predispositions compared to

---

*This chapter was originally published in the special collection on neuroaesthetics in *PsyCh Journal.* Zeki, S., Bao, Y., & Pöppel, E. (2020). Neuroaesthetics: The art, science, and brain triptych. *PsyCh Journal*, 9(4), 427–428. https://doi.org/10.1002/pchj.383.

acquired ones, the extent to which aesthetic judgments are subjective, and the relationship of aesthetic and perceptual judgments to the experience of pleasure, desire, love, and hate, among other topics, with all the complexities that such sentiments involve.

For reasons that are hard to comprehend intellectually, though relatively easy to understand emotionally, there are some who dislike and resent what they see as the intrusion of science into the private world of subjective experiences. This attitude carries with it a widely shared belief among scientists and the public alike, namely, that aesthetic experiences are necessarily subjective and therefore impervious to scientific enquiry — a belief that is not necessarily shared by philosophers like Immanuel Kant and Arthur Schopenhauer nor by those pursuing neuroaesthetics as a discipline; it is in any case an assumption that requires rigorous reappraisal and reexamination. Those who fear or resent this new discipline accuse it and its practitioners of being "reductionists"; they do so without, at the same time, acknowledging that much in the world of art and literature is also reductionist and that there is indeed a splendor and beauty in trying to account for a multiplicity of complex phenomena — in the universe, in our natural world, or in human behavior — in simple terms that explain them all. This is itself something worth studying in terms of the brain's capacity to see connections between disparate fields and approaches and unite them under a single concept. Physicists and mathematicians understand this much better. Interestingly, they never draw the same ire and criticism — of reductionism — when they seek theories of everything; theories that unite, for example, the electromagnetism of James Clerk Maxwell and Albert Einstein's theories of relativity, as the mathematician Hermann Weyl did, or indeed theories that try to account for, under one umbrella, the apparently disparate worlds of quantum mechanics and particle physics, on the one hand, and of the gravitational physics of Isaac Newton and Einstein, on the other, which string theory has been trying to do for the past 30 years. Perhaps the mistake of neuroaesthetics lies in it not having been bold and audacious enough to emulate other disciplines, by coining a term like estheticome to take this process of "reductionism" to the same acceptable levels as the terms genome, cognitome, and connectome have done for their respective subjects (gawk here).

The detractors of neuroaesthetics appear to be fearful of it as well because they imagine neuroaesthetics to be an exercise in "demystification", yet these same critics are strangely silent *vis-à-vis* physicists and astronomers who, for example, try to chart the limits of the universe or try to pinpoint with astonishing precision the moment of the birth of the universe from an infinitesimally small particle of infinite mass (the Big Bang), a scientific account that is almost Biblical in its grandeur but which, unlike the mystery of Biblical stories, carries increasingly the weight of scientific authority. Yet this demystification in physics is a source of wonder even greater than the one that prevailed when the Bible was the sole authority for the origins of the universe. One cannot help but conclude that it is not demystification and reductionism *per se* that are the root causes of these strange apprehensions but specifically what some see as reductionism and demystification that aim better to understand the neural complexities that underlie those atavistic impulses that determine so much that is critical and universal in human behavior. It is perhaps not surprising, therefore, that those critics come mainly from the humanities and seem to fear that neuroaesthetics may trespass into their fields, which is not to say that all in the humanities display the same fear and hostility, far from it. Indeed, some of the most ardent supporters of the discipline of neuroaesthetics come from the humanities and especially prominent among them are artists.

This should not surprise anyone; artists, painters, sculptors, philosophers, writers, and musicians, among others, have all in their way tried their hand at understanding something about not only the mysteries of perception but also the very same experiences of love, beauty, and desire, of pleasure and reward, and of the complexity of both emotional and cognitive factors that guide human behavior that neuroaesthetics addresses. Perhaps more has been written in the humanities about what knowledge can be derived about the human psyche from studies of the character of Shakespeare's *Hamlet*, Tolstoy's *Anna Karenina*, or Flaubert's *Madame Bovary* than from any neurobiological study. Philosophers and humanists have read much into human behavior, the human psyche, and human destiny in Euripidean drama, in the tragedies of Sophocles, or in the spiritual and philosophical discourse between Arjuna and Lord Krishna in the Bhagavad Gita and the teachings of the Dao De Jing. Edward Gibbon's *History of The Decline and Fall of the Roman Empire* may in fact be summarized by saying that it is a voluminous study that leads to a portrait of power, the many and devious ways in which it is sought, and its corrupting influence (sorry for the reductionism here) while Richard Wagner's music is an ineffable incursion into human emotional behavior. These examples of artistic and literary efforts, among many others, to explain and account for human behavior constitute good proof that, for the humanities too, there is something that needs to be explained and accounted for. If so, why exclude neuroscience from this effort, especially since neuroscience is trying to do so by learning more about the source of much, and perhaps all, that dictates human conduct by different techniques, ones which complement those so effectively used by the humanities over the centuries?

Long before neuroaesthetics was formally born as a discipline, the humanities and the world of art were exploring the same problems, though with means and techniques that are different from those of neuroaesthetics; it is because of this long tradition in the humanities that the modern discipline of neuroaesthetics relies so heavily on the writings of philosophers and artists from the time of Plato onwards. And it is for this reason that the reader will find among our pages, in this and in future issues, articles written by artists themselves as well as by philosophers, art historians, and literary experts, in addition to neurobiologists and psychologists. Their contribution is indispensable for our field and therefore the more welcome for it. It nevertheless needs to be emphasized, not only to allay the fears of those who are resentful of neuroaesthetics, that this is a field constituted mainly to understand the brain and brain mechanisms that are engaged during aesthetic and allied experiences. It does not try to trespass into territories that belong more properly to art history and to philosophies of aesthetics, nor does it try to address the question of what beauty is. It is for this reason that the reader will find that a significant number of the articles published here are experimental and belong to the field of neuroaesthetics, but the reader will also find contributions from artists, art historians, and philosophers who have contributed so well to our field and who continue to do so.

In a sense, it could be said that there is, currently, an asymmetry in the relationship between neuroaesthetics and the fields upon which it draws for inspiration and instruction, in that neuroaesthetics has much to learn from them but little to teach them in return. We do not pretend that the knowledge gained from studies in the field of neuroaesthetics will improve in the slightest the work of artists. But in terms of philosophies of aesthetics, the position is perhaps not quite so asymmetric. It would be surprising if the current contributions from neuroaesthetics failed to ultimately influence significantly philosophies of aesthetics. But whether it influences other fields or not, the endeavor to understand, in neural terms, experiences that are so central to the human condition is a rewarding and noble one, and as long as scientists continue to have an interest in such issues, that endeavor will continue. And we, in turn, will continue to contribute to that endeavor by publishing the results of their scientific experiments, artistic descriptions, and theoretical formulations in our pages.

https://doi.org/10.1142/9789811298639_0002

Chapter 2

Interdisciplinarity vs. Intradisciplinarity: A Failure in Neurobiology*

Semir Zeki[†], Yan Bao[‡,§,¶] and Ernst Pöppel[§,¶]

[†]*University College London, London, UK*
[‡]*School of Psychological and Cognitive Sciences, and Beijing Key Laboratory of Behavior and Mental Health, Peking University, Beijing, China*
[§]*Institute of Medical Psychology, Ludwig Maximilian University Munich, Munich, Germany*
[¶]*Parmenides Centre for Art and Science, Pöcking, Germany*

The loud beating of gongs, proclaiming interdisciplinarity and risk-taking as hallmarks of a new attitude to research, one that will reap rich rewards in extravagant cash for further work, would have sounded like a re-invention of the wheel — and a poorly constructed wheel at that — to

---

*This chapter was originally published in the special collection on neuroaesthetics in *PsyCh Journal.* Zeki, S., Bao, Y., & Pöppel, E. (2022). Interdisciplinarity versus intradisciplinarity: A failure in neurobiology. *PsyCh Journal*, 11(5), 617–618. https://doi.org/10.1002/pchj.607.

Polykleitos, Phidias, Michelangelo, and Leonardo, among many others. It would probably have sounded strange even to Einstein (1936), who once wrote that "the whole of science is nothing more than a refinement of everyday thinking. It is for this reason that the critical thinking of the physicist cannot possibly be restricted to the examination of the concepts of his own specific field. He cannot proceed without considering critically a much more difficult problem, the problem of analysing the nature of everyday thinking." This can be taken as a clear invitation to the physicist to enquire more deeply into the knowledge-acquiring system of the brain, through which all knowledge, including that in physics, is acquired. One can look at this re-invented wheel and consider much of the hype behind it as nothing more than guff, which indeed it largely is. Even so, it is nevertheless more than welcome and perhaps will mature eventually into what it promises to be but has so far rarely been.

But neurobiologists should perhaps go beyond interdisciplinarity and look more closely into intradisciplinarity and specifically within the domain of neuroscience itself. This is especially true of neuroaesthetics, which is perhaps the most interdisciplinary branch of neuroscience and one of the most interdisciplinary subjects currently available (Bao *et al.*, 2017; Rentschler *et al.*, 1988). This is not surprising: It addresses a neurobiological problem, of what brain mechanisms are engaged in aesthetic, and allied, experiences such as those of love, desire, sorrow, and moral states. Crucially, it holds that there is no brain mechanism that is devoted uniquely to art, or to beauty, or indeed to knowledge that is derived from them. Rather, it supposes that one of the primordial functions of the brain is the acquisition of knowledge and that knowledge can be acquired through both the cognitive and the emotional systems of the brain. It also believes that the brain — through its systems — can acquire knowledge through the arts, humanities, sciences, and mathematics (Tsukiura & Cabezza, 2011; Zeki, 1999, 2009, 2014). It is for this reason that neuroaesthetics has been among the most transgressive and radical subjects in neurobiology, relying for inspiration and for framing its question upon philosophy, philosophies of aesthetics, art criticism, mathematics, and physics, among other fields. Of course, it also believes what is demonstrably true, that beauty (for example) has many different sources and that these different sources activate different brain systems — music, for example, activating the auditory system, and visual art activating the visual system. But regardless of its source, the experience of beauty appears to correlate with activity in field A1 of the medial orbitofrontal cortex (mOFC) (Ishizu & Zeki, 2011) of the brain; the intensity of the activity there correlates with the intensity of the declared experience of beauty — the more intense the experience, the more intense the activity. That the activity that correlates with beauty can be localized and quantified brings the study of apparently such abstract experiences as beauty or love firmly within the realm of science, no doubt to the dismay of those who dread "de-mystifying" beauty and love but are not nearly as concerned when science tries, through physics, to de-mystify the universe.

Beauty, love, and desire, in the broadest sense, are pleasurable and rewarding experiences. They are also cardinal experiences, which we all seek in life almost continuously. It is this that led Sigmund Freud (1930) to his belief that reward, through satisfaction of the pleasure principle (Lustprinzip), is a main driving force in the conduct of human affairs. In German, the word lust can mean both pleasure and desire, and it is interesting therefore to note that the experience of both correlates with activity in field A1 of the mOFC. Naturally, experiencing something that is pleasurable and rewarding is also intimately linked to decision-making, in favor of the object, or the person, or the situation that is experienced as pleasurable and rewarding. Hence, decision-making is also a critical question for neuroaesthetics or should be so.

Looked at in this way, it is astonishing that the volumes of papers that have been written on decision-making and on the experience of reward and pleasure since 2010 (when some of the major papers on neuroaesthetics had been published many years previously) have almost nothing to say about the importance of aesthetic and romantic experiences as instruments of reward and pleasure and remain completely silent about neuroaesthetics; this is especially noticeable in papers that address the question of "subjective value" without any reference to the output of neuroaesthetics, as if experiences of beauty or of love or of moral values are entirely irrelevant in this domain. It would be unfair, and impractical, to cite a few papers for an error that is made by many. Suffice it to say that ignoring work on neuroaesthetics that has implicated field A1 in aesthetic experiences derived from many sources characterizes the vast majority of papers that have found the same area, field A1 of the mOFC, active in decision-making. Those working on neuroaesthetics fare a little, but not much, better: While they refer to the work done on reward and pleasure and sometimes even decision-making, they are for the most part silent about the relationship of aesthetic experiences to reward and pleasure. Both groups fail to address in any significant way the important neurobiological issue of the relationship of activity in the mOFC that correlates with aesthetic and allied experiences, on the one hand, and with decision-making and reward, on the other. It is presumably not entirely by accident that the mOFC is involved in all of these activities but a neural reflection of the fact that they are all interlinked.

We concentrate in this editorial on the mOFC because activity in it has been found, on the one hand, to correlate with the experience of beauty and desire derived from many different sources, including mathematical beauty, and, on the other, with the experience of reward that leads to decision-making. Given the silence of those working in one of these areas about the work produced by those working in the other, anyone could reasonably assume that these are entirely separate topics, when in fact they are intimately linked.

It is therefore important to emphasize the importance of intradisciplinarity in future research, as opposed to that of interdisciplinarity. The latter has been, and continues to be, broadcast loudly. The former has not had a voice, though it is not less important and deserves one. We give it a voice here and hope to continue doing so in our pages.

References

Bao, Y., von Stosch, A., Park, M., & Pöppel, E. (2017). Complementarity as generative principle: A thought pattern for aesthetic appreciations and cognitive appraisals in general. *Frontiers in Psychology*, 8, 727. https://doi.org/10.3389/fpsyg.2017.00727.

Einstein, A. (1936). Physics and reality. *Journal of the Franklin Institute*, 221, 349–382.

Freud, S. (1930). *Das Unbehagen in der Kultur* (*Civilization and Discontents*). Internationaler Psychoanalytischer Verlag.

Ishizu, T. & Zeki, S. (2011). Toward a brain-based theory of beauty. *PLoS One*, 6, e21852. https://doi.org/10.1371/journal.pone.0021852.

Rentschler, I., Herzberger, B., & Epstein, D. (eds.) (1988). *Beauty and the Brain: Biological Aspects of Aesthetics*. Basel: Birkhäuser Verlag.

Tsukiura, T. & Cabezza, R. (2011). Shared brain activity for aesthetic and moral judgments: Implications for the Beauty-is-Good stereotype. *Social Cognitive and Affective Neuroscience*, 6, 138–148. https://doi.org/10.1093/scan/nsq025.

Zeki, S. (1999). *Inner Vision: An Exploration of Art and the Brain.* Oxford University Press.

Zeki, S. (2009). *Splendors and Miseries of the Brain*. Wiley-Blackwell.

Zeki, S. (2014). The experience of mathematical beauty and its neural correlates. *Frontiers in Human Neuroscience*, 8, 68. https://doi.org/10.3389/fnhum.2014.00068.

https://doi.org/10.1142/9789811298639_0003

Chapter 3

Bridge or Chiasm between Art and Science?

Alexandra von Stosch[*] and Ernst Pöppel[†]

[*]*Barenboim Said Akademie, Berlin, Germany*
[†]*Institute of Medical Psychology, Ludwig Maximilian University Munich, Germany*

Abstract

Art and science have a common root. In spite of occasional misunderstandings between the sciences and the humanities representing different academic disciplines, a bridge between art and science can be built to allow broader perspectives in understanding and foster new creativities both for artists and scientists. We provide examples from the visual arts, music, and poetry. In particular, the spatial configuration in pictures or the temporal structure in music and poetry are addressed. They directly represent a conceptual link between artistic appreciation and mechanisms of information processing in the brain.

Keywords: interdisciplinarity; episodic memory; complementarity; poetry; music; visual art

Is it possible to better understand what happens in the arts when scientific knowledge is used? Can natural scientists learn from the arts for their research? Is a creative exchange in both directions possible? The authors, the art scientist and the neuroscientist, have been in a dialogue on this question for a long time. We are convinced that interactions in both directions are not only valid but even necessary. Impulses for a broader or at least different understanding of the arts can be derived from the natural sciences (Pöppel *et al.*, 2013). The idea of an expanded or shifted perspective may cause resistance among experts in the humanities who may have some distance to natural science. However, the natural scientist as a layperson should also feel free to look at works of art and to do so within his or her conceptual framework, even if he or she comes to unexpected conclusions which may be strange in the eyes of experts coming from the humanities or the arts themselves.

Here is an example to test: The well-known drip paintings of Jackson Pollock result at first sight in visual insecurity or even confusion. In spite of this initial reaction, one should feel invited to associate in episodic memory the interconnection of nerve cells in the brain triggered by the randomly connected drips in the painting; this interconnection cannot be visualized in such a dynamic way with a pictorial representation of nerve cells as is typically used in textbooks of neuroanatomy. Nerve cells in the brain are indeed dynamically interconnected and new functional connections are constantly activated implementing the neural basis of mental processes. Jackson Pollock certainly did not create such drip paintings with the intention that they would be used decades later to symbolize neural processes in the brain. However, what one sees nowadays gives special insight into such dynamic processes; this is much richer than the usual scientific illustration. The inner image in episodic memory matches the outer image in front of the eyes. If this complementarity is initiated, then satisfaction based on coherence arises; complementarity becomes an indicator of aesthetic appreciation (Bao *et al.*, 2017; Zhao *et al.*, 2022). However, this match may also fail, at least on the surface.

With the beginning of modernism, we experience a paradoxical situation: The more complex works of art become, the more they transcend traditional boundaries. But paradoxically, their reception becomes more hermetic by self-appointed interpreters who claim sovereignty of interpretation. The answer of the artist Joseph Beuys "everyone is an artist" is of little use for the reception of art. It would make more sense to say "everyone is a perceiver". Given this situation, the following question arises: Who — or what — determines what the correct or appropriate (or even "true") reception of art is? How do mainstreams emerge? How does a certain trajectory of artistic expression or also a topic in scientific research suddenly become dominant? Why is the interplay between art and science now of particular interest and even fascination, although art and science have always been linked, if we think of Leonardo da Vinci, the invention of oil paintings or photography? Every new technological development has also brought forth new art. It seems that nowadays communicative processes in interdisciplinary networks are triggered that result in high visibility such as the Ars Electronica Center in Linz (Austria) or the scientifically inspired projects by the artist Olafur Eliasson with his Berlin laboratory for spatial experiments.

If we follow the cultural philosopher Ernst Cassirer with his fundamental work "the philosophy of symbolic forms" (1923/1929), art and science have a common root. With this perspective (which we adopt), one can indeed conclude that artists and scientists can learn from each other and also stimulate each other. Scientists explore general principles of experiences or in more general terms potential knowledge of the world within and around us, but this exploration is also true for artistic works, usually on an implicit level. Art provides a rich source of knowledge for gaining insight into what makes us human. Art and science can even be considered "schools of perception". The work of an artist may be expanded through insights from research; scientific research like in the neurosciences and psychology may create a different perspective about the work of art.

Coming back to the question of reception of art, how are works of art evaluated such that some, and not others, are considered as being valuable and significant? Picasso once apparently said that museums only exhibit unsuccessful paintings that represented a break or disruption at the time of their creation. Unpredicted events or marketing strategies certainly play an important role, but in particular the spirit of time, the "Zeitgeist", is essential to create a new perception and appreciation. But what is the spirit of time that dominates communication, and where does the “Zeitgeist” come from? We have to leave the answer open. The communicative mechanism is expressed nicely in a scene of the drama *Faust* by the German poet Johann Wolfgang von Goethe, in which the student Wagner asks how to give a talk and Professor Faust indicates how important the “Zeitgeist” is — but also not giving an answer:

> *And what the spirit of the times men call*
> *Is merely their own spirit after all,*
> *Wherein, distorted oft, the times are glass'd.*
> *(Was ihr den Geist der Zeiten heißt,*
> *Das ist im Grund der Herren eigner Geist,*
> *In dem die Zeiten sich bespiegeln.)*

Given this cultural frame, how can a natural scientist or a layperson in general find a way to the arts anyway? We suggest a detour on this way to start with the elementary basic needs that characterize all living beings. The central question is as follows: What drives us? Every organism, whether human or even unicellular, is concerned with maintaining the inner environment, the homeostatic balance. Every organism asks itself (whether explicitly or implicitly) the following question: “What is good for me, what is bad for me?” Although the biological basis for decision-making processes is different, it is always about making “decisions”. What are the needs whose satisfaction keeps us in balance? Hunger and thirst are at the beginning. It is the search for warmth, security, and shelter. The basic needs also include self-assertion, which — sometimes in active isolation — serves to secure our own identity (Pöppel, 2010). Avoiding pain and longing for union are further drives. We are driven by the desire for duality, for direct communication with others. We are characterized by concern for others and our own future. Activity is another immediate need; the purely sessile way of life is unnatural; the body demands movement. We order our immediate surroundings, we rule over others, and we seek to control time. We hide behind the fear of death; curiosity drives us to decipher the unknown.

All these basic needs do not stand alone. They characterize life and experience in a complementary way, and although they can be thought of separately, they cannot be realized separately. The basic needs also address the central themes of the arts. In which medium are these themes created, and what form do they take? This is our argument: From a biological point of view, the arts are derived from the direct experience of sensory perception. Seeing, hearing, touching, feeling, smelling, tasting, and moving are the starting points for artistic creation. It is striking that all ethnic groups (from the Stone Age cultures to the so-called advanced civilizations) developed similar art forms, and this also applies to cultural groups that were never in contact with each other (Sütterlin and Yu, 2021). These similarities speak for the universality of artistic representation and its rootedness in fundamental mechanisms of sensory and physical experience of the world.

According to the thesis, the direct drive to create art is explained by the effort to communicate to others what cannot be said in explicit language. In the artistic process, the artist experiences communication itself as a direct drive; writing a novel, writing poetry, painting, drawing, sculpting, filming, composing, making music, and dancing are, despite all self-absorption, activities related to others, even if the other is oneself, one's own doppelgänger. Such self-referentiality reveals the complementary relationship between implicit and explicit knowledge, which is fundamental to the definition of personal identity (Bao and Pöppel, 2012; Pöppel, 2010).

A few observations on specific arts: The production of a picture or a drawing is only possible from the primary experience of seeing — a trivial statement but one with consequences. The painter and the draughtsman create a picture on the basis of elementary visual experiences for which the brain naturally provides specific processing mechanisms. This is where recourse to pictorial or episodic memory comes into play: Images that are important to us are stored forever. Lines, surfaces, colors, contrasts, and topographical classifications are the basic building blocks of depicted objects. These elements are emphasized in different ways, and in some modernist art movements, they even take on independent significance, as in minimal art in the 1960s. The building blocks themselves can be elevated to the subject of design, such as The Black Square by Kazimir Malevich, an icon of modern painting. Expressing oneself only through contrasts of topographically arranged surfaces, through contours, or through colors appeals to original sensory experiences for which the brain has developed its own mechanisms.

Listening and speaking as performances of the brain are the starting points for music and poetry. Again, fundamental mechanisms of the brain must be taken into account if one wants to understand the elementary phenomena of these temporal arts. A piece of music can only be understood if the right tempo has been found; musical motifs are experienced in a distorted way if the tempo is wrong. The control of tempo in music results less from traditions or even the piece of music itself but is subject to neuronal mechanisms, especially those that control movement sequences. It is by no means a trivial matter to maintain a tempo or to play half or quarter notes. Such tempo controls require specific programs of the brain (Pöppel, 1989).

As in painting, different components are particularly emphasized in music by preferably varying them. A basic phenomenon in music is the fact that musical motifs are often integrated into a time interval of just a few seconds (Pöppel, 1989, 2009). Experimental studies have shown that this time is limited to around three seconds. Apparently, musicians and composers have an implicit knowledge of neuronal principles: namely, that our brain provides a temporal working platform on which individual contents of consciousness are summarized as a unified form. In the tradition of classical music from Joseph Haydn to Gustav Mahler, we often find musical motifs that reflect this time interval. The well-known Tristan motif by Richard Wagner, which in fact lasts longer than three seconds, may at first sight speak against the temporal limitation of a musical motif to this time interval, but we suppose that it achieves its emotional effect precisely by combining musical elements in successive time windows and contrasting them in a complementarity way. The rubato, the modulation of the tempo, as we know it, for instance, from Frédéric Chopin, is also an example of playing with time: One modulates the tempo but still recognizes the basic pattern.

The neural processes are also fundamental to aesthetic evaluation: What is experienced as aesthetically satisfying cannot be discussed as independent of underlying brain mechanisms (Bao *et al.*, 2017). The brain provides a framework for artistic expression. This is by no means to claim that this must be a rigid framework or even a template. However, it is established that our aesthetic perception is anchored in biologically predetermined conditions (Rentschler *et al.*, 1988), whereby, for example, the basic temporal structure is recognizable but does not have to be slavishly adhered to.

Can the framework for aesthetic evaluations be changed, or are we at our own mercy? Can we move as far away as we like from the biological foundations of the aesthetic, or can we even develop completely new principles for it that no longer have anything to do with our original sensuality? New aesthetic forms can certainly be created, but the removal of biological principles would have a different effect: They would not reach the recipient in his immediate experience but would address a rational framework of evaluation. This is by no means to claim that we are completely rigid in our aesthetic criteria; new aspects will be included in the evaluation repertoire and can then become the basis of a new aesthetic. However, this takes time and can only be achieved through imprinting learning in a defined cultural context. Nevertheless, certain basic structures of neuronal information processing are so constant that they cannot be shaken off with regard to aesthetics (Bao *et al.*, 2022).

In poetry, we also find a direct link to the temporal segmentation of information processing in the brain. In the analysis of poems from different languages, it was found that the same time segmentation applies universally across all languages (Turner and Pöppel, 1988). With longer lines of verse, the speaker automatically makes a caesura, as is the case, for example, with hexameter. The reader is asked to read the three lines of poetry from Goethe's *Faust* quoted above, and he or she will note that each line of verse, spoken loudly at a normal pace, will take about three seconds.

The temporal structure of the poem thus confirms the power of a three-second time window in the temporal arts. The fact that we find a phenomenon in poetry that is derived from integration mechanisms of the brain speaks for the stability of the neural process. If poets always use the same segmentation process in their poems, regardless of the language, this means that different cultural traditions or different syntactic rules of a language do not affect a basic mechanism of the brain that is responsible for temporal segmentation. In addition to the temporal structure, expression naturally plays a prominent role in poetry, for example on stage.

Linguistic expression, combined with the corresponding movement, is determined by various mechanisms of the brain. Emotional expression in language is controlled by neural structures in the right hemisphere of the brain.

There are observations that indicate that the modulation of pitch is made possible by the right hemisphere of the brain. If one temporarily switches off the right hemisphere and asks the examined patient to sing, the temporal structure of singing is unchanged, but the song is only sung at one pitch. Numerous observations suggest that the right hemisphere of the brain is indeed dominant for the emotional evaluation of experiences (Pöppel, 1982). This asymmetry of information processing in our brain is interestingly reflected in the asymmetrical structure of many pictures in art (Stoerig *et al.*, 1983) This anatomical fact must be taken into account: Everything to the left of our line of vision moves to the right hemisphere of the brain, and everything to the right of the line of vision is represented in the left hemisphere. What lies to the left of the line of vision therefore has direct access to emotional evaluation. If one analyzes the structure of pictures that express an emotion, one will find that the focus of the picture is significantly more often on the left side. Pictures that are less emotionally moving tend to be more symmetrical. The preferred compositional form of placing the focal point of the picture on the left when a strong emotion is to be expressed suggests that the artist has an implicit knowledge of how the visual system processes the picture. Incidentally, this characteristic of the design has little to do with our preferred reading direction from left to right. An asymmetrical pictorial structure is also found in cultures with other reading directions.

However, a completely different way of creating a sense of closeness to the self in a painting than the asymmetrical composition can be seen in a large number of Chinese or Japanese landscape paintings: They are characterized by an idiosyncratic perspective composition. The multilayered perspective makes the viewer appear to float above the picture (Bao *et al.*, 2016; Yang *et al.*, 2019). Perhaps this phenomenon can be described with the image of a circle or an ellipse, which is created in the viewer by the structure of the picture. In imagination, a vertical circle (or an ellipse) is created that connects the superimposed elements of the picture; this imaginary circle emerges from the picture and the viewer becomes part of the circle. This participation in the imaginary circle (or ellipse) creates a special closeness to the self, which would not be possible without the multi-layered perspective.

As may have become clear from the examples of different art forms from different areas, the gulf between art and science is not so deep that it cannot be bridged. Above all, the natural scientist must point to the fact that the bridge is crossed in both directions but sometimes at different times. It may be the case that artists know something that researchers only become aware of much later. A historical example would be suprematism; the Black Square of Kasimir Malevich, which was exhibited for the first time in 1915, indicates this in the act of seeing only surfaces and edges are represented on the retina of the eye. Without knowledge of this iconic image, a Chinese researcher developed a theory of vision that uses only these elements as basic topological features when we see almost a century later (Chen, 2005). But there are also direct stimuli from the visual arts to carry out experiments that would not have been possible without the inspiration from art. For example, it was asked in a study whether the human brain reacts differently when the viewer sees either a naturalistic or a surrealistic picture (Silveira *et al.*, 2012). The fact that this is indeed the case shows that the physically impossible is effortlessly separated from the physically possible. Modern imaging techniques with which one can trace the implicit processing procedures of the brain provide information about principles of representation that go beyond the explicit knowledge that manifests itself in a linguistically expressed opinion. Another example is as follows: If we compare aesthetic and moral judgments, we may assume that they are independent of each other. However, if one looks into the brain using advanced imaging technology one finds that they are intertwined. A moral judgment always has an aesthetic component (Avram *et al.*, 2013). But this is nothing new either: The fact that the good and the beautiful (and also the true) are connected was already known thousands of years ago.

References

Avram, M., Gutyrchik, E., Bao, Y., Pöppel, E., Reiser, M., & Blautzik, J. (2013). Neurofunctional correlates of aesthetic and moral judgments: Equal but not the same. *Neuroscience Letters*, 534, 128–132. https://doi.org/10.1016/j.neulet.2012.11.053.

Bao, Y. & Pöppel, E. (2012). Anthropological universals and cultural specifics: Conceptual and methodological challenges in cultural neuroscience. *Neuroscience and Biobehavioral Reviews*, 36(2012), 2143–2146. https://doi.org/10.1016/j.neubiorev.2012.06.008.

Bao, Y., von Stosch, A., Park, M., & Pöppel, E. (2017). Complementarity as generative principle: A thought pattern for aesthetic appreciations and cognitive appraisals in general. *Frontiers in Psychology*, 8, 727. https://doi.org/10.3389/fpsyg.2017.00727.

Bao, Y., Yang, T., Lin, X., Fang, Y., Wang, Y., Pöppel, E., & Lei, Q. (2016). Aesthetic preferences for Eastern and Western traditional visual art: Identity matters. *Frontiers in Psychology*, 7, 1596. https://doi.org/10.3389/fpsyg.2016.01596.

Bao,Y., Zhang, D., Zhao, C., Pöppel, E., & Zabotkina, V. (2022). An aesthetic frame for three modes of knowing. *PsyCh Journal*, 636–644. https://doi.org/10.1002/pchj.588.

Cassirer, E. (1923/1929). *Philosophie der symbolischen Formen (Philosophy of Symbolic Forms).* Berlin, Germany: Bruno Cassirer.

Chen, L. (2005). The topological approach to perceptual organization. *Visual Cognition*, 12, 553–637. https://doi.org/10.1080/13506280444000256.

Pöppel, E. (1982). *Lust und Schmerz (Pain and Pleasure).* Berlin, Germany: Severin & Siedler.

Pöppel, E. (1989). The measurement of music and the cerebral clock: A new theory. *Leonardo*, 22, 83–89.

Pöppel, E. (2009). Pre-semantically defined time windows for cognitive processing. *Philosophical Transactions of the Royal Society B*, 364, 1887–1896. https://doi.org/10.1098/rstb.2009.0015.

Pöppel, E. (2010). Perceptual identity and personal self: Neurobiological reflections. In: T. Maruszewski, M. Fajkowska, & M. W. Eysenck (eds.), *Personality from Biological, Cognitive, and Social Perspectives*, pp. 75–82. New York, NY: Eliot Werner Publications.

Pöppel, E., Avram, M., Bao, Y., Graupmann, V., Gutyrchik, E., Lutz, A., Park, M., Reiser, M., Russell, E., Silveira, S., Smigielski, L., Szymanski, C., & Zaytseva, Y. (2013). Sensory processing of art as a unique window into cognitive mechanisms: Evidence from behavioral experiments and fMRI studies. *Procedia – Social and Behavioral Sciences*, 86, 10–17. https://doi.org/10.1016/j.sbspro.2013.08.517.

Rentschler, I., Herzberger, B., & Epstein, D. (1988). *Beauty and the Brain. Biological Aspects of Aesthetics*. Basel, Switzerland: Birkhäuser Verlag.

Silveira, S., Graupmann, V., Frey, D., Blautzik, J., Meindl, T., Reiser, M., Chen, C., Wang, Y., Bao, Y., Pöppel, E., & Gutyrchik, E. (2012). Matching reality in the arts: Self-referential neural processing of naturalistic compared to surrealistic images. *Perception*, 41, 569–576. https://doi.org/10.1068/p7191.

Stoerig, P., Sütterlin, C., & Pöppel, E. (1983). Rechts und links in Bildwerken: Ein neuropsychologischer Beitrag zum Kunstverständnis (Right and left in artwork: A neuropsychological contribution for an understanding of art). *Umschau in Wissenschaft und Technik*, 83(14), 427–428.

Sütterlin, C. & Yu, X. (2021). Aristotle's dream: Evolutionary and neural aspects of aesthetic communication in the arts. *PsyCh Journal*, 10, 224–243. https://doi.org/10.1002/pchj.416.

Turner, F. & Pöppel, E. (1988). Metered poetry, the brain, and time. In: I. Rentschler, B. Herzberger, & D. Epstein (eds.), *Beauty and the Brain: Biological Aspects of Aesthetics*, pp. 71–90. Basel, Switzerland: Birkhäuser Verlag.

Yang, T., Silveira, S., Formuli, A., Paolini, M., Pöppel, E., Sander, T., & Bao, Y. (2019). Aesthetic experiences across cultures: Neural correlates when viewing traditional Eastern or Western landscape paintings. *Frontiers in Psychology*, 10, 798. https://doi.org/10.3389/fpsyg.2019.00798.

Zhao C., Enriquez, P., Izadifar, M., Pöppel, E., Bao, Y., & Zabotkina, V. (2022). Complementarity of mental content and logistic algorithms in a taxonomy of cognitive functions. *PsyCh Journal*, 973–979. https://doi.org/10.1002/pchj.602.

https://doi.org/10.1142/9789811298639_0004

Chapter 4

An Aesthetic Frame for Three Modes of Knowing*

Yan Bao[†,‡,§], Dongxue Zhang[§], Chen Zhao[§], Ernst Pöppel[†,§] and Vera Zabotkina[¶]

[†]*School of Psychological and Cognitive Sciences, Peking University, Beijing, China*
[‡]*Beijing Key Laboratory of Behavior and Mental Health, Peking University, Beijing, China*
[§]*Institute of Medical Psychology, Ludwig Maximilian University Munich, Germany*
[¶]*Russian State University of the Humanities, Moscow, Russia*

Abstract

This is the claim: Human knowledge expresses itself in three different modes — explicit, implicit, and sensory (e.g., visual) knowledge. The reference to knowledge as explicit knowledge only (as it is usually done) neglects the other modes of knowledge. Instead of using the noun "knowledge", it is suggested that we use "knowing" as a term to stress the active mode in cognition. The three modes of knowing are reflected

---

*This chapter was originally published in the special collection on neuroaesthetics in *PsyCh Journal.* Bao, Y., Zhang, D., Zhao, C.,Pöppel, E., & Zabotkina, V. (2022). An aesthetic frame for three modes of knowing. *PsyCh Journal*, 11(5), 636–644. https://doi.org/10.1002/pchj.588.

in different mechanisms of cognitive and neural processing; these can already be observed in classical documents of humankind. This unity of the different modes of knowing can also be extracted from cultural artifacts, such as poetry. Although possessing different characteristics, the three modes of knowing are related to each other in a complementary way, thus providing an integrative network for cognition. We suggest that a unifying frame of the modes of knowing lies in aesthetic principles. These different modes of knowing with their unifying frame reflect some general features of cognition. We stress that basic concepts in "artificial or machine intelligence" do not match at all human modes of knowing.

Keywords: artificial intelligence; complementarity; aesthetics; knowledge; poetry

Three Modes of Knowing

We begin with an apparently simple question: How is communication possible? How is it possible that the authors whose mother tongues are Chinese, German, and Russian writing a text in English understand each other (or have the impression that they do so) and can thus relate to each other in an apparently shared cognitive world? We do not only speak different languages but are embedded in different cultural traditions with different action patterns, different physical environments, different rituals, different sensory experiences, and different linguistic imprintings. Despite this diversity, we assume implicitly that we can exchange ideas and can trust the meaning of the verbal and nonverbal expressions of the others, and that their facial expressions reflect their emotional states. This may sound obvious and even trivial, but maybe we are dealing with an "unasked question" (Pöppel, Bao, *et al.*, 2013); we accept automatically that we share a common denominator in cognition which presumably reflects anthropological universals (Bao & Pöppel, 2012). This is our hypothesis: Our evolutionary heritage provides a unifying frame for understanding. The common denominator is believed to be represented in three different modes of knowing, and we claim that these modes are universal and therefore shared by everybody, thus bridging the gap between real or suspected differences.

Not surprisingly, these modes of knowing have been recognized since antiquity; in different cultural traditions, they have been "discovered", presumably independently, and they define an encompassing frame of knowing. They express themselves (a) explicitly, semantically, or verbally; (b) implicitly, tacitly, intuitively, and also in control of bodily actions; and (c) visually, i.e., pictorially, topologically, or episodically.

Here are some historical examples (which at first sight may surprise one as being rather far-fetched in making a scientific argument). The Christian Bible in the King James (1769) version begins with the following words: "In the beginning God created the heavens and earth. The earth was without form and void, and darkness was upon the face of the deep; and the Spirit of God was moving over the face of the waters. And God said, 'Let there be light'; and there was light. And God saw that the light was good; and God separated the light from the darkness. God called the light Day, and the darkness he called Night." In this text, reference is made to different kinds of knowing which can be defined as being verbal, explicit (calling and saying), or being implicit, action-oriented (doing, creating), or being pictorial, visual, and sensory (recognizing and seeing). Similarly, in the first six verses of Surah 96 of the Koran, the first revelation that Mohammed received, reference is also made to the three modes of knowing. In the Chinese tradition, we find a similar reference to different knowledge systems in Lun-yu II-18, when Confucius tells his student Zizhang how to get a higher position in the government: "Listen more, put aside those doubtful words and say the rest cautiously (explicit knowledge), then you will make fewer mistakes (action-oriented knowledge). Watch more (visual knowledge), put aside those perilous actions and do the rest carefully, then you will have less regret. With fewer mistakes in your words and less regret in your behavior, you are doing exactly what will bring you to what you want."

As authors, we trust that we can take such traditional documents from different cultures as a confirmation or testimony of what we have been able to extract in science over the last decades or even centuries with empirical methods and reasoning about what we know about the world, or what we refer to as "reality" (Pöppel & Bao, 2011). These distinct pillars of knowing are grounded in the evolutionary heritage which we cannot leave behind (Eibl-Eibesfeldt, 1989; Lorenz, 1943; Mayr, 2001; Skinner, 1981). Our evolutionary heritage determines how information is processed in the sensory and motor systems, and how such processing leads to distinct representations in different areas of the brain and their connectivities (Nauta & Feirtag, 1986). To use three pillars of knowing is inherent to us, although we are usually not aware of this genetic endowment. Awareness arises when functions are lost after a brain injury or another kind of neural pathology. If the self-evident has disappeared, attention is drawn to what is missing (Luria, 1973).

One basic characteristic of sensory information processing and the neural control of activities can be expressed with these three words: easy, effortless, and efficient. The goal is to have easy access to phenomena and events in the environment; the acquired information with respect to saving energy is processed without too much effort; as a consequence, actions will be efficient. These characteristics are critical parameters in our evolutionary heritage because they reflect principles of evolutionary selection (Bao & Pöppel, 2012; Zhou *et al.*, 2014). Hearing words, seeing pictures, and understanding facial expressions are self-evident consequences of neural activities in the brain; reflections that can take quite some time are only secondarily involved when we reflect on hearing or seeing. As these brain activities are automatic, they are beyond conscious attentional control. This unquestioned self-evidence of whatever is mentally represented is the basis of trust in the world; whatever reality is conceived of in different epistemological contexts, reality is taken for granted and is not explicitly monitored continuously. It is as it is, as it has been, and as it will be, and only under pathological conditions is such trust violated.

How could one characterize the three modes of knowing? It is preferable to describe these modes rather than trying to define them at an abstract level. An abstract definition necessarily reduces a richer mental space of associations. This has already been expressed by Baumgarten (1750/1758; §560), the founder of modern aesthetics. He wrote, "quid enim est abstractio, si iactura non est" (What else is abstraction than a loss?). Although the three modes of knowing are described separately in what follows, on the neural basis, they are closely interconnected. Due to this high interconnectivity in the neural machinery, absolute independence of functional systems on a conscious level is not possible (Pöppel, 1989). To make a strong statement, there is no hearing or listening without an emotional flavor, there is no seeing an object or a face outside an aesthetic frame, there is no episodic memory without a perceptual component, and there is no semantic memory without a linguistic reference. Some empirical examples beyond these general statements are, for instance, that touch and visual imagery are interconnected (von Trott zu Solz *et al.*, 2017), as are aesthetic and moral judgments (Avram *et al.*, 2013); it is a trivial statement, but it is documented in empirical studies that music modulates emotions (Park *et al.*, 2015). For the past few years, the studies of emotion and emotive evaluation have become one of the major fields of neuroscience (e.g., Zabotkina, 2018). Our ideas resonate with the works of such scholars as Ekman (2004); Damasio (1999, 2010), and others who have shown that any research of brain and mind should involve the role of emotions. Emotive evaluation plays an important role in the language creativity, specifically in the act of creation of new words and new meaning development. Human creativity requires the combination of previously unconnected mental representations constituted by patterns of neural activity (Thagard & Stewart, 2010). Creative thought also has vital emotional components, including the reaction of pleasure that accompanies novel combinations in the treasured "Aha!" experience. The generation of new representations involves binding together previously unconnected representations in ways that also generate new emotional bindings. Thus, the modes of knowing are connected with each other within a neural frame of interdependence. They reflect complementarity as a generative principle (not a descriptive principle) on the neural level (Bao & Pöppel, 2012; Bao, von Stosch *et al.*, 2017). Physically speaking, knowing is like a tripod: It shows stability on three legs being explicit, implicit, and visual. This stability is, however, endangered if one leg or even two legs of the tripod are broken.

Knowing Explicitly

Explicit knowledge or knowing (EK) refers to information usually expressed in language or symbol systems. EK is reflected in thinking or in "talking to oneself" as Plato had characterized "thinking". EK being consciously represented can be communicated to others by specific symbol systems, such as spoken or written language. If EK is lost, it can be recovered in documents. This implies that EK can be shared by others and, thus, loses individual references only. We accumulate and structure EK by learning, and we construct by doing so a landscape of semantic knowledge. Although this results in the feeling of personal ownership, it is socially shared knowledge.

EK is that kind of knowing that has dominated the Western world during the last centuries, and if one refers to knowledge in this part of the world, EK is usually implied. One of the most important historical examples of EK is the "Discours de la Méthode" by René Descartes (1637); this discourse can be taken as one of the basic texts of rationalism. The four rules of thinking that Descartes promoted are rooted in a strong belief in rationality. He suggested that when dealing with a problem or a question, an important step is to segment the problem before answering it; then one moves from what appears to be easy to what is difficult; of course, one has to take all aspects into account without neglecting or forgetting anything. The most important rule may be the first one: Before coming to a conclusion, one has to make sure not to be a victim of prejudice. In this prescription of thinking about problems or questions, it is obviously assumed that this can in fact be done. However, the rules of segmentation, of going from easy aspects to more difficult ones, imply that one must have prior knowledge of the problem in its complexity. Thus, one is caught logically in a circular argument: One has to know already what one wants to know.

What might from a neuroscientific and cognitive point of view be a specific purpose of EK? We argue within this context that EK is primarily used for the reduction of complexity of information; this allows the construction of a "landscape" of semantic knowledge that can be communicated in an effortless way. The importance of complexity reduction is an issue, if one considers the challenges of information processing at the neural level. Sensory information in all modalities is distributed in neural structures in an unpredictable way both spatially and temporally. At the cellular level, the central representation of visual, auditory, or tactile stimuli is ill-defined. The transduction processes (i.e., the transformation of physical energy into a neural "language") are characterized by different time constants at the receptor level in the different modalities. This leads to temporal indeterminacy of information within the different sensory systems. By a process of complexity reduction using "time windows" of finite duration (Bao *et al.*, 2015; Pöppel & Bao, 2014) and by a complementary use of local and global features, or bottom-up information and top-down schemata (Bao, von Stosch, *et al.*, 2017), the unavoidable complexity to know about the world around us and within us can be overcome. The explicit mode of knowing from a neuroscientific and cognitive point of view can be considered as an adaptive system to extract information from the past to be used in the present and potentially in the future; EK is made available to oneself or others using language or other forms of symbolic representation.

Knowing Implicitly

The second mode of knowledge or knowing refers to knowing implicitly (IK). The difference between EK and IK can be clarified with two short statements going back to antiquity in the Western world. Socrates, in self-defense at court (where he was sentenced to capital punishment), stated that "he knows that he does not know." Rephrasing this statement to characterize IK, one can say, "I do not know that I know." The difference between EK and IK is also apparent in a classical quotation of the philosopher Augustinus (n.d., 397/8); in the 11th book of his *Confessiones* Augustinus wrote, "Quid ergo est tempus? Si nemo ex me quaerat, scio; si quaerenti explicare velim, nescio." ("What then is time? If nobody asks me, I know it; if I want to explain it to somebody who asks me, I do not know it.") In this quotation, "knowing" is used within two conceptual frames or categories; namely, first as IK and then as EK. Thus, this famous citation about time can be considered to be distorted by a categorical error: Two different frames of reference are used from a logical point of view. In fact, one can replace "time" with other terms such as "space", "beauty", "attention", or "consciousness", and one always is seduced to the same answer of obviously knowing what one is referring to but in fact not knowing when asked to be specific.

IK has several facets. It refers, for instance, also to bodily and action-oriented knowledge such as movement patterns or sequential acts which are beyond explicit control. A typical example is playing a musical instrument. IK is dominant in ritual behavior patterns when they become automatized; one does not control explicitly what one is doing even in religious ceremonies or in praying (Silveira *et al.*, 2015). For learned movements as in sports, it is impossible to indicate in detail how the movement is done and which motor components would have to be integrated to create a smooth and successful movement pattern. We are much less rational and dominated by EK than we may believe. The philosopher Friedrich Nietzsche (1886) once remarked that the thesis of René Descartes: "Cogito ergo sum" ("I think therefore I am") is unacceptable. The most one can say could be "It thinks." With such a perspective, thinking can be understood as an implicit process not being supervised continuously by rational parameters; as a result of such hidden activities (one may say "in the dark"), an insight may suddenly become explicit and rise to the conscious surface.

Intuition is also an expression of IK. Intuitive decisions are made without the explicit availability of all the variables that should be taken into account. Decisions are embedded within emotional frames, and they harvest items in the memory systems which in the moment of a decision are not mentally represented on an explicit level; obviously, these components in the decision process cannot be discarded. Intuitive or IK is, however, not irrational (with the exception of outrageous prejudices and evident documentation of stupidity) because, retrospectively, it is in principle possible to analyze the decision within a rational frame. In fact, intuitive decisions are sometimes better and not worse when compared to decisions which are believed to be rational only (Ilg *et al.*, 2007); if one has decided on the basis of an intuition, one has more trust in such an implicit decision as it has apparently a stronger or broader basis in the neural systems.

With respect to aesthetic judgments, it is impossible to make a statement like "this is good art" on a merely explicit basis; as indicated earlier with respect to intuition, too many intervening variables in a judgment about "quality" play a role that remains in the attentional background. Thus, the richness of IK is beyond computability; there are too many intervening factors to be considered. To give an example, let us assume 100 modules in the brain (there are in fact many more), which can be either active or inactive at a given moment. Such a brain would have 10^{30} potential states; looked at from a computational point of view, this number means that each brain state is unique. The number of potential brain states goes beyond computability. It follows that experiences or behavioral patterns are in detail unpredictable. On the other hand, we believe in predictability, at least within limits. How is this possible? Using a metaphor, IK is like an ocean; under its surface, activities are going on that cannot be observed. Sometimes the surface of the ocean is penetrated from below, and an island appears with a stable position for some time. In this metaphor, let the island be a representation of EK; knowing explicitly is characterized by rational analysis and conscious representation. We suggest that the complementarity of the two modes of knowing is crucial for aesthetic judgments (Bao, von Stosch, *et al.*, 2017).

The concept of complementarity as a generative principle, and not as a descriptive principle as in quantum mechanics (Bohr, 2010; Ruhnau & Pöppel, 1991), can be considered to be fundamental in appreciating artwork, although not necessarily using these terms. In his classical book, Worringer (1911) emphasized the importance of abstraction *and* empathy ("Abstraktion und Einfühlung"); these terms indicate that EK and IK are interrelated. Similarly, Arnheim (1979) referred to complementarity in the arts when he wrote that the artist is overcoming disorder by creating order; order is recognized on the explicit level, but the process to reach order remains "silent" or implicit during the creative process. From the point of view of experimental research, artwork provides a unique document of IK, but in addition, musical motifs, poetic lines in different languages, or paintings from different cultures and periods can be used as stimulus material to obtain (in a different methodological way) insight into cognition within distinct levels of information processing (Pöppel, Avram, *et al.*, 2013). With some knowledge background about temporal processing, another perspective is obtained to appreciate, for instance, poetry; presemantic temporal processing within the "time window" of some 3 s represents an element of IK that is one basis of poetic creativity (discussed later; also see Zhao *et al.*, 2018). Similarly, one learns that musical education increases the sensitivity of emotional processing (Park *et al.*, 2015) or results in a confirmation and stabilization of personal identity (Zaytseva *et al.*, 2014). These observations, among many others, indicate neural processing on an implicit level in poetry and music, and also in visual art. Artists from the East and the West are embedded in different cultural traditions when representing the visual world pictorially (Pöppel, 2018; Yang *et al.*, 2019). Western artists since the Renaissance have represented the world mainly using a central perspective focusing on salient objects; Chinese artists, on the other hand, concentrate more on embedding objects into the context and often with a negative perspective in their paintings, or they create a new kind of context such as the "third abstraction" by the Chinese artist LaoZhu (Bao, Yang, *et al.*, 2017). A different cultural imprinting which remains implicit is presumably ritualized and may even contribute to validating cultural identity (Bao, Yang, Lin, Fang, *et al.*, 2016).

Knowing Visually

Seeing is knowing. The third mode of knowing is visual knowledge (VK), which can be subdivided into at least three different domains: sensory knowledge within a subjective present (present VK), mnemonic VK, as in episodic memory, and topological VK, as in geometric figures, visual models, diagrams, or histograms. To refer to seeing as knowing may be at first sight unusual, but historically going back to antiquity, *aisthesis* not only refers to perceiving but also to knowing or understanding (Baumgarten, 1750/1758). We tend to forget that seeing is knowing presumably because present VK is self-evident (Zhou *et al.*, 2016). The importance of seeing as knowing is realized in blindness or whenever specific attributes in the visual modality are lost, such as color or movement perception.

Present VK is created in less than 1 s when we direct our gaze with saccadic eye movements to new visual stimuli, or when the attentional machinery selects what is processed in the visual world (Bao & Pöppel, 2007). We are automatically forced to see something as "something", and in case what is visually represented is geometrically impossible as, for instance, in some surrealistic artwork (e.g., Silveira *et al.*, 2012), meaning is imposed. The impossible cannot be experienced visually as something that is possible; the impossible can only be constructed. What is perceived visually is grounded in an effortless neural process, and importantly, what is seen is considered to be true. In this case, the German language may be helpful as "perception" translates to "Wahrnehmung", which is related to truth ("Wahrheit"); what is perceived is instantaneously without reflection accepted to be true ("wahr"). Thus, seeing is indeed knowing, as has been stressed by Zeki (1999).

In contrast to present VK that relates to the here and now, mnemonic VK refers to the past (Tulving, 2002). Relevant experiences of personal importance from our past going back to early childhood may be anchored visually forever in episodic memory. These images are most of the time stationary and not moving and are anchored in a special location. "One-trial learning" experiences with high emotional load are imprinted in the memory system. When counting how many images in episodic memory are actually available and can actively be recovered later, it is surprising that the number is rather limited (unpublished observations). From the previous day, some 20 to 30 images on average may be activated again (although the interindividual variance is rather high), and this estimate is similar for different genders, ages, or cultures. If one tries to reload from episodic memory the images from previous weeks and months, one experiences the quick loss of images; only what has personal importance and strong emotional relevance is moved into episodic memory. Taken together, only several hundred images from our past can be actively "visited" and can be made consciously available. It is important to stress that the number of recognizable images is much higher. This raises the question of why the number of images in our "inner museum" (as the content of episodic memory can be understood) is rather limited. One may speculate that images from our own past are essential for defining and maintaining our personal identity (Zhou *et al.*, 2014), as in a real museum where cultural identities are maintained by pictorial representations. Not everything we experience and which can be stored in memory systems plays a role in self-related identity. Images from our past can be considered to create a narrative about ourselves that we accept and indeed create ourself, and from a "neuro-economic" perspective or a pragmatic view, the number of accessible images for self-confirmation should indeed be limited.

Topological VK is fundamentally different compared to mnemonic VK or present VK, as it refers to abstractions like those in geometric figures. Conic sections allow the creation of a circle, an ellipse, a hyperbola, or a parabola; importantly, they are also aesthetically pleasing. These aesthetic characteristics apply also to simple squares or rectangles, and to golden sections, Fibonacci spirals, or helical spirals. Possibly because of their order and simplicity, they evoke an aesthetic appreciation. This was one of the great scientific contributions of Descartes (1637), who in the appendix of his *Discours de la méthode* developed analytical geometry. With this mathematical insight, it has become possible to visualize complex phenomena in a simple way such as growth functions. The sigmoid curve is visually extremely simple, but it can indicate phase transitions of high complexity in different domains: It can represent a learning curve in psychology or the time course of chemical reactions. Apparently, transitions between different states can best be described by a sigmoid characteristic. Logarithmic functions, exponential decays, or Gaussian distributions give an effortless insight into data-generating mechanisms which otherwise would remain opaque, or at least more difficult to understand. In particular, oscillations in different frequencies indicate visually functional brain states.

Three Modes of Knowing in Poetry

It may not come as a surprise after what has been described earlier that in poetry the three modes of knowing can be recognized and experienced. In fact, one might even argue that without this complementary frame of knowing, poems would be empty or at least would have a very different aesthetic characteristic. We give only a few examples from different languages; they speak for themselves in integrating the three modes of knowing in explicit verbal knowledge, in triggering implicit action-oriented knowledge, and in creating visual images.

One such example is from *A Man Young and Old* (*The Mermaid*) by William Butler Yeats. The words themselves represent EK, mental images are created reflecting VK, and implicit bodily or action-oriented knowledge gives the poem meaning:

A mermaid found a swimming lad,
Picked him for her own,
Pressed her body to his body,
Laughed: and plunging down,
Forgot in cruel happiness
That even lovers drown.

The poem *In Memory of W.B. Yeats* by W. H. Auden shows in a similar way the three modes of knowing:

He disappeared in the dead of winter:
The brooks were frozen, the airports almost deserted,
And snow disfigured the public statues;
The mercury sank in the mouth of the dying day,
What instruments we have agree.
The day of his death was a dark cold day.

This is another example by Alexander Pushkin from *Eugen Onegin* (strophe 33), translated into English by Nabokov:

Я помню море пред грозою:
Как я завидовал волнам,
Бегущим бурной чередою.
С любовью лечь к ее ногам!
Как я желал тогда с волнами.

I recollect the sea before a tempest:
how I envied the waves.
running in turbulent succession.
with love to lie down at her feet!
How much I wished then with the waves.

And another example is given by the Chinese poet Li Bai from the Tang Dynasty, translated by Yan Bao, in the poem *Early Departure from the White Emperor Town*:

早发白帝城
【唐】李白
朝辞白帝彩云间,
千里江陵一日还°
两岸猿声啼不住,
轻舟已过万重山°

Leaving at dawn the White Emperor town
with clouds surrounded,
Sailing to Jiangling a thousand miles
within just one day.
Ceaseless crying the monkey sounds
along the riverside,
Thousands of mountains left behind
my skiff now far away.

The poetic examples of Yeats, Auden, Pushkin, and Li Bai take us into landscapes of knowing different cultures and different times, but they all reflect fundamental principles of human cognition on the explicit, implicit, and sensory levels.

The Unifying Aesthetic Principle

How are the different modes of knowing connected with each other — if at all? We propose the aesthetic principle; *aisthesis* provides a unifying frame. Using the original meaning of the word *aisthesis* (Baumgarten, 1750/1758/2007), aesthetics is both a theory of sensory perception and knowledge and also a theory of the arts. Within such a frame, the arts can be conceived of as representing particular trajectories of human activities (Cassirer, 1944; Eibl-Eibesfeldt & Sütterlin, 2008; Sütterlin & Yu, 2021). We suggest that the aesthetic principle is the necessary frame of knowing. Thus, *aisthesis* within this concept is fundamental both for knowing about the world within us and around us in general and in the arts in particular. It was the Roman poet Horatius who already more than 2,000 years ago called for an aesthetic principle in the arts: the "simplex et unum" (*De Arte Poetica*); there must be unity and congruence between what is perceived in a piece of art and what has been previously experienced by the person.

What are aesthetic principles in addition to the Horatian demands of unity and simplicity that are typical for the explicit, implicit, and visual modes of knowing and that also apply to different artistic expressions like those in poetry, music, visual art, dance, or architecture? With respect to knowing explicitly, it is useful to go back again to the *Discours de la Méthode* by René Descartes (1637). In his rules of thinking, he stresses precision and clarity in perceiving and reasoning, order in analyzing a problem, observing unity whatever has to be understood. If order is disrupted, it has to be established again, but this can be done only if a sensitivity for order as an aesthetic principle is imprinted in the neurocognitive machinery; this principle is related to the arts as Gombrich (1979) stated. Furthermore, EK is characterized by simplicity in comparison to usually non-transparent complexity; reduction of complexity is one of the hallmarks of *Cognitive Processing*, although it sometimes is exaggerated and results in an illusion of understanding. Laws in science are in fact considered as being true if they correspond to *Occam's Razor* (i.e., the simplest explanation is considered as being correct). An aesthetic pleasure and the sense of beauty are evoked with a clear statement, often in a mathematical language (Zhang & Zeki, 2022). The characteristics of EK, such as order, clarity, simplicity, or unity, are aesthetic principles.

Such aesthetic principles also apply to the other modes of knowing. Present VK is characterized by effortless processing. For instance, a schema is essential to recognize a face, and a schema (or a "Platonic idea") is simple. When perceiving a face or an object ("something" as something in its identity, i.e., a visual image), the percept is just "one" in the "inner theater" being characterized by unity. There is, however, also an abstract frame: Percepts can be conceived of as being realized with an abstract frame of expectations. A visual, auditory, or tactile percept can be looked at as being instantiated within an expected frame. It is the falsification or the verification of a hypothesis represented in the sensory experience that characterizes our access to the physical world. One is invited to reflect on a fundamental statement in the *Critique of Pure Reasons* by Immanuel Kant (1781/1787): "Gedanken ohne Inhalt sind leer, Anschauungen ohne Begriffe sind blind." (Thoughts without content are empty. Sensations without concepts are blind.) With a different kind of terminology, it can be restated that a top-down component is complemented by bottom-up information to create a unitary visual percept.

Complementarity as a generative principle (Bao, von Stosch, *et al.*, 2017) can conceptually be taken as the basis for creating unity of a percept, and unity is an aesthetic principle. The importance of the aesthetic principle in a personal frame of reference applies also for mnemonic VK. Only what as an image from the personal past is emotionally relevant will be engraved in episodic memory. This "inner museum" is the home of events and images that relate to our personal history, and they substantiate our personal self and confirm our identity (Pöppel, 2010). Images in episodic memory representing mnemonic VK are essential building blocks to construct also subjective continuity by relating what has happened in the past to the present and looking into the future. Thus, mnemonic VK can be understood of as being instrumental for the creation of the notion of time. Linear time in its simplicity allows mental time to travel backwards and forwards in the "inner museum" visiting mnemonic representations along the path of continuous time. This brings us again to the aesthetic principles of simplicity and unity being a feature of linear unidimensional time. "Time" is homogeneous "that flows equably without relation to anything external" as Isaac Newton (1686/1980) defined absolute, true, and mathematical time in *Principia Mathematica Philosophiae Universalis*.

Unity and simplicity also apply to topological VK, as indicated earlier. A Gaussian distribution, a logarithmic function, an exponential decay, a sinusoidal oscillation, or a sigmoid curve in their perceptual beauty create an immediate and effortless access on an abstract level to data collected in experiments representing a functional system in its complexity in all domains of science. A visual symmetry or a violation of a symmetry provides an immediate understanding of a data-generating mechanism. This implies that the aesthetic principle can even be used as a research tool — and in fact, it is.

For implicit VK, the reafference principle (Von Holst & Mittelstaedt, 1950) may give an additional cue for the unifying aesthetic principle. Originally, the reafference principle expresses the concept that movements are both regulated by efferent commands to the effectors (like to the eye muscles triggering a saccadic eye movement) and by an "efference copy" of a command. When a movement has been executed, the copy of the efference is compared with incoming information from the sense organs: the "reafference". When both neural representations match in intensity (i.e., the reafference is equivalent to the efference copy), the movement is signaled to be completed. With respect to saccadic eye movements, this concept allows us to understand why visual space is experienced as stable, although the retinal input varies after each saccadic eye movement (Teuber, 1960). This concept of the reafference principle with *comparison* as a central neurocognitive operation needs to be generalized. Such a generalized reafference principle (GRP) provides a theoretical frame for understanding actions and intentions in general as knowing implicitly. Any action triggered by a command (or more generally by a plan) is controlled on the neural level by efferent signals and efferent copies which are stored centrally in neural systems. This implies that the copies of intended actions or plans can provide a basis for monitoring continuously whether targets have already been reached or not, whether an action or a plan has been completed. The GRP offers a concept of neural control with respect to not only short-term activities but also long-term actions and intentions (which may take days, months, or even years). Thus, one owns an endogenous monitoring system that is linked to neural mechanisms that signal satisfaction after the completion of an action or frustration for not having reached the goal. We suggest that the impression of unity of a movement or an action after completion is the consequence of the operative mechanism of comparison as conceived of in the GRP. Many human activities (if not all) that are voluntarily initiated or being ritualized are typically characterized by the implicit expectation of completion and fulfillment. Satisfaction is obtained by the gestalt of a movement, of a performance like playing a musical instrument, preparing a meal, sailing a boat, hitting a golf ball, giving a lecture, passing an examination, or else. It is the retrospective feeling of unity that gives satisfaction. Thus, striving for unity in acts and actions represents the aesthetic principle for implicit knowing; experiencing joy is an instantiation or verification of the aesthetic principle.

Taken together, several indicators confirm *aisthesis* as a unifying principle of the different modes of knowing as well as for the arts in particular. To identify the aesthetic principle directly, one can project oneself into different situations that represent an internal point of view such as remembering and citing a verse from the German poet Friedrich Schiller from his poem *Nänie* (*Auch das Schöne muss sterben* [*also the beautiful has to die....*]), listening to the last fugue of Bach, climbing a mountain, looking at self-portraits of Rembrandt, activating in episodic memory an episode of love, having a sudden insight as a scientist in understanding data, or drinking a glass of wine with friends. In all these activities, in perceiving, acting, and knowing, the essential element is the impression of unity. There is simplicity, there is clarity, there is order with the proper relation of the parts to the whole, there is rhythm and harmony, and there is a conceptual and an emotional frame defining meaning of experiences. Without these attributes in the three modes of knowing or in the arts, we feel frustration, disappointment, and even anger. Such subjective parameters indicate that the aesthetic principle has been violated. The disruption proves the essence of the aesthetic principle.

Three Modes of Knowing and Limits of Artificial Intelligence

How is artificial intelligence (AI) or machine intelligence (MI) related to the three modes of knowing and their interdependence — if at all (Rahwan *et al.*, 2019; Sejnowski, 2018)? How are they related to the aesthetic principle as a unifying frame — if at all? Is it possible (at least at present) to formulate algorithms which mimic knowing in its entirety? Are AI and/or MI models for humans to know? Taking the characteristics of the modes of knowing together, one has to conclude that as long as AI is understood mainly with respect to pattern recognition and learning algorithms, it reflects only a small partial set of human "intelligence" or knowing. The underlying concept in AI is oriented toward EK. This, of course, has its value, and the engineering challenges show impressive solutions. But this successful trajectory is far removed from "understanding" the knowing of humans. One essential feature of human existence is the desire to know; humans are driven by curiosity. Are man-made artifacts curious? When one refers to models to reflect and explain human knowing, one has to conclude that both the "deep learning" paradigm (Sejnowski, 2018) and the classical approach in AI (Minsky & Papert, 1969) are indeed far away from human reality. The critical feature of similarity of a useful model for what has to be modeled is out of reach.

And there are additional limitations if one wants to relate AI to human cognition: What is missing, for instance, in algorithms of AI and MI is the distinction between content functions and logistic functions, as has been suggested in a taxonomy of functions (Pöppel, 1989; Pöppel & Ruhnau, 2011). Temporal processing of neural information has to be considered as a logistic function as it is reflected in the time window of approximately 3 s (e.g., Bao *et al.*, 2015; S. Chen *et al.*, 2020; Pöppel & Bao, 2014; Wang, Bao, *et al.*, 2016; Yu & Bao, 2020; Zhao *et al.*, 2018; Zhou *et al.*, 2014), or the "magic seconds of three." The logistics of the brain can also be seen for a shorter time window in the domain of some tens of milliseconds as it is observed in multimodal reaction time distributions or the temporal order threshold in different sensory modalities (Bao, Szymaszek, *et al.*, 2013; Bao, Yang, Lin, & Pöppel, 2016; Pöppel, 1970). Attentional control also has to be treated conceptually as a logistic function providing the fundamental neural mechanisms to create content. An example is spatial attention with the discovered "eccentricity effect" in the visual field (Bao & Pöppel, 2007; Lei & Bao, 2012), which is modulated temporally within the time window of some 3 s, indicating the functional link between different domains of logistic functions (Bao, Wang, *et al.*, 2013). In addition, activation of neural systems is also a logistic function providing the necessary "power supply" of cognition (Pöppel, 1989) and making sure homeostatic regulation, i.e., to keep all systems, including the cognitive machinery, in balance (Bernard, 1856/1957; Wang, Lin, *et al.*, 2016). Importantly, activation is modulated by the circadian clock (Aschoff, 1965; Bao *et al.*, 2015; Kleitman, 1963); as all physiological, biochemical, and psychological functions show a 24-hr rhythm with maxima and minima at different phases throughout the day, the question arises about how we define or understand personal identity. The "phase-map" of all functions changes continuously, and we are identical with ourselves from this biological point of view only every 24 hr. Apparently, our cognitive memory systems provide a "temporal glue" resulting in the impression of a personal identity across time. The concept of identity, which is crucial for human cognition, is not even a topic in AI or MI.

As stated earlier, knowing as a basic element of human cognition is like a tripod: It is stable only on three legs with knowing explicitly, implicitly, and visually, and the interdependent modes of knowing are related to each other by the aesthetic principle. Furthermore, typical characteristics of the logistic functions of the brain as observed in temporal processing or attentional control indicate the lack of similarity between AI or MI and "knowing" in human beings. It is best that we do not fall into the trap of the illusion that technical systems implement how humans are meant to be with their evolutionary heritage.

Acknowledgments

The concepts developed in this chapter are partly based on "Three modes of knowledge as basis for intercultural cognition and communication: A theoretical perspective" by Ernst Pöppel and Yan Bao published in *Culture and Neural Frames of Cognition and Communication* edited by Shihui Han and Ernst Pöppel, Springer-Verlag, Heidelberg (2011), pp. 215–231. The authors express their gratitude to Evgeny Gutyrchik, Yuelin Liu, and Taoxi Yang for their support in selecting and discussing poetic examples. This work was supported by the Parmenides Foundation, Pöcking (Germany) to Yan Bao, Ernst Pöppel, and Chen Zhao. The National Natural Science Foundation of China (Grants 31771213 and 31371018) supports the work of Yan Bao. Dongxue Zhang is supported by the Chinese Scholarship Council. The work of Vera Zabotkina is supported by the Russian Science Foundation Project 22-18-00594 "Cognitive models of manipulation in media discourse."

References

Arnheim, R. (1979). *Entropie und Kunst. Ein Versuch über Unordnung und Ordnung* [*Entropy and Art: An Essay on Disorder and Order*]. DuMont.

Aschoff, J. (1965). Circadian rhythms in man. *Science*, 148, 1427–1432. https://doi.org/10.1126/science.148.3676.1427.

Augustinus, A. (n.d.). *Confessiones* [*Confessions*]. Klostermann (397/3981993).

Avram, M., Gutyrchik, E., Bao, Y., Pöppel, E., Reiser, M., & Blautzik, J. (2013). Neurofunctional correlates of aesthetic and moral judgments: Equal but not the same. *Neuroscience Letters*, 534, 128–132. https://doi.org/10.1016/j.neulet.2012.11.053.

Bao, Y. & Pöppel, E. (2007). Two spatially separated attention systems in the visual field: Evidence from inhibition of return. *Cognitive Processing*, 8, 37–44. https://doi.org/10.1007/s10339-006-0151-x.

Bao, Y. & Pöppel, E. (2012). Anthropological universals and cultural specifics: Conceptual and methodological challenges in cultural neuroscience. *Neuroscience and Biobehavioral Reviews*, 36, 2143–2146. https://doi.org/10.1016/j.neubiorev.2012.06.008.

Bao, Y., Pöppel, E., Wang, L., Lin, X., Yang, T., Avram, M., Blautzik, J., Paolini, M., Silveira, S., Vedder, A., Zaytseva, Y., & Zhou, B. (2015). Synchronization as a biological, psychological and social mechanism to create common time: A theoretical frame and a single case study. *PsyCh Journal*, 4(4), 243–254. https://doi.org/10.1002/pchj.119.

Bao, Y., von Stosch, A., Park, M., & Pöppel, E. (2017). Complementarity as generative principle: A thought pattern for aesthetic appreciations and cognitive appraisals in general. *Frontiers in Psychology: Theoretical and Philosophical Psychology*, 8, 727. https://doi.org/10.3389/fp-syg.2017.00727.

Bao, Y., Szymaszek, A., Wang, X., Oron, A., Pöppel, E., & Szelag, E. (2013). Temporal order perception of auditory stimuli is selectively modified by tonal and non-tonal language environments. *Cognition*, 129, 579–585. https://doi.org/10.1016/j.cognition.2013.08.019.

Bao, Y., Wang, Z., Liang, W., Wang, Y., Pöppel, E., & Li, H. (2013). Inhibition of return at different eccentricities in the visual field share the same temporal window. *Neuroscience Letters*, 534, 7–11. https://doi.org/10.1016/j.neulet.2012.11.046.

Bao, Y., Yang, T., Lin, X., Fang, Y., Wang, Y., Pöppel, E., & Lei, Q. (2016). Aesthetic preferences for eastern and Western traditional visual art: Identity matters. *Frontiers in Psychology*, 7, 1596. https://doi.org/10.3389/fpsyg.2016.01596.

Bao, Y., Yang, T., Lin, X., & Pöppel, E. (2016). Donders revisited: Discrete or continuous temporal processing underlying reaction time distributions? *PsyCh Journal*, 5(3), 177–179. https://doi.org/10.1002/pchj.142.

Bao, Y., Yang, T., Zhang, J., Zhang, J., Lin, X., Paolini, M., Pöppel, E., & Silveira, S. (2017). The "third abstraction" of the Chinese artist LaoZhu: Neural and behavioral indicators of aesthetic appreciation. *PsyCh Journal*, 6(2), 110–119. https://doi.org/10.1002/pchj.167.

Baumgarten, A. G. (1750). *Aesthetica* [*Aesthetics*]. Felix Meiner 1758/2007.

Bernard, C. (1856). *Introduction à l'étude de la médecine expérimentale* [*An introduction to the study of experimental medicine*]. Dover (1957).

Bohr, N. (2010). *Atomic Physics and Human Knowledge*. Courier Dover Publications.

Cassirer, E. (1944). *An Essay on Man: An Introduction to a Philosophy of Human Culture*. Yale University Press.

Chen, S., Cai, M., & Bao, Y. (2020). The two- to three-second time window of shot durations in movies. *PsyCh Journal*, 9(4), 516–518. https://doi.org/10.1002/pchj.384.

Damasio, A. R. (1999). *The Feeling of What Happens: Body and Emotions in the Making of Consciousness*. Harcourt Brace.

Damasio, A. R. (2010). *Self Comes to Mind: Constricting the Conscious Mind*. Pantheon Books.

Descartes, R. (1637, 1990). *Discours de la méthode* [*Discourse on Method*]. Felix Meiner.

Eibl-Eibesfeldt, I. (1989). *Human Ethology*. Aldine de Gruyter.

Eibl-Eibesfeldt, I. & Sütterlin, C. (2008). *Weltsprache Kunst* [*World Language Art*]. Brandstätter.

Ekman, P. (2004). *Emotions Revealed: Understanding Faces and Feelings*. Phoenix Press.

Gombrich, E. H. (1979). *The Sense of Order: A Study in the Psychology of Decorative Art*. Phaidon Press.

Ilg, R., Vogeley, K., Goschke, T., Bolte, A., Shah, J., Pöppel, E., & Fink, G. (2007). Neuronal processes underlying intuitive coherence judgments as revealed by fMRI on a semantic judgment task. *NeuroImage*, 38, 228–238. https://doi.org/10.1016/j.*NeuroImage*.2007.07.014.

Kant, I. (1781). *Kritik der reinen Vernunft* [*Critique of Pure Reason*]. Hartknoch (1787).

Kleitman, N. (1963). *Sleep and Wakefulness*. University of Chicago Press.

Lorenz, K. (1943). Die angeborenen Formen möglicher Erfahrung [Innate forms of possible experience]. *Zeitschrift für Tierpsychologie*, 5, 235–409.

Luria, A. R. (1973). *The Working Brain. An Introduction to Neuropsychology*. Basic Books.

Lei, Q. & Bao, Y. (2012). The eccentricity effect of inhibition of return is independent of cortical magnification. *Journal of Vision*, 12(9), 674–674. https://doi.org/10.1167/12.9.674.

Mayr, E. (2001). *What Evolution Is*. Basic Books.

Minsky, M. & Papert, S. (1969). *Perceptrons*. MIT Press.

Nauta, W. & Feirtag, M. (1986). *Fundamental Neuroanatomy*. Freeman.

Newton, I. (1686). *Principia Mathematica Philosophiae Universalis*. Josephi Streater.

Nietzsche, F. (1886). *Jenseits von Gut und Böse* [*Beyond Good and Evil*]. dtv (1980).

Park, M., Gutyrchik, E., Welker, L., Carl, P., Pöppel, E., Zaytseva, Y., Meindl, T., Blautzik, J., Reiser, M., & Bao, Y. (2015). Sadness is unique: Neural processing of emotions in speech prosody in musicians and non-musicians. *Frontiers in Human Neuroscience*, 8. https://doi.org/10.3389/fnhum.2014.01049.

Pöppel, E. (1970). Excitability cycles in central intermittency. *Psychologische Forschung*, 34, 1–9.

Pöppel, E. (1989). Taxonomy of the subjective: An evolutionary perspective. In J. W. Brown (ed.), *Neuropsychology of Visual Perception* (pp. 219–232). Lawrence Erlbaum Associates.

Pöppel, E. (2010). Perceptual identity and personal self: Neurobiological reflections. In: T. Maruszewski, M. Fajkowska, & M. M. Eysenck (eds.), *Personality from Biological, Cognitive, and Social Perspectives* (pp. 75–82). Eliot Werner Publications.

Pöppel, E. (2018). East of west, west of east: A matter of global and local identity. *Cognitive Processing*, 19, 115–119. https://doi.org/10.1007/s10339-018-0885-2.

Pöppel, E., Avram, M., Bao, Y., Graupmann, V., Gutyrchik, E., Lutz, A., Park, M., Reiser, M., Russell, E., Silveira, S., Smigielski, L., Szymanski, C., & Zaytseva, Y. (2013). Sensory processing of art as a unique window into cognitive mechanisms: Evidence from behavioral experiments and fMRI studies. *Procedia Social and Behavioral Sciences*, 86, 10–17. https://doi.org/10.1016/j.sbspro.2013.08.517.

Pöppel, E. & Bao, Y. (2011). Three modes of knowledge as basis for intercultural cognition and communication: A theoretical perspective. In S. Han & E. Pöppel (eds.), *Culture and Neural Frames of Cognition and Communication* (pp. 215–231). Springer-Verlag. https://doi.org/10.1007/978-3-642-15423-2_14.

Pöppel, E. & Bao, Y. (2014). Temporal windows as a bridge from objective to subjective time. In D. Lloyd & V. Arstila (eds.), *Subjective Time* (pp. 241–261). MIT Press.

Pöppel, E., Bao, Y., Han, S., Gutyrchik, E., Sozinov, A., Ushakov, D., Kovalev, A., Chernorizov, A. M., Menshikova, G. Y., Zabotkina, V. I., Zaytseva, Y., & Zinchenko, Y. P. (2013). Unasked questions and unused answers in psychology. *Psychology in Russia: State of the Art*, 6(2013), 4–18. https://doi.org/10.11621/pir.2013.0301.

Pöppel, E. & Ruhnau, E. (2011). Psychologie als eine auf Modelle angewiesene Angelegenheit ohne Taxonomie — eine Polemik [psychology as a business using models without a taxonomy — A polemic]. *Nova Acta Leopoldina NF*, 110(377), 213–233.

Rahwan, I., Cebrian, M., Obradovich, N., Bongard, J., Bonnefon, J.-F., Breazeal, C., Crandall, J. W., Christakis, N. A., Couzin, I. D., Jackson, M. O., Jennings, N. R., Kamar, E., Kloumann, I. M., Larochelle, H., Lazer, D., McElreath, R., Mislove, A., Parkes, D. C., Pentland, A. '. S.'., … Wellman, M. (2019). Machine behavior. *Nature*, 568, 477–486. https://doi.org/10.1038/s41586-019-1138-y.

Ruhnau, E. & Pöppel, E. (1991). A directional temporal zones in quantum physics and brain physiology. *International Journal of Theoretical Physics*, 30, 1083–1090. https://doi.org/10.1007/BF00671487.

Sejnowski, T. J. (2018). *The Deep Learning Revolution*. MIT Press.

Silveira, S., Bao, Y., Wang, L., Pöppel, E., Avram, M., Simmank, F., Zaytseva, Y., & Blautzik, J. (2015). Does a bishop pray when he prays? And does his brain distinguish between different religions? *PsyCh Journal*, 4(4), 199–207. https://doi.org/10.1002/pchj.116.

Silveira, S., Graupmann, V., Frey, D., Blautzik, J., Meindl, T., Reiser, M., Chen, C., Wang, Y., Bao, Y., Pöppel, E., & Gutyrchik, E. (2012). Matching reality in the arts: Self-referential neural processing of naturalistic compared to surrealistic images. *Perception*, 41, 569–576. https://doi.org/10.1068/p7191.

Skinner, B. F. (1981). Selection by consequences. *Science*, 213, 501–504. https://doi.org/10.1126/science.7244649.

Sütterlin, C. & Yu, X. (2021). Aristotle's dream: Evolutionary and neural aspects of aesthetic communication in the arts. *PsyCh Journal*, 10(2), 224–243. https://doi.org/10.1002/pchj.416.

Teuber, H. L. (1960). Perception. *Handbook of Physiology — Neurophysiology III* (pp. 1595–1668). Springer Verlag.

Thagard, P. & Stewart, T. (2010). The AHA! Experience: Creativity through emergent binding in neural networks. *Cognitive Science A Multidisciplinary Journal*, 35, 1–33. https://doi.org/10.1111/j.1551-6709.2010.01142.x.

Tulving, E. (2002). Episodic memory: From mind to brain. *Annual Review of Psychology*, 53, 1–25. https://doi.org/10.1146/annurev.psych.53.100901.135114.

The King James Bible (1769). Oxford Edition. https://en.wikisource.org/wiki/Bible_(King_James).

von Holst, E. & Mittelstaedt, H. (1950). *Das Reafferenzprinzip* [*The reafference principle*]. *Naturwissenschaften*, 37, 464–476.

von Trott Zu Solz, J., Paolini, M., & Silveira, S. (2017). Does touch inhibit visual imagery? A case study on acquired blindness. *PsyCh Journal*, 6(2), 139–140. https://doi.org/10.1002/pchj.162.

Wang, L., Bao, Y., Zhang, J., Lin, X., Yang, L., Pöppel, E., & Zhou, B. (2016). Scanning the world in three seconds: Mismatch negativity as an indicator of temporal segmentation. *PsyCh Journal*, 5(3), 170–176. https://doi.org/10.1002/pchj.144.

Wang, L., Lin, X., Zhou, B., Pöppel, E., & Bao, Y. (2016). Rubberband effect in temporal control of mismatch negativity. *Frontiers in Psychology*, 7, 1299. https://doi.org/10.3389/fpsyg.2016.01299.

Worringer, W. (1911). *Abstraktion und Einfühlung* [*Abstraction and Empathy*]. Wilhelm Fink (2007).

Yang, T., Silveira, S., Formuli, A., Paolini, M., Pöppel, E., Sander, T., & Bao, Y. (2019). Aesthetic experiences across cultures: Neural correlates when viewing traditional eastern or Western landscape paintings. *Frontiers in Psychology*, 10, 798. https://doi.org/10.3389/fpsyg.2019.00798.

Yu, X. & Bao, Y. (2020). The three second time window in poems and language processing in general: Complementarity of discrete timing and temporal continuity. *PsyCh Journal*, 9(4), 429–443. https://doi.org/10.1002/pchj.390.

Zabotkina, V. I. (2018). Towards emotion-cognition interface in conceptualization: Interdisciplinary approach [вопросу о соотношении когниции и эмоции в процессе концептуализации: междисциплинарный подход]. *Cognitive Studies of Language/Когнитивные исследования языка*, 35(35), 31–37.

Zaytseva, Y., Gutyrchik, E., Bao, Y., Pöppel, E., Han, S., Northoff, G., Welker, L., Meindl, T., & Blautzik, J. (2014). Self processing in the brain: A paradigmatic fMRI case study with a professional singer. *Brain and Cognition*, 87, 104–108. https://doi.org/10.1016/j.bandc.2014.03.012.

Zeki, S. (1999). *Inner Vision. An Exploration of Art and Brain*. Oxford University Press.

Zhang, H. & Zeki, S. (2022). Judgments of mathematical beauty are resistant to revision through external opinion. *PsyCh Journal*, 11(5), 741–747, https://doi.org/10.1002/pchj.556.

Zhao, C., Zhang, D., & Bao, Y. (2018). A time window of three seconds in the aesthetic appreciation of poems across languages. *PsyCh Journal*, 7(1), 51–52. https://doi.org/10.1002/pchj.194.

Zhou, B., Pöppel, E., & Bao, Y. (2014). In the jungle of time: The concept of identity as an escape. *Frontiers in Psychology*, 5. https://doi.org/10.3389/fpsyg.2014.00844.

Zhou, B., Pöppel, E., Wang, L., Yang, T., Zaytseva, Y., & Bao, Y. (2016). Seeing without knowing: Operational principles along the early visual pathway. *PsyCh Journal*, 5(3), 145–160. https://doi.org/10.1002/pchj.141.

https://doi.org/10.1142/9789811298639_0005

Chapter 5

Burke's "Sensitive" Sublime Rooted in Brain and Body Versus Kant* Transcendental Concept in the German Romantic Discourse

Grazia Pulvirenti and Renata Gambino

Department of Human Studies, University of Catania, Research Center NewHums — Neurocognitive and Human Studies, Catania, Italy
International Network NeuroHumanities Studies (www.neurohumanitiestudies.eu)

Abstract

During the 18th century, debates about what constituted the sublime flourished in Europe, particularly in Germany. These debates were nourished by two different visions: The Kantian concept supposed that the sublime is supra-sensible and rooted in reason (*Logos*) rather than in the object, thus provoking a mental state of tension between nature and art;

*This chapter was originally published in the special collection on neuroaesthetics in *PsyCh Journal*. Gambino, R. & Pulvirenti, G. (2021). Burke's "sensitive" sublime rooted in brain and body versus Kant's transcendental concept in the German Romantic discourse. *PsyCh Journal*, 10(2), 210-223. https://doi.org/10.1002/pchj.414.

Edmund Burke's concept, on the other hand, conceived of the sublime as a bodily immersive experience, which we here define as "sensitive" sublime. In summary, Burke's view of the sublime is rooted in the senses and not in the power of reason, unlike Kant's. This was to disrupt the mainstream ideas of that time, unconsciously anticipating some of the recent neuroaesthetic acquisitions regarding the central role played by the sensory apparatus in the experience of beauty and of the sublime.

***Keywords*:** beauty; ekphrasis; sublime

Edmund Burke's revolutionary view is the core issue of the innovative ekphrasis of Caspar David Friedrich's famous painting *Monk by the Sea* (1808–1810), written by a German author of the Romantic era, Heinrich von Kleist (1810/2013): *Feelings about a Seascape by Friedrich*. This text represents a milestone in the crucial German sublime debate from the early 18th to the 19th century; in our view, Kleist expresses in this text a sharp anti-Kantian position, featuring the concept of the sublime represented in Friedrich's painting as a physiological brain–body experience, deeply rooted in the senses. In fact, the author does not directly describe the art object, according to the traditional ekphrastic strategy, but rather the bodily emotional process engaged in by the reader of the ekphrasis as if they were the viewer of the painting. Kleist stresses the recipient's bodily involvement, allowing the reader to vividly experience the landscape of the painting as if it were real. He places the reader within the specific natural quality of the landscape depicted in the painting, in order to prompt the activation of their emotional and imaginative power through the involvement of the body, more than through the hypothetical stimulation at a supra-sensible level triggering a response rooted in reason. Reconstructing the bodily and mental processes involved by the sublime experience within the painting's landscape, Kleist successfully compares the sublime felt in front of nature with that felt in front of art from a very modern perspective. Kleist's ekphrasis is based on the innovative idea that the sublime experience relies on a bodily perceptual response, which can be triggered by natural, artefactual, as well as linguistic stimuli. In brief, this means that Kleist considers a textual representation to be as effective as a pictorial one — or as a natural landscape — as the trigger of a complex experience that involves perception, emotion, and cognition. This intuitively anticipates Zeki's neuroscientific discoveries about the common origins of different experiences of beauty and of the sublime: Beauty prompts "feelings of pleasure, reward and satisfaction" (Ishizu & Zeki, 2014, p. 1) regardless of the input sources, correlating with the activation of the emotional brain, specifically field A1 of the medial orbitofrontal cortex (Ishizu & Zeki, 2011); the sublime experience on the other hand, although instantiated by natural or artful sources, is associated with "awe, fear, terror" (Ishizu & Zeki, 2014, p. 1) and its experience correlates with a different set of brain areas, such as "in the cortical area the inferior temporal cortex, the posterior hippocampus and the inferior middle frontal gyri; in the subcortical area the basial ganglia and in addition to this the cerebellum" (Ishizu & Zeki, 2014, p. 3). This implies that the big difference between the experience of the beautiful, on the one hand, and the sublime, on the other, relies on the involvement of different emotions. What both experiences have in common is that they may be triggered as well by nature as by art through the activation of the imagination, which has been tested only with regard to represented landscapes (Ishizu & Zeki, 2014). We suppose that the direct experience of nature is only filtered by the senses, while the artistic one involves also simulation of an imagined nature. Therefore, the main difference between nature and art consists in the role played in art by the imagination, which constructs counter-factual representations.

We develop this theme by examining the following issues: (1) the neuroscientific theories about experiencing beauty and the sublime, (2) the 18th-century German debate about the concepts of the Kantian transcendental sublime versus Burke's "sensitive" sublime, and (3) an analysis of Kleist's text in the frame of the genre of ekphrasis as a key example in the modern representation of the sensitive sublime rooted in the body of the human being considered as part of the living biosphere.

The Neuroscientific Theoretical Frame of the Sublime

Before focusing on the German debate about the sublime from the turn of the 18th to the 19th century and before introducing what we refer to as

Edmund Burke's sensitive concept of the sublime, we need to briefly present the results of neurobiological research conducted by Semir Zeki and colleagues (Ishizu & Zeki, 2011, 2014; Zeki, 1998, 1999, 2008, 2019; Zeki, Romaya, Benincasa, & Atiyah, 2014) on the role of the brain in experiencing the beautiful and the sublime (Zeki *et al.*, 2014). Zeki has demonstrated, with brain-imaging techniques, his revolutionary intuition: The experience of beauty derived from different sources, whether sensory (Ishizu & Zeki, 2011), cognitive (Ishizu & Zeki, 2014), or mathematical (Zeki *et al.*, 2014), correlates with "activity in a given, specific, part of the emotional brain, namely field A1 of the medial orbitofrontal cortex" (Zeki, 2019, p. 108).

The philosophical frame of Zeki's theory refers to Burke's concepts of beauty as the output of the interaction among bodily senses and the mind: "Beauty is, for the greater part, some quality in bodies acting mechanically upon the human mind by intervention of the senses" (Burke, 1887, p. 162). The experience of artefactual and natural beauty is not really dependent on the proportion or characteristics of the object but on a vague list of qualities, such as variation and color, intervening on the senses. Zeki points out that "the value of Burke's definition lies … in acknowledging explicitly the critical role played by the brain in the experience of beauty" (Zeki, 2019, p. 108).

Although experience of the sublime is apparently assimilated to that of beauty, Ishizu & Zeki's (2014) experiments suggest that there is a significant difference in the pattern of brain activity that correlates with the experience of the beautiful, on the one hand, and the sublime, on the other.

In particular, field A1 of the medial orbitofrontal cortex, in which activity correlates parametrically with the declared experience of beauty derived from different sources (Ishizu & Zeki, 2011, 2014; Zeki *et al.*, 2014, *inter alia*), was not active during experience of the sublime, while other regions, notably the head of the caudate, the putamen, and the posterior hippocampus, in which activity has been recorded as correlating with experiences of pleasure, hate, and memory, respectively, were also active with the experience of sublimity. Taken together, these results indicate that aesthetic (beauty) experiences and experience of the sublime engage separate and distinct brain systems (Ishizu & Zeki, 2014). These very fascinating results prove the need for an epistemic convergence of literary, artistic, philosophical, and neuroscientific approaches to an aesthetic phenomenon that has been the object of observation and wonder since antiquity and the claim to further cross-disciplinary studies on the different receptions of beauty and the sublime.

These experiments thus suggest that experience of the sublime correlates with activity in brain regions involved during a variety of emotions, like fear and horror, corresponding in general to what Kant defined as "negative pleasures" (Kant, 1790/2009). But the results of the experiments also raise the following issues: The overall pattern of activation during experience of the sublime is significantly different from the overall activity observed in studies dealing with pain, threat, or fear, each one of which qualifies — at least lexically — as constituting an element in the experience of the sublime. It is therefore perhaps futile to give too rigid an interpretation of the functions of regions that were activated and de-activated in our study in relation to past imaging studies that have addressed the individual components (such as fear or surprise) of the experiences that are subsumed under the term (and experience) of the sublime. This is even more so because, although the regions themselves can be fairly accurately demarcated, a search of the literature reveals that there is no unanimity of views on the overall functions of each of these active regions, beyond the somewhat general statement that they are involved in some kind of emotional experience, with some having cognitive functions as well (Zeki *et al.*, 2014).

The results published by Zeki and his colleagues are significant to our enquiry as they confirm Burke's intuition of the fundamental differences in the principles leading to the experience of beauty and those leading to that of the sublime, basing the first on positive feelings (such as "pleasure") and the second on negative ones (such as "fear").

A further relevant issue is the similarity between the sensitive experience of a viewer put in front of an artifact and that of a viewer put in front of a natural landscape: The main factor of difference lies in the role played by the imagination in processing the experience of artefactual sublime, carrying out a process of "completion", as the qualities of sublimity conveyed by fine arts and words lie beyond the senses, and require the creation of a counter-factual experience carried out by the recipient's brain. In particular, we need to reflect on the imagination prompted by language, as we need to contextualize in a scientific frame the process instantiated not only in the image-beholder of Friedrich's painting but particularly in the reader of Kleist's ekphrasis.

In addition to the actions, emotions, and sensations we experience, those we read about also activate bodily representations "as if" we were directly experiencing similar emotions, feelings, or sensations, which are also evoked by representational forms, colors, lines, and words. The aesthetic and emotional response during literary experience is strongly enhanced by the dynamic qualities of literary imagery. It has already been demonstrated through brain-imaging studies that the activity of sensorimotor areas is particularly enhanced by motor imagery triggered both by visual stimuli as well as by literary rhetorical figures (Cuccio, Carapezza, & Gallese, 2013). Particularly in the case of language, it has been shown that "retrieving an object from memory or thinking or imagining or reading generates responses in the body that show similarities to the responses that would occur if the object were present" (Dijkstra & Post, 2015, p. 10).

The multisensory nature of imagination has also been studied through the scientific evidence of the overlapping processes between haptic and visual sensitive substrates (see Kim & James, 2010; Stilla & Sathian, 2008): Information derived from visual and haptic origins converges in the intraparietal sulcus, which is part of not only the somatosensory and visual system but also the motor system. This convergence probably allows, in ways that are not yet clear, for the integration of sensory, visual, and motor processes, which is at the basis of the multisensory nature of imagination (Goebel, Khorram-Sefat, Muckli, Hacker, & Singer, 1998). In particular, looking at a painting representing a physical action or reading about a physical action engages part of the motor system, specifically the mirror-neuron system (Cattaneo & Rizzolatti, 2009; Di Pellegrino, Fadiga, Fogassi, Gallese, & Rizzolatti, 1992; Gallese, Fadiga, Fogassi, & Rizzolatti, 1996; Iacoboni *et al.*, 1999), which responds, both in monkeys and human beings, not only when executing but also when observing, an action, or when reading about it, simulating the execution of that action.

The past 15 years of studies in this field have provided a great deal of evidence suggesting that language comprehension involves sensorimotor representations (Louwerse, Hutchinson, Tillman, & Recchia, 2015; Meteyard, Cuadrado, Bahrami, & Vigliocco, 2012). According to studies of text-processing, the reader's mind creates a counterfactual model of the narrative world described by the text (Johnson-Laird, 1983; van Dijk & Walter, 1983). The mind's construction of a counterfactual world is an important process for understanding narratives because it allows the reader to connect different aspects and sources of information (time, space, motivation, character, and cause-effect) within a coherent and fluid mental representation that can be continually integrated, stabilized, updated, and emotionally experienced (Kintsch, 1988; Kintsch, Welsch, Schmalhofer, & Zimny, 1990; van Dijk & Walter, 1983). Most studies of reading in empirical psychology distinguish three levels of cognitive representation in processing a text: (1) the surface code (wording and syntax), (2) the text base (explicit text propositions and text coherence), and (3) the referential situation model (what the text is about). At this last level, readers can take on an active role by placing themselves in the perspective of the characters and their surroundings (i.e., the time frame of the story, the location in which the events are happening, and the objects that the characters are carrying). Unlike the surface structure or the propositional text base, in which these mental representations rely solely on the information given in the sentences, counterfactual models require the reader's imagination and their interpretation of the information conveyed (Graesser & Zwaan, 1995).

As proposed by Kuzmičová (2014), the reader experiences the actions described in the literary text through their "enactment-imagery", a process that sets the reader virtually within the counterfactual world (see Alber, Caracciolo, & Marchesini, 2018). In the aesthetic experience of a literary text, the intensity of a reader's emotional response is bound to the intensity of the activation of the sensitive motor areas because of the interaction between sensorimotor, emotion–valuation, and meaning–knowledge circuitries during aesthetic experiences (Chatterjee & Vartanian, 2014, p. 370; Shimamura, 2012, 2013). Language has been shown to exploit the pre-existing multimodal character of the sensory-motor system (Gallese & Lakoff, 2005, p. 456). Many studies also prove that some parts of the brain activated by seeing are also engaged in visual imagination (imagining that you are seeing; Farah, 1989; Gallese & Lakoff, 2005, p. 463; Kosslyn, 1980, 1983; Kosslyn, Thompson, & Ganis, 2006). Finally, as Starr (2010, 2013) also argues, referring to empirical studies that have shown the involvement of the motor system in imagined actions, any kind of imagery, because of its multisensory nature, enhances a bodily kinesthetic reaction. Moreover, a large number of references to action and motion within a literary text produces a particularly intense response in sensorimotor circuitries during the act of reading (see Zwaan, Langston, & Graesser, 1995).

With these scientific theories and outputs as a background, we are now in a position to interpret Kleist's ekphrasis as a milestone in the representation of the sensitive sublime: Kleist's ekphrasis strikingly shows textual features that refer to bodily experiences and the author, by doing so, implicitly attests to the validity of Burke's sensitive sublime theory in opposition to the mainstream Kantian one. The textual features strongly trigger the reader's bodily and emotional activation through a mental simulation of visual perception and actions, evoking an experience of the sublime rooted in the body.

Transcendental Sublime Versus Sensitive Sublime

The German debate about the sublime during the 18th century was strongly influenced by the reception of Edmund Burke's theories, leading to a large divide between the positions of those who believed in a transcendental sublime and those endorsing a sensitive one. In the tradition of Wolffian philosophy, Moses Mendelssohn considered the sublime as a sensible expression of perfection, arousing wonder (*Über das Erhabene und das Naive*, 1757), as beauty also did (*Über die Empfindungen*), thrilling body and reason as well. But the idea of negative sublime feelings, like fear and pain, arising from darkness or disorder, claimed by Burke, was contrasted with the assumptions put forward by Wolffian rationalism. It is the sensible perfection of greatness — also if related to a frighteningly sublime experience — that constitutes the core of Mendelssohn's concept. Just as with the premises put forward by Baumgarten (1758) in his *Aesthetica*, for Mendelssohn, too, the greatness is bound to the perfection of the object, which deeply differed from Burke's view, whose theories were well known to Mendelssohn, having read them at the suggestion of Lessing. In 1761, Mendelssohn wrote *Rhapsodie*, rethinking Burke's theories and integrating them in his concept of the sublime as triggered by sensibly unmeasurable qualities, like the Burkian ones of vastness, grandeur, and magnitude. Moreover, Mendelssohn (1771/1972) rewrote *Über das Erhabene und Naive in den schönen Wissenschaften* in 1771, leaning on Burke's theories on the artful way of depicting the immeasurable and on his distinction between the sublime and the beautiful, overcoming his previous vision of the infinite as a result of perfection. Sublime to him now becomes an experience of the brain (Geist) being thrilled by the sensual perception of the immeasurable.

Edmund Burke, in his treatise *A Philosophical Inquiry into the Origin of Our Ideas of the Sublime and Beautiful* (1757/1887), traced back the perception of the sublime to the most natural and elementary feelings of human nature (Meyer-Sickendiek, 2005) that regulate survival in the natural and social world: fear and pain. Burke claimed that the human body is either in a state of pain or in a state of pleasure, and sometimes in neither of them (i.e., in a condition of indifference). The removal or diminution of pain does not provoke pleasure but delight (Burke, 1887, Part 1, Sections 4 and 5). Pain and pleasure, or their modifications, like danger, are oriented or finalized to "self-preservation and society" (Burke, 1887, p. 113). This last is intended to mean not only human society but also "general society, which we have in common with men and other animals, and which we may in some sort be said to have even with the inanimate world" (Burke, 1887, p. 112). Erasing the big divide between culture and nature, Burke established a new conception of natural life, which was taken up by some authors of the German Romantic time (Berlin, 1999), such as Kleist, as in his representation of a cross-species community of humans and animals (Haraway, 2008) in *Penthesilea* (Kleist, 1987, 1993) (Gambino & Pulvirenti, 2017b).

Pain and danger are, therefore, "productive of the strongest emotion which the mind is capable of feeling" and are instantiated by "whatever is in any sort terrible" (Burke, 1887, p. 111). According to Burke, we experience beauty and the sublime in art because of our ability to feel sympathy, i.e., a sort of substitution by which "we are put into the place of another man, and affected in a good measure, as he is affected, so that this passion may either partake of the nature of those which regard self-preservation, and turning upon pain, may be a source of the sublime; or it may turn upon ideas of pleasure" (Burke, 1887, p. 117).

Burke claims that tragic representations in art are a source of pleasure, denying that this pleasure arises from the reasoning faculty but from the "mechanical structure of our bodies" (Burke, 1887, p. 118). While in real life, passions regarding self-preservation turn on pain and danger, they may also turn on delight "when we have an idea of pain and danger, without being actually in such circumstances. … Whatever excites this delight, I call *sublime*" (Burke, 1887, p. 125).

Burke distinguishes two different passions aroused by the sublime: (1) astonishment (i.e., the effect of sublime in its highest degree, which is related to horror in nature and may cause admiration or reverence or respect) and (2) terror, which is related to fear, considered as "an apprehension of pain or death" (Burke, 1887, p. 131). In nature, such passions are induced by the following features: magnificence, greatness in dimensions, vastness of extent, quantity, obscurity, vacuity, darkness, solitude, silence, and infinity (cause of delightful horror). In the arts, they are triggered by the following features: succession of the parts, magnitude of dimension, unfinished and difficult composition that solicits the imagination more strongly, and magnificence and disorder.

As Burke points out, since sight may be a source of fear, sublime feelings may be triggered by paintings in cases where these least resemble nature, depicting it with dark or confused images, or with features of privation, like "vacuity, darkness, solitude, and silence", which strongly influence the imagination (Burke, 1887, p. 146). In his opinion, colors are a powerful source of magnificence, since they convey light and darkness, which is exalted through colors overpowering the senses, and producing an effect of grandeur (Burke, 1887, Part 1, Sections 16–18). Therefore, a dark, gloomy, cloudy, or nightly sky is more solemn than a daily blue sky and this may be obtained by using "sad and fuscous colours, as black, or brown, or deep purple and the like" (Burke, 1887, p. 159). A further source of fear is words, that is, the literary texts that convey "the affections of the mind" (Burke, 1887, p. 134). Therefore, words are conceived as the best way to convey passions, giving rise to imagery, that is always of an obscure kind: "So that poetry, with all its obscurity, has a more general, as well as a more powerful dominion over the passions, than the other arts"

(Burke, 1887, p. 135). Sounds produce sublime passions, particularly the cries of animals in pain or danger: "The angry tones of wild beasts are equally capable of causing a great and awful sensation" (Burke, 1887, p. 160) as they express the universal connection among humans, animals, and the inanimate world, without any further need of interpretation, being non-semantic (Burke, 1887, Part 2, Section 20). Other sources of the sublime may be smells and tastes, particularly if they convey modes and degrees of pain, anguish, or torment. Troubles and danger are at the origin of great virtues, more than any pleasant experience, as strong unpleasant "affections of the body produce such a distinct emotion of mind" (Burke, 1887, p. 208).

These same feelings are at work when sensuously experiencing exceedingly fearful manifestations of nature: its boundlessness, greatness, infinity, obscurity, solitude, silence, and so forth. This is because, in Burke's opinion, the feelings mentioned above provoke a "violent emotion of the nerves" (Burke, 1887, p. 212) and a "violent pulling of the fibres" (Burke, 1887, p. 245), which are positive feelings, as they enhance the vital forces. In this way, Burke highlights a theory of the sublime as rooted in the bodily sensory experience:

But I conceive, if we can discover what affections of the mind produce certain emotions of the body, and what distinct feelings and qualities of body shall produce certain determinate passions in the mind, and no others, I fancy a great deal will be done; something not unuseful toward a distinct knowledge of our passions, so far at least as we have them at present under our consideration (Burke, 1887, p. 298).

For Burke, the sublime does not depend on the intuition of a transcendental realm but on the experience of the body touching its limits and achieving a feeling of contact with the essence of the world. To this sensitive and physiological concept of the sublime, Kant will oppose a transcendental one, focused on the idea of an abstract reason, as Richardson (2010) claims:

> "Kant's influential theory of the sublime, especially as it has been taken up and rehearsed by later writers, notably leaves the body and its brain behind to posit an idealized Reason that, in the course of the sublime moment, succeeds in touching on the noumenal itself, however obliquely,

> however indescribably. Kant dismissed Burke's theory of the sublime, after all, as a "physiological" or "merely empirical" exposition of the subject, unqualified to speak to the universality of aesthetic taste revealed by the sublime experience and so required to give place to a transcendental treatment of the subject such as Kant's own. … [T]he Kantian sublime definitely trends towards valuing the mentalistic over the physiological, the abstract over the empirical. Kant's sublime might itself be described as a cognitive sublime, but a neural sublime would seem to remain, by contrast, tied to the phenomenal, tethered to the body and its limitations — and what kind of a sublime would that be?" (pp. 24–25)

Kant described the sublime as a "negative pleasure" (Kant, 1790/2009, p. 76), leading to an inhibition of life's bodily forces; this inhibition is overcome by a transcendental mental elaboration of this experience. The sublime overwhelms the sensual perception and the imagination, leading to the activation of the rational intellect, which is considered as a "measure of the senses" (Kant, 1790/2009, p. 85). This is particularly enhanced by what Kant defines as the "dynamic sublime" in contrast to the "mathematical sublime" (Kant, 1764): While the mathematical sublime is intended as the feeling of a supersensible faculty apprehending the magnitude of nature, the dynamic sublime is intended as the experience of the dreadful boundless power of nature. The dynamic sublime evokes feelings of limits and nothingness in the viewer and leads them to refuse the identification with their own bodily finiteness and to discover their mental superiority over senses and nature. In this way, the experience of the dynamic sublime leads to the perception of a mental self-experience as empowerment of the rational faculties over nature (Kant, 1790/2009, p. 102). In both cases, Kant relates the sublime to the mind's power of surpassing sensory perception, of feeling pleasure in experiencing the superiority of reason over nature, conveying reason to transcendence.

Kant's theories were successfully reinterpreted by Schiller's works *Vom Erhabenen* (1793/1962) and *Über das Erhabene* (1801/1972): The experience of the sublime was described as provoking a kind of sublimation of the soul with regard to the dangerous and therefore "delightful" experience of the mixed feelings of horror and pleasure, enlarging the human ability to think, to feel one's own self, nature, and the world.

Figure 1. Caspar David Friedrich, Monk by the Sea, Berlin, Alte Nationalgelerie.

Source: Obtained from SMB-digital online collections database (http://www.smb-digital.de/eMuseumPlus?service=ExternalInterface&module=collection&objectId=965511&viewType=detailView) under the Creative Commons Attribution-NonCommercial-ShareAlike 3.0 Germany License.

Caspar David Friedrich's Representation of the Sublime

A masterpiece of the pictorial representation of the Romantic sublime is Caspar David Friedrich's painting, *Monk by the Sea*[1] (1808–1810; Figure 1), marking the radical shift occurring in the pictorial representation of space perception. At the end of the 18th century, theories regarding visual perception were deeply modified by innovations in the field of physics, as exemplified by the circular representations of panoramas (i.e., oversized rounded paintings), such as the famous one by Robert Barker, shown in London in 1792. Panorama exhibitions became a fashion in Europe, as other cases show: In 1808, Karl Friedrich Schinkel exhibited his *Panorama of Palermo* in Berlin, in the round Church of St. Hedwig. This new way of looking at a wide landscape deprived the observer of the traditional central perspective, compelling them to move 360 degrees around the "installation", to change their point of view continuously, and to take part in the space of the representation.

This new experience of the gaze disrupted Alberti's central perspective, which was the expression of a rational elaboration of world perception, as it is to be perceived looking at Friedrich's painting, which subverts almost all the formal rules of landscape painting of that time: in addition to Alberti's perspective construction, the use of the frame, the relationship between foreground and background, the references to proportion, and the function of staffage figures (i.e., human and animals depicted in a landscape). Moreover, the painting renders the violence of nature and the loneliness of the endeavors by depicting a mostly schematic scene, horizontally divided into three areas characterized only by colors, the pale ochre of the dune, the black of the sea, the blue-gray of the cloudy sky. Friedrich's representation exceeds the frame of the human gaze and amplifies the vision toward boundlessness. This is achieved by applying different strategies: First of all, the painting modifies the use of the foreground, introducing a great change in the basic principles of the 18th century's "framed gaze" (Stoichita, 2013). The importance given to the frame in 18th-century theory was related to the implicit rational intention of the gaze: This art of exploring the world emphasized precision and sharpness and therefore the need to cut out apart from the whole. Isolation and sharpness determined by the framed gaze helped give a cognitive structure to the elements in paintings. Step-by-step, the eye was conducted to explore the surface of the canvas, putting single elements together into a sort of complex cluster that gave a sense to the parts and to the whole of the representation. This "framed model" responded to the taste of the time and to the principles of the educational model, called *Anschauungsunterricht* (Gambino, 2010), based on observation, recognition, and narration, as theorized during the 18th century. Friedrich rejected the rules of the framed gaze by opening the sight to immeasurable spaces created by means of color fields composed without any perspective structure (Begemann, 1990) and of staffage figures disproportionately reduced in dimension.

The presence of just the tiny dark figure of a monk, placed under the line of the horizon, almost completely turned away from the viewer, compels the observer of the painting to "physically" enter the depicted natural space, by identifying themself with the monk, who is the only reference for the viewer. Therefore, the viewer is led to experience imaginatively the representation not as a picture but as a real natural space: they become part of the scene, taking over the perspective view of the monk. Therefore, the viewer is "transported" into the natural endeavor, instead of staying in front of it, elaborating a judgment about the artistic quality of the painting. In this way, Friedrich tries to represent the physical and emotional experience of the subject in front of a boundless and menacing nature. According to the groundbreaking study by Koerner (1990), the bodily sensible experience aimed at by Friedrich in his conceptual landscapes and in the conscious creation of the sublime leads to a cathartic/ spiritual effect. This confirms our thesis about the efficacy of the painting compelling the viewer to experience the sublime as rooted in the bodily sensible perception of overwhelming nature, in contrast to the Kantian concept of sublime intended as a metaphysical experience of an abstract intellect.

According to this idea, and as a result of the so-called Kantian crisis (Mandelartz, 2006), in his text about Friedrich's painting, Kleist evokes a sublime experience as a powerful one bound not to a mental empowerment of the self but to an enhanced bodily perception of nature: The human being is driven to perceive more strongly the material force of nature through the senses, by activating their most archaic emotions, bound to a natural, we may say "animal", reaction to primitive feelings of fear and anguish.

Kleist's Sensitive Concept of the Sublime

When Friedrich's painting was first exhibited in October 1810 at the Academy of Fine Arts in Berlin, it became a controversial case because it was a radical departure from the basic principles of classical landscape painting (Begemann, 1990; Müller, 1986, 1995; Rühse, 2013). Heinrich von Kleist published in the journal *Berliner Abendblätter*[2] (Kleist, 2013) a review of Friedrich's painting by Clemens Brentano and Achim von Arnim (1810/2013), but the text had been re-edited by Kleist himself, who maintained the first part (lines 1–12 of the following text) and added a completely new one — starting in line 12 — although he printed it using the initials of Brentano's name ("cb"), leading to a major quarrel with his fellow writers.[3] What is it that struck Kleist so powerfully as to lead him to re-write the text emphasizing in the second totally original part the painting's aesthetic power and novelty, which the other authors had not understood as deeply as he? Why is Kleist's ekphrasis of Friedrich's painting of such great impact and why has it been so difficult to interpret until now?

From its origins, ekphrasis was meant to stimulate the imagination (Webb, 2009, p. 1) and to achieve an alteration in the reader/listener's emotional state. This was to be obtained by enhancing through words the "liveliness" of the mental image representing the described object (Henkel, 1997), consequently illuminating the quality and essence of the object itself. This technique was considered in Latin culture and particularly by Quintilian (*Institutio Oratoria*, 6.2.29–30) as a powerful tool "used to make the audience feel involved in the subject matter, to make them feel as if they were at the scene of a crime, or that they themselves witnessed the achievements for which an emperor is being praised" (Webb, 2009, p. 10). In fact, ekphrasis was best applied to represent not just a scene (i.e., a time, place, or characters) but the action itself (i.e., a murder or a battle).

With regard to the relevance of actions in narrative form, recent neurocognitive studies have empirically shown that the description of an action triggers a more intense emotional response in the reader/listener than the description of a static scene or of objects, as the somatic components of emotion react more strongly to the activation of the solicited sensorimotor circuitry (see Cuccio *et al.*, 2013). Although unaware of these brain processes, ancient writers were well aware of the efficacy of the rhetorical use of the language and aimed to depict the subject matter through actions and events, instead of carefully listing the elements in the scene. In fact, the final purpose of ekphrasis was to "persuade" and emotionally move the audience by creating the fictive and lively illusion of "being present" at the described scene. At a linguistic level, this was achieved through the quality of *energeia*. As analyzed by Webb (2009), Quintilian claimed that *energeia* is to be regarded as an active force (Quintilian, *Institutio Oratoria*, 6.2.29–30) activating the imagination

(Webb, 2009) and the "simulation" of being bodily present at and involved in the space and time of the described situation.

These aspects were significant for the Romantic use of ekphrasis, reflecting the radical innovations occurring in the philosophical debate about perception, thought, and emotion, as pointed out by many scholars focusing their investigations on the regimes of the visible as related to language (Boehm, 2014; Boehm & Pfotenhauer, 1995; Miller, 2002; Osterkamp, 1991; Pfotenhauer, 2000). What is at stake in the ekphrasis of Romanticism is the crisis of the classical aesthetic representation of visible qualities (Gambino & Pulvirenti, 2017a). This is the case of Kleist's ekphrasis: The text does not describe the work of art itself (i.e., the pictorial features of Friedrich's painting) but the process engaged in by the viewer looking at the picture, that is, the reader's/viewer's physical and emotional process of relating to the depicted overwhelming nature by moving into it. In this way, Kleist amplifies Friedrich's intentions of representing the process of seeing and of relating it to a natural scene, as he claimed in a letter to Johannes Schulze in February 1809: *Monk by the Sea* "cannot be defined as a description. It is instead the act of seeing itself" (Friedrich & Zschoche, 1809/2006, p. 45).

We now quote Kleist's text, with numbered lines to make further references clear:

1. It is magnificent to stand in infinite solitude on the seashore, beneath an overcast sky, and to
2. look on an endless waste of water. Part of this feeling is the fact that one has made life's way
3. there and yet must go back, that one would like to cross over but cannot, that one sees
4. nothing to support life and yet senses the voice of life in the sight of the waves, the murmur
5. of the air, the passing clouds and the lonely cry of birds. Part of this feeling is a claim made
6. by the heart and a rejection, if I may call it that, on the part of nature. But this is impossible
7. in front of the picture, and what I should have found in the picture itself I found only

8. between myself and the picture, namely a claim my heart made on the picture and the
9. picture's rejection of me; and so, I myself became the monk, and the picture became the
10. dune, but the sea itself, on which I should have looked out with longing — the sea was
11. absent.
12. There can be nothing sadder or more desolate in the world than this place: the only spark of
13. life in the broad domain of death, the lonely centre in the lonely circle. The picture, with its
14. two or three mysterious subjects, lies there like an apocalypse, as if it were thinking Edward
15. Young's *Night Thoughts* and since it has, in its uniformity and boundlessness, no
16. foreground but the frame, it is as if one's eyelids had been cut off. Yet the painter has
17. undoubtedly broken an entirely new path in the field of his art, and I am convinced that with
18. his spirit, a square mile of the sand of Mark Brandenburg could be represented with a
19. barberry bush, on which a lone crow might sit preening itself, and that such a picture would
20. have an effect that rivaled Ossian or Kosegarten. Why, if the artist painted this landscape
21. using its own chalk and its own water, I believe he would make the foxes and wolves weep:
22. the most powerful praise, without doubt, that could be given to this kind of landscape
23. painting (Kleist, 1810/2013).

The text is structured in three parts: The first (Lines 1–11) describes the magnificent experience of solitude in front of the sea, by physically placing the reader in the depicted landscape "as if" it were a real one, which overwhelms the perceptive limits of the human being, challenging their ability to organize a cognitive experience of nature. In Line 7, the reader discovers that the process that they are engaged in is not prompted by a real landscape but by an artificial one. In fact, we find here the first mention of the painting to which the text refers. In the second part (Lines 12–16), the perceptual perspective changes and the author shifts to a strongly organized rhetorical language, in order to overcome the human limits in the cognitive organization, based on space, time, and cause–effect. Finally, in the third part (Lines 16–23), the experience of the sublime attests itself as a sensual, bodily, and prelinguistic one, according to an organic vision of nature as a material essence bounding all creatures.

The opening lines are characterized by two issues: On the one hand, the author uses words that seem to refer directly to those used by Burke to explain the principles of the sublime (in italics in the following quotation); on the other hand, the reader is prompted to imaginatively create mental images that emotionally transport them to the center of the depicted natural scene, thanks to the use of verbs of motion and sight, spatial deixis, and locative terms (in bold in the following quotation):

> It is *magnificent* to stand in *infinite solitude* on the seashore, beneath an *overcast sky*, and to look on an *endless waste* of water. Part of this feeling is the fact that **one has made his life's way there** and yet must **go back**, that one would like to cross over but cannot, that one sees nothing to support life and yet **senses** the voice of life in the **sight** of the waves, the *murmur* of the air, the **passing** clouds, and the *lonely cry* of birds.

The reader is plunged into the "infinite solitude on the seashore, beneath an overcast sky" and cast in front of "an endless waste of water" (see Lines 1–2), by the use of verbs setting the body in space (e.g., "to stand", Line 1), which solicits the sensation of a stable position and expresses the possible movements made to reach the place: "life's way there" (Line 2), "yet must go back" (Line 3), and "to cross over" (Line 3). The use of such verbs of movement elicits a mental representation strongly bound to space and motion, allowing the reader to situate themself in the represented space of nature.

Kleist triggers the reader's sensorimotor system within the mental representation of natural space, guiding them into an "as if" process by using verbs of bodily motion (see Lines 2–3), and soliciting the reader to establish a dynamical perceptual relationship with the space inside the picture, by using verbs of perception, such as “to look” (Lines 1–2), “to see” (Line 3), and “to sense” (Line 4). In the empathic experience triggered by the first part of the text, the author reproduces the somatic states seen in or implied by the description of the action of getting inside the space, as if the body were present.

The use of verbs related to the act of perceiving real material nature triggers the sensorimotor system, involving the visual, auditory, and proprioceptive systems. The action of standing on the seashore, of moving in this endeavor, and of going back activates with particular efficacy the process of simulation and of empathic identification. The author further leads the reader to perceive the landscape, suggesting the double identification of the reader with the viewer in front of the painting but also with the monk represented in the painting (in bold in the following quotation):

> “[W]hat I should have found in the picture itself **I found only between myself and the picture**, namely a **claim** my heart made on the picture and the picture’s rejection of me; and so **I myself became the monk, and the picture became the dune**, but the sea itself, on which I should have looked out with longing — the sea was absent.”

Again, the verbs of movement drive the experiencing self into a disproportionate landscape, which they can perceive through the senses but not intellectually understand and frame. The focus of the painting — the sea (Lines 10–11) — is revealed to be a vacuum, with no references to cognitive coordinates or visual boundaries. The landscape rejects the viewer/reader, as the space references and the cause–effect references are shifted to an immaterial and boundless panorama. This ambiguous experience overwhelms human cognitive abilities, soliciting a particularly intense sensitive response in the reader.

In the second part (Lines 12–16), the natural landscape is shifted to a metaphorical and meta-representational level: The failed challenge of the cognitive process in front of nature is featured as a rejection of the natural landscape and a return to the artful one. In order to convey in the verbal code this aesthetic experience of the viewer in front of the painting, Kleist shifts to a rhetorical language, intertwining figures like hyperbole ("There can be nothing sadder or more desolate in the world than this place", Line 12), prosopopoeia ("the lonely center in the lonely circle", line 13), metaphor ("the only spark of life in the broad domain of death", lines 12–13), and simile ("like an apocalypse", line 14), which creates the climax of this second part. The boundlessness of the vision represented in the painting leads to a "failure" of the gaze, which is no longer able to "frame" and stabilize the image, leaving the readers in front of the painting with their "eyelids … cut off" (line 16). The "failure" to frame the vision of the natural landscape attests, in Kleist's view, to the power of Friedrich's painting, which is able to solicit the reader's bodily perception in a new way. By compelling the reader to access the picture through an as-if experience, Kleist allows the subject to feel their own being as deeply related to the very essence of nature, which is meant to be the focus of the picture.

In the last part (lines 16–23), Kleist expresses his own opinion about the painting in the first person. Writing as an art critic, he attests to the deep relation existing between nature and art, underlying the fact that the "artful" power of Friedrich's picture consists in its ability to create a bodily perception of the material components of the landscape, pushing the subject to regain their "animal" core. In Kleist's opinion, the material and perceptual experience of the viewer/reader is achieved by the painting because it has dismissed every symbolic and allegorical reference, presenting instead the physical elements in their pure materiality ("chalk" and "water"). The experience of nature becomes a prelinguistic one so that finally the true human emotional response cannot be expressed through semantic language but rather through something similar to the natural cries of animals, which Burke already considered as examples of a powerful expression of the sensitive sublime, stressed by Kleist through the reference to the whine of "foxes and wolves" (Line 21):

If the artist painted this landscape, using its own chalk and its own water, I believe he would make the foxes and wolves weep: the most powerful praise, without doubt, that could be given to this kind of landscape painting (Lines 20–23).

The discourse about the natural and the artful sublime is significant in the German aesthetic discourse of Kleist's epoch because the two forms of experiencing the sublime were not regarded as equally effective since art and nature were thought to affect the human mind in different ways. Burke, on the contrary, considered both equally powerful in affecting the body and the emotions. Kleist apparently denies the possibility of experiencing the powerful pictorial image by relocating the viewer inside the depicted natural scene. But in fact, through the rhetorical construction of the second part of the text, the author puts the reader in conditions of imagining not only the real natural scene but also the powerful effect of the language presenting the painting as a medium of the natural landscape elaborated by the painter in terms of a sensitive sublime, which is stressed in the last part of the text by abolishing every intermedial experience and relocating the reader in the materiality of the natural world.

Kleist was not so much a theorist in aesthetics and poetics but rather an aphoristic declaimer of his ideas and an experimenter with them in his works, essays, and letters so that we do not find the enunciation of a theory of the sublime but many passages that are representations of sublime experiences, where nature and culture do not appear as opposites. A letter from Leipzig about his stay in Dresden, sent to Wilhelmine von Zenge on May 21, 1801 (Kleist, 1801/n.d.), is the best demonstration of this issue: The author describes the masterworks of the Gemäldgalerie in Dresden as natural landscapes evoking the experience of the sublime, which is again and again expressed in relation to the perceptions of the body ("senses and heart") and "not of the intellect". The landscape around Dresden and the natural environment experienced during his wanderings are also described in terms of a sensual sublime, with a handful of expressions concerning the grandeur of the sights, the irregular movement of the river, and the immensity of the landscape, which is compared to a "turbulent sea of earth." The complementariness, and not opposition, of nature to art is achieved through the description of mountains that look "like colossal Pyramids … made by the hand of Angels playing in the sand." In this simile, everything leads to the sublime emotion: nature, art, and the evocation of a superior power in terms of the bodily action of the angel's hand. This comparison between art and nature as a source of the sublime again evokes Burke's theories and their revolutionary implications for a new conception of the human cognitive and imaginative faculties:

> The images in paintings are exactly similar to those in nature, and in nature, dark, confused, uncertain images have a greater power on the fancy to form grand passions (Burke, 1887).

For Kleist, verbal as well as visual arts are able to give rise to the sublime as they both concern the bodily experience of nature. Stretching bodily perception to its limits, Kleist represents man not as a being placed above nature, able to transcend it thanks to reason, but as a living part of nature, experiencing it with body and brain.

The experience of the sublime leads to a stronger perception of nature common to humans and animals. This issue appears again in the short story *St. Cecilia, Or the Power of Music* (published in November 1810; Kleist, 2016). Here, four brothers are deeply struck by an artistic-mystic sublime experience of the music played in the fictional Cathedral of Corpus Christi; they become crazy and are kept in the city's house for the insane. After listening to the composition, probably mysteriously conducted by Saint Cecilia, they begin praying in front of a crucifix and singing the *Gloria in excelsis Deo* at midnight, which deeply moves their senses (Greiner, 1996). The relevant point is the way in which their singing is described, more similar to animals' cries than to human song:

> "[T]hey began to sing the *Gloria in excelsis* in the most hideous and horrible voice. The sound of leopards and wolves, when on an icy winter's night they roar at the sky, may be something like it. The pillars of the house ... were shaken, and the window-panes smitten by the visible breath from their lungs, rattled and threatened to fall in, as if handfuls of heavy sand were dashed against their surface" (Kleist, 2016, p. 303).

Their "fearful and revolting howls" are the expression of an experience that is not elaborated by the rational mind, overcoming the fear and horror of the sublime, as in Kant's theory, but by a surrender of the mind to the body's most archaic feelings. According to Burke, the sublime is a product of physiological feelings of fear and pain, instantiated by the perception of a bodily danger, which is not felt as real, as it is mediated by the imagination. The visual and the linguistic codes are therefore equally powerful in soliciting the sublime experience because they both refer to the basic physiological elements and natural qualities shaping the human interaction with nature. According to Herder's (1778/2002) mainstream theories, language participates in the organic complexity of existence and shares the common morphological roots of the whole living system. In this perspective, the imagination models a totality that the human mind is not able to realize through its logical procedures and references to the laws of vital relations. Both forms of natural phenomena and linguistic figures share the same endless reproductive drive, which reason cannot achieve by only elaborating perceptions.

By using language as a tool to get back to the roots of sensations and emotions, Kleist gives poetic evidence and physical corporeality to his idea of the sublime; he disrupts the transcendental and hegemonic conceptions of aesthetic experience as related to the abstract intellect and of the arts as depending on pre-established rules and principles, opening new modern insights into the human concept of the mind as well as into the concept of linguistic production as an experiment of the mind's adventures (Gambino & Pulvirenti, 2018a, 2018b) in a never-ending process of constructing new meanings for human survival in a world menacing his/her fragile and ephemeral nature.

Notes

- The painting is now exhibited at the Alte Nationalgalerie, Berlin; accession number NG 9/85.
- The journal *Berliner Abendblätter* was edited by Heinrich von Kleist with Adam Müller and published in Berlin by Julius Eduard Hitzig from October 1, 1810 until March 30, 1811.
- The text was followed by six short dialogues, a sort of ironic report of the comments made by the audience during the exhibition. For the text's genesis and von Kleist's rewriting of Brentano and von Arnim's original contribution, see Begemann (1990), Seeba (2003), and Meyertholen (2013).

References

Alber, J., Caracciolo, M., & Marchesini, I. (2018). Mimesis: The unnatural between situation models and interpretive strategies. *Poetics Today*, 39(3), 447–471. https://doi.org/10.1215/03335372-7032690.

Baumgarten, A. G. (1758). *Aesthetica*. Frankfurt, Germany: Kleyb.

Begemann, C. (1990). Brentano und Kleist vor Friedrichs Mönch am Meer. Aspekte eines Umbruchs in der geschichte der Wahrnehmung [Brentano and Kleist in front of Friedrich's painting "Monk by the Sea": Elements of change in the history of reception]. *Deutsche Vierteljahrsschrift für Literaturwissenschaft Und Geistesgeschichte*, 64(1), 54–95. https://doi.org/10.1007/bf03396159.

Berlin, I. (1999). *The Roots of Romanticism*. Princeton, NJ: Princeton University Press.

Boehm, G. (2014). Bildbeschreibung. Über die Grenzen von Bild und Sprache [Image descriptions: Beyond the limits of images and words]. In M. R. Müller, J. Raab, & H.-G. Soeffner (eds.), *Grenzen der Bildinterpretation* [*Limits of the Interpretation of Images*], pp. 15–37. Wiesbaden, Germany: Springer.

Boehm, G. & Pfotenhauer, H. (1995). *Beschreibungskunst-Kunstbeschreibung: Ekphrasis von der Antike bis zur Gegenwart* [*The Art of Description and the Description of Art: The History of ekphrasis from Antiquity to Present Times*]. Munich, Germany: Wilhelm Fink.

Burke, E. & Nimmo, J. C. (ed.) (1887). A philosophical inquiry into the origin of our ideas of the sublime and beautiful. In: *The Works of the Right Honourable Edmund Burke*, Vol. 1, pp. 67–263. London, England (Original work published 1757).

Cattaneo, L. & Rizzolatti, G. (2009). The mirror neuron system. *Archives of Neurology*, 66(5), 557–560. https://doi.org/10.1001/archneurol.2009.41.

Chatterjee, A. & Vartanian, O. (2014). Neuroaesthetics. *Trends in Cognitive Sciences*, 18(7), 370–375. https://doi.org/10.1016/j.tics.2014.03.003.

Cuccio, V., Carapezza, M., & Gallese, V. (2013). *Metafore che risuonano. Linguaggio e corpo tra filosofia e neuroscienze* [*Metaphors that Resonate: Language and Body between Philosophy and Neuroscience*]. Rivista dell'Associazione Italiana Studi Semiotici, 7(17), 69–74.

Dijkstra, K. & Post, L. (2015). Mechanisms of embodiment. *Frontiers in Psychology*, 6, 1525. https://doi.org/10.3389/fpsyg.2015.01525.

Di Pellegrino, G., Fadiga, L., Fogassi, L., Gallese, V., & Rizzolatti, G. (1992). Understanding motor events: A neurophysiological study. *Experimental Brain Research*, 91, 176–180. https://doi.org/10.1007/bf00230027.

Farah, M. J. (1989). The neural basis of mental imagery. *Trends in Neurosciences*, 12(10), 363–416. https://doi.org/10.1016/0166-2236(89)90079-9.

Friedrich, C. D. (1808–1810). *Monk by the Sea* [Painting]. Retrieved from http://www.smb-digital.de/eMuseumPlus?service=ExternalInterface&module=collection&objectId=965511&viewType=detailView.

Friedrich, C. D. & Zschoche, H. (eds.) (1809/2006). *Caspar David Friedrich: Die Briefe* [*Caspar David Friedrich: The Letters*]. Hamburg, Germany: ConferencePoint (Original work published 1809).

Gallese, V., Fadiga, L., Fogassi, L., & Rizzolatti, G. (1996). Action recognition in the premotor cortex. *Brain*, 119(2), 593–609. https://doi.org/10.1093/brain/119.2.593.

Gallese, V. & Lakoff, G. (2005). The brain's concepts: The role of the sensory-motor system in conceptual knowledge. *Cognitive Neuropsychology*, 22(3–4), 455–479. https://doi.org/10.1080/02643290442000310.

Gambino, R. (2010). Auf dem Weg zum "Mittelpunkt des Schönen". Das pädagogische Konzept von Karl Philipp Moritz' italienischer Reisebeschreibung [The core of beauty: Karl Philipp Moritz's pedagogical theory and his Italian journey]. In: A. Krupp (ed.), *Karl Philipp Moritz: Signaturen des Denkens* [*Karl Philipp Moritz's thought and Poetics*], pp. 254–276. Amsterdam, Netherlands: Rodopi. https://doi.org/10.1163/9789042032217_016.

Gambino, R. & Pulvirenti, G. (2017a). The paradox of ekphrasis: Metacritic discourse, perception and imagination in art description. *Metacritic Journal for Comparative Studies and Theory*, 3(1), 151–179. https://doi.org/10.24193/mjcst.2017.3.09.

Gambino, R. & Pulvirenti, G. (2017b). Vom Zweikampf zur Hetzjagd: Die politische Bedeutung der Tiere in Heinrich von Kleists 'Penthesilea' [From duel to hunting: The political value of Parforce in Kleist's Penthesilea]. *Weimarer Beiträge*, 63(3), 367–388.

Gambino, R. & Pulvirenti, G. (2018a). *La mente narrativa di Heinrich von Kleist* [*Heinrich von Kleist's Narrative Mind*]. Milan, Italy: Mimesis.

Gambino, R. & Pulvirenti, G. (2018b). *Storie, menti, mondi: Approccio neuroermeneutico alla letteratura* [*Stories, Minds, Worlds: A Neurohermeneutical Approach to Literature*]. Milan, Italy: Mimesis.

Goebel, R., Khorram-Sefat, D., Muckli, L., Hacker, H., & Singer, W. (1998). The constructive nature of vision: Direct evidence from functional magnetic resonance imaging studies of apparent motion and motion imagery. *European Journal of Neuroscience*, 10(5), 1563–1573. https://doi.org/10.1046/j.1460-9568.1998.00181.x.

Graesser, A. C. & Zwaan, R. A. (1995). Inference generation and the construction of situation models. In C. A. Weaver, S. Mannes, & C. R. Fletcher (eds.), *Discourse Comprehension: Essays in Honor of Walter Kintsch*, pp. 117–140. Mahwah, NJ: Lawrence Erlbaum Associates.

Greiner, B. (1996). Das ganze Schrecken der Tonkunst. Die heilige Cäcilie oder Die Gewalt der Musik: Kleists erzählender Entwurf des Erhabenen [The horror of music. Saint Cecil or the power of music: Kleist's narrative attempt about the sublime]. *Zeitschrift für Deutsche Philologie*, 115(4), 501–520.

Haraway, D. (2008). *When Species Meet*. Minneapolis, MN: University of Minnesota Press.

Henkel, N. (1997). Descriptio [Description]. In: G. Braungart, H. Fricke, K. Grubmüller, J.-D. Müller, F. Vollhardt, & K. Weimar (eds.), *Reallexikon der deutschen Literaturwissenschaft* [*Lexicon of German Literature*], Vol. 1, pp. 337–339. Berlin, Germany: De Gruyter.

Herder, J. G. (2002). On the cognition and sensation of the human soul. In: M. N. Forster (ed.), *Herder: Philosophical Writings*, pp. 187–246. Cambridge, England: Cambridge University Press (Original work published 1778).

Iacoboni, M., Woods, R. P., Brass, M., Bekkering, H., Mazziotta, J. C., & Rizzolatti, G. (1999). Cortical mechanisms of human imitation. *Science*, 286(5449), 2526–2528. https://doi.org/10.1126/science.286.5449.2526.

Ishizu, T. & Zeki, S. (2011). Toward a brain-based theory of beauty. *PLoS One*, 6(7), e21852. https://doi.org/10.1371/journal.pone.0021852.

Ishizu, T., & Zeki, S. (2014). A neurobiological enquiry into the origins of our experience of the sublime and beautiful. *Frontiers in Human Neuroscience*, 8, 891. https://doi.org/10.3389/fnhum.2014.00891.

Johnson-Laird, P. (1983). *Mental Models*. Cambridge, England: Cambridge University Press.

Kant, I. (1764). *Beobachtungen über das Gefühl des Schönen und Erhabenen* [*Observations on Beauty and the Sublime*]. Königsberg, Germany: Kanter.

Kant, I. (2009). Analytik des Erhabenen [Analytics of the sublime]. In: H. F. Klemme (ed.), *Kritik der Urteilskraft* [*Critique of Judgement*], pp. 105–153. Meiner: Hamburg, Germany (Original work published 1790).

Kim, S. & James, T. W. (2010). Enhanced effectiveness in visuo-haptic object-selective brain regions with increasing stimulus salience. *Human Brain Mapping*, 31(5), 678–693. https://doi.org/10.1002/hbm.20897.

Kintsch, W. (1988). The role of knowledge in discourse comprehension: A construction-integration model. *Psychological Review*, 95(2), 163–182. https://doi.org/10.1037/0033-295x.95.2.163.

Kintsch, W., Welsch, D., Schmalhofer, F., & Zimny, S. (1990). Sentence memory: A theoretical analysis. *Journal of Memory and Language*, 29(2), 133–159. https://doi.org/10.1016/0749-596x(90)90069-c.

Von Kleist, H. (n.d.). Letter 45. Briefe [Letters]. *KLEIST.digital*. Retrieved from http://www.heinrich-von-kleist.org/kleistdigital/ (Original work published 1801).

Von Kleist, H. (1987). In: I. M. Barth, K. Müller-Salget, S. Ormanns, & H. C. Seeba (eds.), *Sämtliche Werke und Briefe in vier Bänden* [*The Complete Works and Letters in Four Volumes*]. Frankfurt, Germany: Deutscher Klassiker.

Von Kleist, H. (1993). In: H. Sembdner (ed.), *Sämtliche Werke und Briefe von Heinrich von Kleist* [*The Complete Works and Letters*]. Munich, Germany: Carl Hanser.

Von Kleist, H. (2013). Empfindungen vor Friedrichs Seelandschaft [Feelings in Front of Friedrich's Seascape]. *Berliner Abendblätter*, 12, 47–48 (Original work published 1810).

Von Kleist, H. (2016). Saint Cecilia, or the power of music. In: J. Oxenford & C. A. Feiling (eds.), *Tales from the German, Comprising Specimens from the Most Celebrated Authors*, pp. 298–305. London, England: Chapman and Hall (Original work published 1844).

Von Kleist, H., Bretano, C., & von Arnim, A. (2013). Various expressions experienced before A Seascape with a Monk by Caspar David Friedrich (J. Hunt, Trans.), *Art in Fiction: Anthology of Art in Literature*. Retrieved from https://artinfiction.wordpress.com/2013/07/11/heinrich-von-kleist-clemens-bretano-achim-von-arnim-various-expressions-experienced-before-a-seascape-with-a-monk-by-caspar-david-friedrich-1810/ (Original work published 1810).

Koerner, J. L. (1990). *Caspar David Friedrich and the Subject of Landscape*. New Haven, CT: Yale University Press.

Kosslyn, S. M. (1980). *Image and Mind*. Cambridge, MA: Harvard University Press.

Kosslyn, S. M. (1983). *Ghosts in the Mind's Machine*. New York, NY: W. W. Norton.

Kosslyn, S. M., Thompson, W. L., & Ganis, G. (2006). *The Case for Mental Imagery*. Oxford, England: Oxford University Press. https://doi.org/10.1093/acprof:oso/9780195179088.001.0001.

Kuzmičová, A. (2014). Literary narrative and mental imagery: A view from embodied cognition. *Style*, 48(3), 275–293. https://doi.org/10.5325/STYLE.48.3.275.

Louwerse, M. M., Hutchinson, S., Tillman, R., & Recchia, G. (2015). Effect size matters: The role of language statistics and perceptual simulation in conceptual processing. *Language, Cognition and Neuroscience*, 30(4), 430–447. https://doi.org/10.1080/23273798.2014.981552.

Mandelartz, M. (2006). Von der Tugendlehre zur Lasterschule. Die sogenannte 'Kantkrise' und Fichtes 'Wissenschaftslehre' [From the doctrine of virtue to the school of vice]. In: G. Blamberger, I. Breuer, S. Doering, & K. Müller-Salget (eds.), *Kleist-Jahrbuch 2006* [*Kleist Yearbook 2006*], pp. 120–136. Stuttgart, Germany: J. B. Metzler. https://doi.org/10.1007/978-3-476-00204-4_9.

Mendelssohn, M. (1972). Über das Erhabene und Naive in den schönen Wissenschaften [On the sublime and naive in the beautiful sciences]. In: I. Elbogen (ed.), *Mendelssohn: Gesammelte Schriften, Jubiläumsausgabe* [*Mendelssohn: The Complete Works, Jubilee Edition*]. Stuttgart, Germany: Frommann-Holzboog (Original work published 1771).

Meteyard, L., Cuadrado, S. R., Bahrami, B., & Vigliocco, G. (2012). Coming of age: A review of embodiment and the neuroscience of semantics. *Cortex*, 48(7), 788–804. https://doi.org/10.1016/j.cortex.2010.11.002.

Meyer-Sickendiek, B. (2005). *Affektpoetik. Eine Kulturgeschichte literarischer Emotionen* [*Poetic of Emotions: A Cultural History of Literary Feelings*]. Würzburg, Germany: Königshausen & Neumann.

Meyertholen, A. (2013). Apocalypse now: On Heinrich von Kleist, Caspar David Friedrich, and the emergence of abstract art. *German Quarterly*, 86(4), 404–420.

Miller, N. (2002). *Von Nachtstücken und anderen erzählten Bildern* [*From Nocturnal Pieces to Other Described Paintings*]. Munich, Germany: Carl Hanser.

Müller, G. (1986). Kleists Rhetorik der Innerlichkeit [Kleist's rhetoric of inwardness]. *Studia Neophilologica*, 58(2), 231–242.

Müller, G. (1995). *Kleist und die bildende Kunst* [*Kleist and the Fine Arts*]. Tübingen, Germany: Francke.

Osterkamp, E. (1991). *Im Buchstabenbilde. Studien zum Verfahren Goethescher Bildbeschreibungen* [*Letter Image: Studies about Goethe's Description of Ekphrasis*]. Stuttgart, Germany: J. B. Metzler. https://doi.org/10.1007/978-3-476-03364-2.

Pfotenhauer, H. (2000). Kleists Rede über Bilder und in Bildern. Breife, Bildkommentare und literarische Werke [Kleist's reflection on pictures and in pictures]. In: H. Pfotenhauer (ed.), *Sprachbilder Untersuchung zur Literatur seit des Achzehnten Jahrhunderts* [*Images in Words: Essays on the Eighteenth-Century Literature*], pp. 85–108. Würzburg, Germany: Königshausen und Neumann.

Richardson, A. (2010). *The Neural Sublime: Cognitive Theories and Romantic Texts*. Baltimore, MD: Johns Hopkins University Press.

Rühse, V. (2013). Dies wunderbare Gemählde [This wonderful painting]. In: G. Blamberger (ed.), *Kleist-Jahrbuch 2013* [*Kleist Yearbook 2013*], pp. 238–255. Stuttgart, Germany: J. B. Metzler. https://doi.org/10.1007/978-3-476-01199-2_20.

Schiller, F. (1962a). Über das Erhabene [About the sublime]. In: *Schiller: Sämtliche Werke* [*Schiller: The Complete Works*], 5, pp. 792–809. Munich, Germany: Carl Hanser (Original work published 1801).

Schiller, F. (1962b). Vom Erhabenen [The sublime]. In: *Schiller: Sämtliche Werke* [*Schiller: The Complete Works*], pp. 489–513. Munich, Germany: Carl Hanser (Original work published 1793).

Seeba, H. C. (2003). The eye of the beholder: Kleist's visual poetics of knowledge. In: B. Fisher (ed.), *A Companion to the Works of Heinrich von Kleist*, pp. 103–122. New York, NY: Camden House.

Shimamura, A. P. (2012). Towards a science of aesthetics. In: A. P. Shimamura & S. E. Palmer (eds.), *Aesthetic Science: Connecting Minds, Brains and Experience*, pp. 3–28. New York, NY: Oxford University Press.

Shimamura, A. P. (2013). *Experiencing Art: In the Brain of the Beholder*. New York, NY: Oxford University Press.

Starr, G. G. (2010). Multisensory imagery. In: L. Zunshine (ed.), *Introduction to Cognitive Cultural Studies*, pp. 275–291. Baltimore, MD: Johns Hopkins University Press.

Starr, G. G. (2013). *Feeling Beauty: The Neuroscience of Aesthetic Experience*. Cambridge, MA: MIT Press.

Stilla, R. & Sathian, K. (2008). Selective visuo-haptic processing of shape and texture. *Human Brain Mapping*, 29(10), 1123–1138.

Stoichita, V. (2013). *L'invenzione del quadro* [*The Invention of Meta-Painting*]. Milan, Italy: Il Saggiatore.

van Dijk, T. A. & Walter, K. (1983). *Strategies of Discourse Comprehension*. New York, NY: Academic Press.

Webb, R. (2009). *Ekphrasis, Imagination and Persuasion in Ancient Rhetorical Theory and Practice*. Farnham, England: Ashgate.

Zeki, S. (1998). Art and the brain. *Dædalus*, 127(2), 71–103.

Zeki, S. (1999). *Inner Vision: An Exploration of Art and the Brain*. Oxford, England: Oxford University Press.

Zeki, S. (2008). *Splendors and Miseries of the Brain: Love, Creativity, and the Quest for Human Happiness*. Chichester, England: Wiley-Blackwell.

Zeki, S. (2019). Notes towards a (neurobiological) definition of beauty. *Gestalt Theory*, 41(2), 107–112.

Zeki, S., Romaya, J. P., Benincasa, D. M., & Atiyah, M. F. (2014). The experience of mathematical beauty and its neural correlates. *Frontiers in Human Neuroscience*, 8, 68.

Zwaan, R. A., Langston, M. C., & Graesser, A. C. (1995). The construction of situation models in narrative comprehension: An event-indexing model. *Psychological Science*, 6(5), 292–297.

https://doi.org/10.1142/9789811298639_0006

Chapter 6

Aesthetic Appreciations: Bridges Between Neuroscience, Psychology, and Psychotherapy

Matteo Sesia[*,†] and Chen Zhao[*]

[*]*Institute of Medical Psychology, Ludwig Maximilian University, Munich, Germany*

[†]*Parmenides Foundation, Pöcking, Germany*

Abstract

After reviewing some publications in the field of neuroaesthetics, we find that in experiments on tasks about aesthetic appreciation, many cortical areas are involved. Combining experimental data with theoretical concepts like complementarity as a generative principle, logistic and content functions, and three modes of knowing, i.e., explicit, implicit, and sensory knowledge, we argue that aesthetic appreciation is both an early implicit reaction (*feeling*) and a late explicit evaluation (*thinking*), representing pre-semantic bottom-up and conscious top-town processing, respectively. On a strategic level, we propose that neuroaesthetics can serve as a conceptual bridge to bring together neuroscientific and psychological research, and we can link it to psychoanalytical considerations. A critical parameter of research in neuroaesthetics may be the cultural frame. Thus, to explore aesthetic appreciation in different cultures, we used a semantic differential questionnaire in Chinese, Japanese, Italian, and Farsi with male and female subjects to evaluate the concept of "beautiful". In spite of some similarities in these cultures, important differences in aesthetic appreciation were observed; our result suggests that semantic frames may also be relevant to other theoretical concepts.

Keywords: complementarity; semantic differential; aesthetic appreciation; beauty; cultural frame

Some References in Neuroaesthetics

Do aesthetic appreciation and relevant experiences (such as making moral evaluations and expressing likeness) have a common denominator? Is aesthetic appreciation a psychological singularity independent of other subjective phenomena? On a more general level, does independence of any psychological phenomenon as represented in language exist at all? An answer to the first two questions may be possible if results from experiments in neuroaesthetics are evaluated. We present a selection of some publications in which the aesthetic experiences and underlying neural processes were investigated with special experimental paradigms. We provide a systematic although limited review of experiments in neuroaesthetics and discuss what neural processes we are referring to when talking about "making aesthetic judgments" or "experiencing beauty". Ultimate questions in this field are — as in neuroscientific work in general — whether we make sensible and reliable statements about the relationship between specific behavioral responses and underlying neural activities.

In what follows, a limited review on aesthetic judgments is given. For example, if art from Eastern and Western cultures is compared, it was found with behavioral ratings that subjects preferred (not surprisingly) art from their own culture; in addition, subjects showed higher preference for naturalistic pictures (Bao *et al.*, 2016; Yang *et al.*, 2019). By applying functional magnetic resonance imaging (fMRI), it was found that the neural activation patterns were different when surrealistic compared to realistic paintings were presented (Silveira *et al.*, 2012); this was also observed under differently primed psychological states such as thinking about death (Graupmann *et al.*, 2013; Silveira *et al.*, 2014). Body perception in an artwork elicits more activation in visual-coding areas than in non-artistic photography (Lutz *et al.*, 2013); thus, art is coded differently in brain networks within different frameworks of appreciation. Bao and colleagues (2017) used a work of abstract art originally in red color and its color-changed versions as stimuli to compare the neural activation; the original red elicited less neural reaction which indicates less effort for the original. This result indicates that on an implicit level, the original artwork can be detected which was observed also in another study (Wang and Bao, 2020). Interestingly, when aesthetic appreciation is not related to an object in front of the eyes but to an imaginary environment, subjectively more beautiful environments engage different brain areas compared to less beautiful ones (Vedder *et al.*, 2015). In other imaging studies, commonality and differences between aesthetic, moral, and commercial judgments have been studied; a relationship between the concepts of "the beautiful" and "the good" has been assumed since antiquity. It was found that the neural activation for aesthetic judgments is apparently a common dominator for both moral and commercial judgments, the latter two involving further activations in other brain areas (Avram *et al.*, 2013; Wang *et al.*, 2021).

In addition to the empirical studies, we also review theoretical publications to rethink the role of aesthetic activities in the taxonomy of cognitive functions. We want to emphasize two conceptual domains: The first is complementarity as a generative principle; neurocognitive processes are characterized by complementary activities like bottom-up and top-down ones, and logistic functions (e.g., temporal control) and content functions (e.g., perceptions; Bao *et al.*, 2017). The second conceptual domain refers to the three modes of knowing; explicit, implicit and sensory (e.g., visual) knowledge is interrelated in a complementary way to provide an integrative network for cognition (Bao *et al.*, 2022). For practical reasons, we will not go into details of each publication. We focus on the conceptual findings and further developments to which these studies may lead.

Neural Correlates of Aesthetic Appreciation

After reviewing a limited set of publications, we have indicated the different brain areas that are reported to be involved in aesthetic tasks; they are summarized and listed in Table 1. We highlight activated brain areas in different colors with respect to different cognitive functions. The areas earmarked in red are activated in relatively "purely aesthetic" tasks; this may be related on a general level to a concept developed by Baumgarten (1750), the founder of aesthetics as a philosophical discipline; he states that aesthetics is related to the "art of beautiful thinking" as in the appreciation of beauty or dealing with object recognition in general. The areas

Table 1. Brain activations in neuroaesthetic studies.

Orbitofrontal Cortex (OFC)	The OFC is involved in aesthetic beauty but also in moral and commercial judgments. The OFC may be considered to be the core mechanism for aesthetic appreciation.
Medial Prefrontal Cortex (mPFC)	The mPFC is involved in emotional processing, cognitive control, decision-making, punishment, and self-processing. It shows engagement in moral judgments and is involved in the theory of mind. It is also implicated in dealing with negative emotions.
Anterior Cingulate Cortex (ACC)	The ACC is activated in conscious mental effort, *Cognitive Processing*, and concept retrieval. It also seems to be involved in conceptual integration and word processing.
Orbito-medial Prefrontal Cortex (OmPFC)	The OmPFC is found to be involved in aesthetic judgment, especially in emotional processing and social cognition.
Precuneus	The precuneus is involved in integrating external information into a self-referential context.
Primary Visual Area	The primary visual area is closely related to the bottom-up visual information processing.
Temporoparietal Junction (TPJ)	The TPJ is involved in moral and commercial judgments, theory of mind (ToM), conceptual integration, and word processing. It is associated with a conceptual thinking mode (especially the left TPJ) and binding spatial information.
Middle Frontal Gyrus (mFG)	The mFG is related to top-down negative emotion regulation.

Table 1. (*Continued*)

Insula	The right anterior insula is activated in hedonic response, norm violation, empathy, and emotional processing, while the left insular cortex is related to the mobilization of resources for potential aversive events. The antero-dorsal sector of the insula is a core of aesthetic experiences.
Supplementary Motor Area (SMA)	The SMA is the motor-limbic interface, which is activated when emotions trigger motor action.
Middle and Inferior Temporal Gyrus (mTG,iTG)	The mTG and iTG activation is observed in aesthetic, moral, and commercial judgment; mTG shows higher activation in viewing for instance naturalistic paintings.
Posterior Cingulate Cortex (PCC)	The PCC plays a role in integrating external information into a self-referential context and top-down processes. It is also activated in commercial judgment and tasks with emotional memories.
Inferior Frontal Gyrus (IFG)	The IFG shows higher activation when looking at paintings that are rated as highly emotional.
Hippocampus	The hippocampus is activated when retrieving information from episodic memory. Some activation was observed in commercial judgments.
Ventral Striatum	The ventral striatum is activated when thinking of rewards and positive emotions.
Default Mode Network (DMN)	The DMN is involved in processing artwork of high aesthetic preference or high predictability in inferring meaningful content. It can be considered a "delayed aesthetic network".
Fusiform Gyrus	The fusiform gyrus is involved in object recognition (including faces and landscapes).
Left Lingual Gyrus	The left lingual gyrus is involved in encoding complex images.
Cuneus	The cuneus is activated when giving aesthetic appreciation.
Superior Parietal Lobule	The superior parietal lobule is involved in the manipulation of information in working memory.
Parahippocampal Cortex	The parahippocampal cortex is activated in commercial judgments.

(*Continued*)

Table 1. (*Continued*)

Nucleus Accumbens	The nucleus accumbens is involved in the generation of positive affective experiences, the detection of emotional values, and the depiction of future rewards.
Anterior Prefrontal Cortex	The anterior prefrontal cortex is involved in conceptual processing.
Medial Frontal Gyrus	The medial frontal gyrus is activated in higher mental functions, such as executive control.
Ventromedial Prefrontal Cortex (vmPFC)	The vmPFC shows higher activation in viewing pictures, and it is also involved in moral judgments.

Color coding of the table

Areas mainly related to aesthetic tasks
Areas mainly related to emotions
Areas mainly related to conceptual thinking
Areas mainly related to "self"

in blue are those correlated with emotional elaboration; they are activated when emotions are particularly involved. In green, the areas are related to conceptual thinking, taking the aesthetic appreciation to a higher level of abstraction, such as retrieval of information from working memory. Finally, in yellow, we highlight the areas related to the concept of "self" when self-identity is addressed. The ones that are not highlighted can be either polyfunctional or they are not significantly activated in a specific task. It should be noted that the indicated brain areas are most likely also involved in other cognitive functions.

The survey shows that aesthetic appreciation is extremely multifaceted, given the different activation patterns of up to 30 brain areas that have been identified in published studies. It is next to impossible to identify a specific brain area or a typical network mainly responsible for aesthetic appreciation. On that basis, we take another perspective to think about the relationship between aesthetic appreciation with other psychological functions.

Aesthetic Appreciation: A Core Mechanism for Higher Mental Functions

As previously mentioned, *aesthetic evaluation* and *judgment* can be considered as core mechanisms underlying other higher and complex mental functions. For instance, Wang and colleagues (2021) investigated the similarities and differences between aesthetic, moral, and commercial judgments. They asked participants whether they found a poetic line beautiful or not, whether they thought that a statement was right or wrong (from a moral standpoint), and whether a commercial slogan was effective or not. The participants were given two alternative forced choices while their brain activation was measured with fMRI. They found a core aesthetic appreciation center in the middle temporal gyrus (mTG) and inferior frontal gyrus (IFG) together with the orbital frontal cortex (OFC). The first two areas seemed to be related to a semantic center and the second one was highly involved in aesthetic appreciation. By looking at brain activation patterns in commercial and moral judgments, they found out that these three areas seemed to be activated as well. Other than that, moral judgment enrolled the temporoparietal junction (TPJ) and the anterior cingulate cortex (ACC), whereas commercial judgment seemed to involve also the hippocampus, the parahippocampal cortex (PhC), and the posterior cingulate cortex (PCC).

Conceptually similar results were obtained by Avram and colleagues (2013) when comparing aesthetic and moral judgments. It seems that an aesthetic core mechanism stays beneath more complex and abstract functions. It makes sense also from an evolutionary standpoint; it is unfortunate that evolutionary aesthetics is a rather underdeveloped field, having come to a stall in the last few years (Bartalesi, 2013; MacLean, 1990). Nevertheless, such studies may help to speculate about an evolutionary basis for aesthetic appreciation. The results provided by Bao and colleagues (2017), in which participants showed a remarkable preference for landscape artworks regardless of their cultural background, can be marked as an indication supporting the so-called *preference model* (Ulrich, 1986; cit. in Bartalesi, 2013). The preference model states that humankind developed a sense of what is beautiful judging whether a certain location could be considered convenient or not. Of course, this is not exhaustive and a lot still has to be understood, if not discovered, regarding aesthetic evaluation. Nonetheless, by looking at different beautiful standards in different cultures, we could investigate whether aesthetic appreciation can be considered an anthropological universal.

If an aesthetic principle is indeed so widely shared (or at least deeply rooted in the human brain), it can be expected to affect also other domains of experiences and cognition. There might exist an aesthetic principle guiding how we perceive, build memories, and even how we approach others. We may open to new perspectives also in other psychological fields, such as psychopathology: maybe also look at phenomena like anorexia, anti-social personality disorder, or narcissism from new points of view. We may even develop a different perspective, embedded in a new epistemological framework, within an aesthetic frame on how humans perceive time capable of including very different and far domains, such as the function of time windows (Pöppel, 1997, 2009), or the concept of continuity of time. Put differently, the aesthetic principle at the core of mental functioning may even revolutionize our understanding of basic cognitive functions and their underlying mechanisms, We may come to new conclusions even in unexpected fields as in the case of the time window of some three seconds found to be relevant in poetry (Turner & Pöppel, 1988; Yu & Bao, 2020).

In spite of the indicated opportunities, it is necessary to take a critical look from the methodological point of view at the results as the tasks in the studies were not always comparable. Moreover, as the experimental design is usually to compare two conditions like aesthetic judgment against moral judgment, the analysis of the fMRI data usually adopts a subtractive comparison. Therefore (at least for some of the results), it is not certain whether the subtractive activation is directly involved in a purely aesthetic process. Due to the limitation in interpreting fMRI data, it is always difficult to say whether the activation of a brain area is necessarily responsible for a certain behavior (Logothetis, 2008). It should also be noted that we cannot even claim whether the stronger activation in defined brain area is a sign for intense or less appreciation, although there may be exceptions for specific areas (Zhang *et al.*, 2024, in preparation).

In the table above, almost all the cortical areas have been reported to be part of the process of either passive aesthetic experience or active aesthetic judgment. In the discussion parts of the publications reviewed, the authors also referred to other studies to demonstrate how the activated areas are involved in other higher cognitive functions, such as decision-making, memory retrieval, or motor control (see details in Table 1). The implication is that the processes of memory retrieving or self-referring are part of an aesthetic experience or judgment. Aesthetic appreciation is regarded explicitly or implicitly as a result of cognitive content. However, if we reverse the thinking, the questions that should be asked are as follows: Which brain areas are NOT activated during aesthetic tasks? And is there a cognitive process that is completely unrelated to aesthetic experience? One possibility based on the theoretical reviews is that aesthetic experience is embedded in many higher cognitive functions, not the other way around as usually implied. It is not because higher cognitive functions are involved that the aesthetic judgment is formed, but aesthetics is a pre-semantic and implicit reaction before memory retrieving, thinking, and motor actions that leads to explicit semantic expression and judgments (Bao *et al.*, 2022; Zhao *et al.*, 2022). In other words, whether the task is to evaluate the beauty or not, when people encounter something/someone that they need to response to, maybe the first logistic reaction is "whether it is of my favor"; this automatic aesthetic experience is not determined by the intellectual analysis of the content, which comes actually later.

Aesthetics: An Intertwining of Knowing

Of course, we are not denying that higher cognitive functions serve in a top-down way to modulate explicit expressions and behaviors, and we want to emphasize that the pre-semantic aesthetic reaction is distinct from judgments after consciously thinking. Given such taxonomical classification, we can also divide the aesthetic processes into two clusters of psychological meaning: the processes that make us *feel* an artwork and the processes that make us *think* about the artwork. This kind of classification highlights the connection between different fields in the psychological domain: a neuroscientific declination of it and a psychological and with a psychotherapeutical nuance. Indeed, we think that neuroaesthetics can be considered a conceptual bridge between different concepts and theories in both psychodynamic and neuroscientific fields (see also Rentschler *et al.*, 1988). The distinction mentioned above can be declined in Bionian terms following explications as proposed for instance by some authors (Ferro, 2002, 2007, 2013; Granieri, 2011; Nespoli, 2019); at the same time, they indicate a way of exemplifying an implicit and explicit way of knowing (Bao *et al.*, 2022), logistical and content functions (Zhao *et al.*, 2022) or a good way of seeing an object (the aesthetic appreciation) as the result of the complementarity of bottom-up processes (feeling the art) and top-down processes (thinking the art), in a reciprocal interaction (Bao *et al.*, 2017).

Some functions are meant to register data at a lower level of analyses impressing them as β elements (Bion, 1962), which is also called the first degree of representation of reality; later on, more complex and specialized areas of the brain are involved to build up a more complex representation of an aesthetic impression that can be mentally manipulated and combined with other concepts and knowledge: α elements (*ibidem*). At the same time, we can see how different areas of the brain are interested when dealing with an aesthetic stimulus: some of them can be thought of as dedicated to encoding and registering different stimuli or giving an emotional nuance to it. At a higher level of abstraction, some other areas appear to be involved either when a self-referential frame is needed to give a frame to the incoming information or when a more specific conceptual knowledge is needed, showing a kind of top-down modulation. Indeed, when asked to give an appreciation of different Western and Eastern pieces of art, Western and Eastern participants showed a preference for artworks of their own culture; this proves how relevant the cultural background is even when providing a naïve and spontaneous judgment about what is beautiful. Once again, we can see how top-down modulations influence the bottom-up perception of artworks, even when we have a piece of strong evidence for the bottom-up salience of the process (Bao *et al.*, 2016, 2017). Finally, these areas can be seen as devolved to provide the functions of arousal and emotional reaction (logistical function) and some of them to create and deal with mental content (content functions) (Zhao *et al.*, 2022).

Psychoanalysis has developed through this century pulling together many different theories and voices, with little attachment to neuroscience and to "harder" sciences. Some remarkable attempts were made (e.g., Schore, 2015; Cozolino, 2021), but they still got highly criticized; all of psychoanalysis was put under heavy pressure given the scarce possibility of proving it right or wrong (Popper, 1935/1959). Nevertheless, apart from some attempts to defend it, even recently (e.g., Grant & Harari, 2005), it might be a good moment for bringing the related fields together, using neuroscience and also evolutionary psychology as tools to give some solid basis not only to psychoanalysis but to psychology as a whole (Mitchell, 1993). In this sense, neuroaesthetics might serve as a field in which to observe how different perspectives happen to be combined. Only when combined do they create a common and fruitful frame and narration to best describe what is happening in our minds when approaching an aesthetic stimulus. Some might say that aesthetic appreciation is highly subjective; we would like to argue that the core experience of aesthetic appreciation could be objective and universal, and just the description of aesthetic phenomena is subjective: making explicit what we intuitively know might be the only subjective part of the process. It has been pointed out that there might be a profound gap between implicit knowledge of beauty and explicit one (Bao *et al.*, 2022). Just like Augustinus said about time: «quid enim est tempus? Si nemo ex me quaerat, scio; si quaerenti explicare velim, nescio» (What then is time? If no one asks me, I know; if I want to explain to who's asking, I do not know). We advocate that the subjectiveness of some knowledge comes from this simple transformation and the switch between modalities of knowing. As asserted by Ferro (2002, 2007), when we have an encounter with the external reality (or "O", in Bionian terms), we register it into β elements, according to our senses and the emotional nuance we give to this experience, and then again into α elements, thanks to the α function of our mind, and then again in "narrative derivatives", the ones to be communicated. In these multi-stage transformations, the contact with the original experience might be lost, and when we try to give meaning and communicate what we feel and felt, we may be aware of the gap between our original knowledge and the way of depicting it (first) and then communicating it (later).

Semantic Aesthetic Appreciation: An Intercultural Study

Taking the perspective of explicit knowledge of aesthetics, we investigated whether the concept of what is *beautiful* can be universally shared or whether it is culturally specific. The study came up with a most basic question in the Western culture: "Are the *true*, the *good*, and the *beautiful indeed* three main concepts at the basis of any philosophy, or are they relevant only in Western culture?". From these basic questions, we started developing a study focusing on just the dimension of "the beautiful".

To address the cultural difference in the construction of the concept of "the beautiful", we applied a semantic differential (Osgood, 1957). It consists of a list of word pairs, usually antonyms, and asks participants to indicate which pole is associated with the target term. In this study, members with various cultural backgrounds (German, British, Italian, and Chinese) first brainstormed over fifty pairs of words as candidates of antonyms in the semantic differential questionnaire. For example, pairs of "Day versus Night", "Simple versus Complex", or "Good versus Bad". The pairs were derived from self-introspection and quick oral surveys in social circles of the coauthors regarding the experience and imagination of beautiful things. The final version of the list has the consensus of coauthors and can be found in the Appendix. Then, to distribute the questionnaire in different cultural environments, we invited several corresponding local scientists related to the field of psychology to help translate the English version of the list into local languages. During the translation, the translators and we discussed the meaning of the words back and forth to make sure that they deliver as same possible meaning across cultures. Then the questionnaire to be answered was sent to students in different universities. The students participated voluntarily without receiving any compensation.

With answers from Italian (n = 36), Chinese (n = 26), Japanese (n = 30), and Farsi students (n = 29), a statistical analysis was done with JASP 0.18.1. The data were first reformatted; the ratings of pairs are originally from 1 to 7 on a Likert scale; by minus 4 with the whole scale, an index scale was created from –3 to 3, where 0 means that this pair is not related to the concept of "the beautiful". The negative index suggests a preference toward the left word in the pair, and the positive index suggests a preference toward the right. On that basis, one can use the index value to summarize cultural group features, i.e., a cultural group mean for each pair as preference and a cultural group variance for each pair indicating variability within the group.

Further statistical analysis indicates that the preference ratings and the variance of the four cultures follow a normal distribution, and the box plot indicates no outliers. Then both dependent variables were rendered to one-way analysis of variance (ANOVA). The results are visualized in the following figure. For preference, the main effect is not significant ($F = 0.331$, $p = 0.803$, $\eta_p^2 = 0.005$), while for variance, the main effect is highly significant ($F = 30.706$, $p < 0.001$, $\eta_p^2 = 0.311$). Further *post hoc* analysis reveals that the variance in the Chinese group is significantly smaller than the Japanese group ($t = –8.003$, $p < 0.001$) and the Farsi group ($t = –7.539$, $p < 0.001$); also, the variance in the Italian group is significantly smaller than the Japanese group ($t = –5.577$, $p < 0.001$) and the Farsi group ($t = –5.113$, $p < 0.001$), as indicated in Figure 1.

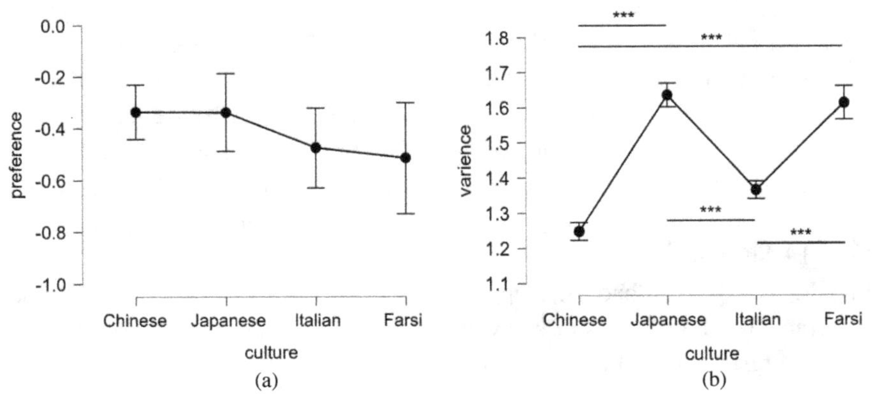

Figure 1. Comparing preference and variance in the semantic differentials in the four cultures. Panel A shows the mean of the preference of the semantic differentials. The error bar stands for 1 standard error. Similarly, panel B shows the mean of the variances, and the error bar stands for 1 standard error.

Note: '***' indicates a significant *post-hoc* comparison.

The analysis suggests that even though overall the preference for the semantic differential does not differ significantly across the four cultures, it should be noted that many quantitative details were hidden behind the averaging processes, which means that the pattern, i.e., the rating distribution for each semantic differential, is different between the cultures. In a culture with a larger rating variance, the rating pattern is more likely to be polarized in a bimodal way. Thus, it can be concluded that the Chinese and Italian subjects evaluate "the beautiful" differently than the Japanese and Farsi subjects. In addition to the result of intercultural differences in judgment behavior, the methodology of semantic differential can be considered to be a usable tool to investigate cultural differences in cognitive concepts in general.

References

Avram, M., Gutyrchik, E., Bao, Y., Pöppel, E., Reiser, M., & Blautzik, J. (2013). Neurofunctional correlates of esthetic and moral judgments. *Neuroscience Letters*, 534, 128–132. https://doi.org/10.1016/j.neulet.2012.11.053.

Bao, Y., von Stosch, A., Park, M., & Pöppel, E. (2017). Complementarity as generative principle: A thought pattern for aesthetic appreciations and cognitive appraisals in general. *Frontiers in Psychology*, 8, 245718. https://doi.org/10.3389/fpsyg.2017.00727.

Bao, Y., Yang, T., Lin, X., Fang, Y., & Pöppel, E. (2016). Aesthetic preferences for Eastern and Western traditional visual art: Identity matters. *Frontiers in Psychology*, 7, 214446. https://doi.org/10.3389/fpsyg.2016.01596.

Bao, Y., Yang, T., Zhang, J., Zhang, J., Lin, X., Paolini, M., Pöppel, E. & Silveira, S. (2017). The "third abstraction" of the Chinese artist LaoZhu: Neural and behavioral indicators of aesthetic appreciation. *PsyCh Journal*, 6(2), 110–119. https://doi.org/10.1002/pchj.167.

Bao, Y., Zhang, D., Zhao, C., Pöppel, E., & Zabotkina, V. (2022). An aesthetic frame for three modes of knowing. *PsyCh Journal*, 11(5), 636–644. https://doi.org/10.1002/pchj.588.

Bartalesi, L. (2013). Quale storia naturale per l'estetico? L'ipotesi darwiniana rivisitata (What is natural history for the aesthetic? The Darwinian hypothesis revisited). *Rivista di estetica*, 54, 7–25. https://doi.org/10.4000/estetica.1433.

Baumgarten, A. G. (1750). *Aesthetica*. Traiecti cis Viadrum (Frankfurt an der Oder), Johannis Christian Kleyb.

Bion, W. R. (2018). *Elements of Psychoanalysis*. London: Routledge.

Cozolino, L. (2021). *Neuroscienze per clinici* (*Neuroscience for Clinicians*). Milano: Raffaello Cortina.

Ferro, A. (2002). *Illness Factors, Healing Factors: Genesis of Suffering and Psychoanalytic Treatment*. Raphael Cortina.

Ferro, A. (2007). *Avoid Emotions, Experience Emotions*. Milan: Raffaello Cortina.

Ferro, A. (2013). *Psicoanalisi oggi* (*Pyschoanalysis Today*) Roma: Carrocci editore.

Granieri, A. (2011). *Bodily Affects and Thought. Intertwining Psychoanalysis and Neurobiology*, pp. XV-136. UTET University.

Grant, D. C. & Harari, E. (2005). Psychoanalysis, science and the seductive theory of Karl Popper. *Australian & New Zealand Journal of Psychiatry*, 39(6), 446–452. https://doi.org/10.1080/j.1440-1614.2005.01602.x.

Graupmann, V., Peres, I., Michaely, T., Meindl, T., Frey, D., Reiser, M., Pöppel, E., Fehse, K., & Gutyrchik, E. (2013). Culture and its neurofunctional correlates when death is in mind. *Neuroscience Letters*, 548, 239–243. https://doi.org/10.1016/j.neulet.2013.05.062.

Logothetis, N. K. (2008). What we can do and what we cannot do with fMRI. *Nature*, 453(7197), 869–878. https://doi.org/10.1038/nature06976.

Lutz, A., Nassehi, A., Bao, Y., Pöppel, E., Sztrókay, A., Reiser, M., Fehse, K., & Gutyrchik, E. (2013). NeuroCognitive Processing of body representations in artistic and photographic images. *NeuroImage*, 66, 288–292. https://doi.org/10.1016/j.NeuroImage.2012.10.067.

MacLean, P. D. (1990). *The Triune Brain in Evolution: Role in Paleocerebral Functions*. New York, NY: Springer Science & Business Media.

Mitchell, S. A. (1993). *Gli orientamenti relazionali. Per un modello integrato* (*Relational Concepts in Psychoanalysis; An Integration*). Torino: Bollati Boringhieri.

Nespoli, G. (2019). *Constructs and Paradigms of Contemporary Psychoanalysis: The Theoretical Assumptions of Dynamic Psychology*. Mondadori University.

Osgood, C. E., Suci, G. J., & Tannenbaum, P. H. (1957). *The Measurement of Meaning*, No. 47. University of Illinois Press.

Pöppel, E. (1982). *Lust and Schmerz. (Pain and Pleasure)*. Berlin: Severin & Siedler.

Pöppel, E. (1997). A hierarchical model of temporal perception. *Trends in Cognitive Sciences*, 1(2), 56–61. https://doi.org/10.1016/S1364-6613(97)01008-5.

Pöppel, E. (2009). Pre-semantically defined temporal windows for Cognitive Processing. *Philosophical Transactions of the Royal Society B: Biological Sciences*, 364(1525), 1887–1896. https://doi.org/10.1098/rstb.2009.0015.

Pöppel, E., Avram, M., Bao, Y., Graupmann, V., Gutyrchik, E., Lutz, A., Park, M., Reiser, M., Russell, E., Silveira, S., Smigielski, L., Szymanski, C., & Zaytseva, Y. (2013). Sensory processing of art as a unique window into cognitive mechanisms: Evidence from behavioral experiments and fMRI studies. *Procedia-Social and Behavioral Sciences*, 86, 10–17. https://doi.org/10.1016/j.sbspro.2013.08.517.

Popper, K. (1935/1959). *Logik der Forschung* (*The Logic of Scientific Discovery*). Wien: Springer, 1935/London: Hutchinson, 1959.

Rentschler, I., Herzberger, B., & Epstein, D. (1988). *Beauty and the Brain. Biological Aspects of Aesthetics*. Basel: Birkäuser Verlag.

Schore, A. N. (2015). *Affect Regulation and the Origin of the Self: The Neurobiology of Emotional Development*. New York, NY: Routledge. https://doi.org/10.4324/9781315680019.

Silveira, S., Graupmann, V., Frey, D., Blautzik, J., Meindl, T., Reiser, M., Chen, C., Wang, Y., Bao, Y., Pöppel, E., & Gutyrchik, E. (2012). Matching reality in the arts: self-referential neural processing of naturalistic compared to surrealistic images. *Perception*, 41(5), 569–576. https://doi.org/10.1068/p7191.

Silveira, S., Gutyrchik, E., Wetherell, G., Bao, Y., Pöppel, E., Blautzik, J., Reiser, M., Frey, D., & Graupmann, V. (2014). Ceci n'est pas la mort: Evidence for the recruitment of self-reference from surrealistic art under mortality salience. *European Journal of Social Psychology*, 45(2), 255–266. https://doi.org/10.1002/ejsp.2076.

Turner, F. & Pöppel, E. (1988). Metered poetry, the brain, and time. In: I. Rentschler, B. Herzberger, & D. Epstein (eds.), *Beauty and the Brain: Biological Aspects of Aesthetics*, pp. 71–90. Basel: Birkhäuser Verlag. https://doi.org/10.1007/978-3-0348-6350-6_4.

Ulrich, R. S. (1986). Human responses to vegetation and landscapes. *Landscape and Urban Planning*, 13, 29–44. https://doi.org/10.1016/0169-2046(86)90005-8.

Vedder, A., Smigielski, L., Gutyrchik, E., Bao, Y., Blautzik, J., Pöppel, E., Zaytseva, Y., & Russell, E. (2015). Neurofunctional correlates of environmental cognition: an fMRI study with images from episodic memory. *PLoS One*, 10(4), e0122470. https://doi.org/10.1371/journal.pone.0122470.

Wang, X. & Bao, Y. (2020). Rembrandt portraits: Implicitly detecting the original perspective. *PsyCh Journal*, 9, 513–515. doi:10.1002/pchj.376.

Wang, X., Simmank, F., Zhang, D., Zeng, Y., Silveira, S., Sander, T., Bao, Y., & Paolini, M. (2021). Aesthetics as a common denominator for moral and commercial judgments. *PsyCh Journal*, 10(2), 187–189. https://doi.org/10.1002/pchj.407.

Yang, T., Silveira, S., Formuli, A., Paolini, M., Pöppel, E., Sander, T., & Bao, Y. (2019). Aesthetic experiences across cultures: Neural correlates when viewing traditional Eastern or Western landscape paintings. *Frontiers in Psychology*, 10, 294675. https://doi.org/10.3389/fpsyg.2019.00798.

Yu, X. & Bao, Y. (2020). The three second time window in poems and language processing in general: Complementarity of discrete timing and temporal continuity. *PsyCh Journal*, 9(4), 429–443. https://doi.org/10.1002/pchj.390.

Zhang, D., Formuli, A., Pöppel, E., & Paolini, M. (2024). The occipital lobe as a neural reference in an fMRI study on imagery (in preparation).

Zhao, C., Enriquez, P., Izadifar, M., Pöppel, E., Bao, Y., & Zabotina, V. (2022). Complementarity of mental content and logistic algorithms in a taxonomy of cognitive functions. *PsyCh Journal*, 1–7. https://doi.org/10.1002/pchj.602.

Appendix: Semantic Differential

1	Day	Night
2	Predetermined	Developed
3	Stable	Changing
4	Simple	Complex
5	Majestic	Insignificant
6	Symmetric	Asymmetric
7	Good	Bad
8	Ambiguous	Precise
9	Straight	Crooked
10	Past	Future
11	Slow	Fast
12	Sincere	Fake
13	Thrilling	Peaceful
14	Positive	Negative

(*Continued*)

(*Continued*)

15	Profound	superficial
16	Lazy	Hardworking
17	Big	Small
18	Virtuous	Dissolute
19	Regular	Random
20	Mysterious	Evident
21	Natural	Artificial
22	Pleasant	Annoying
23	Green	Red
24	Inside	Outside
25	Happy	Sad
26	Empty	Full
27	Drunk	Sober
28	Intentional	Coincidental
29	True	False
30	Pure	Polluted
31	Bright	Dark
32	Concrete	Abstract
33	Theoretical	Empirical
34	Cheerful	Grim
35	Rare	Common
36	Young	Old
37	Intense	Sparse
38	Innovative	Traditional
39	East	West
40	Calm	Agitated
41	Now	Later
42	Fragile	Tough
43	Dirty	Clean
44	Harmonious	Chaotic
45	Essential	Extra
46	Implicit	Explicit

(*Continued*)

47	Scientific	Artistic
48	Thin	Thick
49	Pessimistic	Optimistic
50	Introverted	Extraverted
51	Open	Closed
52	Holistic	Discrete

https://doi.org/10.1142/9789811298639_0007

Chapter 7

A Hidden Message: Decoding Artistic Intent*

Jacek Rogala[†], Beata Bajno[‡] and Andrzej Wróbel[§]

[†]*Bioimaging Research Center, World Hearing Center, Institute of Physiology and Pathology of Hearing, Warsaw, Poland*

[‡]*Independent Artist, Warsaw, Poland*

[§]*Department of Epistemology, Institute of Philosophy, University of Warsaw, Warsaw, Poland*

Abstract

Understanding how art makes impressions upon the perceiver has been a fundamental topic of philosophical interest since the time of ancient Greece. However, the extent of artistic perception and aesthetic appreciation has been the topic of empirical studies only recently, following the emergence of psychology as an independent field of science. This chapter discusses the hypothesis that the impression created by an artwork on the viewer can be predicted by examining the activity of neuronal networks. In particular, we focus on neural activity evoked by abstract stimuli that matches elements of the viewers' previously learned

*This chapter was originally published in the special collection on neuroaesthetics in *PsyCh Journal*. Rogala, J., Bajno, B., & Wróbel, A. (2020), A hidden message: Decoding artistic intent. *PsyCh Journal*, 9(4), 507–512. https://doi.org/10.1002/pchj.374.

conceptual dictionary. We show that artistic appreciation fundamentally depends on how easily the author's intent expressed in his or her artwork can be abstracted and decoded, on a neuronal level, into new or merged concept networks. More diverse intellectual and personal experiences — and their corollary neural networks — may facilitate the creation of new networks. These new networks, in turn, modulate the extent to which art can be apprehended and appreciated.

Keywords: art appreciation; beauty; concept networks; information theory; neuroaesthetics

Since classic antiquity, philosophers have sought to address fundamental questions regarding the nature of beauty and artistic appreciation. However, artistic perception and aesthetic appreciation emerged as a topic of empirical investigation relatively recently, following the recognition of psychology as a distinct scientific field. The pioneering experimental psychologist Fechner (1876) was one of the first to consider aesthetics as a distinct area of scientific inquiry. Beauty and other aesthetic properties, however, do not need to be identified within the art. On one hand, nature itself possesses aesthetic values; on the other hand, art may express other attributes, such as balance, dynamism, and harshness (Zangwill, 2003). In addition, the creative intent of the artist does not necessarily express itself through the language of beauty and aesthetics (Zangwill, 1999). Artistic examples that do not invoke aesthetic value include the majority of contemporary art from the 20th and 21st centuries (Kimball, 1997). Moreover, the ubiquity and popularity of art pertaining to negative emotions (Sachs, Damasio, & Habibi, 2015) and the intercultural differences in aesthetic perception (Marcus & Myers, 1995) preclude the identification of common mechanisms of the art experience that are based solely on aesthetic values.

At the same time, the same masterpiece can evoke similar sensations across different continents and cultures. For example, the popularity of Chopin's music in both Europe and East Asia highlights the existence of some common perceptual mechanisms across continents and cultures (Fung, 2007; Mlodawska-Bronowska, 2015). Indeed, contemporary psychological investigations show that art can elicit both cognitive and emotional reactions (Silvia, 2006) and that understanding art can enhance an individual's experience and enjoyment (Millis, 2001). As demonstrated in research by Silvia (2005), the experience of interest in a given artwork is highly correlated with its apprised understanding. Subsequent research on the interdependence between perception and subsequent decoding of art was conducted by Russell and Milne (1997). In their work, Russell and Milne demonstrated that the ability to understand abstract paintings is enhanced by providing titles as well as the artist's biography or the work's context of creation (Russell, 2003). In this chapter, we posit that the original assumption of Gestalt psychology is that every subjective phenomenon embedded within a frame of emotional evaluation (Pöppel, 1982; Zeki, 1998) is likely accompanied by cognitive processes.

Sending the Message

Psychological investigations have indicated that the experience evoked by an artwork may evolve during the decoding of the encrypted artist's message (Leder, Belke, Oeberst, & Augustin, 2004). This decoding occurs in the same way that information is passed from sender to receiver via an information channel, as described in the information theory proposed by Shannon (1948). The first applications of the information theory to art research date back to the middle of the 20th century. At that time, Bense (1969) described artwork using definitions of chaos and order that were derived from Shannon's information theory (Shannon, 1948). Both Bense and Shannon applied this theory to the creative process to characterize the transfer of information between the author and the artwork. In particular, Bense (1969) combined the information theory with Birkhoff's (1933) mathematical analysis of aesthetic measure as well as with Noam Chomsky's (1966) generative grammar. Bense's (1969) new theory stated that in contrast to the physical world, the world of art progresses from disorder (i.e., a low level of information) toward order (i.e., a high level of information). Thus, according to this theory, aesthetic value is defined by the relationship between chaos (or complexity) and order.

Note that Bense's theory does not define how the information encoded in the artwork is received by the audience and subsequently affects aesthetic and emotional judgments. Furthermore, contemporary studies that adopt information theory measures to art research have not included decoding, which is inherent to the process of perceiving art. For instance, Rigau, Feixas, and Sbert (2008) applied elements of the information theory to analyze paintings created by different artists in the period 1880–1939. Using the approach proposed originally by Bense, the authors interpreted the creative process as encoding of information, which can be described by the transition from the initial repertoire of colors to its selection and distribution in artistic objects (i.e., canvas). By applying three different measures of order and complexity (i.e., Shannon's entropy, Kolmogorov complexity, and Shannon's mutual entropy) to the progressively partitioned image, the researchers were able to capture the image's spatial order. In another information-theoretic study, Lopes and Tenreiro Machado (2019) adopted complexity measures (e.g., entropy, Kolmogorov complexity, and fractal dimensions) to analyze 457 works by graphic artist Escher. The researchers' work showed that the evolution of the complexity of the artworks correlated with the periods of Escher's artistic career. None of these studies, however, examined the perception and decoding of the artwork by the audience.

Receivers: Understanding the Message

The transfer of information from the author to the audience is critical for the reception of artistic intent as well as the full understanding of the experience of art. To our knowledge, however, no studies have examined the transfer of information from author to audience. The information embedded in art, along with its reception, is often implicit ("implicit learning;" Reber, 1989) or tacit ("tacit knowledge;" Reber, 1989). Neuroimaging studies have provided some evidence for the presence of implicit information in abstract art. One study by Kawabata & Zeki (2004) examined brain activity while participants viewed representational (e.g., landscape or portrait) versus abstract art. These researchers found no difference in brain activity in response to viewing representational versus abstract art (Kawabata & Zeki, 2004). This result may suggest that the same features and compositional elements are present in both representational and abstract paintings (Kawabata & Zeki, 2004).

A study of electrical brain activity carried out by Umilta, Berchio, Sestito, Freedberg, and Gallese (2012) suggests that the information embedded in artwork may also be implicitly decoded by the audience. The results of this research are consistent with the notion that traces of the artist's gestures on the canvas (e.g., brush strokes) activate motor areas in the viewer's brain (Freedberg & Gallese, 2007). In an experiment conducted by Umilta *et al.* (2012)), researchers compared the activity of the motor cortex in participants viewing either high-resolution digitized images or simplified versions of artworks by Lucio Fontana by showing cuts on the canvas. The experiment was conducted with two groups: (1) participants who were already familiar with Fontana's artworks and (2) participants who were exposed to Fontana's paintings for the first time. Results revealed in both groups a decrease in the amplitude of the mu rhythm (8–14 Hz), as compared to baseline, recorded by electrodes clustered around sites C3 and C4 located over the motor cortex. This amplitude decrease occurred only when subjects were viewing high-resolution digitized images of the original artworks and not when they viewed simplified versions. Importantly, previous studies have reported a suppression of mu rhythm on the same clusters of electrodes when subjects viewed hand grasping (Muthukumaraswamy & Johnson, 2004a, 2004b; Streltsova, Berchio, Gallese, & Umilta, 2010) or meaningless gestures (Babiloni *et al.*, 2002; Streltsova *et al.*, 2010). Together, these similarities suggest that the viewer's brain (intentionally or not) decodes detailed information about the artist's strokes in the original artwork.

According to the information theory, successful decoding of transferred information can occur only if the same encoding/decoding mechanism is at work in both the sender and the receiver (Shannon, 1948). However, it is unlikely that developmental rules that are embodied in the nervous network are sufficient for encoding and/or decoding culturally conditioned information. Mechanisms of individual plasticity and contextual information should also be included in this process. The notion that individual plasticity and contextual information play a role in encoding and decoding processes is supported by several psychological studies on contextual information processing, conducted in Eastern Asia, Western Europe, and the United States. Behavioral investigations have shown that people from East Asia detected changes in contextual information more often than individuals from Europe and the United States. East Asians also paid attention to a broader display spectrum and were more capable of detecting color changes in the periphery of a scene, whereas Westerners detected mostly central changes (Boduroglu, Shah, & Nisbett, 2009). Furthermore, in studies of emotional perception, Japanese subjects used facial expressions of the surrounding people to establish the target person's emotions, whereas Westerners did not use this approach (for a review, see Mesquita, Boiger, & De Leersnyder, 2016). Intercultural differences in brain information processing have also been detected using brain imaging. One neuroimaging study by Ma *et al.* (2014) has compared brain activity of subjects from China and Denmark while they made attribute judgments across three dimensions: social roles, personality traits, and physical attributes. The analyses showed greater activity in the medial prefrontal cortex during self-reflection in Western subjects as compared to subjects from East Asia. These results suggest greater encoding of self-relevant stimuli and less interdependence among Western individuals versus individuals with East Asian cultural experiences.

Decoding Mechanisms

Based on the aforementioned studies, the proposed encoding/decoding mechanism for understanding an artwork should also include sociocultural factors, as these factors are necessary for the emotional processing and enjoyment of art. Here, we posit that this encoding/decoding mechanism could be based on the set of internal concepts shaped by common sociocultural factors, such as shared values, education level, social role, culture, and personal experience. This approach is similar to "conceptual schemes," as proposed by Davidson (1973), that can be used to organize personal experience into points of view. This notion is consistent with the previously reported correlation between understanding and appreciating an artwork (Silvia, 2005, 2006). Similarly, psychological and sociological investigations have shown that more sophisticated and complex musical genres (e.g., folk, bluegrass, blues, jazz, opera, and classic) that require more experienced composers with more musical education are preferred by audiences with higher social or professional positions (Mellander, Florida, Rentfrow, & Potter, 2018; Peterson, 1992), higher education (Mellander *et al.*, 2018), and greater intellectual openness (Rentfrow & Gosling, 2003). An analogous relationship also has been found between age and musical preferences; in particular, older people with more experience have been shown to prefer more complex genres (North & Hargreaves, 1997; Rentfrow & Gosling, 2003). Thus, the reported high popularity of Chopin's music in Europe, China, and Japan may be explained by a relatively large ratio of highly educated individuals in all three of these regions (https://ourworldindata.org/quality-of-education; http://worldpopulationreview.com/countries/education-rankings-by-country; https://en.wikipedia.org/wiki/Education_Index).

One key question should be investigated: What is the neural basis of the aforementioned concepts? Further, it is unknown how the information embedded in artwork is decoded by the perceiver. One key candidate mechanism is the idea of concept cells or networks, as proposed by Quiroga (2012). Quiroga found that concept cells integrate and store all information pertaining to one concept, regardless of the modality of the inducing stimulus. The concept described by multimodal stimuli could include the encoding of people, events, or more complex cases (e.g., family and television show). Neurons of the same concept network are activated regardless of the modality of the stimulus, as long as the stimulus is associated with that concept. For example, take the case of a *Star Wars* movie (Kurtz, Lucas & Lucas, 1977). Neurons encoding the concept of *Star Wars* will be activated by a movie excerpt, a picture of a movie character, or even the spoken or written name of the film (Quiroga, 2012). We hypothesize that concept networks, which integrate multimodal information, may also store abstract stimuli that are common across less culturally dependent concepts. These abstract stimuli may also constitute part of a conceptual dictionary that is composed of putative concept networks that store information common to many concepts.

Premises for concept networks and dictionaries were laid out in an article by Mitchell *et al.* (2008), who investigated brain activity associated with the meaning of nouns. To examine brain activity related to the meaning (or the "associated concept") of a given noun, researchers created a map of activations for 60 word-picture pairs of selected nouns. Then, each map was decomposed into activations that were associated with intermediate features of that noun, modeled as the activations of its corresponding n-grams. The n-grams were composed of contiguous sequences of n items from a large text corpus, in which a given text can be found in a large corpus of text. Such meaning maps were then used to predict brain activation for newly presented nouns. Results showed a mean prediction accuracy of 70%, suggesting that brain activity associated with specific concepts can be identified using neuronal constituents of elements from a conceptual dictionary. We hypothesize that this conceptual dictionary is comprised of brain areas activated by *n*-grams of a given noun.

We propose that elements of the conceptual dictionary are encoded in concept networks, which can induce similar subjective sensations that are mediated by common brain networks. Support for this notion comes from the work of Vessel, Starr, and Rubin (2012), who asked participants to apprise works of art from a variety of cultural traditions and historical periods, as aesthetically moving or not. The researchers found large individual differences in aesthetic ratings; for example, an image that was highly recommended by one observer was given a low recommendation by another. However, the brain activation maps showed a consistent pattern, such that there was a linear association between recommendation level and blood-oxygen level-dependent response in occipitotemporal regions (left inferior temporal sulcus, left parahippocampal cortex, and right superior temporal gyrus) and hubs of the default mode network.

A more diverse dictionary would allow for more connections between elements, and thus a greater ability to describe more complex objects and relations. We hypothesize that higher levels of education and more varied personal experiences encourage a more diverse conceptual dictionary. Support for this notion comes from the previously cited investigations that have examined the popularity of different genres of music among individuals with different levels of education and of different ages (Mellander *et al.*, 2018; Peterson, 1992).

The hypothesis presented here proposes that the impact of art depends on the activation of neuronal networks that encode elements of the conceptual dictionary of one or many concepts. When many different and originally disconnected concept networks are simultaneously activated, a new network may emerge. This new network may, in turn, lead to new associations or concepts, which is similar to a recent model of memory consolidation (Klinzing, Niethard, & Born, 2019). A diverse conceptual dictionary that allows for many new associations (including abstract associations) could elucidate sensations evoked by abstract or complex art. In contrast, a poor dictionary, associated with lower education and/or a narrower personal experience, could limit perception to less complex art (e.g., pop, rap, and representational paintings). Thus, perception is directly associated with the experience and knowledge of a given individual. A poor dictionary may impede comprehension of the artistic intent embedded in more sophisticated art or in art that originates from different cultures, thus undermining artistic value. These data imply that the recognition of an artwork depends upon the individual's conceptual dictionary.

The Sender's Perspective

Interestingly, our hypothesis regarding the value of the presence of an audience is in line with the opinion of artists across different genres (e.g., film, ballet, or theater). In a letter to Edward Alden Jewell, an art critic from *The New York Times* (1943), Mark Rothko and Adolph Gottlieb wrote:

> "There is no such thing as good painting about nothing. We do not intend to defend our pictures. They make their own defense. We consider them clear statements. Your failure to dismiss or disparage them is prima facie evidence that they carry some communicative power. It is our function as artists to make the spectator see the world our way — not his way" (as cited in Clearwater, 2006, p. 181).

In another piece by Stanisław Dróżdż, a creator of concrete poetry, stated the following in a review:

> "Concrete poetry works are informative. I send a message and leave it for the audience to read. [...] Information is not always designed by me from the beginning to the end. It sometimes surfaces by itself. And I don't know how this happens" (as cited in Dawidek, 2012, p. 312).

Finally, the following statement was delivered by the sculptor Tony Cragg during the 2016 "Sculpture Today" conference:

> "The point of exhibiting a pissoir may well have initially been to shock a bourgeois French public, but giving an industrial object a context in which it was given other meanings did have another far-reaching significance. We know that all things have their physical four-dimensional extension in space and time but they also exist in the form of memories and mental pictures in our minds every object or thing not only has a physical existence but also has a cloud of meaning — words and pictures — associated to it. Artists realized that it was possible to manipulate the terms associated to a thing so as to effect changes in its meaning. Enabling banal objects, like a soup can, to take on a new existence. In fact it was possible to completely transform something without even having to change its form" (Cragg, 2016, p. 152).

To summarize, we hypothesize that the impression created by an artwork on the audience is based on the process of decoding the message imparted by the author. For the successful decoding of transferred information, the authors and the audience must share similar encoding/decoding mechanisms. These mechanisms are modulated by sociocultural factors, such as personal experience, education, and cultural tradition. The neurophysiological basis of this mechanism may include activation of the neuronal networks evoked by the set of encoded abstract stimuli. These mechanisms are common for many concepts that comprise an individual's conceptual dictionary.

Interestingly, abstract drawings prepared by any type of automated process do not evoke the interest of the audience. Similarly, experiencing natural phenomena (e.g., mountains and stormy ocean views) does not require an encoding/decoding process, which appears to be an indispensable component of experiencing an artwork. These observations suggest that what makes art fascinating is the possibility of insight into the mind of an artist.

References

Babiloni, C., Babiloni, F., Carducci, F., Cincotti, F., Cocozza, G., Del Percio, C., *et al.* (2002). Human cortical electroencephalography (EEG) rhythms during the observation of simple aimless movements: A high resolution EEG study. *NeuroImage*, 17(2), 559–572. https://doi.org/10.1006/nimg.2002.1192.

Bense, M. (1969). *Introduction to the Information-Theoretical Aesthetics. Foundation and Application in the Text Theory*. Stuttgart: Rowohlt Taschenbuch Verlag GmbH.

Birkhoff, G. D. (1933). *Aesthetic Measure*. Cambridge: Harvard University Press.

Boduroglu, A., Shah, P., & Nisbett, R. E. (2009). Cultural differences in allocation of attention in visual information processing. *Journal of Cross-Cultural Psychology*, 40(3), 349–360. https://doi.org/10.1177/0022022108331005.

Chomsky, N. (1966). *Topics in the Theory of Generative Grammar.* The Hague: Mouton.

Clearwater, B. (2006). *The Rothko Book*. London, UK: Tate Publishing.

Cragg, T. Centrum Rzeź by Polskiej w Orońsku (2016). *Sculpture Today*, pp. 145–154. Orońsko, Poland.

Davidson, D. (1973). On the very idea of a conceptual scheme. *Proceedings and Addresses of the American Philosophical Association*, 47, 5–20. https://doi.org/10.2307/3129898.

Dawidek, M. (2012). *Odprysk poezji. Stanisław Dróżdż mówi/A Piece of Poetry*. Conversations with Stanisław Dróżdż. Kraków, Poland: Korporacja Ha!art.

Fechner, G. T. (1876). *Vorschule der Aesthetik.* [*Preschool of Aesthetics*]. Hildesheim, Germany: Olms.

Freedberg, D. & Gallese, V. (2007). Motion, emotion and empathy in esthetic experience. *Trends in Cognitive Sciences*, 11(5), 197–203. https://doi.org/10.1016/j.tics.2007.02.003.

Fung, A. Y. H. (2007). The emerging (national) popular music culture in China. *Inter-Asia Cultural Studies*, 8(3), 425–437. https://doi.org/10.1080/14649370701393824.

Leder, H., Belke, B., Oeberst, A., & Augustin, D. (2004). A model of aesthetic appreciation and aesthetic judgments. *British Journal of Psychology*, 95(4), 489–508. https://doi.org/10.1348/0007126042369811.

Kawabata, H. & Zeki, S. (2004). Neural correlates of beauty. *Journal of Neurophysiology*, 91(4), 1699–1705. https://doi.org/10.1152/jn.00696.2003.

Kimball, R. (1997). Art without beauty. *The Public Interest*, 127, 44–59.

Klinzing, J. G., Niethard, N., & Born, J. (2019). Mechanisms of systems memory consolidation during sleep. *Nature Neuroscience*, 22, 1598–1610. https://doi.org/10.1038/s41593-019-0467-3.

Kurtz, G., Lucas, G. (Producers), & Lucas, G. (Director). (1977). *Star Wars* [*Motion Picture*]. United States: Twentieth Century Fox.

Lopes, A. M. & Tenreiro Machado, J. A. (2019). Complexity analysis of Escher's art. *Entropy*, 21(6), 553. https://doi.org/10.3390/e21060553.

Ma, Y., Bang, D., Wang, C., Allen, M., Frith, C., Roepstorff, A., & Han, S. (2014). Sociocultural patterning of neural activity during self-reflection. *Social Cognitive and Affective Neuroscience*, 9(1), 73–80. https://doi.org/10.1093/scan/nss103.

Marcus, G. E. & Myers, F. R. (1995). *The Traffic in Culture: Refiguring Art and Anthropology*. California, CA: University of California Press.

Mellander, C., Florida, R., Rentfrow, P. J., & Potter, J. (2018). The geography of music preferences. *Journal of Cultural Economics*, 42, 593–618. https://doi.org/10.1007/s10824-018-9320-x.

Mesquita, B., Boiger, B., & De Leersnyder, J. (2016). The cultural construction of emotions. *Current Opinion in Psychology*, 8, 31–36. https://doi.org/10.1016/j.copsyc.2015.09.015.

Millis, K. (2001). Making meaning brings pleasure: The influence of titles on aesthetic experience. *Emotion*, 1(3), 320–329. https://doi.org/10.1037/1528-3542.1.3.320.

Mitchell, T. M., Shinkareva, S. V., Carlson, A., Chang, K., Malave, V. L., Mason, R. A., & Just, M. A. (2008). Predicting human brain activity associated with the meanings of nouns. *Science*, 320(5880), 1191–1195. https://doi.org/10.1126/science.1152876.

Młodawska-Bronowska, J. (2015). Corporate culture. In: J. Młodawska-Bronowska (ed.), *Japan — A Western View in New Opportunities for Polish-Japanese Cooperation: Diagnosis and Prospects*, pp. 95–106. Łódz, Poland: Wydawnictwo Uniwersytetu Łódzkiego.

Muthukumaraswamy, S. D. & Johnson, B. W. (2004a). Changes in rolandic mu rhythm during observation of a precision grip. *Psychophysiology*, 41(1), 152–156. https://doi.org/10.1046/j.1469-8986.2003.00129.x.

Muthukumaraswamy, S. D. & Johnson, B. W. (2004b). Primary motor cortex activation during action observation revealed by wavelet analysis of the EEG. *Clinical Neurophysiology*, 115(8), 1760–1766. https://doi.org/10.1016/j.clinph.2004.03.004.

North, A. C. & Hargreaves, D. J. (1997). *Music and Consumer Behaviour. The Social Psychology of Music.* Oxford: Oxford University Press.

Peterson, R. A. (1992). Understanding audience segmentation: From elite and mass to omnivore and univore. *Poetics,* 21(4), 243–258. https://doi.org/10.1016/0304-422X(92)90008-Q.

Pöppel, E. (1982). *Lust und Schmerz. Grundlagen menschlichen Erlebens und Verhaltens (Pleasure and Pain. Bases of Human Experience and Behaviour).* Berlin, Germany: Verlag Severin & Siedler.

Quiroga, R. Q. (2012). Concept cells: The building blocks of declarative memory functions. *Nature Review Neuroscience*, 13, 587–597. https://doi.org/10.1038/nrn3251.

Rentfrow, P. J. & Gosling, S. D. (2003). The do re mi's of everyday life: The structure and personality correlates of music preferences. *Journal of Personality and Social Psychology*, 84(6), 1236–1256. https://doi.org/10.1037/0022-3514.84.6.1236.

Reber, A. S. (1989). *Implicit Learning and Tacit Knowledge.* New York, NY: Oxford University Press.

Rigau, J., Feixas, M., & Sbert, M. (2008). Informational aesthetics measures. *IEEE Computer Graphics and Applications*, 28(2), 24–34. https://doi.org/10.1109/MCG.2008.34.

Russell, P. A. (2003). Effort after meaning and the hedonic value of paintings. *British Journal of Psychology*, 94(1), 99–110. https://doi.org/10.1348/000712603762842138.

Russell, P. A. & Milne, S. (1997). Meaningfulness and hedonic value of paintings: Effects of titles. *Empirical Studies of the Arts*, 15(1), 61–73. https://doi.org/10.2190/EHT3-HWVM-52CB-8QHJ.

Sachs, M. E., Damasio, A., & Habibi, A. (2015). The pleasures of sad music: A systematic review. *Frontiers in Human Neuroscience*, 9, 404. https://doi.org/10.3389/fnhum.2015.00404.

Shannon, C. E. (1948). A mathematical theory of communication. *Bell System Technical Journal*, 27(4), 623–656. https://doi.org/10.1002/j.1538-7305.1948.tb00917.x.

Silvia, P. J. (2006). Artistic training and interest in visual art: Applying the appraisal model of aesthetic emotions. *Empirical Studies of the Arts*, 24(2), 139–161. https://doi.org/10.2190/DX8K-6WEA-6WPA-FM84.

Silvia, P. J. (2005). Cognitive appraisals and interest in visual art: Exploring an appraisal theory of aesthetic emotions. *Empirical Studies of the Arts*, 23(2), 119–133. https://doi.org/10.2190/12AV-AH2P-MCEH-289E.

Streltsova, A., Berchio, C., Gallese, V., & Umilta, M. A. (2010). Time course and specificity of sensory-motor alpha modulation during the observation of hand motor acts and gestures: A high density EEG study. *Experimental Brain Research*, 205, 363–373. https://doi.org/10.1007/s00221-010-2371-7.

Vessel, E., Starr, G. G., & Rubin, N. (2012). The brain on art: Intense aesthetic experience activates the default mode network. *Frontiers in Human Neuroscience*, 6, 66. https://doi.org/10.3389/fnhum.2012.00066.

Umilta, M. A., Berchio, C., Sestito, M., Freedberg, D., & Gallese, V. (2012). Abstract art and cortical motor activation: An EEG study. *Frontiers in Human Neuroscience*, 6, 311. https://doi.org/10.3389/fnhum.2012.00311.

Zangwill, N. (2003). *The Oxford Handbook of Aesthetics*, p. 2003. Oxford, UK: Oxford University Press.

Zangwill, N. (1999). Art and audience. *The Journal of Aesthetics and Art Criticism*, 57(3), 315–332. https://doi.org/10.2307/432197.

Zeki, S. (1998). Art and the brain. *Daedalus*, 127, 71–103.

https://doi.org/10.1142/9789811298639_0008

Chapter 8

Judgments of Mathematical Beauty Are Resistant to Revision through External Opinion*

Haoxuan Zhang and Semir Zeki

Laboratory of Neurobiology, Division of Cell &Developmental Biology, University College London, UK

Abstract

We here address the question of the extent to which judgments of mathematical beauty (which we categorize as biological beauty) are resistant to revision through external opinion. A total of 100 mathematicians of different national and ethnic origins were asked to rate 60 mathematical equations for their beauty; after being presented a fictitious "expert rating," they were asked to re-rate the same equations. Results showed that the judgments of mathematical beauty had a high level of resistance to external opinion. This is in line with the resistance to revision of judgments for other categories of biological beauty.

***Keywords*:** aesthetic judgments; biological beauty; neuroaesthetics; mathematical beauty

*This chapter was originally published in the special collection on neuroaesthetics in *PsyCh Journal*. Zhang, H. & Zeki, S. (2022). Judgments of mathematical beauty are resistant to revision through external opinion. *PsyCh Journal*, 11(5), 741–747. https://doi.org/10.1002/pchj.556.

Introduction

Mathematicians commonly exalt the beauty of mathematical formulae; some, such as Bertrand Russell (1919), have declared that mathematics possesses "not only truth but supreme beauty," whereas others such as Herman Weyl, if pressed to choose between truth and beauty in a mathematical formula, have declared a preference for beauty. The physicist Paul Dirac (1940) developed a theory of mathematical beauty and proposed that it was the beauty rather than the simplicity of a mathematical formula that was a guide to its veracity. But what does the beauty of a mathematical formula consist of? Immanuel Kant supposed that a mathematical formula is experienced as beautiful because "it makes sense" (Breitenbach, 2013), which of course begs the question of what it makes sense to. One possibility is that it makes sense because it obeys the rules of the brain's logical systems (Zeki *et al.*, 2014); because one assumes that these are the same for all brains, regardless of ethnic and cultural groupings, it is reasonable to assign mathematical beauty to the category of biological beauty, just like the beauty of human faces or bodies, as opposed to artifactual beauty (Zeki *et al.*, 2018; Zeki *et al.*, 2014). This may seem strange; unlike the experience of beauty in human faces or figures, mathematical beauty is perhaps among the most extreme examples of the experience of beauty that is dependent upon culture and learning; no one who is not conversant with the language of mathematics is likely to be able to judge mathematical formulae aesthetically. One test of this would lie in psychophysical experiments to determine the extent to which aesthetic judgments in mathematics are hostage to peer opinion. In general, a characteristic of biological beauty is that, unlike artifactual beauty, it is more resistant to change in light of external opinion; hence, the Bayesian system of belief-updating should be relatively less effective for mathematical beauty than for aesthetic experiences that belong in the artifactual category (Zeki & Chén, 2020). We have tested this hypothesis here, in light of recent psychophysical studies which have demonstrated that there is indeed greater resistance to modifying aesthetic judgments of stimuli belonging to the biological category compared to ones belonging to the artifactual one (Bignardi *et al.*, 2020; Glennon & Zeki, 2021; Vessel *et al.*,

2018). To undertake our study, we asked mathematicians to rate the beauty of 60 mathematical equations collected for a previous study (Zeki *et al.*, 2014) and studied the extent to which they changed their rating in light of external "peer" opinions.

We have used the same approach here as we did for studying the appreciation of the beauty of human faces and of abstract paintings, where we found that a revision of opinion about the aesthetic status of a face was about twice as unlikely as a revision of opinion about the aesthetic status of abstract paintings, which fall firmly in the artifactual category (Bignardi *et al.*, 2020).

Materials and Methods

Participants

One hundred participants (60 female) were recruited from mathematical departments of universities worldwide or directly from our knowledge of them; their ages ranged from 20 to 51 ($M = 23.0$, $SD = 5.68$), and all had majored in mathematics, applied mathematics, or statistics. The majority were nationals of the UK ($n = 40$) and various European countries ($n = 38$), whereas the rest were from Asia ($n = 13$), the Middle East ($n = 5$), and South America ($n = 4$). All participants gave their consent after being informed of the objectives of the study, which had been approved by the Ethics Committee of University College London.

Stimuli

We used the 60 mathematical formulae that were used in a study of the neural correlates of the experience of mathematical beauty (Zeki *et al.*, 2014). As in that previous study, the equations were displayed on a screen in white font on a dark gray background (CMYK 0.00, 0.00, 0.00, 0.80). The base size of the image was 600×200 pixels; the width and height of each image were adjusted to fit the different sizes of the equations to produce more visually uniform stimuli.

Procedure

The experimental procedure was designed using PsychoPy 3.0 software (Peirce *et al.*, 2019), and the experiment was conducted through the Pavlovia.org online research platform. Before starting the experiment, demographic information was collected from the participants regarding their gender, age, and nationalities. Subjects first viewed a sample equation and were asked to complete a sample trial; they were told that the objective of the experiment was to assess the degree to which they would revise their rating of the beauty of the formulae in light of feedback from external experts about the aesthetic status of these formulae. They were not, however, told that the expert ratings were fictitious.

The experiment consisted of 60 trials, each with one mathematical formula (the formulae are available in the supplementary details of the article by Zeki *et al.*, 2014). These were displayed in random order to prevent order effects. Each trial had two components: (1) a rating of how beautiful the equations were judged by the subject and (2) a rating for how well the equations were understood. Both ratings were given separately, using a scale of 0 (*not beautiful at all*) or (*not understood at all*) to 10 (*very beautiful*) or (*very well understood*). In Part 1 of the experiment, the beauty and understanding ratings were collected via the slider provided in PsychoPy 3.0; to prevent any misclick, the slider was activated 5 s after the equations appeared on the screen, where they remained until ratings were given, which ensured that sufficient time was provided for participants to study the equations and respond. In Part 2, participants were given a fictitious "expert beauty rating" and were told that it was the average beauty rating given to that equation by mathematical experts; in reality, the ratings had been computer-generated. The original rating given by the participants was not displayed to them when they were asked to re-rate the equation in light of external opinion. Following the previous work of Zaki *et al.* (2011), Bignardi *et al.* (2020), and Glennon & Zeki (2021), the fictitious "expert" ratings were derived from the beauty ratings given by the actual participants (discussed in the following).

We used fictitious rather than real ratings because we consider mathematical beauty to fall under the category of biological beauty (see Zeki & Chén, 2020; Zeki *et al.*, 2014). We therefore expected ratings to be

Table 1. The probability distribution of fictitious ratings, ΔE.

	First beauty rating										
Input	**First beauty rating < 4**			**4 < First beauty rating < 6**					**First beauty rating > 6**		
Possible ΔE	0	2	3	0	−2	−3	2	3	0	−2	−3
Probability of the ΔE	50%	25%	25%	50%	12.5%	12.5%	12.5%	12.5%	50%	25%	25%
Output	Fictitious rating = first beauty rating + ΔE										

consistent across individuals; hence, the use of real ratings would have been less effective in swaying opinion of the aesthetic status of the equations. When the participants' first beauty rating was less than 4 (*not beautiful*), the displayed fictitious "feedback" rating had a 50% chance of being the same (*no change*) and a 25% chance of being 2 or 3 points higher. Similarly, where participants had rated the equation higher than 6, the "fictitious" rating had a 50% chance of being the same and a 25% chance of being 2 or 3 points lower. Finally, when the initial rating given was between 4 and 6, the fictitious rating had a 50% chance of being 2 or 3 points higher and 2 or 3 lower (with a 12.5% chance for each), whereas there was a 50% chance of it being the same as the initial rating given by the subject. Table 1 shows how the fictitious ratings were generated using the aforementioned probabilities. In summary, there was on any given occasion always a 50% chance for the fictitious rating to be the same as the initial beauty rating given by the participant. The rating was rounded up to the nearest integer before being displayed to the participants.

Participants were asked to rate the same equation again immediately after seeing the fictitious "peer" ratings. They were free to either maintain their original rating or give a new one, to both questions.

Analysis

To ascertain whether external feedback opinion influences the beauty ratings that subjects gave to the equations, responses were divided into two groups: In one group (Group A) were responses in which the feedback rating was identical to the rating given by the subjects; Group A served as a control for the second group (Group B), which was divided into two subgroups according to whether the feedback rating was above or below the initial rating given by the subjects. There were 3,016 ratings in Group A and 2,984 ratings in Group B. To test whether the rating changes (RCs) were derived from normally distributed populations, we submitted the data from the three groups (Group A and the two subgroups of B) to the Lilliefors test to verify whether the RC (second beauty rating–first beauty rating) came from normally distributed populations. The possible effect of gender on the rating was also inspected via a Kruskal–Wallis one-way analysis of variance.

To investigate the magnitude of RC induced by external feedback opinion, two variables were defined: the RC itself and the difference between the first beauty rating and the external feedback rating (ΔE). To facilitate the analysis and grouping, we introduce the term "|RC|" which defines the absolute magnitude of RC; values of RC and ΔE were calculated according to the following formula:

RC = second beauty rating–first beauty rating

ΔE = fictitious "expert" rating–first beauty rating

Our main prediction was that external information will not influence aesthetic judgments of mathematical beauty because we hypothesized that mathematical beauty belongs to the biological category (see Zeki *et al.*, 2014). This was tested by the Kruskal–Wallis analysis; an interval plot and multiple comparison tests were applied to both the RC and to within-subject analyses to determine whether external information affects the magnitude and direction of RC. The RCs in the three groups were compared to check if the fictitious feedback altered the distribution and the mean ranks of the RC. Next, a Spearman rank correlation coefficient was calculated between the two variables: RC (which refers to how much subjects changed their rating) and ΔE (which refers to how much we tried, through the fictitious rating, to manipulate their judgment). This allowed us to learn to what extent RC and the feedback are monotonically related and establish the difference between the fictitious "expert rating" and the initial beauty rating (ΔE). The result, the correlation coefficient [r_s], reflected how well the two variables were monotonically correlated (i.e., how the magnitude of the rating change correlated with the magnitude of the external fictitious information).

In addition to the main analysis, a regression analysis was undertaken to check if the levels of understanding of the mathematical equations influence the magnitude of RC; that is, whether better understanding of an equation improves the robustness of the initial aesthetic judgment of mathematical formulae judgments, thus making it more resistant to outside influence.

We used the mean minus one (MM1) analysis (Vessel *et al.*, 2018) to investigate the average agreement in mathematical beauty rating among subjects. To calculate the MM1, the correlation between two variables — (1) the beauty rating given by one subject for each equation and (2) the average beauty rating given by all other subjects for the same equation — was computed. This was repeated for all subjects and all equations and produced an individual preference score for each subject; the latter reflects whether an individual subject's beauty ratings for all the equations were in agreement with the mean rating derived from all other subjects for the same equations (Germine *et al.*, 2015; Sutherland *et al.*, 2020). The across-observer average MM1 score was acquired by applying Fisher r to z transformation to the individual preference score (r value), calculating the mean, and transforming back to the r value again. Compared to just taking the mean of individual r values, the transformations result in less biased estimates of between-subject average agreement (Bronstad & Russell, 2007), with a higher MM1 value suggesting better agreement between beauty ratings of mathematical equations.

The more widely used interclass correlation coefficient (ICC) (Bi & Kuesten, 2012) was used as an additional check on the results derived from the MM1 analysis. ICC indicates the level of agreement in aesthetic rating given to the formulae by different subjects. In this analysis, the degree of the absolute agreement in beauty ratings, expressed as the average of 100 independent ratings per equation made by 100 participants, was evaluated. A higher ICC value represents a higher uniformity among subjects in beauty ratings for each mathematical equation and vice versa.

All analyses were conducted in MATLAB R2020b and RStudio Desktop 1.4.1106.

Results

Mean beauty and understanding ratings

The initial and second beauty ratings had means of 5.73 and 5.68 (*SD*s = 2.72 and 2.69), respectively; this shows that there was a high correlation between the second and first beauty ratings, $r_s(5{,}998) = 0.91$, $p < 0.001$. The mean understanding rating for the mathematical equations was 5.86 ($SD = 3.34$), suggesting that equations were well distributed according to the ability to understand. As in our previous study (Zeki *et al.*, 2014), a significant positive correlation between understanding level and beauty rating was observed, $r(5{,}998) = 0.60$, $p < 0.001$. The average absolute difference between the first and second beauty ratings was 0.40 ($SD = 0.74$), suggesting a low level of RC across all trials. The feedback Group B (n = 2,984) had a significantly higher mean RC (0.56), $p < 0.001$, compared to the RC of the control Group A (0.28) (n = 3,016), for which the external feedback was identical to their original rating. The *SD* of RC in the feedback Group B (0.80) was also numerically higher than that in Group A (0.66). These results show that external feedback resulted in RCs in 21% of the examples in that they revised their judgment of the beauty of the equations. In these 21% of cases, external feedback led to an increase in the magnitude of the RC. In the remaining 79%, the RC was less than 0.77; we regard this as insignificant because it can be the result of errors from inaccurate clicks or rounding-up processes during fictitious rating generation. The criterion of 0.77 was adopted from the 90% quartile of RC produced by Group A, which reflects 90% of the error generated during the process.

The Kruskal–Wallis test on gender led to the rejection of the null hypothesis that the beauty rating given by the two genders came from the same population distribution, $H(1) = 9.552$, $p = 0.002$. In other words, gender had an effect on the first beauty rating, which was also reflected by the mean beauty ratings as female participants tended to give lower beauty ratings (5.62 vs 5.80). Next, the same analysis was applied to the changes in beauty ratings made by the two genders to check if the magnitude of RC was influenced by gender, with the result that it was not, $H(1) = 49.8$, $p < 0.001$. It appears that female participants tended to have larger RCs compared to male participants, with a mean value of 0.46 and 0.36, respectively.

Tests for the normality of rating change distributions

The Lilliefors test was applied to the RC in all three groups (Group A and the two subgroups of B) for ratings given positive and negative feedback, ΔE. The test indicated that none of these groups followed a normal distribution at the 95% CI, $h = 1$, $p < .001$, in that responses were more concentrated centrally; in other words, subjects had been less likely to change their opinion than would have been expected from a normally distributed population. This led us to use a nonparametric analysis for evaluating the effects of external opinion on RC.

Effect of external opinion on mathematical aesthetic judgments

In Figure 1, where the x-axis represents the RC, it is observable that, for all three groups, the great majority of responses had their RCs clustered around zero. This indicates that judgments of mathematical beauty are very resistant to change in light of external opinion; where external opinion had an effect (i.e., where ΔE existed) (Figure 1(b) and (c)), a portion of the responses was observed to have more positive or negative RCs, as compared to the responses in Group A (Figure 1(a)).

To evaluate the effect of the external information on the RC, we applied the Kruskal–Wallis test to all three groups. The result showed a significant difference between the means among the three groups, $H(2) = 613.02$, $p < 0.01$; in other words, where there was a rating change, this was the result of the presence of ΔE and not due to random factors. This conclusion was further supported by the multiple comparison test (see Figure 2(b)), which shows that both positive and negative ΔE feedback groups had mean ranks significantly different from Group A and were also

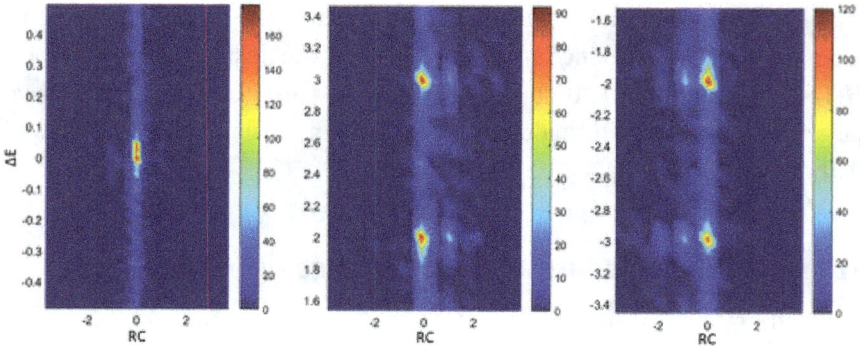

(a) (b) (c)

Figure 1. Heat maps of the distribution of rating change (RC) versus ΔE in the three groups. Warmer colors represent more RCs at a given value. (a) Distribution in Group A ($-0.5 < \Delta E < 0.5$), (b) distribution in Group B positive ΔE subgroup ($1.5 < \Delta E < 3.5$), and (c) distribution in Group B negative ΔE subgroup ($-3.5 < \Delta E < 1.5$).

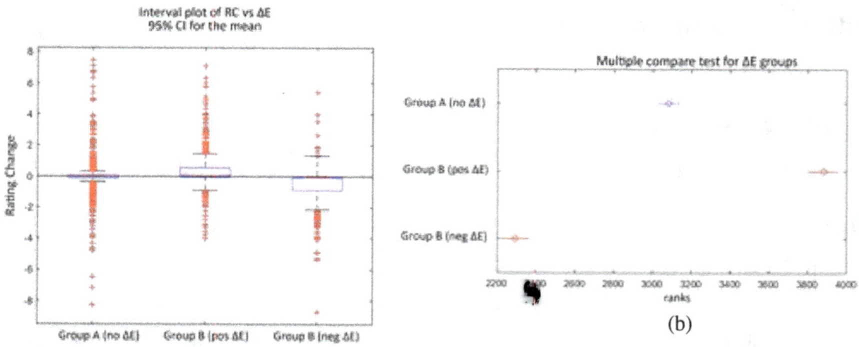

(a) (b)

Figure 2. (a) Interval plot of rating change (RC) versus ΔE for the three groups and (b) multiple comparison test results for the three groups; RCs in Group A have mean ranks significantly different from two subgroups of Group B.

different from each other. The interval plot of Figure 2(a) shows that in Group A, the upper and lower quartiles were densely packed within a 1-point range, with mean RCs close to 0. For the positive and negative ΔE groups, the means were higher and lower than 0, respectively, with significantly wider interquartile ranges. Thus, where ΔE was effective in inducing an RC, it affected both the occurrence and magnitude of the RC, though the overall RC was small for both.

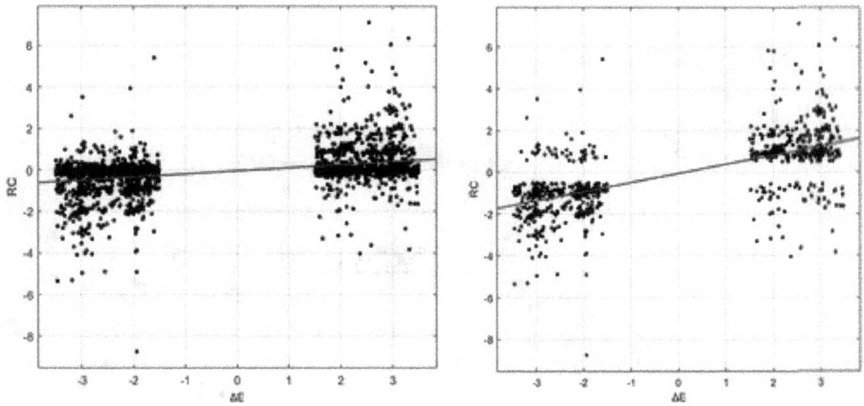

(a) (b)

Figure 3. (a) Correlation of rating change (RC) and ΔE in the feedback group and (b) same correlation, with trials without significant |RC|<0.5 excluded.

The Spearman correlation analysis also showed that within the feedback group, RC and ΔE had a significant positive correlation, $r_s(2{,}984) = 0.391$, $p < 0.001$. In other words, a more positive ΔE rendered participants less resistant to external opinion, and vice versa (Figure 3(a)); considering that there was no significant RC, $|\text{RC}| < 0.77$, for over two-thirds of trials (2,068 in 3,047), the exclusion of such trials will render the above correlation numerically higher, $r_s(978) = 0.672$, $p < 0.001$.

There was a weak negative correlation r_s between understanding level and RC, $r_s(2{,}984) = -0.112$, $p < 0.01$, suggesting that a better understanding may lead to an even smaller RC.

MM1

The MM1 score shows the degree of agreement among subjects in their ratings of stimuli, with an MM1 score above 0.5 indicating good agreement. In this study, the across-observer average MM1 score showed a good degree of agreement for mathematical beauty, MM1 = 0.509, 95% CI [0.455, 0.528]. This MM1 value is higher than those for abstract works, which were 0.3 in the works by Bignardi *et al.* (2020) and Vessel *et al.* (2018); however, it is lower in value than that obtained for other biological

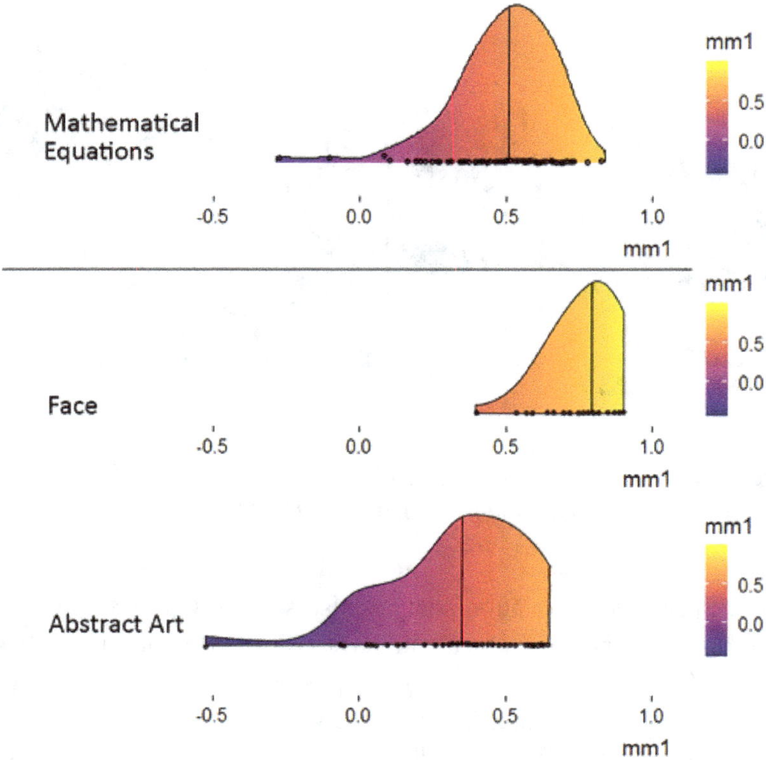

Figure 4. Mean minus one (MM1) plots to show the degree of agreement in beauty ratings across subjects. Data for faces and abstract artworks were adapted from Bignardi *et al.* (2020). Black dots represent individual preference scores (how well a certain subject's beauty rating agrees with the average rating given by others). Vertical line represents the across-observer average MM1 score.

categories, which stand at *c* 0.7 for facial beauty and *c* 0.6 for landscapes in the study by Vessel *et al.* (2018) and at 0.81 in the studies by Bignardi *et al.* (2020) and Glennon & Zeki (2021) (Figure 4).

The extent of agreement between individuals' aesthetic (or other) judgments is probably a reliable pointer to the extent to which they would revise their judgments in light of external opinion; the higher the agreement, the more resistant they would probably be to revision. The extent of the agreement between individuals' ratings is better shown by the more widely used ICC, where a value above 0.75 is excellent, whereas a value below 0.4 is poor (Cicchetti, 2001). In our study, we obtained a two-way random, average ICC value of 0.967 at a 95% CI level for beauty ratings of mathematical equations. This shows that there is a very high level of absolute agreement in the aesthetic rating of mathematical equations, thus further supporting the view that mathematical beauty can be categorized as a form of biological beauty. Taken together, these values would appear to justify the correlation between high agreement and high resistance to revision of opinion.

Discussion

In this study, we extend our previous work on the extent to which opinions formed through different experiences are susceptible to modification or revision through external opinion. At the basis of this work lies our belief that all experiences are interfaced through brain concepts, which we broadly divide into two categories: inherited concepts and postnatally acquired synthetic ones (Zeki, 2009; Zeki & Chén, 2020). In developing this line, we have suggested that in the Bayesian context, inherited biological concepts generate posteriors which are either very or relatively more resistant to modification through further experience or feedback. By contrast, the posteriors generated from acquired concepts are much more hospitable to external influence and opinion (Zeki & Chén, 2020). We furthermore conjectured that within the biological category, the resistance to modification through external opinion is not rigid but graded, with the aesthetic rating for some categories of biological stimuli being more resistant to change through external opinion than others. The studies that we and others have conducted (Bignardi *et al.*, 2020; Glennon & Zeki, 2021; Vessel *et al.*, 2018) support this broad general subdivision. One of the extremes that falls into the biological category is that of color or, more precisely, color categories; it is difficult, if not impossible, to get individuals to change the color category into which they place a surface, even when there are wide changes in the wavelength-energy composition of the light reflected from them (Zeki *et al.*, 2019). Another, less extreme, category is that of facial beauty, a category for which subjects are about twice as resistant to changing their aesthetic ratings in light of received opinion, compared to artifactual beauty (e.g., abstract paintings) (Bignardi *et al.*,

2020). From the weak correlation between understanding of equations and the rating change, we conclude that the rating change is not necessarily dictated by a lack of understanding.

We have based our experiments on the belief that the aesthetic judgments of mathematical formulations belong in the biological category (Zeki *et al.*, 2014); indeed, the very fact that the veracity of mathematical formulations demands universal assent among mathematicians of all races points to that end. And, though not explicitly enunciated, this is implicit in Plato's belief that mathematical beauty is the highest form of beauty. Paul Dirac's principle of beauty in mathematics is also based on the supposition of a common agreement between mathematicians: "It is the same in all countries and at all periods of time" (Kragh, 1990, p. 288).

It could be argued that because our subjects were professional mathematicians, they may have been in one way or another familiar with these equations, the judgments related to them, or their contexts (as coming, e.g., from famous mathematicians). Although we cannot discount this possibility, we consider it to be unlikely. If mathematicians use (or have used) similar rules to understand and appreciate these formulae, then our point becomes more emphatic because it implies that the rules are general ones, despite the diversity of our mathematicians; this, in turn, would imply that these rules are dictated by brain logic. Moreover, we have shown that understanding had little effect on the RC. Finally, such an interpretation would be contradicted by the fact that in 21% of the cases, the mathematicians did revise their aesthetic judgment in light of external opinion, which suggests that although the resistance to revision of opinion is high, it is not absolute. This leads us to conclude that it is largely the logical system of the brain that is the determining factor in making aesthetic judgments in mathematics but that there is nevertheless room for doubt in the making of such judgments because there was an RC in 21% of the examples.

We are truly sorry to have taken the reader on such a complex statistical journey to prove what is essentially a simple point — that the resistance to revision of aesthetic judgment of mathematical formulae in light of external opinion is very high; this in turn justifies our categorization of mathematical beauty in the biological category.

Acknowledgment

This work was supported by a grant (RPG-2017-341) to S. Zeki from The Leverhulme Trust, London.

References

Bi, J. & Kuesten, C. (2012). Intraclass correlation coefficient (ICC): A framework for monitoring and assessing performance of trained sensory panels and panelists. *Journal of Sensory Studies*, 27(5), 352–364. https://doi.org/10.1111/j.1745-459x.2012.00399.x.

Bignardi, G., Ishizu, T., & Zeki, S. (2020). The differential power of extraneous influences to modify aesthetic judgments of biological and artifactual stimuli. *PsyCh Journal*, 10, 190–199. https://doi.org/10.1002/pchj.415.

Breitenbach, A. (2013). Beauty in proofs: Kant on aesthetics in mathematics. *European Journal of Philosophy*, 23(4), 955–977. https://doi.org/10.1111/ejop.12021.

Bronstad, P. M. & Russell, R. (2007). Beauty is in the "we" of the beholder: Greater agreement on facial attractiveness among close relations. *Perception*, 36(11), 1674–1681. https://doi.org/10.1068/p5793.

Cicchetti, D. V. (2001). The precision of reliability and validity estimates revisited: Distinguishing between clinical and statistical significance of sample size requirements. *Journal of Clinical and Experimental Neuropsychology*, 23(5), 695–700. https://doi.org/10.1076/jcen.23.5.695.1249.

Dirac, P. A. M. (1940). The relation between mathematics and physics. *Proceedings of the Royal Society of Edinburgh*, 59, 122–129. https://doi.org/10.1017/s0370164600012207.

Germine, L., Russell, R., Bronstad, P. M., Blokland, G. A. M., Smoller, J. W., Kwok, H., Anthony, S. E., Nakayama, K., Rhodes, G., & Wilmer, J. B. (2015). Individual aesthetic preferences for faces are shaped mostly by environments. Not genes. *Current Biology*, 25(20), 2684–2689. https://doi.org/10.1016/j.cub.2015.08.048.

Glennon, M. J. & Zeki, S. (2021). The power of external influences to modify judgments of facial and moral beauty. *PsyCh Journal*, 1–8. https://doi.org/10.1002/pchj.492.

Kragh, H. (1990). *Dirac: A Scientific Biography*, p. 288. New York, NY: Cambridge University Press.

Peirce, J., Gray, J. R., Simpson, S., MacAskill, M., Höchenberger, R., Sogo, H., Kastman, E., & Lindeløv, J. K. (2019). PsychoPy2: Experiments in behavior made easy. *Behavior Research Methods*, 51(1), 195–203. https://doi.org/10.3758/s13428-018-01193-y.

Russell, B. (1919). *Mysticism and Logic and Other Essays*. London: Allen and Unwin.

Sutherland, C. A. M., Burton, N. S., Wilmer, J. B., Blokland, G. A. M., Germine, L., Palermo, R., Collova, J. R., & Rhodes, G. (2020). Individual differences in trust evaluations are shaped mostly by environments, not genes. *Proceedings of the National Academy of Sciences of the USA*, 117(19), 10218–10224. https://doi.org/10.1073/pnas.1920131117.

Vessel, E. A., Maurer, N., Denker, A. H., & Starr, G. G. (2018). Stronger shared taste for natural aesthetic domains than for artifacts of human culture. *Cognition*, 179, 121–131. https://doi.org/10.1016/j.cognition.2018.06.009.

Zaki, J., Schirmer, J., & Mitchell, J. P. (2011). Social influence modulates the neural computation of value. *Psychological Science*, 22(7), 894–900. https://doi.org/10.1177/0956797611411057.

Zeki, S. (2009). *Splendors and Miseries of the Brain: Love, Creativity, and the Quest for Human Happiness*. Chichester, UK and Malden, MA: Wiley-Blackwell.

Zeki, S. & Chén, O. Y. (2020). The Bayesian-Laplacian brain. *European Journal of Neuroscience*, 51(6), 1441–1462. https://doi.org/10.1111/ejn.14540.

Zeki, S., Chén, O. Y., & Romaya, J. P. (2018). The biological basis of mathematical beauty. *Frontiers in Human Neuroscience*, 12, 467. https://doi.org/10.3389/fnhum.2018.00467.

Zeki, S., Javier, A., & Mylonas, D. (2019). The biological basis of the experience and categorization of colour. *European Journal of Neuroscience*, 51(2), 670–680. https://doi.org/10.1111/ejn.14557.

Zeki, S., Romaya, J. P., Benincasa, D. M. T., & Atiyah, M. F. (2014). The experience of mathematical beauty and its neural correlates. *Frontiers in Human Neuroscience*, 8, 68. https://doi.org/10.3389/fnhum.2014.00068.

https://doi.org/10.1142/9789811298639_0009

Chapter 9

Aesthetics as a Common Denominator for Moral and Commercial Judgments*

Xuanyu Wang[†], Fabian Simmank[‡,§], Dongxue Zhang[†,‡], Yifan Zeng[†], Sarita Silveira[‡,§], Tilmann Sander[¶], Yan Bao[†,‡,§,||] and Marco Paolini[**]

[†]*School of Psychological and Cognitive Sciences, Peking University, Beijing, China*
[‡]*Institute of Medical Psychology, Ludwig Maximilian University, Munich, Germany*
[§]*Parmenides Foundation, Pöcking, Germany*
[¶]*Physikalisch-Technische Bundesanstalt, Berlin, Germany*
[||]*Beijing Key Laboratory of Behavior and Mental Health, Peking University, Beijing, China*
[**]*Department of Radiology, University Hospital, Ludwig Maximilian University, Munich, Germany*

*This chapter was originally published in the special collection on neuroaesthetics in *PsyCh Journal*. Wang, X., Simmank, F., Zhang, D., Zeng, Y., Silveira, S., Sander, T., Bao, Y., & Paolini, M. (2021). Aesthetics as a common denominator for moral and commercial judgments. *PsyCh Journal*, 10(2), 187–189. https://doi.org/10.1002/pchj.407.

Abstract

Using fMRI, a core evaluation mechanism was found for aesthetic judgments with add-on neural activities for moral and commercial judgments. We propose that aesthetic evaluations serve as a basic core mechanism implicitly for moral and commercial judgments.

***Keywords*:** aesthetic evaluation; commercial judgment; fMRI; implicit knowledge; moral judgment

The relationship between beauty and goodness has been a topic since antiquity; Aristotle mentions in *Nicomachean Ethics* that virtue aims at the beautiful. Indeed, beauty has been regarded to define moral goodness for a long time, but beauty may also play a role in economic activities like in advertisements to convince customers. However, few empirical studies have dealt with this problem. Using fMRI, we compared activation patterns related to aesthetic, moral, and commercial judgments to identify neural processes underlying these types of judgments and in doing so to test the concept of complementarity as a generative principle (Bao *et al.*, 2017). Seventeen right-handed healthy German male students were asked to make two alternative forced choices with respect to aesthetic evaluations (whether a poetic line was beautiful or not), moral evaluations (whether a statement was right or wrong), and commercial evaluations (whether an advertisement slogan was convincing or not) while their brains were scanned with 3T fMRI. The study applied a block design with five conditions, namely, the aforementioned three judgment categories plus two control conditions: a scramble control with pseudosentences and a test of general knowledge (whether a statement was correct or wrong). All stimuli were sentences in German. Each condition involved 10 blocks, and each block contained three different stimuli. For each condition, 15 different stimuli were selected from a pool of 50 stimuli, and they were presented twice with prebalanced valence and arousal. For each trial, a sentence was presented for 5000 ms followed by a 200 ms dark screen. Subsequent to each block, a central fixation was presented for 7500 ms. Technical details of the fMRI setup and the preprocessing procedure were identical to the study by Simmank *et al.* (2020), and the same dataset was shared. For first-level analyses to control for visual components, each experimental condition was compared to the scramble control. Subjective t-statistic maps for each of these contrasts were then fed into a second-level group analysis with cluster-level family-wise error correction, $p(\text{FWE}) < 0.05$; only clusters larger than 200 voxels were reported. The study was conducted in accordance with the Declaration of Helsinki and approved by the ethics committee of the Medical Faculty of Ludwig Maximilians University, Munich, Germany. Written informed consent was collected from all participants.

The behavioral measurements showed that subjects rated 50.1% of the aesthetic stimuli as beautiful, 54.3% of the moral stimuli as right, and 51.1% of the commercial slogans as convincing; there was no difference between the three (ANOVA, $F(2, 32) = 0.19$, $p > 0.05$). However, the reaction time (RT) was different (ANOVA, $F(2.28, 34.22) = 99.29$, $p < 0.001$) with a longer RT for aesthetic (M = 2.50 s, SD = 0.76 s) compared to moral (M = 2.31 s, SD = 0.78 s) and commercial judgments (M = 2.16 s, SD = 0.71 s), $p < 0.001$ each; RT for moral judgment was also statistically longer than commercial judgment, $p = 0.0084$.

With respect to fMRI, common areas of activations were found by conjunction analysis for the aesthetic, moral, and commercial judgments with the scramble control extracted. These were clusters in the temporal and frontal lobes in the left hemisphere, mainly composed of the "semantic circuit" that processes abstract word meanings (Della Rosa *et al.*, 2018), including the middle temporal gyrus (MTG) and the inferior frontal gyrus (IFG). Additionally, common activation was found in the left orbitofrontal cortex (OFC) (see Figure 1, red). The OFC has been found to be involved in both aesthetic and moral decision-making (Avram *et al.*, 2013; Cheng *et al.*, 2020), and it correlates with positive outcomes or subjective values in general (Stalnaker *et al.*, 2015). We observed also common activations for moral and commercial evaluations, i.e., moral and commercial judgments both elicit higher activations in the temporoparietal junction (TPJ) as well as in the anterior cingulate cortex (ACC) all lateralized in the left hemisphere (i.e., a conjunction between Moral > Aesthetics and Commercial > Aesthetics; see Figure 1, green). No significant cluster was found in the reverse contrasts Aesthetics > Moral/

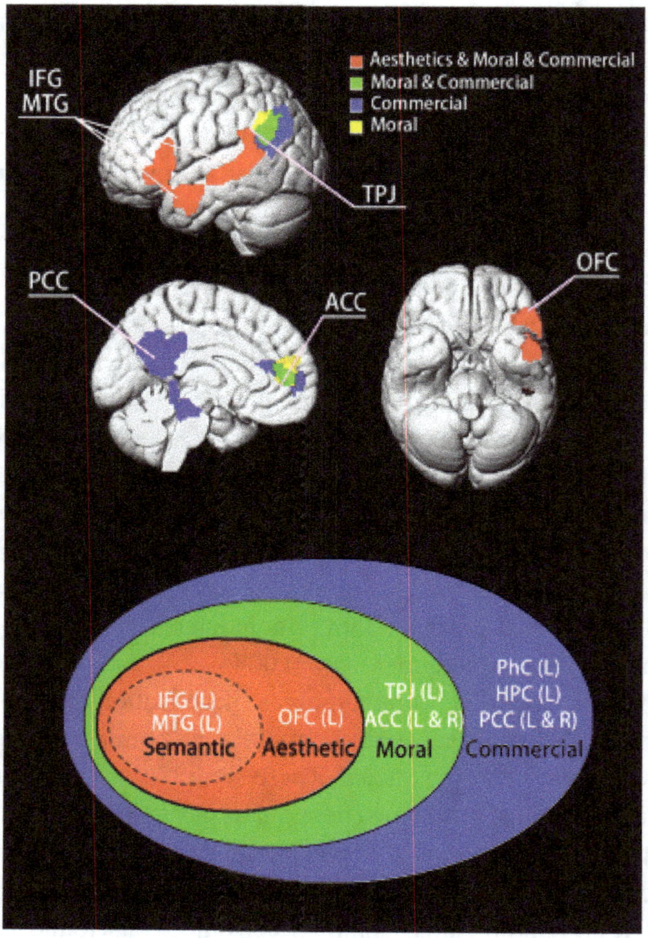

Figure 1. Brain activities related to three types of judgments. In the upper part, conjunctions between Aesthetics > Scramble, Moral > Scramble, and Commercial > Scramble are shown in red; conjunctions between Moral > Aesthetics and Commercial > Aesthetics are shown in green. Blue marks indicate Commercial > Aesthetics, and yellow marks indicate Moral > Aesthetics. In the lower figure, each type of judgment is summarized with a Venn diagram.

Note: IFG = inferior frontal gyrus; MTG = middle temporal gyrus; TPJ = temporoparietal junction; ACC/PCC = anterior/posterior cingulate cortex; OFC = orbitofrontal cortex; PhC = parahippocampal cortex; HPC = hippocampus; L = left hemisphere; R = right hemisphere.

Commercial. Furthermore, commercial-specific activations were found in the posterior cingulate cortex (PCC). When comparing commercial to moral judgments, a higher activation was also found in the parahippocampal cortex and the hippocampus (see Figure 1, blue).

Concerning the overlapping brain activities across three categories in addition to the semantic network, the OFC was involved in all three experimental conditions presumably representing an aesthetic core mechanism of evaluations underlying all three categories of judgment. Most likely, the semantic component was a result of the scramble control lacking a semantic meaning, as the same areas were also found when comparing the general knowledge condition to the scramble control (Simmank *et al.*, 2020). In addition, moral and commercial judgments both recruited the theory of mind network, i.e., the ACC cortex as well as the left TPJ, associated with concept retrieval, conceptual integration, and more generally with word processing (Cabeza *et al.*, 2012). This may imply a more conceptual thinking mode for moral and commercial judgments. Commercial-specific activations were more widespread, involving the PCC, the hippocampus, and the parahippocampal cortex, possibly relating to its acquired nature.

Taken together, a central mechanism can be suspected underlying all three categories of judgment, with aesthetic judgment as the core. Moral and commercial judgments are recruiting additional areas during the decision-making processes. A hierarchical relationship between the three judgment categories is visualized in a Venn diagram (Figure 1, bottom). An aesthetic-like evaluation mechanism presumably operates on an implicit level as it is also typical in art appreciation (Bao *et al.*, 2016). These results support the hypothesis that beauty as one component of aesthetic judgments is an intrinsic component of moral goodness and commercial convincingness. However, one has to deal with a paradoxical observation: How can the dominance of aesthetic judgments be understood given the fact that reaction time to aesthetic statements was the longest? We argue that the cognitive processes for aesthetic judgments make it necessary to be moved to an explicit level which usually remains implicit; this shift process requires more time. In comparison, commercial and moral judgments are categories which are often overlearned in everyday activities enrolling circuits related to memory, and this process may function faster on an explicit level. In conclusion, both the data of fMRI and the behavioral results support the hypothesis of an aesthetic-like core mechanism with additional neural activations for moral and economic judgments, also emphasizing the concept of complementarity in *Cognitive Processing*.

Acknowledgments

This study was supported by Grant 31371018 and 31771213 from the National Natural Science Foundation of China for Y.B., the Parmenides Foundation for F.S., and the Hanns-Seidel Foundation for S.S.; the authors are grateful for the support of Mihai Avram. Open access funding was enabled and organized by Projekt DEAL.

References

Avram, M., Gutyrchik, E., Bao, Y., Pöppel, E., Reiser, M., & Blautzik, J. (2013). Neurofunctional correlates of esthetic and moral judgments. *Neuroscience Letters*, 534, 128–132. https://doi.org/10.1016/j.neulet.2012.11.053.

Bao, Y., von Stosch, A., Park, M., & Pöppel, E. (2017). Complementarity as generative principle: A thought pattern for aesthetic appreciations and cognitive appraisals in general. *Frontiers in Psychology*, 8, 727. https://doi.org/10.3389/fpsyg.2017.00727.

Bao, Y., Yang, T., Lin, X., Fang, Y., Wang, Y., Pöppel, E., & Lei, Q. (2016). Aesthetic preferences for Eastern and Western traditional visual art: Identity matters. *Frontiers in Psychology*, 7, 1596. https://doi.org/10.3389/fpsyg.2016.01596.

Cabeza, R., Ciaramelli, E., & Moscovitch, M. (2012). Cognitive contributions of the ventral parietal cortex: An integrative theoretical account. *Trends in Cognitive Sciences*, 16(6), 338–352. https://doi.org/10.1016/j.tics.2012.04.008.

Cheng, Q., Cui, X., Lin, J., Weng, X., & Mo, L. (2020). Neural correlates of moral goodness and moral beauty judgments. *Brain Research*, 1726, 146534. https://doi.org/10.1016/j.brainres.2019.146534.

Della Rosa, P. A., Catricalà, E., Canini, M., Vigliocco, G., & Cappa, S. F. (2018). The left inferior frontal gyrus: A neural crossroads between abstract and concrete knowledge. *NeuroImage*, 175, 449–459. https://doi.org/10.1016/j.NeuroImage.2018.04.021.

Simmank, F., Avram, M., Fehse, K., Sander, T., Zaytseva, Y., Paolini, M., … Silveira, S. (2020). Morality in advertising: An fMRI study on persuasion in communication. *PsyCh Journal*. 9(5), 629–643. https://doi.org/10.1002/pchj.358.

Stalnaker, T. A., Cooch, N. K., & Schoenbaum, G. (2015). What the orbitofrontal cortex does not do. *Nature Neuroscience*, 18(5), 620–627. https://doi.org/10.1038/nn.3982.

https://doi.org/10.1142/9789811298639_0010

Chapter 10

(Neuro)Aesthetics: Beauty, Ugliness, and Ethics*

Philip Dietrich and Thomas Knieper

Digital and Strategic Communication, University of Passau, Germany

Abstract

The objective of this chapter is a holistic view of aesthetics, ethics, and neuroaesthetics. After a few introductory case studies, aesthetics is systematically introduced as a philosophical subdiscipline. This perspective is then expanded from aesthetics to neuroaesthetics. Using various art forms as well as current media formats, the aspects of beauty and ugliness are discussed, and aesthetic properties are expanded to include ethical implications. These can be expressed through ideals of beauty and the compulsion for body transformation. This perspective is then expanded from aesthetics to neuroaesthetics. From this point of view of art, the so-called golden ratio will play a central role. It is shown how representations affect people and what ethical implications are associated with the effects. Therefore, this chapter first has to look at art from the perspective of neuroaesthetics and then consider the ethical aspects of the beautiful and the ugly. The considerations lead to a brief discussion of Socrates's three sieves.

---

*This chapter was originally published in the special collection on neuroaesthetics in *PsyCh Journal*. Dietrich, P. & Knieper, T. (2022). (Neuro)Aesthetics: Beauty, ugliness, and ethics. *PsyCh Journal*, 11(5), 619–627. https://doi.org/10.1002/pchj.478.

***Keywords*:** aesthetics; art; beauty; ethics; film; neuroaesthetics; photography; terror; ugliness; video

Aesthetics is all about the assessment of sensual perception but not only sensual delights. If it becomes ugly, there will also be a question of ethics included. Representations of the horrible, atrocity, hatefulness, disgust, mutilation, torture, and further body horror are increasingly taken up by various art forms and manage to fascinate more and more people. However, the coverage of catastrophes and the human abyss is not new. On the other hand, the use of detailed and revealing images is indeed a new phenomenon. It is a creeping process that may follow developments within the visual arts. Graphic images were always used to depict and denounce war crimes and other atrocities. Examples are artistic works such as the etching cycles *Les Misères et les Malheurs de la Guerre* (*Horror and Sorrows of War*) by Jacques Callot (17th century) or *Desastres de la Guerra* (*Horrors of War*) by Francisco José de Goya y Lucientes (19th century). Such inhuman and dehumanizing images are now — besides art — also part of news photography, movies, and computer games (Knieper & Saleh, 2015; Müller, 2005; Müller & Knieper, 2019; Wilke, 2005). For example, remember the iconic photograph of Eddie Adams from February 1, 1968, depicting an execution in the streets of Saigon or the photographical documentation of the self-immolation of a Buddhist monk in Saigon by Malcom Browne (Grewenig, 1999). Both photographers were awarded high prizes for their photographs. Cruelties and horror are also shown in movies. This is true across all genres (action, martial arts, fantasy, horror, thriller, war, etc.). For example, movies such as *The Killing Fields* (Joffé, 1984), *The Deer Hunter* (Cimino, 1978), or *City of Life and Death* (Lu, 2009) allow a glimpse into the depths of the war and its atrocities. At the same time, war becomes increasingly aestheticized. As an example, the opening scene of the film *Saving Private Ryan* (Spielberg, 1998) depicts the storming of Omaha Beach in its cruelty and complements it with elements of beauty (Röwekamp, 2011).

Even if their status as an art form is still controversial, video games cannot remain unmentioned. For example, the Museum of Modern Art in New York has been exhibiting video games in its permanent exhibition since 2012 (Feige, 2017). Regarding their interactive character, depictions of violence in video games are under particular scrutiny. In the game *Grand Theft Auto V* (Rockstar Games, 2013), the player is forced to torture a character to continue playing the game. The developers reacted to criticism by stating that they wanted to draw attention to the absurdity of torture as an instrument (Hern, 2013). Another extreme example is the game *Rape Day* (Desk Plant, 2019). The setting is a zombie apocalypse where the player slips into the role of a serial killer who rapes women while they are on the run from zombies. Although the game was very quickly banned from official sales sites, such as Steam, it continued to be distributed through unofficial sites (Evans, 2019).

In addition to visuals, music should not go unmentioned. Various genres, such as jazz, rock and roll, or rap, have taken up the grievances of their time and processed them musically. Thus, Kurt Pahlen (1996) stated in his history of music that there would have been no jazz without violence and slavery. Horror and beauty cannot be portrayed in music the way that they can in pictures. However, there are adequate forms of representation in music to evoke visual images. This includes, for example, the so-called *Tristan Chord*, which has a high dissonance (Zeki, 2004). In his opera *Tristan und Isolde*, Wagner is said to have processed the idea of romantic love because the ideal conveyed through art could not be achieved in reality (Zeki, 2004). Such representations are frequently represented, especially in classical music. Thus, there are numerous links between what is heard and visual stimuli. There are, for example, tone sequences that remind one of the flowing of water and the chirping of birds. An aesthetic image of beauty in the form of consonances and of ugliness in the form of dissonances can also be created through music.

Representations of horror have the potential not only to inform and to fascinate but also to emotionally disturb and frighten. In addition, people may become dull in relation to cruelty, and so-called *habitualization effects* may occur. It is questionable how such an aesthetic experience can be placed between the obligation to inform the public, the economic compulsion to present, and an ethical responsibility (Isermann & Knieper, 2009). In this context, it is necessary to clarify whether such presentations must be shown or whether they may be shown at all.

In this chapter, therefore, the connections between aesthetics and ethics are discussed and connected to the laws of visual perception from the neuroaesthetics. The objective is to show an interdisciplinary approach. For this purpose, elements of ethics, aesthetics, and neuroaesthetics are combined to show commonalities.

Aesthetics as a Philosophical Subdiscipline

Aesthetics as a philosophical subdiscipline is comparatively young. The first explanations can be found in Alexander Gottlieb Baumgarten's (1750) *Aesthetica.* The term *aesthetics* is derived from the ancient Greek word "aisthesis" and can be translated as "sensual perception" (Reicher, 2015). Baumgarten described aesthetics as a form of cognition through sensual perception. What is interesting about Baumgarten's remarks is that he sees aesthetics as both a science and an art. Here, the term *ars* is used, suggesting the concept of art in its old meaning as the ability to do something (Schneider, 2021). Sensual perception helps to find truth because the senses have their own quality of cognition. It is noticeable that Baumgarten often mixed the concepts of truth and beauty (Schneider, 2021) — a concept that we already know from antiquity because Plato already postulated the triad of goodness, beauty, and truth (Büttner, 2006). More modern approaches have moved away from Baumgarten and describe aesthetics as a theory of beauty or art. Against this, however, it can be argued that aesthetics is not only concerned with beautiful appearances but with all forms of perception (Welsch, 1960). The objective of aesthetics is not to expand conceptual knowledge but the sensual appropriation of reality (Waibl, 2009). Therefore, following Baumgarten, more recent approaches have focused on aesthetic events (Bubner, 1989; Düwell, 1999) or the aesthetic experience (Reicher, 2015). This refers to the experience of sensual qualities of perceived phenomena (Fenner, 2000; Reicher, 2015). The advantage of relying on aesthetic experience is that it can capture a variety of aesthetic qualities such as the ugly, the exciting, the funny, or the boring as well as numerous aesthetic objects such as natural phenomena.

Art from the Perspective of Neuroaesthetics

An important scientific milestone in the question of how the brain perceives beauty was a conference theme of the study group *Biologische Grundlagen der Ästhetik* (*Biological Basis of Aesthetics*). The resulting anthology *Beauty and the Brain* (Rentschler *et al.*, 1988) contains the reflections of neuroscientists, psychologists, anthropologists, and philosophers and also artists, musicians, and writers. With philosophical, biological, and considerations about the effect of art in the human brain, an important fundament was laid for today's neuroaesthetic considerations. From today's perspective of neuroaesthetics, aesthetics is mainly related to the perception and evaluation of art. Therefore, it has a close connection to cognitive psychology, affective psychology, and neuroscience (Brielmann & Pelli, 2018). The aesthetic experience is also part of neuroaesthetics, which investigates which neurobiological processes lead to aesthetic experiences (Zeki, 2001). Empirical aesthetics defines the discipline as the perception of, production of, and response to art as well as the engagement with objects that elicit strong feelings such as pleasure (Brielmann & Pelli, 2018). Concepts such as desire, beauty, or pleasure always play a major role. From the perspective of philosophy, these terms can be more narrowly defined. The first question to be answered here is how art is to be understood from a neuroaesthetic point of view. On one hand, "mimesis" defines art as the realistic representation of what is seen. On the other hand, art can be understood as a form of complexity reduction. In an overview of the history of art, Sütterlin &Yu (2021) advocated the second understanding of art. The authors justify this with the fact that when looking at a wide variety of works of art, a formal reduction and schematization can be felt. The special thing about this design is that art creates a common basis regardless of culture. From a neuroaesthetic point of view, characteristics such as beautiful or ugly can also be examined from the concept of art on the basis of the perception of art. Ethical implications of aesthetic experiences can be derived. Thus, experiences of beauty may be associated with activity within the medial orbitofrontal cortex and the overlapping ventromedial prefrontal cortex

(Sütterlin & Yu, 2021). In general, the brain takes on the important task of creating a picture of the outside world from many experiences, especially visual ones. This includes ordering, separating, and spatially distributing information such as color, shape, movement, and depth (Zeki, 2000). The processing of visual knowledge in the human brain follows three basic principles. The first principle states that the visual brain has several parallel systems of information processing that are spatially distributed across the brain. Second, the final result reaches the parallel systems at different times. Therefore, third, there is an asynchrony in perception (Zeki, 2000). Zeki (2000) therefore came to the assumption that especially colors and movements are perceived spatially distributed in the brain whereas form and depth are processed in the same place. This aspect is exciting insofar as it is shown later that shape in particular plays a central role in the perception of beauty. Building on this understanding, art is also subject to the laws of the brain (Zeki, 2004). In this context, art takes the form of a special kind of knowledge and is thus subject to the biological foundations of knowledge. Building on brain research, laws of art and aesthetics are therefore to be formulated in neuroaesthetics (Zeki, 2004). The greatest challenge for the brain in this context is that change is a central cultural element in both Western and Eastern cultures. Therefore, the brain must take on the task of retaining the important properties of objects. This is done by the ability of abstraction, which enables the brain to recognize regularities from numerous examples. Zeki (2004) assumed that this process can also be taken over by an artist by creating an abstracted image of reality. For the experience of art and the neurobiological implementation, Ramachandran & Hirstein (1999) developed eight heuristics to help explain it. Among them is the form of exaggeration. This is primarily intended to map the central characteristics of an object and thus to evoke strong responses. This occurs especially in caricatures, where physical features are deliberately exaggerated to be better recognized and to trigger a reaction on the part of the viewer (Knieper, 2015). The form of isolation as a visual model is also intended to cause an effect in the viewer. Since attention is basically very limited, isolated aspects in particular stand out. Symmetry also plays a central role in the laws of art experience. Symmetry is presented as an important aspect of beauty. Within the laws, this is explained from evolution–biological perspective with the fact that with the help of the symmetry, potential sexual partners are to be recognized better. The selection of the heuristics already shows that through elements such as exaggeration, isolation, and symmetry, abstractions of reality are conveyed to the brain and trigger reactions there. Especially with regard to the evaluation of beauty and the connection to form and symmetry, aesthetic aspects also come into play. The special importance of symmetry suggests a mathematical basis in the perception of beauty. Zeki *et al.* (2014) therefore investigated whether the experience of beauty from mathematics activates the same emotional part of the brain as the experience of beauty from sensory and perception sources. Their results showed that mathematical beauty activates the same areas of the brain (A1 of the medial orbitofrontal cortex) as the experience of visual, musical, and moral beauty. The connection between symmetry, art, and moral beauty suggests that the perception of beauty and ugliness deserves an ethical consideration in addition to aesthetic and neuroaesthetic ones.

The fact that neuroaesthetics has been able to observe processes in the brain that arise from aesthetic experiences makes it all the more important to deal with the aesthetical and ethical implications of both the beauty and the ugliness. Yet, the importance of the topic in current research discourse is well evident. The metastudy by Chuan-Peng *et al.* (2020) has provided a detailed overview of research on the beauty of faces and visual art. In addition, research by Zhao *et al.* (2018), von Brück (2021), and Yang *et al.* (2021) has shown that neuroaesthetic considerations of beauty and ethics are a very current topic in art, commercial judgments, and culture. This justifies an interdisciplinary look at the beautiful and the ugly as well as their ethical implications.

Ethics of the Aesthetics

In everyday life, aesthetic is often associated with the art or the beautiful. This idea is not far from aesthetics as a philosophical subdiscipline (Majetschak, 2019). Since antiquity, philosophers have thought about the beautiful (Platon, 1988), the ugly, and the art (Aristoteles, 1994). It is true that aesthetics as a subdiscipline of philosophy did not emerge until much later. Yet, many philosophical texts contain comments on beauty, ugliness, and art. These individual aspects are particularly relevant with regard to a relation to ethics.

The first connections between art and ethics and their impacts on human beings can be found in Plato's idea of an ideal state (Platon, 1991). In this context, Plato reproaches artists for not telling the truth and thereby corrupting people's souls. The ideas of goodness, beauty, and truth could thereby not be recognized in their original form; only the superficially perceptible is depicted. Plato criticized above all the representation of sensuality, which destroys the reason of the recipients. Thereby, such art should be censored, and the artists should be banished from the ideal state. Plato saw an exception to this in the educational functions of art. Thus, artists in the ideal state may pursue their work only if they depict the virtues by means of noble characters. Art should exclusively depict the moral virtues of society (Fenner, 2013). Even if the factor of censorship must of course be questioned, more recent approaches also have shown that certain music does have an impact on people's emotional world (Rinderle, 2011). Aristotle also picked up on the educational and formative nature of art (Aristoteles, 1994). In his *Poetics*, he postulates that the source of all art derives from the desire for imitation. Through this, the ugly can be experienced at all. He assumed that thereby things, at which one would turn away in reality, can be represented by art. He named two places of ugliness in art in particular: pain and death in tragedy and the ridiculous in comedy. Thus, certain art forms, especially Greek musical dramas, can induce a catharsis in the audience. For example, Aristotle described that certain keys, rhythms, and instruments can trigger strong reactions in the audience. Building on the idea of catharsis, artistic therapy also attempts to activate emotions (Heimes, 2010). Overall, Aristotle argued that by imitating the ugly, affection becomes possible, and educational effects are associated with it. Here, Lessing (1996) contradicted Aristotle by asking whether there is a limit to what can be tolerated until the viewer also turns away from representations of art. Using the sculpture of Laokoon as an example, Lessing discussed why Laokoon's face shows so little sign of pain, whereas in literature, Laokoon's pain is described excessively. His assumption is that the pictorial representation is too cruel and would lead to aversion. Therefore, he argued, the artist depicted a softened version. Transferred to the present time, this would mean that representations of the ugly could be shown only if the spectators did not turn away. However, the degree to which images of horror can be endured shifts. Nowadays, society is voyeuristic to its core (Taylor, 1998). Through habituation effects, an even greater fascination with horror develops, and in some cases, viewers even experience pleasure (Knieper & Saleh, 2015). At this point, the concept of aesthetic sensation by Friedrich Schlegel (1988) must be mentioned. He presented shock as the central aesthetic experience because it can arouse the interest of the audience. He divided the interests into the adventurous, the disgusting, and the ghastly. All three aspects stand for aesthetic borderline experiences, which can cause a strong effect in people and thus contribute to the experience of reality (Liessmann, 2009). Modern versions take up the educational aspect of art and transfer it to newer media formats, such as novels, films, or videos. Within the so-called *ethical turn*, the idea of a narrative ethics was raised. This refers to the communication of ethical aspects through the form of narration (Fenner, 2013). Through successful narration, ethical experiences can be conveyed and help gain one's own insights into general moral principles. Narration helps put oneself in a situation that is not commonplace and to understand the actions of the persons portrayed (Nussbaum, 1986). Besides the experience of general moral principles, narration can also sensitize to the perception of cruelty and discrimination, as well as vulnerability and need for help (Rorty, 1992). The emergence and increasing popularity of the video game gives new significance to narrative ethics in this regard. Through the active actions of the player, a deeper understanding of new situations can emerge.

As more and more areas of life are being aestheticized (e.g., fashion, living, and eating), other areas are coming into focus alongside art. Wolfgang Welsch (1993) picked up on this and postulated that ethics is turning into a "subdiscipline of aesthetics" (p. 44). For Welsch, the focus is especially on individual ethical tendencies. These are increasingly oriented toward the aesthetic perfection of one's own body (Welsch, 1993). Art takes on the role of broadening the perspectives of the viewer and sensitizing the sense for plurality and contradictions. In this way, the individualistic actions of others are to be comprehended. Through the theoretical history of aesthetics, it is evident that art has always been ascribed an educational function. Particular aesthetic experiences trigger strong feelings in people and thereby sensitize them to general moral principles. Both beauty and ugliness always take on a special role. Therefore, the ethical implications of beauty and ugliness will first be discussed in more detail before the concepts are linked to the idea of neuroaesthetics.

Beauty in Ethics and Aesthetics

The concept of beauty can be argued about in everyday life as well as in philosophical debates. Since ancient times, theorists have been discussing what the beautiful in itself actually is. It is interesting to note that the fundamental questions have hardly changed (Liessmann, 2009). In the style of his Socratic conversations, Platon (1988) reproduced a conversation between his teacher Socrates and Hippias. In it, Socrates asked his interlocutor whether he can define the beautiful. By means of various examples, it is shown that all attempts at definition are refuted. However, the individual questions are also relevant for today's reflections on beauty. In his first attempt, Hippias defined beauty by the beauty of a girl. For it is clear that beauty in itself can be recognized when looking at a beautiful girl. Even today, we see that beauty is often equated with the attractiveness of people, which has an impact on how they are perceived in society. People who are perceived as beautiful are generally assigned more positive characteristics (Renz, 2007). The other attempts of definition by Hippias also had significance in the debate until today. Subsequently, Hippias tried to define the beautiful as the decent. Here, too, we see the ethical elements already mentioned, which are connected to an aesthetic in that the beautiful is the morally expectable. Although Socrates and Hippias merely concluded in their dialogue that they now know how difficult it is to define the beautiful, more concrete approaches can be found in other works of Plato that still have an impact today. As mentioned earlier, in Plato's ancient doctrine of ideas, the beautiful is equated with the truth and the goodness. These are original images of a perfection from which various individual aspects are derived. According to Plato, this perfection can only be achieved through the concept of harmony. Harmony stands for a compulsive connection of two elements which actually do not fit together. Harmony consists of appropriate proportions and a symmetry

(Büttner, 2006). What is immediately recognizable here is that beauty as perfection is captured with the interaction of numbers, and thus a connection to mathematics can be established.

The so-called *golden ratio approach*, in which the ratio of the whole to its larger part is equal to the ratio of the larger part to the smaller part, is therefore particularly well known (Kessler, 2007). This form of representation can be found in the fine arts (e.g., Leonardo da Vinci and Albrecht Dürer), in architecture (e.g., Notre Dame in Paris and Florence Cathedral), and in music (e.g., Gérard Grisey). This so-called *Grand Theory* (Tatarkiewicz, 2003) determined ideas about beauty until the Renaissance. Additionally, in modern psychological studies on attractiveness research, a focus on the symmetry of faces can be found (Armann & Bülthoff, 2009). It becomes clear that this idea of perfection must be subjected to an ethical consideration in the further course. From this, a compulsion to perfection can be derived through modern ideals of beauty. Following the grand theory, natural phenomena, which Kant (1790) grasped as the sublime, were also included in the consideration in addition to art. More and more, the ideas of beauty were oriented toward individualistic tendencies. Thus, beauty became increasingly subjectivized and, above all, popular culture was determined as an important influencing factor (Maase, 2008). These individualistic tendencies, however, do not stand for a great plurality of beauty but are, depending on the cultural imprint, connected with a certain uniformity and synchronization in which the ideas of the political society and one's own peer group serve as a standard (Mentges & Richard, 2005).

Both in the reception of art and in research on attractiveness, however, the aspect of symmetry is referred to again and again. William Hogarth (1997) showed in his book *The Analysis of Beauty* aspects that can trigger aesthetic sensations. Besides uniformity, size, proportion, and coloration, symmetry also plays a major role. These elements are important in the design of lines and, in his view, are considered to be the most universally beautiful object (Hogarth, 1997). Colin Martindale (1988) also investigated the effect of aesthetic sensations on neural processes. Here, too, symmetry is one of the most important influencing factors. Follow-up studies have shown that symmetry and a higher degree of complexity lead to a more positive beauty judgment in most people. Among other things, it was recognized that ugliness is detected very early in the judgment process and activates the frontal area of the brain, whereas beauty, in the form of mathematical-logical structures, is detected later and activates the rear area of the brain (Jacobsen & Höfel, 2003). It can be seen that elements of the golden ratio are still represented in the observation of art. Because art, as shown, has an educational function and especially in modern times images of beauty are conveyed, the question arises how the forms of symmetry affect the perception of beauty in people. At international beauty contests, the most beautiful people in the world are chosen by a jury, and magazines regularly designate celebrities as the "sexiest man/woman alive." In this context, it is important to clarify which influencing variables have an impact on the aesthetic perception when judging people. Many studies have referred to the face (Dunn *et al.*, 1996; Langlois *et al.*, 1994) or the body (Furnham *et al.*, 2003; Singh, 1993; Singh & Luis, 1995). Two elements will be highlighted that are related to the way we look at art. In an experimental trial by Dunn *et al.* (1996), 297 subjects were asked to rate the smiles of 8 men and 8 women. The photographs that were shown differed in the symmetry of the teeth, the color of the teeth, the number of visible teeth, the condition of the teeth, and the height of the upper lip. The test subjects judged the most beautiful smile on the basis of the symmetry of the teeth and the color of the teeth. The same can be applied to the entire face. Langlois & Roggman (1990) designed an average and artificial face from numerous natural faces. The artificial faces showed a higher degree of symmetry, and the more natural faces were included in the model. In 300 test subjects, the artificial faces were always rated as more beautiful than the natural faces. This also shows that the elements of the golden ratio still play a role in perceived beauty and judgment.

These elements can also be found in beauty judgments relating to the body. Singh (1993) studied the waist-to-hip ratio and its influence on beauty judgments. As a result, it was found that especially female bodies with a waist-to-hip ratio of 0.7 were judged as attractive. It can therefore be seen that proportions also play a central role here. However, another aspect of the study becomes ethically relevant. Not only were these women rated as more attractive but they were also rated as healthier and more fertile. This leads to ethical questions about how perceived beauty affects many aspects of everyday life. Ideals of beauty take a central role in many decisions, and the question is how to respond to deviations. These areas can influence the choice of a partner and career opportunities and also one's own psyche. To what extent do beauty ideals have to be achieved by the individual to be successful in society?

Therefore, a question remains: Why does man strive to fulfill the respective cultural ideals of beauty? As already shown, modern attractiveness research examines the connection between perceived beauty and the attribution of positive characteristics. In a notable study, T. Wang *et al.* (2015) investigated whether moral beauty is different from facial beauty. Two experiments were carried out for this purpose. In the first experiment, participants were asked to make aesthetic judgments about faces as well as moral actions. The second experiment examined the implicit perception of facial and moral beauty. The results showed that aesthetic and moral judgments are based on the same emotional and cognitive components (orbitofrontal cortex, inferior temporal gyrus, and medial superior frontal gyrus). Furthermore, the researchers concluded that reward regions in the brain are involved in the perception of facial beauty. Another study has already shown that there is a connection between the aesthetic assessment of beauty, moral, and economic evaluation. Beauty can be a component for moral goodness or commercial conviction (X. Wang *et al.*, 2021). The effect of aesthetic qualities on moral goodness and commercial persuasiveness can be combined, especially in modern media formats. In this context, ideals of beauty are increasingly manifested through the socialized appearance of heroes and advertisers. The connection between morality and beauty has already been discussed from a philosophical perspective. Philosophers, such as Nietzsche (1980), also have confirmed a connection between physical beauty and ascribed moral qualities. This supports Plato's considerations that the beauty, the goodness, and the truth form a unity, thus creating an urge to perfect one's own body. Although it can be argued that the triad can already be derived from a combination of the texts "Phaidros," "Philebos," and "Symposion," there are other opinions that locate a unity of the beautiful, the good, and the true only in Neo-Platonism (Martin, 2017). Nevertheless, this unity of ideas convincingly reflects the connection between aesthetics and ethics. In this context, the societal conceptions of beauty and the associated implications play a major role, leading to an examination of the concept of beauty from an ethical perspective. Beauty plays a central role especially in the romantic search for a partner. In this context, young people increasingly orient themselves to the popular beauty ideals of the respective society.

The fact that ideals of beauty are prescribed by pop culture icons and reinforced in social media by influencers creates a pressure to conform that leads to various mechanical or virtual interventions to conform to the ideal. From an ethical perspective, people can and must ask themselves how far these changes should go.

The Ugly in Ethics and Aesthetics

Analogous to the considerations of beauty, there are also numerous considerations of ugliness in the history of ideas. Here, ugliness as a counterpart to beauty initially takes on a central role. Plato described ugliness as the imperfection directed against perfect beauty. By this, he means the absence of everything divine and thus of the true, beautiful, and good (Platon, 2013). Hereby, also the preceding assumption after the perfection of one's own body can be taken up again. However, this raises the question of whether ugliness really has to be the opposite of beauty, or whether it can even be a part of beauty. One thinks here of the fear when watching a horror film which is, however, at the same time horrifying as pleasant. The philosopher Nelson Goodman (1995) described the phenomenon as the paradox of ugliness. He described that similar to the beautiful, aesthetic effects arise also through perceived ugliness. This includes, for example, asymmetries in design or the examples from photography mentioned at the beginning of the article. This is also consistent with the early reflections of Friedrich Schlegel (1988), who named the effect of shock as a central aesthetic experience. By breaking with expectations, interest is awakened in an aesthetic object. In this way, the ugly can, among other things, trigger pleasure by playing with aesthetic borderline experiences. Here, people consciously break with their norms, values, and ideas, and an inner sense of self-preservation is addressed. Thus, a claim to show artistic representations of horror can be derived because shocking moments of fate are informed and enlightened. In this way, people can be sensitized to gruesome events outside their everyday lives. It is precisely through elements of narrative ethics that a change of perspective can take place here, which makes it possible within an aesthetic experience to comprehend the views of others and to sensitize people to other actions and injustices. Besides this positive aspect of the representation of ugliness, however, other effects must not go unmentioned. As already shown, culturally determined ideals of beauty prevail in societies. However, these are beginning to merge in the context of globalization, after countries from the Asian region have increasingly participated in Western and international beauty contests since 2012. The pressure exerted by the ideals of beauty on one's own body is also intensified by the increasing media portrayal of features that are not considered beautiful. More and more real-life formats are emerging whose protagonists deviate from the ideals of beauty (e.g., *The Biggest Loser*, *Radical Body Transformations*, and *My 600-lb Life*). In all these formats, the objective is to achieve the ideal of a desirable body; therefore, the participants must be willing to transform themselves. These trigger cultivation effects in the recipients, which can result in self-doubt, developmental impairments, depression, and/or eating disorders. For changes to occur here, moral appeals must be made to viewers because little will change in the media market through self-control (Hoffmann, 2019; Rossmann & Brosius, 2005). Thus, it can be seen that aesthetic effects can also result from perceived ugliness. On one hand, these can sensitize through possibilities of narrative ethics; on the other hand, they can trigger media-generated constraints through the presentation of beauty ideals.

The Three Sieves of Socrates

It is fundamental to distinguish between the outrage in the picture and the outrageous picture. For example, the German Press Council explained in its ethical guidelines (1973) that the depiction of a dying or physically or mentally suffering person in a manner that goes beyond the public interest is inappropriate. Persons affected by an accident, assault, or calamity must not become a victim for a second time as a result of the depiction. The assessment depends strongly on the nature of the picture. Does the enlightening, journalistic view come to the fore or the voyeuristic or propagandistic character? This means that the ethical classification can no longer be restricted to the image content, *per se*. For a final value judgment, the context of the picture must also be taken into account (Bock *et al.*, 2011; Knieper & Müller, 2019). A possible aid for classification is the principle of the so-called *three sieves of Socrates*: (a) Sieve of truth: Does the picture correspond to the truth in its form of a subjective eyewitness? (b) Sieve of goodness: Does the impression of the picture correspond to the good? (c) Sieve of necessity and usefulness, respectively: Is there a need to publish the picture? Or in short, if the story behind a picture you are about to show is not true, good, or necessary, do not show it. It is as simple as that. In this respect, an ethics of image usage corresponds with the professional standards of ethics in journalism. The principles are accuracy, reality, truth, and fairness (Isermann & Knieper, 2009). However, conflicts, disasters, and wars are reported in the media day after day. In the digital age with its photo- and video-compatible mobile devices, production of pictures was never this easy (Knieper & Saleh, 2015). Numerous snapshots and videos of every conceivable atrocity in public spaces will be made. The horror picture becomes evidence of being an eyewitness, a marker of authenticity, and a trophy. However, this image production by the masses defies both political and ethical control. In the age of mobile devices, the uncontrolled production of horror images can no longer be held back. Those pictures are characterized by the charm of an amateur aesthetic that puts authenticity above professional composition. In this case, it is all about the evidential value. However, it becomes problematic at the latest, when the control of the image motifs and their distribution and reception becomes a strategic weapon of war and terror. Modern terrorism works with the manufacturing and distribution of horror images. With this special kind of terror aesthetics, political enemies are to be frightened and also potential supporters are to be recruited through heroic images. Both aesthetically appealing and cruel and ugly pictures are remembered equally. Regarding the underlying mnemonic function, a picture is remembered even if the presented content is morally reprehensible (Goertz, 2018; Müller & Knieper, 2019).

Conclusion

All in all, the complexity of this interdisciplinary approach is evident. However, some exciting and highly relevant questions can be derived from the combination of aesthetics and ethics as well as in relation to neuroaesthetics. It could be shown that in particular, the aesthetic effect of art has always been connected with an educational factor. This is often expressed through the representation of virtues and also moral border areas. In this way, moral action can be conveyed, and immoral action can be experienced and understood. Especially modern media and art forms, such as photography and film and also video games, have a strong immersive character and can convey ethical ideas through aesthetics. It could be shown that especially representations of the beauty and the ugliness play a central role. These categorizations are based in particular on ideas, such as the golden ratio. Neuroaesthetics has shown that aesthetic experiences trigger special effects in the brain and that aesthetic experiences influence judgment. The perceived attractiveness of a person thus also affects the judgment about this person. Thus, attractiveness is often equated with positive characteristics and high moral qualities. From an ethical perspective, this combination therefore becomes interesting in terms of the extent to which this effect creates constraints on people. These include the compulsion to fulfill cultural ideals of beauty to increase one's chances at work, finding a place to live, or finding a partner. It opens up the question of the extent to which one's own body becomes an artificial reproduction of the beauty ideals and the extent to which one's own body can be modified to this end. This also includes the question of the extent to which horror images or morally reprehensible acts such as torture or rape should be presented. Is a limit of acceptable depiction crossed, as in *Grand Theft Auto V* or *Rape Day*, or can images like the self-immolation of the monk be revealing? Closely related to this is the question of how to deal with a terrorist war of graphic images. All these questions hold great potential to be pursued further in empirical studies. Overall, this chapter showed how fruitful a holistic view of aesthetics, ethics, and neuroaesthetics can be.

References

Aristoteles [Aristotle] (1994). *Poetik* [*Poetics*] *Griechisch/Deutsch.* Reclam (in German).

Armann, R. & Bülthoff, I. (2009). Gaze behavior in face comparison: The roles of sex, task, and symmetry. *Attention, Perception, & Psychophysics*, 71, 1107–1126. https://doi.org/10.3758/APP.71.5.1107.

Baumgarten, A. G. (1750). *Aesthetica.* Kleyb (in Latin).

Brielmann, A. & Pelli, D. (2018). Aesthetics. *Current Biology*, 28, R847–R870. https://doi.org/10.1016/j.cub.2018.06.004.

Bock, A., Isermann, H., & Knieper, T. (2011). Ikonologische Kontextanalyse [Iconological context analysis]. In T. Petersen & C. Schwender (eds.), *Die Entschlüselung der Bilder: Methoden zur Erforschung visueller Kommunikation. Ein Handbuch* [*The Decoding of Images: Methods for the Study of Visual Communication. A Handbook*, pp. 56–71. Halem Verlag (in German).

Bubner, R. (1989). *Ästhetische Erfahrung* [*Aesthetic Experience*]. Suhrkamp (in German).

Büttner, S. (2006). *Antike Ästhetik. Eine Einführung in die Prinzipien des Schönen* [*Ancient Aesthetics. An Introduction to the Principles of Beauty*]. Beck (in German).

Callot, J. (1633). *Malheurs de la Guerre* [*Painting*]. New York, NY: The Metropolitan Museum of Art.

Chuan-Peng, H., Huang, Y., Eickhoff, S. B., Peng, K., & Sui, J. (2020). Seeking the "Beauty Center" in the brain: A meta-analysis of fMRI studies of beautiful human faces and visual art. *Cognitive, Affective, & Behavioral Neuroscience*, 20(6), 1200–1215. https://doi.org/10.1101/081539.

Cimino, M. (Director) (1978). *The Deer Hunter*. EMI Films, Universal Studios.

de Goya y Lucientes, F. (1863). *Desastres de la Guerra* [*Aquatint Etching*], Pasadena, CA: Norton Simon Museum.

Dunn, W., Murchison, D., & Broome, J. (1996). Esthetics: Patients' perceptions of dental attractiveness. *Journal of Prosthodontics*, 5, 55–71. https://doi.org/10.1111/j.1532-849X.1996.tb00292.x.

Düwell, M. (1999). *Ästhetische Erfahrung und Moral. Zur Bedeutung des Ästhetischen für die Handlungsspielräume des Menschen* [*Aesthetic Experience and Morality. On the Importance of the Aesthetic for the People's Freedom of Action*]. Alber (in German).

Evans, P. (2019). Rape day game pulled by steam platform after outcry. *BBC News*, March 7 https://www.bbc.com/news/blogs-trending-47484397.

Feige, D. M. (2017). Videospiele im Spannungsfeld von Kunst und Kulturindustrie [Video games in the field of tension between the art industries and the culture industries]. In M. Niederauer & G. Schweppenhäuser (eds.), *Kritische Theorie der Kulturindustrie nach 70 Jahren. Interpretationen aus Philosophie, Soziologie und Medientheorie* [*Critical Theory of the Culture Industry After 70 Years. Interpretations from Philosophy, Sociology, and Media Theory*], pp. 201–219. Springer Fachmedien Wiesbaden (in German).

Fenner, D. (2000). *Kunst — Jenseits von gut und Böse? Kritischer Versuch über das Verhältnis von Ästhetik und Ethik* [*Art — Beyond Good and Bad? Critical Attempt on the Relationship between Aesthetics and Ethics*]. Francke (in German).

Fenner, D. (2013). Ethik und Ästhetik [Ethics and aesthetics]. *Kulturelle Bildung Online* [*Cultural Education Online*]. https://www.kubi-online.de/artikel/ethik-aesthetik (in German).

Furnham, A., McClelland, A., & Omer, L. (2003). A cross-cultural comparison of ratings of perceived fecundity and sexual attractiveness as a function of body weight and waist-to-hip ratio. *Psychology, Health & Medicine*, 8(2), 219–230. https://doi.org/10.1016/S0191-8869(01)00073-3.

Goertz, S. (2018). *Der Neue Terrorismus. Neue Akteure, neue Strategien, neue Taktiken und neue Mittel* [*The New Terrorism. New Actors, New Strategies, New Tactics, New Resources*]. Springer VS (in German).

Goodman, N. (1995). *Sprachen der Kunst* [*Languages of Art*]. Suhrkamp (in German).

Grewenig, M. M. (ed.) (1999). *Augenblicke des Jahrhunderts. Meisterwerke der Reportagefotografie von Associated Press* [*Moments of the Century. Masterpieces of Reportage Photography from the Associated Press*]. Hatje Cantz (in German).

Heimes, S. (2010). *Künstlerische Therapien. Ein intermedialer Ansatz* [*Artistic Therapies. An Intermedia Approach*]. Vandenhoeck & Ruprecht (in German).

Hern, A. (2013). Grand theft auto 5 under fire for graphic torture scene. *theguardian.com*, September 18, https://www.theguardian.com/technology/2013/sep/18/grand-theft-auto-5-under-fire-for-graphic-torture-scene.

Hoffmann, D. (2019). Körperbilder in der Kommunikationsforschung [Body images in communication research]. In K. Lobinger (ed.), *Handbuch Visuelle Kommunikationsforschung* [*Handbook of Visual Communication Research*], pp. 421–439. Springer VS (in German). https://doi.org/10.1007/978-3-658-06508-9_13.

Hogarth, W. (1997). *The Analyses of Beauty*. New Haven, CT: Yale University Press.

Isermann, H. & Knieper, T. (2009). Tod und Leid auf Seite eins [Death and Suffering on Page One]. *message: Internationale Fachzeitschrift für Journalismus* [*Message: International Trade Journal for Journalism*], 11(2), 26–29 (in German).

Jacobsen, T. & Höfel, L. (2003). Descriptive and evaluative judgment processes: Behavioral and electrophysiological indices of processing symmetry and aesthetics. *Cognitive, Affective, & Behavioral Neuroscience*, 3, 289–299. https://doi.org/10.3758/CABN.3.4.289.

Joffé, R. (Director). (1984). *The Killing Fields*. Goldcrest Films, Columbia Pictures.

Kant, I. (1790). *Kritik der Urteilskraft* [*Critique of Judgment*]. Lagarde and Friedrich (in German).

Kessler, E. (2007). Die Proportionen der Schönheit [The proportions of beauty]. In C. Gutwald & R. Zons (eds.), *Die Macht der Schönheit* [*The Power of Beauty*], pp. 133–160. Wilhelm Fink (in German).

Knieper, T. & Müller, M. G. (2019). Zur Bedeutung von Bildkontexten und Produktionsprozessen für die Analyse visueller Kommunikation [On the importance of image contexts and production processes for the analysis of visual communication]. In K. Lobinger (ed.), *Handbuch Visuelle Kommunikationsforschung* [*Handbook Visual Communication Research*], pp. S.515–S.526. Springer Fachmedien (in German).

Knieper, T. & Saleh, I. (2015). Die audiovisuelle Medienberichterstattung über das Massaker von Marikana: Bestimmt die Weltanschauung die Sicht auf die Dinge? [Audiovisual media coverage of the Marikana massacre: Does the worldview determine how things are viewed?]. In F. Reer, K. Sachs-Hombach, & S. Schahadat (eds.), *Krieg und Konflikt in den Medien. Multidisziplinäre Perspektiven auf mediale Kriegsdarstellungen und deren Wirkungen* [*War and Conflict in the Media. Multidisciplinary Perspectives on Media Depictions of War and Their Effects*], pp. 241–278. Halem (in German).

Knieper, T. (2015). Caricature. *The International Encyclopedia of Communication*. https://doi.org/10.1002/9781405186407.wbiecc005.pub2.

Langlois, J. & Roggman, L. (1990). Attractive faces are only average. *Psychological Science*, 1(2), 115–121. https://doi.org/10.1111/j.1467-9280.1990.tb00079.x.

Langlois, J., Roggman, L., & Musselman, L. (1994). What is average and what is not average about attractive faces? *Psychological Science*, 5(4), 214–220. https://doi.org/10.1111/j.1467-9280.1994.tb00503.x.

Lessing, G. E. (1996). In H. G. Göpfert (ed.), *Six volumes Werke* [*Works*]. WGB (in German).

Liessmann, K. P. (2009). *Schönheit* [*Beauty*]. Wien: UTB (in German).

Lu, C. (Director). (2009). *City of Life and Death*. China Film Group Corporation, NewKSM.

Maase, K. (2008). *Die Schönheit des Populären. Ästhetische Erfahrung der Gegenwart* [*The Beauty of the Popular. Aesthetic Experience of the Present Age*]. Campus Verlag (in German).

Majetschak, S. (2019). *Ästhetik. Zur Einführung* [*Aesthetics. An Introduction*]. Junius (in German).

Martin, J. L. (2017). The birth of the true, the good, and the beautiful: Toward an investigation of the structures of social thought. *Current Perspectives in Social Theory*, 35, 3–56. https://doi.org/10.1108/S0278-120420160000035001.

Martindale, C. (1988). Aesthetics, psychobiology, and cognition. In F. H. Farley & R. W. Neperud (Hrsg.) (eds.), *The Foundations of Aesthetics, Art and Art Education*, S. 7–42. New York, NY: Praeger.

Mentges, G. & Richard, B. (2005). *Schönheit der Uniformität. Körper, Kleidung, Medien* [*Beauty of Uniformity. Body, Clothing, Media*]. Campus Verlag (in German).

Müller, M. G. (2005). >Burning Bodies<: Visueller Horror als strategisches Element kriegerischen Terrors — eine ikonologische Betrachtung ohne Bilder [>Burning bodies<: Visual horror as a strategic element of warlike terror — An iconological view without images]. In T. Knieper & M. G. Müller (eds.), *War Visions: Bildkommunikation und Krieg* [*War Visions: Picture Communication and War*], pp. 405–423. Herbert von Halem (in German).

Müller, M. G. & Knieper, T. (2019). Terror der Bilder — Visuelle Kommunikation in Krieg und Terrorismus [Terror of images — Visual communication in war and terrorism]. In K. Lobinger (ed.), *Handbuch Visuelle Kommunikationsforschung*, pp. 145–179 (in German). Springer Nature. https://doi.org/10.1007/978-3-658-06508-9_9.

Nietzsche, F. (1980). In G. Colli & M. Montinari (eds.), (KSA) *Sämtliche Werke. Kritische Studienausgabe in 15 Bänden* [*Complete Works. Critical Study Edition in 15 Volumes*] dtv (in German).

Nussbaum, M. C. (1986). *The Fragility of Goodness: Luck and Ethics in Greek Tragedy and Philosophy*. University Press.

Pahlen, K. (1996). *Die große Geschichte der Musik* [*The Great History of Music*]. Coromoran Verlag (in German).

Platon [Plato] (1988). In O. Apelt (ed.), *Der größere Hippias. Sämtliche Dialoge, Bd. III* [*The Greater Hippias. All Dialogues, Vol. III*]. Meiner (in German).

Platon [Plato] (1991). Der Staat [The state]. In K.-H. Hülser (ed.), *Sämtliche Werke in zehn Bänden. Bd. 5* [*Complete Works in Ten Volumes, Vol. 5*]. Insel Verlag (in German).

Platon [Plato] (2013). In M. Holzinger (ed.), *Phaidros* [*Phaedrus*]. Berlin Edition, Holzinger (in German).

Ramachandran, V. & Hirstein, W. (1999). The science of art: A neurological theory of aesthetic experience. *Journal of Consciousness Studies*, 6, 15–51.

Reicher, M. E. (2015). *Einführung in die philosophische Ästhetik* [*Introduction to Philosophical Aesthetics*]. WGB (in German).

Rentschler, I. Herzberger, B. & Epstein D. (eds.) (1988). *Beauty and the Brain. Biological Aspects of Aesthetics*. Birkhäuser Verlag.

Renz, U. (2007). *Schönheit. Eine Wissenschaft für sich* [*Beauty. A Science in Itself*]. Berlin Verlag (in German).

Rinderle, P. (2011). *Musik, Emotionen und Ethik* [*Music, Emotions, and Ethics*]. Karl Alber (in German).

Rorty, R. (1992). *Kontingenz, Ironie und Solidarität* [*Contingency, Irony, and Solidarity*]. Suhrkamp (in German).

Rossmann, C. & Brosius, H.-B. (2005). Vom hässlichen Entlein zum schönen Schwan? Zur Darstellung und Wirkung von Schönheitsoperationen im Fernsehen [From ugly duckling to beautiful swan? Depicting cosmetic surgeries on television and the resulting impacts]. *M&K Medien & Kommunikationswissenschaft [Media & Communication Studies]*, 53(4), 507–532 (in German). https://doi.org/10.5771/1615-634x-2005-4-507.

Röwekamp, B. (2011). *Antikriegsfilm. Zur Ästhetik, Geschichte und Theorie einer filmhistorischen Praxis* [*Anti-War Film. On the Aesthetics, History, and Theory of a Film Historical Practice*]. Edition Text und Kritik (in German).

Schlegel, F. (1988). In E. Behler & H. Eichner (eds.), *Kritische Schriften und Fragmente. Studienausgabe in sechs Bänden* [*Critical Writings and Fragments. Study Edition in Six Volumes*]. Brill/Schöningh (in German).

Schneider, N. (2021). *Geschichte der Ästhetik von der Aufklärung bis zur Postmoderne* [*History of Aesthetics from Enlightenment to Postmodernism*]. Reclam (in German).

Singh, D. (1993). Adaptive significance of female physical attractiveness: Role of waist-to-hip ratio. *Journal of Personality and Social Psychology*, 65, 293–307. https://doi.org/10.1037/0022-3514.65.2.293.

Singh, D. & Luis, S. (1995). Ethnic and gender consensus for the effect of waist-to-hip ratio on judgment of women's attractiveness. *Human Nature*, 6, 51–65. https://doi.org/10.1007/BF02734135.

Spielberg, S. (Director) (1998). *Saving Privat Ryan*. DreamWorks Home Entertainment.

Sütterlin, C. & Yu, X. (2021). Aristotle's dream: Evolutionary and neural aspects of aesthetic communication in the arts. *PsyCh Journal*, 10, 224–243. https://doi.org/10.1002/pchj.416.

Tatarkiewicz, W. (2003). *Geschichte der sechs Begriffe. Kunst, Schönheit, Form, Kreativität, Mimesis, Ästhetisches Erlebnis* [*History of the Six Terms. Art, Beauty, Form, Creativity, Mimesis, Aesthetic Experience*]. Suhrkamp (in German).

Taylor, J. F. A. (1998). *Body Horror. Photojournalism, Catastrophe and War*. New York, NY: New York University Press.

von Brück, M. (2021). Aesthetic principles, the arts, and the interpretation of culture. *PsyCh Journal*, 10(2), 200–209. https://doi.org/10.1002/pchj.406.

Waibl, E. (2009). *Ästhetik und Kunst von Pythagoras bis Freud* [*Aesthetics and Art from Pythagoras to Freud*]. UTB (in German).

Wang, T., Mo, L., Mo, C., Tan, L. H., Cant, J. S., Zhong, L., & Cupchik, G. (2015). Is moral beauty different from facial beauty? Evidence from an fMRI study. *Social Cognitive and Affective Neuroscience*, 10(6), 814–823. https://doi.org/10.1093/scan/nsu123.

Wang, X., Simmank, F., Zhang, D., Zeng, Y., Silveira, S., Sander, T., Bao, Y., & Paolini, M. (2021). Aesthetics as a common denominator for moral and commercial judgments. *PsyCh Journal*, 10(2), 187–189. https://doi.org/10.1002/pchj.407.

Welsch, W. (1960). *Ästhetisches Denken* [*Aesthetic Thinking*]. Reclam (in German).

Welsch, W. (1993). *Die Aktualität des Ästhetischen* [*The Timeliness of Aesthetics*]. Wilhelm Fink (in German).

Wilke, J. (2005). Kriegsbilder in der historischen (Bild-)Publizistik [Images of war in historical (visual) journalism]. In T. Knieper & M. G. Müller (eds.), *War Visions: Bildkommunikation und Krieg* [*War Visions: Picture Communication and War*], pp. 22–56. Herbert von Halem (in German).

Yang, T., Formuli, A., Paolini, M., & Zeki, S. (2021). The neural determinants of beauty. *BioRxiv*. https://doi.org/10.1101/2021.05.21.444999.

Zeki, S. (2000). Farbe, Form, Bewegung — Zur Verarbeitung des visuellen Wissens im menschlichen Gehirn [Color, shape, movement — The processing of visual knowledge in the human brain]. In C. Maar, H. U. Obrist, & E. Pöppel (eds.), *Weltwissen Wissenswelt. Das globale Netz von Text und Bild* [*World Knowledge — Knowing World. The Global Net of Text and Images*], pp. 170–174. Du Mont (in German).

Zeki, S. (2001). Artistic creativity and the brain. *Science*, 293, 51–52. https://doi.org/10.1126/science.106233.

Zeki, S. (2004). Dante, Michelangelo und Wagner. Das Gehirn als Konstrukteur genialer Kunstwerke [Dante, Michelangelo, and Wagner. The brain as a designer of ingenious art]. In C. Maar & H. Burda (eds.), *Iconic Turn. Die neue Macht der Bilder* [*Iconic Turn. The New Power of Pictures*], pp. 77–102. Du Mont (in German).

Zeki, S., Romaya, J. P., Benincasa, D. M. T., & Atiyah, M. F. (2014). The experience of mathematical beauty and its neural correlates. *Frontiers in Human Neuroscience*, 8, 68. https://doi.org/10.3389/fnhum.2014.00068.

Zhao, C., Zhang, D., & Bao, Y. (2018). A time window of 3 s in the aesthetic appreciation of poems. *PsyCh Journal*, 7(1), 51–52. https://doi.org/10.1002/pchj.194.

https://doi.org/10.1142/9789811298639_0011

Chapter 11

On the Necessity of Importing Neurobiology into Mathematics*

Mikhail Filippov[†] and Semir Zeki[‡]

[†]*National University of Singapore, Singapore*
[‡]*University College London, UK*

Abstract

In Gödel's incompleteness theorems, for every mathematical system, there are correct statements that cannot be proven to be correct within that system. We here extend this to address the question of axiomatic statements that are *perceived* (or known) to be correct but which mathematics, as presently constituted, cannot prove. We refer to these as *perceptual axioms*.

Keywords: "illusions"; incompleteness theorems; logic; perceptual axioms; perceptual reality

*This chapter was originally published in the special collection on neuroaesthetics in *PsyCh Journal*. Filippov, M. & Zeki, S. (2022). On the necessity of importing neurobiology into mathematics. *PsyCh Journal*, 11(5), 755–756. https://doi.org/10.1002/pchj.524.

Introduction

> *The opposite of a correct statement is a false statement. But the opposite of a profound truth can be another profound truth.*
>
> — Niels Bohr

In a famous encounter at a meeting entitled *The Second Conference on the Epistemology of the Exact Sciences (2. Tagung für Erkenntnislehre der exakten Wissenschaften)*, held in Königsberg in 1931, Kurt Gödel confronted the famous mathematician David Hilbert with regard to his incompleteness theorems; Gödel argued that, for a mathematical system to be consistent, it cannot be complete, in the sense that it cannot prove all correct statements as correct and incorrect ones as incorrect (Gödel, 1931); moreover, that the consistency of the axioms cannot be proven within its own particular axiomatic system. The implication is that, within mathematics, there is an important difference between truth and proof. But Gödel was writing only about the limitations of the mathematical axiomatic theories. Here, we address what amounts to the same question in a more neurobiological context. We formulate our question as follows: Are there any perceived truths that are universally known but which mathematics cannot prove? We believe that we can address this question within the context of the organization of the perceptual brain and what it allows us to experience.

Truths Derived from Perception that are Unprovable Mathematically

The concept of mathematical axioms is well known; it refers to statements which the logical system of the brain accepts without proof; these statements constitute the building blocks which allow us to construct mathematical theories. We use the following examples to propose the concept of another axiomatic system and introduce the term *perceptual axiomatic* system to do so; this is also universally accepted by the human brain and works in parallel but independently of the logical system of the brain underlying mathematics.

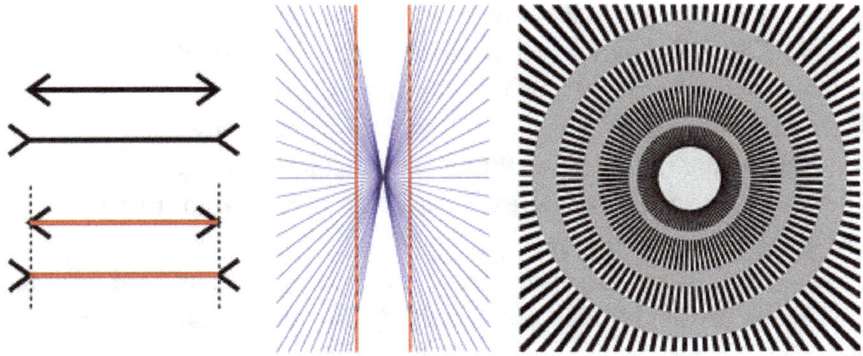

Figure 1. Müller-Lyer "illusion" (left), Hering "illusion" (center), and a modified (achromatic) version of Isia Leviant's enigma (right).

The perceptual truths of the three examples given in the following cannot be doubted and also cannot be proven mathematically. These perceptual truths are so obvious that we refer to them as *perceptual axioms*.

Consider first the Müller-Lyer "illusion" (Figure 1, left). In it, there are two lines of the same length; both have arrows directed inwards for one line and outwards for the other. That they are of equal length can be proven mathematically. However, no normal person sees these lines to be of the same length, even when they know them to be mathematically equal; instead, the line with the outward arrows is perceived to be shorter than the one with the inward arrows. This is true even if, when trying to convince the viewer of the equality of lengths, the lines are made more conspicuous by being highlighted, for example, in red, as in the lower left panel.

A second example is the Hering "illusion" (Figure 1, center). Here, two lines which can be proven to be parallel mathematically are perceived as being bent in a concave way. The *perceptual* fact of them being so bent is one that has universal assent: All humans with a normal visual apparatus (including the visual brain) see these lines as slightly bent and not as parallel; it can therefore be said to be perceptually axiomatic. But mathematical formulae cannot prove these two lines to be so bent, unless a perceptual axiom is taken into consideration, which amounts to incorporating perceptual axioms into mathematics. The only conclusion that mathematics can reach about them is that they are parallel, thus contradicting a profound perceptual axiomatic truth. For normal individuals, it is the latter truth that prevails and which they have to live with while the mathematical truth is the one that is rejected perceptually.

Here again, the perceptual truth is contradictory to the mathematical one; this constitutes another example of mathematics not being able to prove a simple perceived reality.

In our third and last example, we consider a different type of "illusion": one in which motion is perceived in a static picture, at least for some individuals (Figure 1, right); the motion is very rapid and reverses direction with prolonged viewing. The reality of this cannot be doubted by individuals who perceive the motion in the rings (not all individuals do). The "illusion" is one prepared by the French artist, Isia Leviant, who has written about the steps in its construction (Leviant, 1996). By sole use of mathematics, which expresses the occurrence of motion in terms of displacement, distance, velocity, acceleration, speed, and time, for example, in the form

$$v = \frac{dr}{dt}, a = \frac{dv}{dt} = \frac{d^2r}{dt^2},$$

it would be impossible to prove the existence of motion in such a picture, and even less so the reversal in its direction. In other words, it will be impossible to prove mathematically to a person who experiences the motion that there is motion. This is another example of perceptual truths which mathematics cannot prove to be true.

Conclusion

We have consistently used the term "illusion" above in quotes because, by definition, illusion is a departure from "reality," in the examples given here from "mathematical reality". However, mathematics depends upon the healthy functioning of brain mechanisms just as perception does; where the two clash, it is the perceptual reality that prevails. It is worth emphasizing that the only perceptual reality that we are capable of experiencing is that which is allowed and made possible by the organization of our brains. Viewed in this context, it is the mathematical reality that is the "illusion" and the perceptual reality that is the truth, something which no argument in mathematics can overcome.

We have given only three examples; there are no doubt many more that fall under the category of perceptual axioms. We use these three examples to make the general point that, without incorporating an understanding of neurobiological mechanisms, the whole field of mathematical logic is limited in its application, as are its results. The only truths that we are sure of, whether mathematical or otherwise, are the ones that our brain organization allows us to experience. Therefore, mathematics should include in its repertoire an understanding of the neurobiological mechanisms of perceptual reality in order to address not only the problem of incompleteness but also the entire problem of knowledge.

Acknowledgments

This work was supported by the Leverhulme Trust London. We are grateful to members of the Laboratory of Neurobiology at UCL and members of The Platonic Academy London for useful discussion.

References

Gödel, K. (1931). Über formal unentscheidbare Sätze der Principia Mathematica und verwandter Systeme I. *Monatshefte für Mathematik*, 38(1), 173–198. https://doi.org/10.1007/BF01700692.

Leviant, I. (1996). Does "brain-power" make enigma spin? *Proceedings: Biological Sciences*, 263(1373), 997–1001. https://doi.org/10.1098/rspb.1996.0147.

Part II

Art in Space

https://doi.org/10.1142/9789811298639_0012

Chapter 12

Aristotle's Dream: Evolutionary and Neural Aspects of Aesthetic Communication in the Arts*

Christa Sütterlin[†] and Xinchi Yu[‡,§]

[†]*Human Ethology Group, Max-Planck-Institute for Ornithology, Seewiesen, Germany*
[‡]*School of Psychological and Cognitive Sciences, Peking University, Beijing, China*
[§]*Department of Chinese Language and Literature, Peking University, Beijing, China*

Abstract

Art in general perception is something that transcends our notion of reality. In view of the earliest findings in Paleolithic sites, their abstract appearance and sometimes ceremonial context increased their status of a secret language. Even the first figurative cave paintings remained in the context of an encoded semantic whole. The highly symbolic value of

*This chapter was originally published in the special collection on neuroaesthetics in *PsyCh Journal*. Sütterlin, C. & Yu, X. (2021), Aristotle's dream: Evolutionary and neural aspects of aesthetic communication in the arts. *PsyCh Journal*, 10(2), 224–243. https://doi.org/10.1002/pchj.416.

art seemed invulnerable. It was just the claim for *mimesis* in Greek antiquity (Plato) that urged artists to "realistically" depict what can be seen — as to stay in track of eternal messages behind. This devaluated the artistic oeuvre to a purely imitating craft and had to overcome at once several inherent obstacles. First, that reality (the phenomenal world) is in general only a pale reflection of what lays behind (Platonic ideas) and second, that the human eye, unlike the human mind, cannot penetrate to more than ephemeral impressions. Moreover, it mixed up reality with what we are able to see (i.e., visual perception), thus supposing a pinpoint representation of the world by our senses. Aristotle was the first to qualify art as picturing more than we usually are meant to see, filling the gap between the sensual and the spiritual world. Aristotelian aesthetics includes concepts of reduction and selection of composition and emotion, thus a summarized view within any performance of poetics or painting. And it took centuries to close the gap between natural and aesthetic perception or art. Life sciences in the 20th century discovered the evolutionary basis of sensory perception as being highly biased and organized, concept as emotion-driven and thus, mentally equipped as well. This sets a new approach in our understanding of perception, art, and aesthetics as an ongoing communication in process on common bases. Art may cooperate or disagree but never can cut the nexus with its perceptual prejudices and substrate.

***Keywords*:** art; communication; complexity reduction; evolution; mimesis; realism; template formation

Aristotle and the Fallacies of the Platonic Mimesis Concept

Art enjoys the nimbus of excellence and mystery, of something sublime up to the present day (Burke, 1757/1989; Gambino & Pulvirenti, 2020; Ishizu & Zeki, 2014; Schoeller & Perlovsky, 2016). Many art scientists relentlessly aim to defend its qualities of difference, even autonomy against everything we call nature, life, or reality. More than ever before, museums and exhibitions are visited like sacred places in a mood of awe: Psychological evidence supports that real museum experience may provide an "aura" for art appreciation (Brieber, Nadal, & Leder, 2015;

Brieber, Nadal, Leder, & Rosenberg, 2014; Pelowski, Forster, Tinio, Scholl, & Leder, 2017; but see Brieber, Leder, & Nadal, 2015). What kind of message do people hope to gather from pictures and objects belonging to the most sumptuous collections in the world? It seems that all these indefinable features take part in the enigmatic prestige art covers in our minds. Furthermore, our knowledge about the origins of artistic production in the context of early rituals, dance, and music in paleolithic caves contributes to the mysterious, even religious embedding of art (e.g., Dissanayake, 1992, 2008; Savage *et al.*, 2020; Sütterlin, 2003). What are the conditions that support such comparisons? Although art could be considered as a symbolic system (just like language), the learning process of art and (written or verbal) language might be different. And in the oldest human documents, pictures appear earlier than writing. It seems that it is accessible more easily by intuition and emotion than by the more analytical and "reasoning" parts of cognition.

From the perspective of the earliest comments and texts on art, however, there is no clear evidence that supports the sublime estimate of art, at least from pre-Socratic thinking up to Plato. Affinities to eternal qualities, like the structures and harmonies behind the visible world — Pythagorean proportions and Platonic ideas — are attributed to the notion of *beauty* alone, which participates in cosmogonic theories (Grassi, 1980). Art, by contrast, is suggested to skillfully depict what is physically present and visible: a world of change and contingency. But as the divine laws are potentially inherent to all sorts of natural things, the artist finds their role and dignity in imitating nature.

Art as *mimesis (imitation)* is discussed by Xenophon in *Memorabilia* (III, 9, 10) and eminently taken up by Plato. Art appears reduced to a craft that can be learned and has no real dignity on its own. It copies what can be perceived as existent and, according to Plato, is just a copy and weaker reflection of the essence (*ideas*), as he exposed in his *Politeia* (X, 598 b; Schumacher, 1983). The dichotomy of Platonic thinking in two worlds, a spiritual world of ideas (essence) and a physical world of matter (existence) that contains only a pale and accidental reflection of the ideas behind, is an early top-down argument, and the concepts of art and beauty seem to assimilate quite seamlessly into this groundwork, whereas the intermediate link of aesthetics is lurking rather between the lines. The notion of mimesis in these early terms marks the onset of a misunderstanding of the conditions under which information about the phenomenal world is acquired and can be realistically translated. The main problem was the underestimation of what is underlying the concept of mimesis: human perception. Imitation depends on what we perceive. If Plato's recommendation for artists was to simply picture what is there and can be seen, he had to suppose a purely mimicking power of the human senses. However, art does not picture reality but our perception of reality. And human visual perception is, however, not necessarily camera-like (cf. Schober & Rentschler, 1972) and a faithful reflection of all the details falling on the retina: Attention is a selective process, and sometimes our visual perception is biased (e.g., Carrasco, 2011; Simons & Levin, 1997; Whitney & Levi, 2011).

The problem seems in fact to be the one of visibility. If already pre-Platonic thinkers had spoken of the hidden laws and harmonies of beauty, Heraclitus even of the invisible harmony prevailing over the visible harmony, how should we then be able to perceive beauty, if not by our eyes? Xenophon, as a disciple of Socrates, was one of the first to speak of the sensory preconditions of beauty (Grassi, 1980), and in his *Symposium* we are witnesses of his fascination for beautiful male appearances. Even in Plato's *Phaidros* and *Timaios*, we read about the thrill in view of physical beauty (Plato, 340 B.C/1990). Actually, beauty is the sheer visible. The arts, however, seem to be limited in their aim to rebuild dull copies of living appearances.

There was a gap to close between the conditions of art and beauty, and this touches on the role of aesthetics. Aristotle, the most distinguished Platonian scholar, was the first to give a critique of the classical concept of mimesis and to reflect on the role of art on new bases. He can be called the first art critic altogether. He asked how it should be possible to get access to the beautiful if Platonian *mimesis* is the only pathway. How can artists then give more than bits and pieces, if they cannot perceive and penetrate the structures underlying beauty? Art implies seeing beyond physical reality and perceiving (implicitly or explicitly) the hidden structures or intents lying behind (Bao, Yang, *et al.*, 2017; Rogala, Bajno, & Wróbel, 2020): This could be demonstrated by the observation that viewers are able to identify the original color or orientation of abstract paintings (Bao, Yang, *et al.*, 2017; Lelièvre & Neri, 2019; Lindauer, 1987; Mather, 2012) and even the original perspective of portraits (Wang & Bao, 2020). Hidden syntax-like structures (or more generally, hierarchical structures) have been proposed for music (London, 2002; Patel, 2003; Pöppel, 1989; Wittmann & Pöppel, 1999), poetry (Teng *et al.*, 2020; Turner & Pöppel, 1988; Yu & Bao, 2020), and even visual arts (Ganczarek, Pietras, & Rosiek, 2020). This process would find its resemblance in perceiving the hierarchical structure of language and making pragmatic inferences.

Platonic theory of forms separated essence (idea) from existing appearances (eidola) and, for Plato, it was the human mind only that could advance to the inner laws and coherences of the physical world — in sublime moments also dreams or states of memory (*Phaidros*, 250 a, b). Both make no part of our senses but certainly of Plato's metaphysics. Aristotle by contrast proceeds, bottom-up, from concrete reality and develops the more general issues and ideas from prevailing circumstances. Humans have to find and filter out the essence from the existing. His view of unifying principles being inherent to our phenomenal world can be exploited by the artist and throws a new light on everything associated with mimesis (Fuhrmann, 1973). Art, according to Aristotle, can complete the incomplete.

In his *Poetica*, for example, he explains that the poet of a drama cannot unfold the story in its natural length of time, but he should shorten and condense it into a plot: It must satisfy the claim of wholeness and unity. Rather than a mimetic depiction, art presents a purified version of reality in which perceptual tools are included and the essential becomes transparent. The pathway to cognition leads via perception (Müller, 1999). But, in contrast to sensualism in the 18th century, the road is oriented toward leading ideas. In addition, the artist could enter events of a possible outcome: This also means fiction. What Plato in his *Politeia* still called "lies" — for instance Hesiod's *Theogony* — is put against the idea that Gods are fundamentally good or at least should be represented like this. Aristotle pleaded for the drama, unfolded by the clash of different human or divine characters. The form of a drama has to be established against its material. The cases or examples are representative models that should apply to all (Fuhrmann, 1973). Aristotle is the first to talk of composition, transformation, and fantasy in art. Art splits from physical reality. In art, catharsis (purification by high emotion) is a desired affective outcome. This notion resonates with modern fMRI experiments on aesthetic perception, indicating that the emotion-related regions are also activated in this process (Bohrn, Altmann, Lubrich, Menninghaus, & Jacobs, 2013; Di Dio & Gallese, 2009; Kawabata & Zeki, 2004; Vartanian & Skov, 2014).

For the *spatial* expression of art, there are similar problems: A painter or sculptor cannot reproduce things in their natural dimensions, they have to shorten as well, to reduce the measures and to give a pooled effect or even essence. A painting or sculpture should be oriented towards an ideal, in a pinch, taking out of various examples the best parts. This clearly means a selection and was the advice of Xenophon earlier in his *Memorabilia* (Grassi, 1980). The notion of the summarized lays inside the artist. The crucial thing about it is in fact that, for these achievements, the eye of the artist is as well prepared as the optical demands of the public.

At an early moment of human history, the mental pad from Plato to Aristotle is leading from the concept of preexisting (divine) ideas, called *idealism* (top-down), to the naturally acquired perceptions and intuitions, called *naturalism* (bottom-up), guiding human's notion of reality.

The Earliest Art Material

The first known drawings by homo erectus, in this case, were found in Bilzingsleben (Thuringia, Germany) and are dated to around 280,000 B.C. A bone artifact shows 21 lines of about the same length and distance from each other (Figure 1). This fact supports the idea of the lines being probably the oldest manifestations of a more than technically intended activity (Feustel, 1987). They do not aspire to depict anything possibly concrete in the world but present pure proportions. We do not know, of course, what kind of message the producer tried to convey in the lines; whether they referred to a number of recordings in their head, or whether they were markings for themselves or for others. But evidently the lines did not contribute directly to the scope of the whole object — an elephant tibia used as a club — but they set signs that were clearly visible against the random

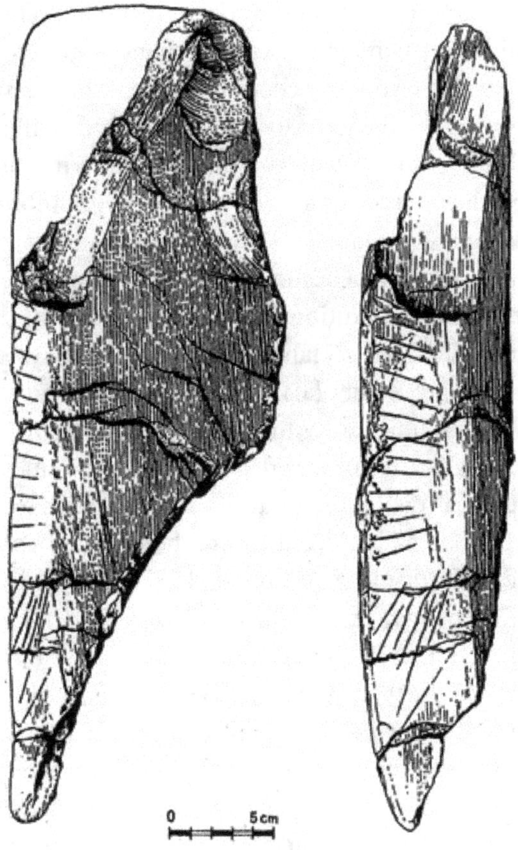

Figure 1. Bilzingsleben, Thuringia, Germany. Bone artifact with 21 cutlines. Ca. 250,000 B.P. (Behm-Blancke, 1983, p. 307) (Courtesy: Behn-Blancke).

texture of the bone. Obviously, their aim goes beyond the straight scope of the tool. The pattern gives evidence that somebody tried deliberately to cut these 21 lines in the same length and in a regular sequence and we can presume that they were proud of the success. To oppose a visible structure against the texture of the bone, something that would surmount the impression of randomness, a persistent incision was needed (i.e., *intentional* engraving).

Beyond the gratification of leaving signs that would probably outlast their own life, it must have been clear to the producer that these signs increase the visibility of the object as a whole and attract the attention of others. The addition of perceptible signs enhances not only the convenience of the object but also its exceptional note, which rises its status as well as the status of its owner. Dissanayake (1992) called this practice "making something special." And certainly, the pattern satisfied a sense of order that the producer shares with the beholders (Gombrich, 1984). The findings of many more examples of similar line patterns on bones in the closer area of Thuringia in Germany show the expansion of a cultural convention that became a tradition. The new and eminent thing with art is that humans created a kind of language, a symbolic system (or at least some elements that are symbolic) that they could read and understand themselves, that communicates with others and themselves, yet stimulates others and themselves (of course, these stages may emerge at different times in the course of evolution). Early humans may have possessed the instinct of expression (as indicated by the behavior of infants and other primates; Boysen, Berntson, & Prentice, 1987; Hodgson, 2006; Morris, 1962; Tanaka, Tomonaga, & Matsuzawa, 2003) and the fact that these expressions (e.g., in terms of signs) would attract attention of conspecifics provides a basis for further stages in the evolution of communication (Eibl-Eibesfeldt & Sütterlin, 2008; Tomasello, 2010). Our brains are actively participating in the sensations we draw from the visual information of our environment, especially the enhancing of edge contrasts favors our basic capacity to discriminate and highlight something against its background — an important condition for the perception of objects at all; this processing of an object as an object, in the field of psychology, is widely referred to as "figure–ground discrimination" (e.g., Likova & Tyler, 2008; E. Rubin, 1921; N. Rubin, 2001; Schnabel *et al.*, 2018). Before the first highlights of art around 35,000 years B.P., humans *collected* natural materials in the forms of faces, animals, or vulvas — what they later manufactured themselves (Herrmann & Ullrich, 1991). Some researchers have attempted to explain the productional outcome from the neural mechanism of perception (for instance, see a recent debate on the so-called neurovisual resonance theory by Hodgson, 2019a, 2019b; Mellet, Salagnon, *et al.*, 2019; and Mellet, Colagè, *et al.*, 2019). However, one should be cautious in this approach, as the neural mechanisms of perception and action are not exactly the same (Goodale & Milner, 1992;

Goodale & Westwood, 2004; Schenk & McIntosh, 2010). Productional outcomes (in this case, the observed pattern of engravings through archaeological excavation) are also constrained by other factors, for example, the texture of the background and the tool that archaic humans employed to engrave. Therefore, it is important also to note that one should not attempt to account for the productional outcomes *merely* by the cognitive mechanisms of perception; however, we could say, in a careful manner, that perception precedes formation.

The fact that we associate some meaning to material structures, such as lines and signs, that subjectively promise to close the gap between the visible and the invisible world, may have played an important role in the evolution of signs. That we attribute some meaning only to mimetically reproduced figures was the error of early theories on art. Chimpanzees and infants have also been found to possess this potential for symbolicity (Eibl-Eibesfeldt, 1984; Gardner & Gardner, 1978; Wolf, 1987). Through symbolicity, the void between the present and the absent is bridged. The attribution of sense seems without need of affirmation at this age and fantasy plays a big role, but the need for consent in a social group is increased. At a later stage, called the "schema" phase, lines are joined to form larger figures. Rectangles are called "houses" and circles, the "sun" (Al-Shamma-Gerber, 1984). The figures are still a token, but they are understood by a larger social group. In another context, the signification has to be learnt. When the Yalenang, Mek-speakers in Western New Guinea, paint their holy shields with red and white parallel lines, every line is given a special meaning (a pig, a marsupial, etc.) that members have to internalize (Eibl-Eibesfeldt & Sütterlin, 2008). This special significance is a shared knowledge in the group and is handed over to the next generation.

The deliberate creation of graphic and plastic structures that hide and contain an invisible meaning represents a new step in evolution. Animals communicate with colored feathers and stripes, songs, and chemical signs that tell others "I am a male" or "This place is mine" (Eibl-Eibesfeldt, 1967; Noble & Bradley, 1933). The understanding is part of the species-specific outfit. To invent new connections between the visible and the invisible is a new and risky task in human evolution. Story-telling, the invention of events that never happened or may have happened once, has been a crucial step in human cultural evolution (Harari, 2015; Junker,

2013; Tooby & Cosmides, 2001). And it was just for this option that Aristotle was applying in his new agenda for the arts.

The leaving of messages in oral or written form to the following generations is what we call culture. It was a tradition for millennia to garnish bones, and later stones and megaliths, with linear patterns, parallel, circular, as bundles, and fan-shaped. It became a pictorial language system long before the written one appeared. And it arose in many different places without cultural contact between the groups. The crossed regular lines on an ochre stone of Blombos Cave in South Africa (ca. 75,000 B.P.) were certainly not inspired by the earlier findings in paleolithic middle Europe. Linear engravings are not only a form of ornamentation around the sites of Bilzingsleben, Oldisleben, and those around Pavlov in the Czech Republic (all from the middle to late Paleolithic eras) but have remained a means of marking or messaging throughout cultural history around the globe (Figure 2). The evident pleasure of setting lines in a clear and distinct order against a ground with fuzzy patterns seems to be an urge and ornamental power cross-culturally until today. Note that a recent experiment with modern humans has pointed out a possibility of increased aesthetic appraisal for the engravings in different stages from the Blombos Cave and the Diepkloof Rock Shelter (Tylén *et al.*, 2020). However, the conclusions may be problematic, due to the mismatch between the target hypothesis and its experimental design (Yu, Pöppel, & Bao, 2020).

Given the evidence of neural predispositions, the artistic cultivation of lines appears in the light of a highly intrinsic form of communication between the artist's work and their own visual system. Jung (1971) has pointed out in unsurpassable terms the parallels between the sensory mechanisms of contrast-perception and features of artistic drawing by delineating the highly selective processes underlying both activities:

Figure 2. Pavlov, Czech Republic. Bundled lines on ivory (12.5 cm). Gravettian, ca. 25,000 B.P. (Archaeological Institute of Brno, Czech Republic, with friendly permission).

Vision evidently is based on an active choice and processing of images from the external world. These images are primarily reduced to formal features defined by their contours, originating in physiological interconnections of visual nervous systems from the eye up to the brain performing active information processing. Such reduced and simplified figures are stored in our memory and facilitate the recognition of similar forms. Art is using an analog selection process in omitting the accidental by emphasizing contours.

This idea should not be confused with the recent concept of the neurovisual resonance theory (Hodgson, 2006, 2019a, 2019b; for an idea with a similar flavor, see Changizi, Zhang, Ye, & Shimojo, 2006), as it is not attempting to make a strong link between the perception machinery and productional outcomes (as in the neurovisual resonance theory); rather, Jung delineated the whole process of how we learn and store the information from our experience and employ it in art creation.

Humans evidently are able to stimulate themselves in a way that their sensory system becomes activated in optimal terms and they can exploit new worlds in their imagination and emotions. They may manipulate artistically their perception and feelings as to transcend sheer satisfaction. Moreover, if somebody scratches lines into a bone, evoking satisfaction or amazement for themself, they can presume that others seeing the marking might find this amazing or powerful as well and share those feelings. This form of communication seems to be a new step in evolution. Chimpanzees, when starting to paint, show excitement. It seems to be satisfying to leave visible traces, but first of all, it seems to be satisfying for themselves and not so much for others. Children might also scribble and paint for their own joy including others. The urge to communicate something to conspecifics, however, to build a community of understanding by means of visual signs and images, a resonance that transcends the purely subjective aim and meaning, is another new thing. Humans can conceive visual worlds that do not necessarily exist physically but are perceived by others (Figure 3).

Proto-Aesthetic Perception

From the moment we understand that not only art but also perception is manipulating what we call "reality," we can integrate the neurophysiological groundworks easier in the question of art and aesthetics (Sütterlin, 2003; Zeki, 1998, 1999). The demand for aesthetics, for instance, in terms

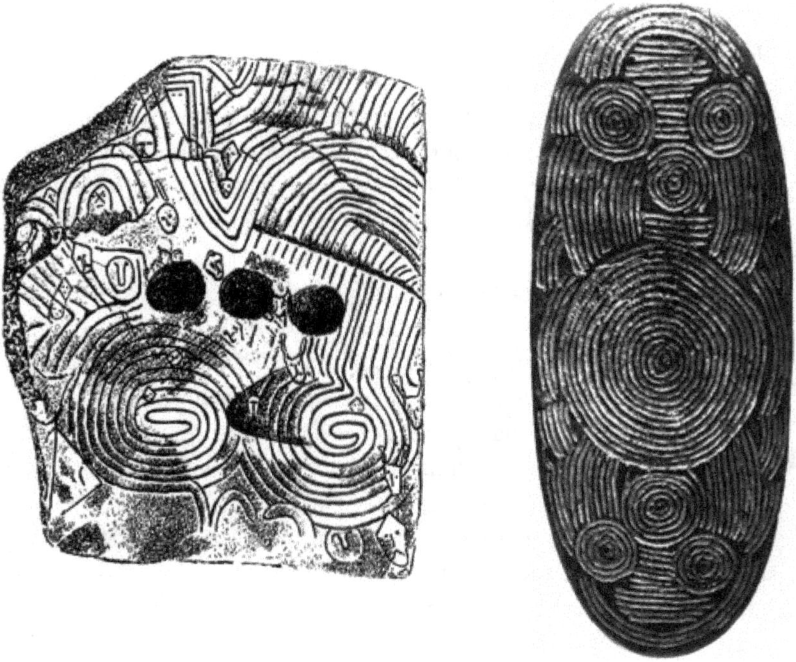

(a) (b)

Figure 3. (a) Decorated orthostat in a tomb at Gavrinis Island, Bretagne, France. Ca. 500 B.P. (Nay-Scheibler, 1981, p. 11). (b) Stone Churinga. Ritual Clan emblem with circled lines from the Walbiri, Mount Allen, Australia, 20th century. Photo: I. Eibl-Eibesfeldt, with friendly permission.

of an "evaluative perception" or "sensory cognition" arises much earlier than in art and much further down toward the bottom: in our everyday perception. The first to promote this idea in an extended philosophical form was Baumgarten (1750/2007) who introduced "Aesthetica" as a "gnoseologia inferior" and "analogon rationis." In his concept, sensitive, cognitive, evaluative, and emotional components were guiding humans' sensory and critical approaches to reality and the arts. The first line of empirical evidence followed only in the 20th century.

This enterprise started again on simple grounds. If we look at an arrangement of lines as presented by Metzger (1953), we feel a constraint to organize these lines within our visual field. Lines in a closer vicinity are summarized into a common figure, whereas new figures arise if some of the lines are interconnected. The same is true for the similarity of form and brightness: similar elements of a random arrangement of basic forms are picked out at once and comprised as one common figure (Arnheim, 1965). If we have to decide if some arrangement of lines is more beautiful than another, we may refer to specific size illusions, such as the Botti type (Störig & Rentschler, 1984). We are prone to attribute dynamic and warmth qualities to certain lines, and even Kandinsky (1926/1973) associated the line with differential mental power.

We preferentially see composed figures rather as a whole and not in their components (Arnheim, 1965); we even see lines and figures that are not physically there, but virtually present by omission in the stimulus, as in the famous Kanisza figures, we tend to correct visually slightly distorted figures into correct ones by memory (Wulf, 1922) and complete them if presented in short-time exposure as incomplete (Schuster & Beisl, 1978). Our perception is an active process, always in search of order and regularities (Bao, von Stosch, Park, & Pöppel, 2017). Evidently, our brains just need some regular ground information to fill up if there are gaps. This proceeds after the inbuilt hypothesis of homogeneity and redundancy of our environment (Zhou *et al.*, 2016), which somehow sounds like a missing link to Aristotelian aesthetics. And, if art or decoration meets one of these measures and attitudes, it communicates in optimal ways with our senses. This match seems to create aesthetic (i.e., sensual) pleasure. There is no region of this world that does not develop its own culture of ornaments in daily and ritual life (Eibl-Eibesfeldt & Sütterlin, 2008; Gombrich, 1984; Figure 4). The political abuse of its suggestive power is discussed as well (Vogel, 2009). The similarity between aesthetic perception and other cognitive processes has also been picked up by recent neuroimaging studies, indicating a common brain machinery (Avram *et al.*, 2013; Tsukiura & Cabeza, 2011; Wang *et al.*, 2020). In this sense, the cognitive and neural machinery for aesthetic perception and other parts of perception and cognition may be intertwined. Therefore, some claims that (proto-)aesthetic perception is "useless" in the course of evolution (e.g., Mellet, Colagè, *et al.*, 2019) are not well founded.

In *The Psychology of Consciousness*, Ornstein (1972) proposed a new approach to conceptualizing the activity of our senses. Rather than "windows to the world," he proposes the metaphor of strongholds or filters

(a) (b)

Figure 4. (a) Raphia-Shoowa of the Kuba, Kongo, Africa, 20th century. Collection and photo: C. Sütterlin. (b) Adornment towel (Mola) of the Kuna Indians, San Blas Islands, Panama, 20th century. Photo: C. Sütterlin.

against the overflow of information. Referring to Aldous Huxley's (1954) *Doors of Perception*, he sees the senses as "systems of data reduction." The notion of filters and valves, thus of confinement, tried to replace that of a limited participation in our sensory environment. Whenever we aim to reduce and simplify information, we have to construct identifiable schemes and symbols out of relevant occurrences and these schemes should contain the features under which we can *recognize* (i.e., compare them with the ones already stored; Zeki, 1998, 1999). Analogy is an intended result. Regular features may be more easily memorized, and this could be explained by the fact that our brains may not be able to provide defined neurons for every visual event in its different appearances (Olshausen & Field, 2004; Quian Quiroga, Kreiman, Koch, & Fried, 2008; Thomas & French, 2017). Simplification and schematization in artistic portrayal hence seem to answer to a strict urge of our sensory outfit.

Categorial Perception and Schema Representation

The early line engravings of archaic humans were followed by the emergence of more complex figures. In late Paleolithic caves, we find the first representations that we can label as figures. Big animals are depicted in

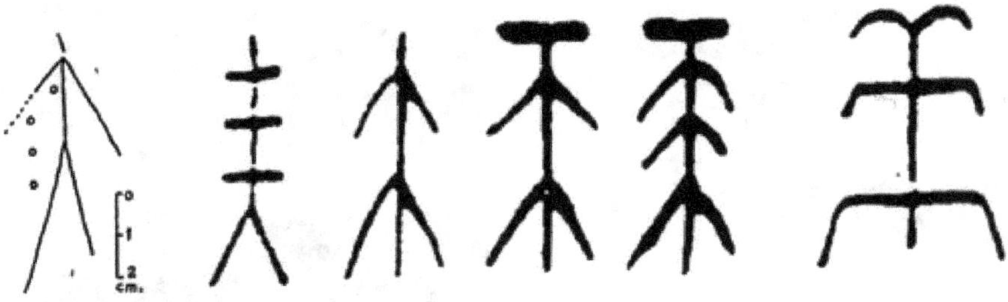

(a) (b)

Figure 5. (a) Schematized anthropomorphic engraving. Magdalenian, ca. 13,000–9000 B.P. Grotte Sainte Eulalie, Ariège, France. (Glory, 1948. 20). (b) Stylized representations of the human figure. Bronze Age, Almaden, Spain. (Glory, 1948, p. 96).

recognizable contours and proportions with even some shadings evoking form. They are admired with good reason as the first masterpieces of art history and attract visitors from all over the world. The human figures, however, appear definitely as underrepresented and less remarkable. In an amazing consistency, they are set in the most vigorous formal reduction of a stick figure. No further attributes are given except sometimes a cane or weapon. Mostly, they are shown in a frontal position. This appears like a stereotyped graphic formula in many different cultures (Figure 5).

Again, both types of motives underly a formal reduction and schematization. But whereas the animal representations include contours and contrasts that come close to a real model, the human ones appear as ideograms, containing the utmost general features characterizing the physical genus homo: the vertical erection of paired limbs along a symmetrical axis with the small sign of a head on top. The limbs are shown in a distinctly spread position. All in all, this formula contains a minimum of what all humans have in common. It is a pure formal schema.

Surprisingly enough, the same pictorial schema arose when unexperienced members of traditional cultures were asked to draw their first design of a human figure. Members of the Yanomami Indians (Venezuela), the Eipo (Western New Guinea), and the Himba (Namibia) groups produced the same reduced formula of the human body on a sheet (Figure 6). The individuals did not take the opportunity to draw after concrete body models of group members being around but conceived this very plain model. As the faces and bodies of individual humans provide a lot of unique

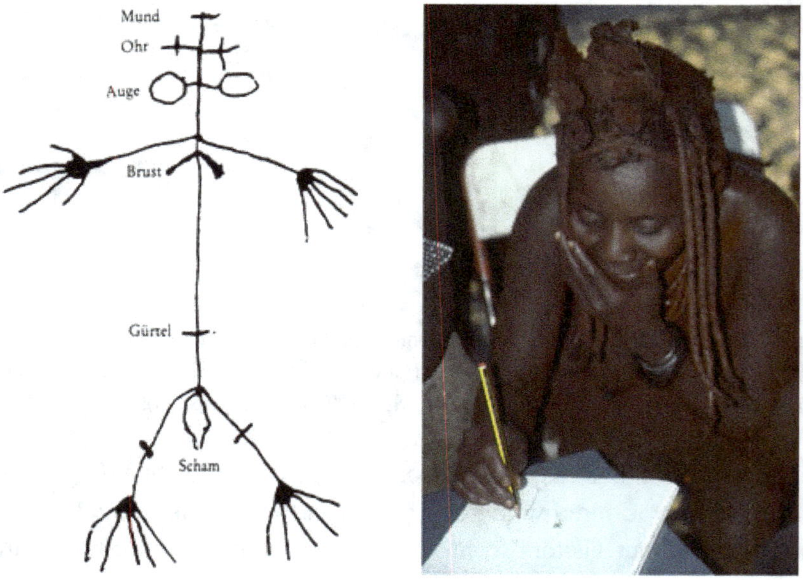

(a) (b)

Figure 6. (a) First design of a woman by an Eipo man, Western New Guinea. Private collection. I. Eibl-Eibesfeldt. (b) First drawing of a human figure by a Himba woman, Namibia. Photo: C. Sütterlin.

information, the body schema appears as a high-grade graphical abstraction. Even in modern art, this human formula is carried on in the works of Paul Klee, Joan Miro, A. R. Penck, and Ugo Dossi (Figure 7). There seems to be a very distinctive and consistent idea of what makes a pattern of lines look like a human figure and what should be the relevant features for its recognition: frontality, paired limbs, symmetry along a central axe. It is a universal concept.

It is really amazing that when humans create a token or logogram of the subject, it looks everywhere the same, allowing for the immediate recognition of the schema. There exists no evident cultural or political necessity to create an abstract or reduced picture of something in a compatible way. (Artists could potentially omit whatever detail.) Actually, this schema is something we cannot see as such since it is underlying the concrete and visible individual bodies as a hidden plot and structure. It seems to produce a kind of cognitive map for the recognition of the human

Figure 7. Ugo Dossi, I am. 2004. So-called automatic drawing with the result of a human figure in the form of a reduced body schema, with friendly permission of the artist.

body. Therefore, Aristotle was right in expecting art to reproduce more than bits and pieces: It reproduces what we really are able to perceive and what comes much closer to the "ideas" than Plato expected. It is not just the mind or art that transcends and explains reality but our perception (Sütterlin, 2014).

The same is true for children's drawings. The phase mentioned earlier as the "scribbling phase" is replaced at the age of about 3 years by the so-called schema phase, which predominates until school age. These schemas or prototypes are not necessarily influenced by learning. As Poeck &Orgass (1964) demonstrate, most of the 4-year-old children who are able to discriminate verbally neck and human torso draw the so-called *Kopffüssler* (i.e., the human figure consisting of a head with feet annexed without neck and torso). This means that the prototypes are more conservative than is possible for a child to know (Piaget & Inhelder, 1971; Serjouie, 2011). It has also been demonstrated that the ability to distinguish biological motion (in point-light) for humans and non-humans appears as early as 3 years of age (Pavlova, Krägeloh-Mann, Sokolov, & Birbaumer, 2001). If there is any addition to the schema, we observe big spread hands, the male sexual organ, or a stick, in one hand. All these graphic attributes, again, are quite stereotyped and can be called expressive signs or signals. Big hands, especially when upheld and with fingers spread, are perceived as an avoidance gesture in all cultures and are performed even in deaf- and blind-born children as a refusal gesture (Eibl-Eibesfeldt, 1973). Phallic display and the carrying of sticks are classical male-imposing gestures to impress other males. Even in the very abstract graphic reduction, some expressive cues of communication are involved.

Picking out what is typical and repeatable in a given environment follows the way our brain simplifies and schematizes incoming information. But the purely cognitive aspects of data reduction do not satisfy the affective and appealing ones. When information theory in the 20th century first tried to quantify the amount of data reduction adequate for a sender and a receiver system (Shannon, 1948), this sounded like an opening chance for new aesthetic theories (Bense, 1965; Moles, 1958). But the theory was originally adapted for passive transmission systems and was ignorant of how visual and auditive channels would weigh the substantial importance of incoming information. Perception, however, acts like actively choosing and bringing out messages from interfering signals masking the message. At this point, ethological research could accommodate the discussion.

All organisms seem to be highly selective to species-specific properties. Stimulus-specific underpinnings for schema have been identified in the field of animal ethology and could also possibly be identified in humans (Eibl-Eibesfeldt, 1967, 1984; Lorenz, 1935). For animals, Gross, Bruce, Desimone, Fleming, and Gattass (1981) and Desimone, Albright, Gross, and Bruce (1984) provide evidence of a differential score of responses to stimuli resulting from single-cell deduction of a macaque's brain. Faces of conspecifics and humans, simple and individualized, with eyes mapped and without, were mixed with other motives, like hands and nonsense patterns. Faces of conspecifics and humans with marked eyes obtained significantly higher neural firings compared to other stimuli. The frog's eye answers preferentially to small black moving stimuli resembling a fly (Lettvin, Maturana, McCulloch, & Pitts, 1959). The body has also been found to be represented in selective regions in macaques (Popivanov, Jastorff, Vanduffel, & Vogels, 2014; Wachsmuth, Oram, & Perrett, 1994). Similar adaptations were expected in humans too, and psychology and human ethology followed the tracks. Research, for example, has demonstrated that the processing of human bodies may be represented in selective regions in the brain (Downing, Jiang, Shuman, & Kanwisher, 2001; Downing & Peelen, 2011; Peelen & Downing, 2005; Pourtois, Peelen, Spinelli, Seeck, & Vuilleumier, 2007), and the representations of point-light bodies and more real-like bodies are significantly similar to each other (Peelen, Wiggett, & Downing, 2006).

Categorical Feature Sensitivity and Semantic Schemata

Categorization serves as a clear discrimination and identification of perceived matter in terms of a valid interpretation. In order to recognize immediately whether an impending object is simply a figure or an animal, a human being, a plant, or a tree, internalized reference patterns (i.e., prototypes, templates) are required that contain general information of all comparable events (e.g., Dittrich, Lea, Barrett, & Gurr, 1998; Goldstone & Hendrickson, 2010; Pavlova *et al.*, 2001; Rosch, 1973, 1975; Thompson & Oden, 2000; Wittgenstein, 1953). These templates are generally acquired by statistical learning (e.g., Yin & McBride, 2015). They encode all information necessary for identification, retaining the congruent features of stored figures according to the hypothesis of similarity. Against this prototype incoming information is compared (Eibl-Eibesfeldt, 1984; Eibl-Eibesfeldt & Sütterlin, 2008; Sütterlin, 2003). What we learn from the scores of Gross *et al.* (1981) and Desimone *et al.* (1984) is that sensory processing serving reduction of data complexity is not simply working in quantitative — or physical — terms as a random process. Perception does not function like the discharge of a dustbin, which would be an indifferent choice; rather selective sensitivity is a process tuned to semantic qualities and thus highly discriminative and undemocratic. Analyzing and categorizing information means transforming it into *meaningful* information. It is highly feature-selective also, in terms of semantic content.

Early ethological writings, like that of von Uexküll (1909), describe these perceptual constraints and limitations as "schemata" that are "no products of the environment, but instruments of the brain." Involved receptors are neuronally connected with the higher motoric centers to release an adequate action and need to be as unambiguous as possible. They are stored in the brain as simplified neuronal images of objects in the environment (Lorenz, 1935; Schleidt, 1962). All this applies to animals but in parts could be explored and extended to human perception and behavior as well. Lorenz (1943) described different perceptual schemata in humans, from chemical (odor) and acoustical to visual ones, like the famous "*Kindchenschema*" (baby schema), referring to certain features and proportions of the small child's head and face. A bulged forehead, big eyes, chubby cheeks, and so forth are easily adapted for dummy experiments and are even prone to exaggeration (Hückstedt, 1965). A whole industry of dolls and comics profits from the exaltation of sensations provoked by this visual schema whose biological significance lays in the close caring relation of a mother to her small child in need of protection. This preference for baby schema has been corroborated by lines of later research, indicating that the baby schema is related to higher perceived cuteness (Borgi, Cogliati-Dezza, Brelsford, Meints, & Cirulli, 2014; Little, 2012; Sanefuji, Ohgami, & Hashiya, 2007), more motivation for caretaking (Glocker, Langleben, Ruparel, Loughead, Gur, & Sachser, 2009), and possibly beauty (Kuraguchi, Taniguchi, & Ashida, 2015); viewing the baby schema also potentially activates the reward system of the brain (Glocker, Langleben, Ruparel, Loughead, Valdez, *et al.*, 2009). We can call this feature preference "aesthetics."

Other perceptual schemata refer to body contours in terms of sexual differentiation. Skrzipek (1978) showed in repeated experiments that body contours are clearly correlated to male or female bodies by sheer proportions (i.e., without marking the secondary sexual characteristics) by children and adults. The male body schema thereby was presented with large shoulders and slim hips, and the female schema with narrow shoulders and large hips. The preferences for one or the other changed during the age of puberty and stayed invariant (Singh, 1993). Our perception seems easily trapped by dummies. Again, internal concepts of a male or female body schema seem to serve as prototypes (templates) against which real occurrences are aligned (Lutz *et al.*, 2013).

The differential body schema seems to represent a complex stimulus category. From the earliest figures in paleolithic art up to recent examples, we can observe the exaltation of this schema in two-dimensional and three-dimensional art. A strong paleolithic example is the one found on a mammoth bone from Predmost (Moravia, 25,000–30,000 B.P.). The high degree of stylization prevails, but the sex of the figure is clearly marked by the breasts and the wide emphasis of the hip section: two hot spots of female body signals (Herrmann & Ullrich, 1991). We may think that artistic simplification and abstraction need the precursor model of a full version of the realistic picture from which they can reduce. But the abstract prototype is evidently stored in the brain. Other plastic figurines, like those from Dolni-Vestonice (Czech Republic), Laussel, and Lespugue (France) from the late Paleolithic era, are sculpted block-like as female schemata and called Venuses (Figure 7). Emphasized again are the primary and secondary sexual attributes. They function as optical triggers.

Since the Neolithic, the female waistline is highly laced into small and smallest curves. Famous are the "Violin Idols" from Hissarlik (Figure 8)

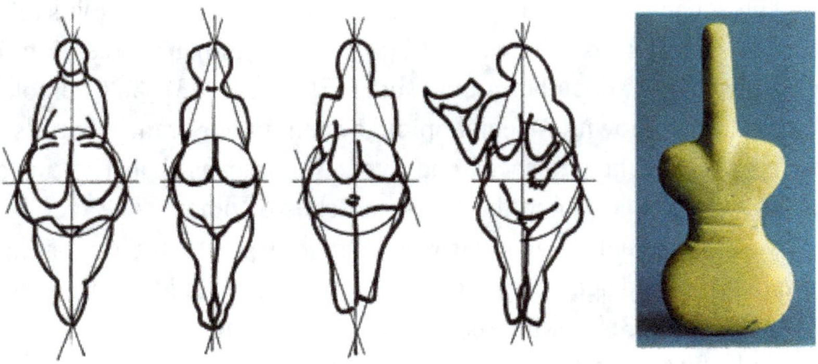

(a) (b)

Figure 8. (a) Paleolithic female body-contour schema. So-called "Venuses" from France, Russia, and the Czech Republic. From left to right: Lespugue, Kostenki, Dolni Vestonice, Laussel. (Leroi-Gourhan, 1971, p. 112). (b) Neolithic female idol (violin form) from Pelos-culture, Cyclades, Greece. 3000–2650 B.C. Marble. 11.5 cm. Badisches Landesmuseum Karlsruhe. Photo: Thomas Goldschmidt, with friendly permission.

and the multi-curved lines of a female contour presented in sophistically twisted body positions we admire in Giovanni da Bologna's "Astronomy" (1572). Cultural history played with this model partly depending on the needs or inhibitions of sexual selection pressures (Sütterlin, 2018). Heroic male representations on the other side profit from the overstatement of shoulder width, even in the highly abstract incisions on cave walls. Statues of Gods and athletes in Ancient Greece show the imposing male torso dimensions as well as portraits of kings and emperors, like in the one of Charles V of Habsburg by Jakob Seisenegger (1532), thereby imitating the real dress code of the court at that time (Eibl-Eibesfeldt & Sütterlin, 2008).

Amazingly enough, the human face did not play a similar role in art for a very long time. As the face belongs to the most individual, revealing, and communicative part of the human body, the suppression or neglect of this feature in early art must be founded on reasons of cultural history. As humans, we like to look into other faces to check the individual, their mood, and maybe their intentions. Face-to-face is the most relevant factor of social communication (Eibl-Eibesfeldt, 1984). A special cortical region is selectively activated for face perception in the human brain (Furey *et al.*, 2006; Kay & Yeatman, 2017; Tsao, Freiwald, Tootell, & Livingstone, 2006; Zhu, Zhang, Luo, Dilks, & Liu, 2011) and in macaque brains (Tsao, Schweers, Moeller, & Freiwald, 2008). Also, human perception has adapted perfectly to decipher mimic signs (Ekman, 1973). As Chelnokova *et al.* (2016) has shown, a special opioid system in our brains controls our social attention to human faces and eyes. This means that it is, in fact, rewarding to see and watch the faces and eyes of others. The face bears the most subtle signal of what our genes cannot show to others: for example, fitness, sex, and age (among others; e.g., Han, 2018; Ma & Han, 2009; Scheele *et al.*, 2013; Tsukiura & Cabeza, 2008). The appearance of individualized human representations arose late in art history, roughly with the first mummy portraits in Egypt. But from then, the evolution of this genre was unstoppable.

Another factor of the human face, attractivity, is scored widely in our brain and evolves with the idealized representations of gods in Ancient Greece. Whatever is associated with standardization and typecasting can be found in the galleries of sculptures in all museums. And this form of facial standardization corresponds again with the way we perceive attractive faces. Johnston and Franklin (1993) have demonstrated that the *beau* ideal coincides significantly with the average of faces in a population. This is the way, again, our brain stores the many different faces seen. Attractive faces also activate the human reward system (Cloutier, Heatherton, Whalen, & Kelley, 2008; Etcoff, 1999). The opioid system may mediate the perceived attractiveness of faces (Chelnokova, Laeng, Eikemo, Riegels, & Loseth, 2014). Thus, the need to create idealized faces wherever humans can manipulate reality up to its energizing measures seems to have inspired the arts for centuries in European and non-European cultures. Facial beauty rewards the beholder. Carved heads by the Yoruba (Benin) or the Baule (Ivory Coast) show the very same ambition for regularity and symmetry as the Athena of Phidias (Figure 9). Even individualized portraits are in general straightened into more regular traits. The allure of sophisticated and interesting individual physiognomies was the interest of 17th- and 18th-century European art history (Jean Antoine Houdon, Quentin de La Tour). Art can explore our sensitivity to specific cues and, even more, contributes to the extension of our common visual habits. Individual faces showing all the interesting characteristics

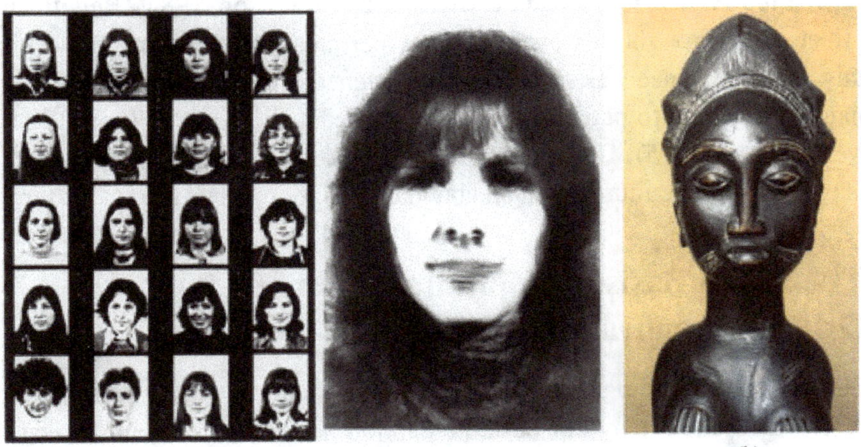

(a) (b)

Figure 9. (a) Female average face (right), morphed from 20 individual faces (left) by H. Daucher, University of Munich, with friendly permission). (b) African sculpture of the Baule, Ivory Coast, Africa Photo: I. Eibl-Eibesfeldt, with friendly permission.

(e.g., asymmetry) we issue as extraordinary are more likely to be stored in our biographic memory (Light, Hollander, & Kayra-Stuart, 1981; Wiese, Altmann, & Schweinberger, 2014; see also Lin, Lendry, & Ebner, 2016; and Lin, Fischer, Johnson, & Ebner, 2020, for a possible enhancing effect of attractiveness of faces on the memory, indicating a U-shape pattern). Exaggerated mimic expressions, however, were always restrained if not for expressive movements in apotropaic figures (Eibl-Eibesfeldt & Sütterlin, 1992). In focusing on a few selective features in a repeated or exaggerated way, art even repeats strategies of behavioral ritualization at ornamental and expressive levels (Sütterlin, Schiefenhövel, Lehmann, Forster, & Apfelauer, 2014). Examples are found in the exaltation of the eye motive through the ages to which humans like many other organisms respond highly attentively (Koenig, 1975; Sütterlin & Eibl-Eibesfeldt, 2015). The eye, even as an autonomous motive, can be taken as a ritualized gazing behavior.

Another feature we pay attention to is certainly nature itself with its green plants and colored flowers. There is probably no culture that ignores the subject — in decoration as well as in the painted or sculpted arts. Plants and flowers are regarded as indicators of a healthy environment in which a good life is possible. Ethologists call it phytophilia or biophilia (Eibl-Eibesfeldt, 1984; Kellert & Wilson, 2013; Wilson, 1984). Beneficial effects were brought into evidence by experiments with patients in hospitals. Convalescence was significantly shortened if there was sight of green nature compared to patients in rooms without such a view (Ulrich, 1984; Vedder *et al.*, 2015). The optical surrogate of such a soothing environment is even found in cultures with an abundance of indigenous tropical nature.

The Social Basis of Communication and its Drive for Innovation

Selective attention to special features in our environment, as said before, is provided already by our basic perception. From the moment we learn that the question of aesthetics begins much earlier than in our elevated taste and emotions, and much earlier than in art, we may understand that in art two systems are talking to each other in a most differentiated and subtle way: our perception and artistic production. All these preferences can also be called visual prejudices in the sense that no prior experiences are necessary for the acquirement of such a perceptual bias. Small children crawling on a glass top laid over the edge of a table stopped before the edge, even though they could feel the continuation of the glass with their hands (Gibson & Walk, 1960). Evidently, they interpreted the end of the table as a "visual cliff" without any foregoing plunge. And others, only 2 weeks old and fixed in a small chair, turned their heads away, blinking their eyes when they saw a shade heading symmetrically toward them. They seemed to anticipate a collision (Ball & Tronick, 1971). The term "prejudice" could easily evoke the notion of something narrowly subjective in an ontogenetic sense. The term, however, as the examples given above can show, in a psychophysical or philosophical sense, indicates not an individualized subject, but a *species-specific* subject, as opposed to an object called environment or reality. The term includes phylogeny, a collective mind, and reflects anthropological universals (Bao & Pöppel, 2012).

It seems that, whenever humans are able to, they like to stimulate themselves, their senses, moods, and emotions. But, as the artist does not talk exclusively to themself but to a community of others as well, they move in a kind of public space. Returning once more to the first "artist" of Bilzingsleben, an early ancestor of modern homo, we can simply suppose that the regular pattern on the elephant bone attracted the attention of many more people, as it attracted the attention of those who excavated and saved it. As already Plato pointed out in *Politeia* (3, 401 ff), art should have an educative effect on people aspiring for higher functions in a political system — this being possible on grounds of common values. Aristotle proposed that art can address effects quite directly as a target and actually should do so. It can trigger emotions by perceptual cues. And his notion of catharsis lives from emotions we all share. Invention and artistic freedom — innovation — are allowed to a certain degree of plausibility.

For a common understanding or communication, some basic agreement about the signification of signs and codes is required (Cassirer, 1923/1980; Hockett, 1960/1982; Pöppel & Bao, 2011; Tomasello, 2010). Regarding this, there is a high consent in how we analyze factors of our environment in our sensory perception. In the first step, an impression is simplified and generalized. Only in further steps it proceeds to analyze more precisely if there is time. In creating a pictorial language, we meet a pre-existing perceptive understanding. This coherence is important. Accordance of similar concepts about reality in art as well as in our perception reduces uncertainty (i.e., creates emotional security and trust; Bao *et al.*, 2016; Halcour, 2002; Sütterlin, 2013) and the discovery of similar meanings evokes pleasure.

For a very long time, this circular connection ensured an accordance between art and the public that cemented social cohesion. If every individual would just cultivate their own concepts about the visual environment, there could be no basic consensus about a common and collective understanding. As mentioned before, neurons in the brain do not represent single events in a one-to-one manner for most of the time; rather, events are represented by the activation pattern of neuron assemblies. It is for sure, however, that single exceeding moments are stored in our biographical mind. Art can even try to restore such exceeding moments to a collective memory. Privacy is becoming public. Art took up to account for change and for learning to see more than we can share easily. To fit optimally into an established circuit of communication feeds our rewards system. But art is playing with both the known and comprehensible as well as with new insights, cultural challenges, and surprises.

How to turn the screw of cultural evolution on both ends? Responsiveness to a single stimulus can be complex in individual adults, and art has evolved to highly elaborated cultural and individual styles as well that seem to transcend any defined common view, especially in modern times. Think of how Caravaggio guided our eyes to perceive bodies and colors in the most dramatic mise-en-scenes of light contrasts. How would we appreciate a painting of Claude Monet without accepting that his view of Reims Cathedral was directed by the changing lights under which the façade was exposed to his eyes? And who would ever recognize Dora Maar in the squeezed deformations in which Picasso liked to portray her if we would not have in mind that a profile and a frontal view of the face usually do not appear at the same time? As Janet Flanner reports, "Cubism was the very occult way of regarding what you could not see" (Adato, 2010). But this might be true for many aspects of vision per se. As the

Italian artist Giorgio Morandi commented "I believe that nothing can be more abstract, more unreal, than what we actually see. We know that all that we can see of the objective world as human beings, never really exists as we see and understand it." (Roditi, 1990). Perception can transcend reality. Certainly, it took years for the appreciation of modern art. And there seemed not much space for vigorous innovations over a long time.

In the 19th century, the fugitive subjective sensation of the moment seemed to adopt all aesthetic considerations. "From the eye to the hand" without reflecting on more general issues of the motive became the head-note of impressionism — as if all Platonic prejudices of the ephemeral enterprise of mimesis would finally come true. But again, this was just a new mode of visual communication among the artists and the beholders. Mimesis does not perform between art and reality but between art and our perception of reality in the visual brain. And it turned out that our shared visual concepts are quite resistant to the seducements of artistic innovation. The return to abstract painting (Bao, Yang, *et al.*, 2017) seemed to be absorbed by humans' visual Gestalt competences. We adapted to monochrome painting, to "Informel," conceptual art, and so forth without losing our normal visual habits. Even if communication between art and the public seemed to run away at one side, we seem to expect more crazy events in the art market. How come?

Communication is not only a system to answer questions but to question answers. Differences in the habitual performance of perceived things break the rules but still brush our brains — even against the fur. How else would it be possible to talk of distortions, divergences, and deviations if not measured in reference to stable neuronal frames of human vision? Deviation can even activate our search for proto- or archetypical information enclosed and reward the system in a different way. It may moreover activate the models and templates themselves. Templates' formation stays active even if culture and art may establish contradictions. It is not the templates that are changing but the cultural answers to them. The brain is a stratified system. We do not forget the opulent appearance of buffalos in the cave of Lascaux when looking at a picture by Miriam Cahn in 2018. It might be an evolutionary stable strategy to follow-up new ways of communication. Training in active art design could affect neural connections in the brain (Bolwerk, Mack-Andrick, Lang, Dörfler, & Maihöfner, 2014).

And a visit to an art gallery or museum may also reduce stress and bring us to a different mood (Clow & Fredhoi, 2006; Mastandrea *et al.*, 2019; Mastandrea, Fagioli, & Biasi, 2019; Silvia, 2005).

For a long time, art seemed to assist the visual system, intensify contrasts, establish and activate three-dimensional spatiality in paintings, play with visual illusion effects, and so forth. The onset of modern art gradually loosened the nexus. Body contours in paintings of Pierre Bonnard's *Bathing Women* got blurred and melted with the pink flowery wallpaper in the background, the pictorial space began to disorganize and pile onto the surface. And, maybe the landscape behind the hospital of Saint-Paul by Vincent Van Gogh just wanted to restitute the many green colors of a meadow after a storm that, in our own experience or memory, would maybe amalgamate into a more general or trivial green after a timespan. If consent between our everyday way of perceiving and what we perceive in paintings reduces uncertainty and creates pleasure (formally or semantically), artistic innovations may increase uncertainty and cause productive tensions. But human vision remains in search of sense and order even in view of art that denies it. This can be demanding but also stimulating. Art is certainly challenging our sensory system but can contribute to its extension and diversification, even enlarging our empathy and spectrum of emotions (Shapiro, Rucker, & Beck, 2006). It has become the subject of new research on happiness (de Muynck, 2019). Art will never be able to destroy what is the result of a long evolution. The frictions between the two levels of adaptations, natural and cultural, inspire new attitudes toward our environment and the arts.

Conclusion

In this chapter, we intended to bring readers through a journey from the earliest art form and (probably also one of the earliest) art theories to modern arts, through the lens of ethology, anthropology, art theory, and neuroscience. The Platonian idea that art is *mimesis*, according to our reasoning, was problematic. On the contrary, the Aristotelian theory of art as a product of complexity reduction, as we have argued, may reflect an anthropological universal for art in different cultures in the course of evolution. Schema and categorical perception could be considered as manifestations of this argument. Besides, aesthetics and perception are deeply connected, which provides us with a unique window for connecting theories of perception and art. Modern art seems to have raised challenges for the Aristotelian theory, yet we argue that the biological basis of perception and aesthetic perception, as a result of long evolution, is unlikely to be destroyed quickly by modern arts.

References

Adato, P. M. Writer, Director, and Producer (2010). *Paris: The luminous years: Towards the making of the modern* [Motion Picture, USA: Thirteen for WNET.org, The Elegant Image LLC, and Ina & Arte France].

Al-Shamma-Gerber, N. (1984). *Der formale Bestand der Schemazeichnung und der Aufbau von Bedeutungsträgern aus definierten Grundmustertypen* [*The Formal Existence of the Schematic Drawing and the Structure of Meaning Carriers from Defined Basic Pattern Types*], Unpublished Doctoral Thesis, LMU, Munich, Germany.

Arnheim, R. (1965). *Kunst und Sehen. Eine Psychologie des schöpferischen Auges* [*Art and Visual Perception: A Psychology of the Creative Eye*]. Berlin, Germany: de Gruyter.

Avram, M., Gutyrchik, E., Bao, Y., Pöppel, E., Reiser, M., & Blautzik, J. (2013). Neurofunctional correlates of esthetic and moral judgments. *Neuroscience Letters*, 534, 128–132. https://doi.org/10.1016/j.neulet.2012.11.053.

Ball, W. & Tronick, F. (1971). Infant responses to impending collision, optical and real. *Science*, 171, 818–820. https://doi.org/10.1126/science.171.3973.818.

Bao, Y. & Pöppel, E. (2012). Anthropological universals and cultural specifics: Conceptual and methodological challenges in cultural neuroscience. *Neuroscience and Biobehavioral Reviews*, 36, 2143–2146. https://doi.org/10.1016/j.neubiorev.2012.06.008.

Bao, Y., von Stosch, A., Park, M., & Pöppel, E. (2017). Complementarity as generative principle: A thought pattern for aesthetic appreciations and cognitive appraisals in general. *Frontiers in Psychology*, 8, 727. https://doi.org/10.3389/fpsyg.2017.00727.

Bao, Y., Yang, T., Lin, X., Fang, Y., Wang, Y., Pöppel, E., & Lei, Q. (2016). Aesthetic preferences for Eastern and Western traditional visual art: Identity matters. *Frontiers in Psychology*, 7, 1596. https://doi.org/10.3389/fpsyg.2016.01596.

Bao, Y., Yang, T., Zhang, J., Zhang, J., Lin, X., Paolini, M., Pöppel, E., & Silveira, S. (2017). The "third abstraction" of the Chinese artist LaoZhu: Neural and behavioral indicators of aesthetic appreciation. *PsyCh Journal*, 6(2), 110–119. https://doi.org/10.1002/pchj.167.

Baumgarten, A. G. (2007). *Ästhetik* (D. Mirbach, Ed. & Trans.). Hamburg, Germany: Felix Meiner (Original Work Published 1750).

Behm-Blancke, G. (1983). Altpaläolithische Gravierungen von Bilzingsleben, Kr. Artern [Lower paleolithic engravings of Bilzingsleben]. *EAZ Ethnographisch-Archäologische Zeitschrift*, 24, 304–320.

Bense, M. (1965). *Aesthetica: Einführung in die neue* [*Aesthetics: Introduction into a New Aesthetics*]. Baden-Baden, Germany: Agis.

Bohrn, I. C., Altmann, U., Lubrich, O., Menninghaus, W., & Jacobs, A. M. (2013). When we like what we know: A parametric fMRI analysis of beauty and familiarity. *Brain and Language*, 124(1), 1–8. https://doi.org/10.1016/j.bandl.2012.10.003.

Bolwerk, A., Mack-Andrick, J., Lang, F. R., Dörfler, A., & Maihöfner, C. (2014). How art changes your brain: Differential effects of visual art production and cognitive art evaluation on functional brain connectivity. *PLoS One*, 9(7), e101035. https://doi.org/10.1371/journal.pone.0101035.

Borgi, M., Cogliati-Dezza, I., Brelsford, V., Meints, K., & Cirulli, F. (2014). Baby schema in human and animal faces induces cuteness perception and gaze allocation in children. *Frontiers in Psychology*, 5, 411. https://doi.org/10.3389/fpsyg.2014.00411.

Boysen, S. T., Berntson, G. G., & Prentice, J. (1987). Simian scribbles: A reappraisal of drawing in the chimpanzee (Pan troglodytes). *Journal of Comparative Psychology*, 101(1), 82. https://doi.org/10.1037/0735-7036.101.1.82.

Brieber, D., Leder, H., & Nadal, M. (2015). The experience of art in museums: An attempt to dissociate the role of physical context and genuineness. *Empirical Studies of the Arts*, 33(1), 95–105. http://doi.org/10.1177/0276237415570000.

Brieber, D., Nadal, M., & Leder, H. (2015). In the white cube: Museum context enhances the valuation and memory of art. *Acta Psychologica*, 154, 36–42. http://doi.org/10.1016/j.actpsy.2014.11.004.

Brieber, D., Nadal, M., Leder, H., & Rosenberg, R. (2014). Art in time and space: Context modulates the relation between art experience and viewing time. *PLoS One*, 9(6), e99019. https://doi.org/10.1371/journal.pone.0099019.

Burke, E. (1989). *Philosophische Untersuchungen über den Ursprung unserer Ideen vom Erhabenen und Schönen*. (F. Bassenge, Trans. & W. Strube, Ed.). Hamburg, Germany: Meiner (Original Work: *A Philosophical Inquiry into the Origin of Our Ideas of the Sublime and Beautiful; with an Introductory Discourse Concerning Taste*. Published 1757 in London).

Carrasco, M. (2011). Visual attention: The past 25 years. *Vision Research*, 51(13), 1484–1525. https://doi.org/10.1016/j.visres.2011.04.012.

Cassirer, E. (1980). *The Philosophy of Symbolic Forms (Volume One: Language)*. (R. Manheim, Trans.). New Haven, CT: Yale University Press (Original Work Published 1923).

Changizi, M. A., Zhang, Q., Ye, H., & Shimojo, S. (2006). The structures of letters and symbols throughout human history are selected to match those found in objects in natural scenes. *American Naturalist*, 167(5), E117–E139. https://doi.org/10.1086/502806.

Chelnokova, O., Laeng, B., Eikemo, M., Riegels, J., & Loseth, G. E. (2014). Rewards of beauty: The opioid system mediates social motivation in humans. *Molecular Psychiatry*, 19, 746–747. https://doi.org/10.1038/mp.2014.1.

Chelnokova, O., Laeng, B., Løseth, G., Eikemo, M., Willoch, F., & Leknes, S. (2016). The μ-opioid system promotes visual attention to faces and eyes. *Social Cognitive and Affective Neuroscience*, 11(12), 1902–1909. https://doi.org/10.1093/scan/nsw116.

Cloutier, J., Heatherton, T. F., Whalen, P. J., & Kelley, W. M. (2008). Are attractive people rewarding? Sex differences in the neural substrates of facial attractiveness. *Journal of Cognitive Neuroscience*, 20(6), 941–951. https://doi.org/10.1162/jocn.2008.20062.

Clow, A. & Fredhoi, C. (2006). Normalisation of salivary cortisol levels and self-report stress by a brief lunchtime visit to an art gallery by London City workers. *Journal of Holistic Healthcare*, 3(2), 29–32.

de Muynck, R. (2019). *A Plea for the Emotional in Art. New Insights of Behavioral Sciences and Brain Research*. Unpublished Oral Presentation, Bavarian Academy of Fine Arts, Munich.

Desimone, R., Albright, T. D., Gross, C. G., & Bruce, C. (1984). Stimulus-selective properties of inferior temporal neurons in the macaque. *Journal of Neuroscience*, 4(8), 2051–2062. https://doi.org/10.1523/JNEUROSCI.04-08-02051.1984.

Di Dio, C. & Gallese, V. (2009). Neuroaesthetics: A review. *Current Opinion in Neurobiology*, 19, 682–687. https://doi.org/10.1016/j.conb.2009.09.001.

Dissanayake, E. (1992). *Homo Aestheticus: Where Art Comes From and Why*. Toronto, Canada: Macmillan.

Dissanayake, E. (2008). If music is the food of love, what about survival and reproductive success? *Musicae Scientiae*, 12(Suppl. 1), 169–195. https://doi.org/10.1177/1029864908012001081.

Dittrich, W. H., Lea, S. E., Barrett, J., & Gurr, P. R. (1998). Categorization of natural movements by pigeons: Visual concept discrimination and biological motion. *Journal of the Experimental Analysis of Behavior*, 70(3), 281–299. https://doi.org/10.1901/jeab.1998.70-281.

Downing, P. E., Jiang, Y., Shuman, M., & Kanwisher, N. (2001). A cortical area selective for visual processing of the human body. *Science*, 293(5539), 2470–2473. https://doi.org/10.1126/science.1063414.

Downing, P. E. & Peelen, M. V. (2011). The role of occipitotemporal body-selective regions in person perception. *Cognitive Neuroscience*, 2(3-4), 186–203. https://doi.org/10.1080/17588928.2011.582945.

Eibl-Eibesfeldt, I. (1967). *Grundriss der vergleichenden Verhaltensforschung. Ethologie* [*Foundation of Comparative Behavioral Research: Ethology*]. Munich, Germany: Piper.

Eibl-Eibesfeldt, I. (1973). *Der vorprogrammierte Mensch* [*The Preprogrammed Human*]. Vienna, Austria: Molden.

Eibl-Eibesfeldt, I. (1984). *Die Biologie menschlichen Verhaltens. Grundriss der Humanethologie* [*Biology of Human Behavior: Foundation of Human Ethology*]. Munich, Germany: Piper.

Eibl-Eibesfeldt, I. & Sütterlin, C. (1992). *Im Banne der Angst. Zur Natur- und Kunstgeschichte menschlicher Abwehrsymbolik* [*Under the Spell of Fear: On the Natural and Art History of Human Defense Symbolism*]. Munich, Germany: Piper.

Eibl-Eibesfeldt, I. & Sütterlin, C. (2008). *Weltsprache Kunst. Zur Natur- und Kunstgeschichte bildlicher Kommunikation* [*World Language Art: On the Natural and Art History of Visual Communication*]. Vienna, Austria: Brandstätter.

Ekman, P. (1973). Cross-cultural studies of facial expression. In P. Ekman (ed.), *Darwin and Facial Expression*, pp. 169–222. New York, NY: Academic Press.

Etcoff, N. (1999). *Survival of the Prettiest: The Science of Beauty*. New York, NY: Doubleday.

Feustel, R. (1987). Eiszeitkunst in Thüringen [Ice age art in Thuringia]. In H. Müller-Beck & G. Albrecht (eds.), *Die Anfänge der Kunst vor 30'000 Jahren* [*The Beginnings of Art 30,000 Years Ago*]. Stuttgart, Germany: Theiss.

Fuhrmann, M. (1973). *Einführung in die antike Dichtungstheorie* [*Introduction to Ancient Poetry Theory*]. Darmstadt, Germany: WBG.

Furey, M. L., Tanskanen, T., Beauchamp, M. S., Avikainen, S., Uutela, K., Hari, R., & Haxby, J. V. (2006). Dissociation of face-selective cortical responses by attention. *Proceedings of the National Academy of Sciences*, 103(4), 1065–1070. https://doi.org/10.1073/pnas.0510124103.

Gambino, R. & Pulvirenti, G. (2020). Burke's "sensitive" sublime rooted in brain and body vs Kant transcendental concept in the German romantic discourse. *PsyCh Journal*, 10(3), 414.

Ganczarek, J., Pietras, K., & Rosiek, R. (2020). Perceived cognitive challenge predicts eye movements while viewing contemporary paintings. *PsyCh Journal*, 9(4), 490–506. https://doi.org/10.1002/pchj.365.

Gardner, R. A. & Gardner, B. T. (1978). Comparative psychology and language acquisition. *Annals of the New York Academy of Sciences*, 309(1), 37–76. https://doi.org/10.1111/j.1749-6632.1978.tb29441.x.

Gibson, E. J. & Walk, R. D. (1960). The visual cliff. *Scientific American*, 202, 64–71. https://doi.org/10.1038/scientificamerican0460-64.

Glocker, M. L., Langleben, D. D., Ruparel, K., Loughead, J. W., Gur, R. C., & Sachser, N. (2009). Baby schema in infant faces induces cuteness perception and motivation for caretaking in adults. *Ethology*, 115(3), 257–263. https://doi.org/10.1111/j.1439-0310.2008.01603.x.

Glocker, M. L., Langleben, D. D., Ruparel, K., Loughead, J. W., Valdez, J. N., Griffin, M. D., Sachser, N., & Gur, R. C. (2009). Baby schema modulates the brain reward system in nulliparous women. *Proceedings of the National Academy of Sciences*, 106(22), 9115–9119. https://doi.org/10.1073/pnas.0811620106.

Glory, A. (1948). Les peintures de l'age du métal en France méridionale [The paintings of the metal period in South of France. *Préhistoire*, 10, 7–135.

Goldstone, R. L. & Hendrickson, A. T. (2010). Categorical perception. *Wiley Interdisciplinary Reviews: Cognitive Science*, 1(1), 69–78. https://doi.org/10.1002/wcs.26.

Gombrich, E. H. (1984). *The Sense of Order*. London, England: Phaidon Press.

Goodale, M. A. & Milner, A. D. (1992). Separate visual pathways for perception and action. *Trends in Neurosciences*, 15(1), 20–25. https://doi.org/10.1016/0166-2236(92)90344-8.

Goodale, M. A. & Westwood, D. A. (2004). An evolving view of duplex vision: Separate but interacting cortical pathways for perception and action. *Current Opinion in Neurobiology*, 14(2), 203–211. https://doi.org/10.1016/j.conb.2004.03.002.

Grassi, E. (1980). *Die Theorie des Schönen in der Antike [The theory of the beautiful in antiquity]*. Köln, Germany: DuMont.

Gross, C. G., Bruce, C. J., Desimone, R., Fleming, J., & Gattass, R. (1981). Cortical visual areas of the temporal lobe. In C. N. Woolsey (ed.), *Multiple Visual Areas*, pp. 187–216. Totowa, NJ: Humana Press.

Halcour, D. (2002). *Europäische Hochschulschriften: Wie wirkt Kunst? Zur Psychologie ästhetischen Erlebens* [*European University Studies: Series VI. On the Psychological Impact of Aesthetic Experience*]. Frankfurt, Germany: Lang.

Han, S. (2018). Neurocognitive basis of racial ingroup bias in empathy. *Trends in Cognitive Sciences*, 22(5), 400–421. https://doi.org/10.1016/j.tics.2018.02.013.

Harari, Y. N. (2015). *Eine kurze Geschichte der Menschheit* [*A Short History of Humankind*]. Munich, Germany: Pantheon.

Herrmann, J. & Ullrich, H. (1991). *Menschwerdung. Millionen Jahre Menschheitsentwicklung — natur — und geisteswissenschaftliche Ergebnisse* [*Incarnation: Millions of Years of Human Development — Results from the Natural Sciences and Humanities*]. Berlin, Germany: Akademie.

Hockett, C. (1982). The origin of speech. In S. Y. Wang (ed.), *Human Communication: Language and Its Psychobiological Bases*, pp. 4–12. San Francisco, CA: W.H. Freeman (Original Work Published 1960).

Hockett, C. (1960). The origin of speech. *Scientific American*, 203(3), 88–111. https://doi.org/10.1038/scientificamerican0960-88.

Hodgson, D. (2006). Understanding the origins of paleoart: The neurovisual resonance theory and brain functioning. *PaleoAnthropology*, 2006, 54–67. https://doi.org/10.1017/s0959774306000023.

Hodgson, D. (2019a). Response to the critique by Mellet *et al.* of Hodgson's neurovisual resonance theory. *Journal of Archaeological Science: Reports*, 28, 102041. https://doi.org/10.1016/j.jasrep.2019.102041.

Hodgson, D. (2019b). The origin, significance, and development of the earliest geometric patterns in the archaeological record. *Journal of Archaeological Science: Reports*, 24, 588–592. https://doi.org/10.1016/j.jasrep.2019.02.025.

Hückstedt, B. (1965). Experimentelle Untersuchungen zum "Kindchenschema" [Experimental studies on the "baby schema"]. *Zeitschrift für Experimentelle und Angewandte Psychologie*, 12, 421–450.

Huxley, A. (1954). *The Doors of Perception*. London, England: Flamingo.

Ishizu, T. & Zeki, S. (2014). A neurobiological enquiry into the origins of our experience of the sublime and beautiful. *Frontiers in Human Neuroscience*, 8, 891. https://doi.org/10.3389/fnhum.2014.00891.

Johnston, V. & Franklin, M. (1993). Is beauty in the eye of the beholder? *Ethology and Sociobiology*, 14, 183–199. https://doi.org/10.1016/0162-3095(93)90005-3.

Jung, R. (1971). Kontrastsehen, Konturbetonung und Künstlerzeichnung [Perception of contrast, emphasis on contour, and artistic drawing]. *Studium Generale*, 24, 1537–1538.

Junker, T. (2013). *Die Evolution der Phantasie. Wie der Mensch zum Künstler wurde* [*The Evolution of the Imagination: How Man Became an Artist*]. Stuttgart, Germany: Hirzel.

Kandinsky, W. (1973). *Punkt und Linie zur Fläche* [*Point and Line to Plane*]. Bern, Switzerland: Benteli (Original Work Published 1926).

Kawabata, H. & Zeki, S. (2004). Neural correlates of beauty. *Journal of Neurophysiology*, 91(4), 1699. https://doi.org/10.1152/jn.00696.2003.

Kay, K. N. & Yeatman, J. D. (2017). Bottom-up and top-down computations in word- and face-selective cortex. *eLife*, 6, e22341. https://doi.org/10.7554/eLife.22341.

Kellert, S. R. & Wilson, E. O. (2013). *The Biophilia Hypothesis*. Washington, DC: Island Press.

Koenig, O. (1975). *Urmotiv Auge* [*The Eye as a Primordial Motive*]. Munich, Germany: Piper.

Kuraguchi, K., Taniguchi, K., & Ashida, H. (2015). The impact of baby schema on perceived attractiveness, beauty, and cuteness in female adults. *SpringerPlus*, 4(1), 164. https://doi.org/10.1186/s40064-015-0940-8.

Lelièvre, P. & Neri, P. (2019, August). A deep learning model of orientation perception and judgement. Paper presented at the *European Conference on Visual Perception*, Leuven, Belgium. http://pierrelelievre.free.fr/ftp-plelievre-static/ECVP2019.pdf.

Leroi-Gourhan, A. (1971). *Prähistorische Kunst. Die Ursprünge der Kunst in Europa* [*Prehistoric Art: The Origins of Art in Europe*]. Stuttgart, Germany: Herder.

Lettvin, J. Y., Maturana, H. R., McCulloch, W. S., & Pitts, W. H. (1959). What the frog's eye tells the frog's brain. *Proceedings of the Institute of Radio Engineers*, 47, 1940–1951. https://doi.org/10.1109/JRPROC.1959.287207.

Light, L. L., Hollander, S., & Kayra-Stuart, F. (1981). Why attractive people are harder to remember. *Personality and Social Psychology Bulletin*, 7(2), 269–276. https://doi.org/10.1177/014616728172014.

Likova, L. T. & Tyler, C. W. (2008). Occipital network for figure/ground organization. *Experimental Brain Research*, 189(3), 257. https://doi.org/10.1007/s00221-008-1417-6.

Lin, T., Fischer, H., Johnson, M. K., & Ebner, N. C. (2020). The effects of face attractiveness on face memory depend on both age of perceiver and age of face. *Cognition and Emotion*, 34(5), 875–889. https://doi.org/10.1080/02699931.2019.1694491.

Lin, T., Lendry, R., & Ebner, N. C. (2016). Face likeability mediates the memory-enhancing effect of face attractiveness in young but not older adults. *Memory*, 24(10), 1396–1406. http://doi.org/10.1080/09658211.2015.1117109.

Lindauer, M. S. (1987). Perceived and preferred orientations of abstract art. *Empirical Studies of the Arts*, 5(1), 47–58. https://doi.org/10.2190/K1X2-X4VJ-6YN9-BKD8.

Little, A. C. (2012). Manipulation of infant-like traits affects perceived cuteness of infant, adult and cat faces. *Ethology*, 118(8), 775–782. https://doi.org/10.1111/j.1439-0310.2012.02068.x.

London, J. (2002). Cognitive constraints on metric systems: Some observations and hypotheses. *Music Perception*, 19(4), 529–550. https://doi.org/10.1525/mp.2002.19.4.529.

Lorenz, K. (1935). Der Kumpan in der Umwelt des Vogels [The mate in the environment of birds]. *Journal of Ornithology*, 83, 137–213. https://doi.org/10.1007/BF01905355.

Lorenz, K. (1943). Die angeborenen Formen möglicher Erfahrung [The innate conditions of possible experience]. *Zeitschrift für Tierpsychologie*, 5(2), 235–409. https://doi.org/10.1111/j.1439-0310.1943.tb00655.x.

Lutz, A., Nassehi, A., Bao, Y., Pöppel, E., Sztrókay, A., Reiser, M., Fahse, K., & Gutyrchik, E. (2013). Neuro*Cognitive Processing* of body representations in artistic and photographic images. *NeuroImage*, 66, 288–292. https://doi.org/10.1016/j.NeuroImage.2012.10.067.

Ma, Y. & Han, S. (2009). Self-face advantage is modulated by social threat: Boss effect on self-face recognition. *Journal of Experimental Social Psychology*, 45(4), 1048–1051. https://doi.org/10.1016/j.jesp.2009.05.008.

Mastandrea, S., Fagioli, S., & Biasi, V. (2019). Art and psychological well-being: Linking the brain to the aesthetic emotion. *Frontiers in Psychology*, 10, 739. https://doi.org/10.3389/fpsyg.2019.00739.

Mastandrea, S., Maricchiolo, F., Carrus, G., Giovannelli, I., Giuliani, V., & Berardi, D. (2019). Visits to figurative art museums may lower blood pressure and stress. *Arts & Health*, 11(2), 123–132. https://doi.org/10.1080/17533015.2018.1443953.

Mather, G. (2012). Aesthetic judgement of orientation in modern art. *i-Perception*, 3(1), 18–24. https://doi.org/10.1068/i0447aap.

Mellet, E., Colagè, I., Bender, A., Henshilwood, C. S., Hugdahl, K., Lindstrøm, T. C., & d'Errico, F. (2019). What processes sparked off symbolic representations? A reply to Hodgson and an alternative perspective. *Journal of Archaeological Science: Reports*, 28, 102043. https://doi.org/10.1016/j.jasrep.2019.102043.

Mellet, E., Salagnon, M., Majkić, A., Cremona, S., Joliot, M., Jobard, G., Mazoyer, B., Mazoyer, N. T., & d'Errico, F. (2019). Neuroimaging supports the representational nature of the earliest human engravings. *Royal Society Open Science*, 6(7), 190086. https://doi.org/10.1098/rsos.190086.

Metzger, W. (1953). *Gesetze des Sehens* (*The Laws of Seeing*). Frankfurt, Germany: Kramer.

Moles, A. (1958). *Théorie de l'information et perception esthétique* [*Information Theory and Aesthetic Perception*]. Paris, France: Flammarion.

Morris, D. (1962). *The Biology of Art*. London, England: Methuen.

Müller, P. (1999). *Die Kunstphilosophie bei Aristoteles unter besonderer Berücksichtigung der Malerei* [*The Philosophy of Art in Aristotle with Special Consideration of Painting*]. Munich, Germany: Grin.

Nay-Scheibler, E. (1981). *Gesichter der Steinzeit* [*Faces of the Stone Age*]. Stuttgart, Germany: Belser.

Noble, G. K. & Bradley, H. T. (1933). The mating behavior of lizards: Its bearing on the theory of sexual selection. *Annals of the New York Academy of Sciences*, 36, 25–100. https://doi.org/10.1111/j.1749-6632.1933.tb55365.x.

Olshausen, B. A. & Field, D. J. (2004). Sparse coding of sensory inputs. *Current Opinion in Neurobiology*, 14(4), 481–487. https://doi.org/10.1016/j.conb.2004.07.007.

Ornstein, R. E. (1972). *The Psychology of Consciousness*. London, England: Penguin.

Patel, A. D. (2003). Language, music, syntax and the brain. *Nature Neuroscience*, 6(7), 674–681. https://doi.org/10.1038/nn1082.

Pavlova, M., Krägeloh-Mann, I., Sokolov, A., & Birbaumer, N. (2001). Recognition of point-light biological motion displays by young children. *Perception*, 30(8), 925–933. https://doi.org/10.1068/p3157.

Peelen, M. V. & Downing, P. E. (2005). Selectivity for the human body in the fusiform gyrus. *Journal of Neurophysiology*, 93(1), 603–608. https://doi.org/10.1152/jn.00513.2004.

Peelen, M. V., Wiggett, A. J., & Downing, P. E. (2006). Patterns of fMRI activity dissociate overlapping functional brain areas that respond to biological motion. *Neuron*, 49(6), 815–822. https://doi.org/10.1016/j.neuron.2006.02.004.

Pelowski, M., Forster, M., Tinio, P. P., Scholl, M., & Leder, H. (2017). Beyond the lab: An examination of key factors influencing interaction with 'real' and museum-based art. *Psychology of Aesthetics, Creativity, and the Arts*, 11(3), 245. http://doi.org/10.1037/aca0000141.

Piaget, J. & Inhelder, B. (1971). *Mental Imagery in the Child*. London, England: Routledge & Kegan Paul.

Plato (1990). In G. Eigler (Ed.), *Platon Werke: Band 5: Phaidros, Parmenides, Epistolai (Briefe)* [*Plato's Works: Vol. 5. Phaedrus, Parmenides, Epistolai* (*Letters*)]. Darmstadt, Germany: WBG (Original Work Published ca. 380 B.C.).

Poeck, K. & Orgass, B. (1964). Die Entwicklung des Körperschemas bei Kindern im Alter von 4–10 Jahren [The development of the body schema in children aged 4–10 years]. *Neuropsychologia*, 2(2), 109–130. https://doi.org/10.1016/0028-3932(64)90016-8.

Popivanov, I. D., Jastorff, J., Vanduffel, W., & Vogels, R. (2014). Heterogeneous single-unit selectivity in an fMRI-defined body-selective patch. *Journal of Neuroscience*, 34(1), 95–111. https://doi.org/10.1523/JNEUROSCI.2748-13.2014.

Pöppel, E. (1989). The measurement of music and the cerebral clock: A new theory. *Leonardo*, 22(1), 83–89. https://doi.org/10.2307/1575145.

Pöppel, E. & Bao, Y. (2011). Three modes of knowledge as basis for intercultural cognition and communication: A theoretical perspective. In S. Han & E. Pöppel (eds.), *Culture and Neural Frames of Cognition and Communication*, pp. 215–231. Berlin, Germany: Springer. https://doi.org/10.1007/978-3-642-15423-2_14.

Pourtois, G., Peelen, M. V., Spinelli, L., Seeck, M., & Vuilleumier, P. (2007). Direct intracranial recording of body-selective responses in human extrastriate visual cortex. *Neuropsychologia*, 45(11), 2621–2625. https://doi.org/10.1016/j.Neuropsychologia.2007.04.005.

Quian Quiroga, R., Kreiman, G., Koch, C., & Fried, I. (2008). Sparse but not 'grandmother-cell' coding in the medial temporal lobe. *Trends in Cognitive Sciences*, 12(3), 87–91. https://doi.org/10.1016/j.tics.2007.12.003.

Roditi, E. (1990). *Dialogues: Conversations with European Artists at Mid-Century*. London, England: Lund Humphreys.

Rogala, J., Bajno, B., & Wróbel, A. (2020). A hidden message: Decoding artistic intent. *PsyCh Journal*, 9, 507–512. https://doi.org/10.1002/pchj.374.

Rosch, E. H. (1973). On the internal structure of perceptual and semantic categories. In T. E. Moore (ed.), *Cognitive Development and Acquisition of Language*, pp. 111–144. New York, NY: Academic Press.

Rosch, E. H. (1975). Cognitive representations of semantic categories. *Journal of Experimental Psychology: General*, 104, 192–233. https://doi.org/10.1037/0096-3445.104.3.192.

Rubin, E. (1921). *Visuell wahrgenommene Figuren* [*Visually Perceived Figures*]. Copenhagen, Denmark: Gyldendal.

Rubin, N. (2001). Figure and ground in the brain. *Nature Neuroscience*, 4(9), 857–858. https://doi.org/10.1038/nn0901-857.

Sanefuji, W., Ohgami, H., & Hashiya, K. (2007). Development of preference for baby faces across species in humans (Homo sapiens). *Journal of Ethology*, 25(3), 249–254. https://doi.org/10.1007/s10164-006-0018-8.

Savage, P. E., Loui, P., Tarr, B., Schachner, A., Glowacki, L., Mithen, S., & Fitch, W. T. (2020). Music as a coevolved system for social bonding. *Behavioral and Brain Sciences*. Advance online publication. https://doi.org/10.1017/S0140525X20000333.

Scheele, D., Wille, A., Kendrick, K. M., Stoffel-Wagner, B., Becker, B., Güntürkün, O., Maier, W., & Hurlemann, R. (2013). Oxytocin enhances brain reward system responses in men viewing the face of their female partner. *Proceedings of the National Academy of Sciences*, 110(50), 20308–20313. https://doi.org/10.1073/pnas.1314190110.

Schenk, T. & McIntosh, R. D. (2010). Do we have independent visual streams for perception and action? *Cognitive Neuroscience*, 1(1), 52–62. https://doi.org/10.1080/17588920903388950.

Schleidt, W. (1962). Die historische Entwicklung der Begriffe "Angeborenes auslösendes Schema" und "Angeborener Auslösemechanismus" in der Ethologie [The historical development of the terms "Angeborenes auslösendes Schema" and "Angeborener Auslösemechanismus" in ethology]. *Zeitschrift für Tierpsychologie*, 19, 697–722. https://doi.org/10.1111/j.1439-0310.1962.tb00800.x.

Schnabel, U. H., Bossens, C., Lorteije, J. A., Self, M. W., de Beeck, H. O., & Roelfsema, P. R. (2018). Figure-ground perception in the awake mouse and neuronal activity elicited by figure-ground stimuli in primary visual cortex. *Scientific Reports*, 8(1), 1–14. https://doi.org/10.1038/s41598-018-36087-8.

Schober, H. & Rentschler, I. (1972). *Das Bild als Schein der Wirklichkeit. Optische Täuschungen in Wissenschaft und Kunst* [*The Picture as the Appearance of Reality: Optical Illusions in Science and Art*]. Munich, Germany: Moos.

Schoeller, F. & Perlovsky, L. (2016). Aesthetic chills: Knowledge-acquisition, meaning-making, and aesthetic emotions. *Frontiers in Psychology*, 7, 1093. https://doi.org/10.3389/fpsyg.2016.01093.

Schumacher, F. (1983). *Der Geist der Baukunst* [*The Spirit of Architecture*]. Stuttgart, Germany: DVA.

Schuster, M. & Beisl, H. (1978). *Kunst-Psychologie. Wodurch Kunstwerke wirken* [*Art Psychology: How Works of Art Work*]. Köln, Germany: DuMont.

Serjouie, A. (2011). *Children's Understanding of Pictures and Expression of Emotion in Their Drawings. A Comparative Study of Articulation of Happiness and Sadness by Iranian Children Living in Islamic Republic of Iran with Iranian Children Living in German Federal Republic*, Unpublished Doctoral Thesis, University of Erfurt, Erfurt, Germany.

Shannon, C. A. (1948). A mathematical theory of communication. *Bell System. Technical Journal*, 27(379–423), 623–656.

Shapiro, J., Rucker, L., & Beck, J. (2006). Training the clinical eye and mind: Using the arts to develop medical students' observational and pattern recognition skills. *Medical Education*, 40(3), 263–268. https://doi.org/10.1111/j.1365-2929.2006.02389.x.

Silvia, P. J. (2005). Emotional responses to art: From collation and arousal to cognition and emotion. *Review of General Psychology*, 9(4), 342–357. https://doi.org/10.1037/1089-2680.9.4.342.

Simons, D. J. & Levin, D. T. (1997). Change blindness. *Trends in Cognitive Sciences*, 1(7), 261–267. https://doi.org/10.1016/S1364-6613(97)01080-2.

Singh, D. (1993). Adaptive significance of female physical attractiveness: Role of waist-to-hip ratio. *Journal of Personality and Social Psychology*, 65(2), 293. https://doi.org/10.1037/0022-3514.65.2.293.

Skrzipek, K. H. (1978). Menschliche "Auslösermerkmale" beider Geschlechter. I. Attrappenwahl-Untersuchungen der Verhaltensentwicklung [Human "trigger characteristics" of both sexes. I. Dummy choice studies of behavioral development]. *Homo*, 29, 75–88.

Störig, P. & Rentschler, I. (1984). Brain and beauty: Bioscientific approaches to aesthetic phenomena. In R. Haller (ed.), *Proceedings of the 8th International Wittgenstein Symposium; 1: Ästhetik. Schriftenreihe der Wittgenstein Ges. 10/1*, pp. 91–94. Vienna, Austria: Hölder-Pichler-Tempsky.

Sütterlin, C. (2003). From sign and schema to iconic representation: Evolutionary aesthetics of pictorial art. In E. Voland & K. Grammer (eds.), *Evolutionary Aesthetics*. Berlin, Germany: Springer.

Sütterlin, C. (2013). Explizite und implizite Symmetrie. Präzises und diffuses Gleichgewicht in der Kunst [Explicit and implicit symmetry: Precise and diffuse balance in art]. In O. Bender, S. Kanitscheider, & A. Treml (eds.), *Gleichheit und Ungleichheit. Symmetrie und Asymmetrie. Erscheinungs- und Verlaufsformen, Funktion und Folgen in Natur und Kultur* [*Equality and Inequality, Symmetry and Asymmetry in Manifestations and Developments in Nature and Art*], pp. 231–284. Norderstedt, Germany: BoD.

Sütterlin, C. (2014). Platonic vision: On art, the nature of beauty and the history of aesthetics. In C. Sütterlin, W. Schiefenhövel, C. Lehmann, J. Forster, & G. Apfelauer (eds.), *Art as Behaviour. An Ethological Approach to Visual and Verbal Art, Music and Architecture*, pp. 169–216. Oldenburg, Germany: BIS-Verlag.

Sütterlin, C. (2018). Weibliche und männliche Körperdarstellungen im Neolithikum. Zur evolutiven Revision eines Körperbildes [Female and male body representations in the Neolithic: The evolutionary revision of body image]. In B. Lange, C. Schwender, & S. Schwarz (eds.), *Geschlecht und Verhalten aus evolutionärer Perspektive* [*Gender and Behavior from an Evolutionary Perspective*], pp. 147–160. Lengerich, Germany: Pabst Science.

Sütterlin, C. & Eibl-Eibesfeldt, I. (2015). Urmotiv Auge revisited [The eye as primordial motive, revisited]. In O. Bender, S. Kanitscheider, & A. Treml (eds.), *100 Jahre Otto Koenig* [*100 Years of Otto Koenig*]. Norderstedt, Germany: BoD.

Sütterlin, C. Schiefenhövel, W. Lehmann, C. Forster, J. & Apfelauer G. (eds.) (2014). *Art as Behaviour: An Ethological Approach to Visual and Verbal Art, Music and Architecture. Hanse Studies*, Vol. 10. Oldenburg, Germany: BIS-Verlag.

Tanaka, M., Tomonaga, M., & Matsuzawa, T. (2003). Finger drawing by infant chimpanzees (Pan troglodytes). *Animal Cognition*, 6(4), 245–251. https://doi.org/10.1007/s10071-003-0198-3.

Teng, X., Ma, M., Yang, J., Blohm, S., Cai, Q., & Tian, X. (2020). Constrained structure of ancient Chinese poetry facilitates speech content grouping. *Current Biology*, 30, 1299–1305. https://doi.org/10.1016/j.cub.2020.01.059.

Thomas, E. & French, R. (2017). Grandmother cells: Much ado about nothing. *Language, Cognition and Neuroscience*, 32(3), 342–349. https://doi.org/10.1080/23273798.2016.1235279.

Thompson, R. K. & Oden, D. L. (2000). Categorical perception and conceptual judgments by nonhuman primates: The paleological monkey and the analogical ape. *Cognitive Science*, 24(3), 363–396. https://doi.org/10.1207/s15516709cog2403_2.

Tomasello, M. (2010). *Origins of Human Communication*. Cambridge, MA: MIT Press. https://doi.org/10.7551/mitpress/7551.001.0001.

Tooby, J. & Cosmides, L. (2001). Does beauty build adapted minds? Toward an evolutionary theory of aesthetics, fiction, and the arts. *SubStance*, 30(1), 6–27. https://doi.org/10.1353/sub.2001.0017.

Tsao, D. Y., Freiwald, W. A., Tootell, R. B., & Livingstone, M. S. (2006). A cortical region consisting entirely of face-selective cells. *Science*, 311(5761), 670–674. https://doi.org/10.1126/science.1119983.

Tsao, D. Y., Schweers, N., Moeller, S., & Freiwald, W. A. (2008). Patches of face-selective cortex in the macaque frontal lobe. *Nature Neuroscience*, 11(8), 877. https://doi.org/10.1038/nn.2158.

Tsukiura, T. & Cabeza, R. (2008). Orbitofrontal and hippocampal contributions to memory for face–name associations: The rewarding power of a smile. *Neuropsychologia*, 46(9), 2310–2319. https://doi.org/10.1016/j.Neuropsychologia.2008.03.013.

Tsukiura, T. & Cabeza, R. (2011). Shared brain activity for aesthetic and moral judgments: Implications for the beauty-is-good stereotype. *Social Cognitive and Affective Neuroscience*, 6(1), 138–148. https://doi.org/10.1093/scan/nsq025.

Turner, F. & Pöppel, E. (1988). Metered poetry, the brain, and time. In I. Rentschler, B. Herzberger, & David Epstein (eds.), *Beauty and the Brain*, pp. 71–90. Basel, Switzerland: Birkhäuser.

Tylén, K., Fusaroli, R., Rojo, S., Heimann, K., Fay, N., Johannsen, N. N., Riede, F., & Lombard, M. (2020). The evolution of early symbolic behavior in Homo sapiens. *Proceedings of the National Academy of Sciences*, 117(9), 4578–4584. https://doi.org/10.1073/pnas.1910880117.

Ulrich, R. S. (1984). View through a window may influence recovery from surgery. *Science*, 224(4647), 420–421. https://doi.org/10.1126/science.6143402.

Vartanian, O. & Skov, M. (2014). Neural correlates of viewing paintings: Evidence from a quantitative meta-analysis of functional magnetic resonance imaging data. *Brain and Cognition*, 87, 52–56. https://doi.org/10.1016/j.bandc.2014.03.004.

Vedder, A., Smigielski, L., Gutyrchik, E., Bao, Y., Blautzik, J., Pöppel, E., Zaytseva, Y., & Russell, E. (2015). Neurofunctional correlates of environmental cognition: An fMRI study with images from episodic memory. *PLoS One*, 10(4), e0122470. https://doi.org/10.1371/journal.pone.0122470.

Vogel, S. (2009). Vom Widerspruch im Ornament [On the contradiction in ornament]. In A. Husslein-Arco & S. B. Vogel (eds.), *Die Macht des Ornaments* [*The Power of Ornaments*]. Vienna, Austria: Belvedere Museum Vienna.

von Uexküll, J. (1909). *Umwelt und Innenwelt der Tiere* [*Environment and Inner World of Animals*]. Berlin, Germany: Julius Springer.

Wachsmuth, E., Oram, M. W., & Perrett, D. I. (1994). Recognition of objects and their component parts: Responses of single units in the temporal cortex of the macaque. *Cerebral Cortex*, 4(5), 509–522. https://doi.org/10.1093/cercor/4.5.509.

Wang, X. & Bao, Y. (2020). Rembrandt portraits: Implicitly detecting the original perspective. *PsyCh Journal*, 9, 513–515. https://doi.org/10.1002/pchj.376.

Wang, X., Simmank, F., Zhang, D., Zeng, Y., Silveira, S., Sander, T., Bao, Y., & Paolini, M. (2020). Aesthetics as a common denominator for moral and commercial judgments. *PsyCh Journal*, 10(2), 407. https://doi.org/10.1002/pchj.407.

Whitney, D. & Levi, D. M. (2011). Visual crowding: A fundamental limit on conscious perception and object recognition. *Trends in Cognitive Sciences*, 15(4), 160–168. https://doi.org/10.1016/j.tics.2011.02.005.

Wiese, H., Altmann, C. S., & Schweinberger, S. R. (2014). Effects of attractiveness on face memory separated from distinctiveness: Evidence from event-related brain potentials. *Neuropsychologia*, 56, 26–36. https://doi.org/10.1016/j.Neuropsychologia.2013.12.023.

Wilson, E. O. (1984). *Biophilia*. Cambridge, MA: Harvard University Press.

Wittgenstein, L. (1953). *Philosophical Investigations* (G. E. M. Amnscombe, Trans.). Oxford, England: Blackwell.

Wittmann, M. & Pöppel, E. (1999). Temporal mechanisms of the brain as fundamentals of communication — with special reference to music perception and performance. *Musicae Scientiae*, 3(Suppl. 1), 13–28. https://doi.org/10.1177/10298649000030S103.

Wolf, D. P. (1987). Drawing conclusions: Insights into the nature of art from children's drawings. In J. G. Hohenzollern & M. Liedtke (eds.), *Vom Kritzeln zur Kunst. Stammes und individualgeschichtliche Komponenten der künstlerischen Fähigkeiten* [*From Doodling to Art: Tribal And Individual Historical Components of Artistic Skills*], pp. 186–198. Bad Heilbrunn, Germany: Klinkhardt.

Wulf, F. (1922). Über die Veränderung von Vorstellungen [About the changing of ideas]. *Psychologische Forschung*, 1, 333–373.

Yin, L. & McBride, C. (2015). Chinese kindergartners learn to read characters analytically. *Psychological Science*, 26(4), 424–432. https://doi.org/10.1177/0956797614567203.

Yu, X. & Bao, Y. (2020). The 3 second time window in poems and language processing in general: Complementarity of discrete timing and temporal permanence. *PsyCh Journal*, 9(4), 429–443. https://doi.org/10.1002/pchj.390.

Yu, X., Pöppel, E., & Bao, Y. (2020). Aesthetic appreciation is more than seeing, and symbolicity is more than discriminating: A comment on Tylén *et al.* (2020). *PsyArXiv*. https://doi.org/10.31234/osf.io/5quya.

Zeki, S. (1998). Art and the brain. *Daedalus*, 127(2), 71–103.

Zeki, S. (1999). *Inner Vision: An Exploration of Art and the Brain*. Oxford, England: Oxford University Press.

Zhou, B., Pöppel, E., Wang, L., Yang, T., Zaytseva, Y., & Bao, Y. (2016). Seeing without knowing: Operational principles along the early visual pathway. *PsyCh Journal*, 5(3), 145–160. https://doi.org/10.1002/pchj.141.

Zhu, Q., Zhang, J., Luo, Y. L., Dilks, D. D., & Liu, J. (2011). Resting-state neural activity across face-selective cortical regions is behaviorally relevant. *Journal of Neuroscience*, 31(28), 10323–10330. https://doi.org/10.1523/JNEUROSCI.0873-11.2011.

https://doi.org/10.1142/9789811298639_0013

Chapter 13

The Golden Ratio as an Ecological Affordance Leading to Aesthetic Attractiveness*

Daniela De Bartolo[†,‡], Maria De Luca[†], Gabriella Antonucci[†,‡], Stefan Schuster[§], Giovanni Morone[†], Stefano Paolucci[†] and Marco Iosa[†,‡]

[†]*Santa Lucia Foundation, Rome, Italy*
[‡]*Department of Psychology, Sapienza University of Rome, Italy*
[§]*Department of Bioinformatics, Friedrich Schiller University Jena, Germany*

Abstract

The golden ratio (GR) is an irrational number (close to 1.618) that repeatedly occurs in nature as well as in masterpieces of art. The GR has been considered a proportion perfectly representing beauty since ancient times, and it was investigated in several scientific fields but with conflicting results. This study aims at investigating if this proportion is associated with a judgment of beauty independently of the type of the

*This chapter was originally published in the special collection on neuroaesthetics in *PsyCh Journal*. De Bartolo, D., De Luca, M., Antonucci, G., Schuster, S., Morone, G., Paolucci, S., & Iosa, M. (2022). The golden ratio as an ecological affordance leading to aesthetic attractiveness. *PsyCh Journal*, 11(5), 729–740. https://doi.org/10.1002/pchj.505.

stimulus, and the factors that may affect this aesthetic preference. In Experiment 1, an online psychophysical questionnaire was administered to 256 volunteers asked to choose among three possible proportions between the parts of the same stimulus (GR, 1.5, and 1.8). In Experiment 2, we recorded eye movements of 15 participants who had to express an aesthetic judgment on the same stimuli as Experiment 1. The results revealed a slight overall preference for GR (53%, $p<0.001$), with higher preferences for stimuli representing humans, anthropomorphic sculptures, and paintings, regardless of the cultural level. In Experiment 2, a shorter dwell time was significantly associated with a better aesthetic judgment ($p = 0.005$), suggesting the possibility that GR could be associated with easier visual processing, and it could be hence considered as a visual affordance.

Keywords: aesthetics; beauty; eye tracking; golden ratio; neuroaesthetics; psychophysics

Introduction

The sense of beauty represents a very addressed and discussed topic in the field of psychology (Ramachandran & Hirstein, 1999), and in the last two decades, neuroaesthetics have become an important field of research for neurophysiologists and neuropsychologists (Bundgaard *et al.*, 2017; Zeki, 1999a, 1999b) interested in the cognitive and perceptual processes involved in the perception of beauty (Bundgaard, 2015). Recent studies exploited the eye-tracking methodology to objectively analyze the "salient" properties of an object that may capture the beholder's attention (Bundgaard, 2015; Chassy *et al.*, 2015; Chenier & Winkielman, 2018; Kumar & Garg, 2010; Reber, 2012), using electroencephalography (Kong *et al.*, 2015; Zhang *et al.*, 2020; Zhang & Deng, 2012) and functional magnetic resonance imaging (Di Dio *et al.*, 2007, 2011) to analyze the brain areas and relevant processes involved in the perception of attractive and beautiful stimuli. Among the properties associated with attractiveness, researchers found symmetry, harmonic proportions, balance, contrast, and a proper easiness/complexity relationship (Fink & Penton-Voak, 2002).

A particular proportion linked to harmony and studied in aesthetics is the golden ratio (GR; Iosa *et al.*, 2018). From a geometric point of view, it is the mathematical solution of dividing a segment in a manner that the ratio between the shorter part to the longer one of the segments is equal to the ratio between the longer part and the whole segment. This problem has only a solution in which the above ratio is the irrational number φ = 1.6180... (Figure 1).

Since the 5th century BCE, this number has occured in the artworks of many artists who have considered the GR as a beautiful and pleasant proportion (Gamwell, 2015). Polykleitos sculptured the Doryphoro with

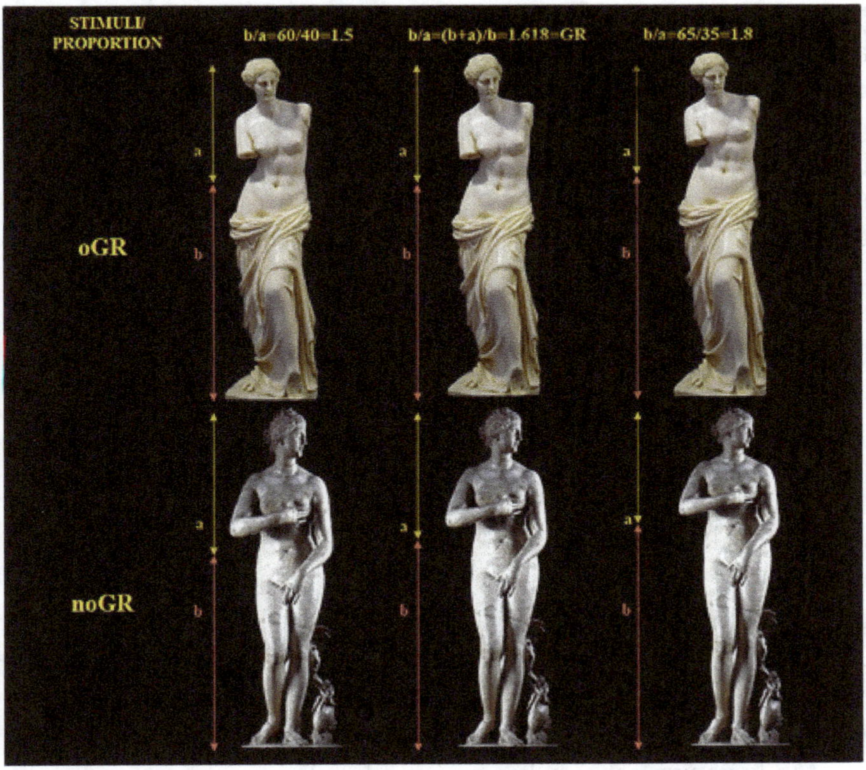

Figure 1. The stimuli modification procedure reported for two exemplificative stimuli: The Venus of Milo (above, originally in golden rule [GR], oGR) and the Venus de' Medici (below, not originally in GR, noGR). The GR refers to the ratio between b (from the ground to the navel) and a (from the navel to the top of the head). In the case of Venus of Milo, the original version was compared to two experimental altered versions in which b/a was 1.5 (on the left) or 1.8 (on the right). The Venus of Medici has originally a ratio b/a = 1.5, and the versions in GR and R1.8 were digitally created.

the ratio between the total height and the distance of the navel from the ground equal to the GR; the same ratio was respected by Alexandros of Antioch when he sculptured the Venus de Milo. Phidias supervised the creation of the façade of the Parthenon as inscribed in a golden rectangle, namely, with the ratio of basis and height of the rectangle equal to the GR (Iosa *et al.*, 2018).

Since the 19th century, psychologists have been testing if the GR led to a subjective experience of beauty supported by objective evidence. However, conflicting results were found among studies: GR seemed to be the preferred proportion if the stimulus was a rectangle (Fechner, 1865), a triangle (Witmer, 1893), partially if it was an ellipse (Fechner, 1865), a sculpture (Di Dio *et al.*, 2007, 2011), or related to human faces (Danikas & Panagopoulos, 2004; Kwart *et al.*, 2012). Other researchers questioned the different methodologies with which these results were obtained (Godkewitsch, 1974; Piehl, 1976). In particular, it was questioned whether GR often emerged as the average response and median ranking but not as the modal, most selected, preferred ratio (Angier, 1903; Nienstedt Jr. & Ross, 1951).

Till now, only reviews compared the potential preferences for GR among different types of stimuli but combining data obtained in different studies, carried out with different methodologies, and using different parameters to assess aesthetic preferences (Bejan, 2009; Green, 1995; Iosa *et al.*, 2018). Russell (2000) compared in the same experiment geometric stimuli versus paintings, finding GR was preferred in the former and not in the latter, in contrast with the fact that GR has been found as a feature of many artistic paintings (de Campos *et al.*, 2015; Iosa *et al.*, 2018). Some studies also suggested a possible confounding role played by the cultural level of the interviewed subjects (Berlyne, 1970; Thompson, 1946).

The literature lacks studies investigating the preference for the GR comparing different classes of stimuli and analyzing if this preference may depend on education, artistic, or mathematical experiences.

The aims of this study were hence to investigate if the golden proportion is or is not preferred, and if this eventual preference could depend on the type of stimulus or on the cultural level of observers (Experiment 1). Then, we aimed at investigating if some parameters of gaze movements, quantitatively assessed by eye tracking, could be associated with aesthetic judgments when the stimulus has a golden proportion (Experiment 2).

This study was conducted according to the Declaration of Helsinki for research on human beings. It was approved by an independent local ethics committee of the Santa Lucia Foundation (CE /PROG660). For the first experiment, the participants were recruited online, and their data were treated anonymously according to the Legislative Decree of 30 June 2003, n. 196, art. 13. The participants read all the information relating to the questionnaire administered online and gave their consent before proceeding with the compilation. For the second experiment, the participants were enrolled in the Laboratory of Developmental Dyslexia Research Laboratory located at the Santa Lucia Foundation in Rome. The participants gave their written consent before proceeding with the experiment.

Materials and Methods of Experiment 1

Participants

A sample of 256 participants accepted to complete an online administered questionnaire on a personal computer. All of them were normal or corrected-to-normal vision. The average age of subjects was 34.2 ± 10.8 years, their average schooling was 16.8 ± 2.9 years, and 167 participants were female.

Procedure

A computerized online questionnaire was administered to participants, according to psychosocial and clinical research methodology (Regmi *et al.*, 2016). Data were treated anonymously. The questionnaire was based on the method of "absolute judgment in psychophysics": It consists of administering stimuli divided in couples, asking the subject to select the most liked. This method of choice is commonly used to identify, if it exists, a relative order of pleasantness (Wever & Zener, 1928). Participants were asked to complete the questionnaire in a comfortable position in front of a PC monitor rather than a smartphone to avoid small displays. The online questionnaire consisted of two parts: In the first one, participants were asked to report demographic data (age and sex), their education (highest degree), and to self-evaluate their cultural skills (because GR could be studied in the fields of mathematics, visual arts, and music, the self-assessed knowledge of these disciplines were separately assessed). The second part contained the experimental paradigm formed by 17 stimuli, each one replicated in three different proportions, forming 51 items of coupled stimuli. In each couple, two images were simultaneously presented on the screen and subjects were asked to choose the image they liked the most by clicking with the mouse on the preferred picture. The left–right positions of the three proportions on the screen were randomly balanced among the trials.

Stimuli

The GR has been traced in the human body proportions, artistic sculptures, and paintings with which a sense of beauty is usually associated; the visual stimuli included in this study therefore included colored images of human figures (Di Dio *et al.*, 2007), images of anthropomorphic sculptures (Di Dio *et al.*, 2011), and images of paintings (Iosa, 2018). In addition, stimuli on which a preference for the GR has been reported in previous studies have been included, such as geometric figures (Berlyne, 1970) and bisected lines (Benjafield & Adams-Webber, 1976). Then, the image of Leonardo's Vitruvian Man drawing was included given its notorious golden proportions. Finally, virtual humanoid avatars were also inserted to provide more recent replications of humanoid figures based on virtual tools (designed using an open-source program Make HumanTM for 3D humanoid prototypes) (Bastioni *et al.*, 2008). The stimuli were chosen by balancing the numbers of those already known to be originally in GR (oGR, such as Venus de Milo; Iosa *et al.*, 2018) and stimuli of the same categories not originally in GR (noGR, such as Venus de' Medici, Figure 1), while the geometric and virtual stimuli, being created *ad hoc*, had no original proportion. Each figure was divided according to a specific ideal line that was the horizontal line passing for the navel of the anthropomorphic figures (photographs, paintings, sculptures, or virtual reproductions of humans) or the vertical line dividing the scene into two parts for horizontal paintings. The proportions between the vertical dimensions of the obtained two parts for human figures and those between the two horizontal dimensions for horizontal paintings were altered to obtain in total three stimuli in proportions of the GR, 1.5, or 1.8. The same proportions were recreated for geometric stimuli selecting the cut point in the line section or altering the proportion of the basis and height of the rectangle. This procedure is shown graphically in Figure 1.

As in previous studies (Di Dio *et al.*, 2007, 2011; Fechner, 1865), each stimulus was presented in three possible versions having the following main proportion between two main parts of the picture (with a tolerance of 0.5%): 61.8:38.2% (ratio = 1.618, approximately equal to GR), 65%–35% (ratio = 1.8, furtherly called R1.8), or 60:40% (ratio = 1.5, R1.5). The other two proportions were chosen in accordance with previous studies (Haines & Davies, 1904) and with the fact that subjects are often exposed also to them (Benjafield & Adams-Webber, 1976). In fact, R1.8 is close to the ratio of television and computer monitors (16:9 = 1.78, or 1920:1080 pixels = 1.78), and it was also one of the ratios used in previous studies (Di Dio *et al.*, 2007, 2011); whereas R1.5 was chosen because some studies suggested that the actual preferred proportion is 3:2 (=1.5) and not GR (Davis, 1933), as also occurred for the ellipses of the study of Fechner (1865).

The 17 stimuli are reported in Table 1. Anthropomorphic figures (humans, humanoids, sculptures, and Vitruvian Man) were shown as

Table 1. List of stimuli.

Stimulus	Category	Main proportion	Original proportion
1. Woman	Photographs of real humans	Ratio between the distance of the navel to the ground and that between the navel and the top of the head	62–38% (oGR)
2. Man			62–38% (oGR)
3. Woman			58–42% (noGR)
4. Man			65–35% (oGR)
5. Venus de Milo	Photographs of sculptures		62–38% (oGR)
6. Doryphoros			62–38% (oGR)
7. Venus de' Medici			60–40% (noGR)
8. David			59–41% (noGR)

(Continued)

Table 1. (*Continued*)

Stimulus	Category	Main proportion	Original proportion
9. Virtual woman	Virtual avatars		—
10. Virtual man			—
11. Vitruvian Man	Drawing		62–38% (oGR)
12. The Flagellation of Christ	Paintings	Ratio between right and left parts with respect to a main element of the scene	62–38% (oGR)
13. The Creation of Adam			62–38% (oGR)
14. The Calling of St. Matthew			60–40% (noGR)
15. Undergrowth with two figures			64–36% (noGR)
16. Section of a horizontal line	Geometric figures	Ratio between two parts	—
17. Rectangle		Basis/height	—

Note: For the artistic stimuli the authors were (5) Alexandros, (6) Polykleitos, (7) Cleomenes, (8 and 13) Michelangelo Buonarroti, (11) Leonardo da Vinci, (12) Piero della Francesca, (14) Caravaggio, and (15) van Gogh.

entire standing figures, and the main proportion considered was the ratio between the total height and the distance of the navel from the ground, according to previous studies (as shown in Figure 1; Davis & Altevogt, 1979; Di Dio *et al.*, 2007). For paintings, the main proportion was that related to GR (de Campos *et al.*, 2015) or another clear horizontal division between the two main parts of the painting. The three couples of comparisons (GR vs. R1.8, GR vs. R1.5, R1.8 vs. R1.5) multiplied for the 17 stimuli resulting in the 51 items of the questionnaire.

Statistical analysis

All analyses were performed using IBM SPSS Statistics Version 23.0. The sample size for Experiment 1 was determined by computing estimated statistical power, based on the results of a previous study (Di Dio *et al.*, 2007). Fixing the alpha level at 0.05 and the power of the test at 0.80, we were required to enroll at least 255 subjects.

The percentage preference for GR, R1.5, and R1.8 proportions were computed for each stimulus and averaged for each stimulus category.

Comparisons were performed using the chi-squared test. Since each proportion was compared to another two, the possibility of finding a statistically significant results was divided by 2 applying Bonferroni correction. So, the alpha level of statistical significance was set at 0.025. Odds ratio (OR) and 95% confidence interval (95% CI) were also computed. The correlations of the aesthetic preference with demographic variables and cultural skills were tested using Spearman's rank correlation coefficient.

Results of Experiment 1

Table 2 shows the percentages of participants' choices for the three presented comparisons. As shown in the last columns of Table 2, GR preferences overcame 50% in 5 out of 6 categories and in 14 out of 17 stimuli. Among these, the significant preferences found in favor of GR were related to photographs of humans (54%, $p = 0.005$), sculptures (53%, $p = 0.029$), paintings (55%, $p = 0.004$), and Vitruvian Man (58%, $p = 0.015$). Preference was only close to significance for virtual humanoids (54%, $p = 0.057$).

Table 2. Percentages of choices and *p*-values related to chi-square among the three comparisons between the ratios (*p*-values are in bold and are lower than alpha-level reduced for Bonferroni's correction; special character Φ was added to significant results in favor of GR ratio).

	GR vs. R1.8		**GR vs. R1.5**		**R1.5 vs. R1.8**		**Mean % for GR**	
Stimulus	**%**	***p***	**%**	***p***	**%**	***p***	**Stimulus**	**Category**
1. oGR woman	62.5%	**<.001**$^{\Phi}$	64.5%	**<.001**$^{\Phi}$	38.3%	**<.001**	63.5%	54.4%
2. oGR man	62.9%	**<.001**$^{\Phi}$	52.0%	.532	53.5%	.261	57.5%	
3. noGR woman	40.6%	**.003**	39.1%	**<.001**	52.0%	.532	39.9%	
4. no GR man	60.5%	**.001**$^{\Phi}$	53.1%	.317	55.1%	.104	56.8%	
5. oGR Venus	64.5%	**<.001**$^{\Phi}$	49.6%	.901	57.0%	**.024**	57.1%	53.4%
6. oGR Doryphoros	52.3%	.453	53.9%	.211	58.6%	**.006**	53.1%	
7. noGR Venus	47.8%	.532	55.5%	.08	66.4%	**<.001**	51.7%	
8. noGR David	67.2%	**<.001**$^{\Phi}$	36.7%	**<.001**	60.2%	**.001**	52.0%	
9. Neutral virtual woman	63.3%	**<.001**$^{\Phi}$	45.3%	.134	65.0%	**<.001**	54.3%	54.2%
10. Neutral virtual man	76.6%	**<.001**$^{\Phi}$	31.6%	**<.001**	71.5%	**<.001**	54.1%	
11. oGR Vitruvian Man	73.0%	**<.001**$^{\Phi}$	42.2%	**.012**	71.5%	**<.001**	57.6%	57.6%

(Continued)

Table 2. (*Continued*)

	GR vs. R1.8		**GR vs. R1.5**		**R1.5 vs. R1.8**		**Mean % for GR**	
Stimulus	%	*p*	%	*p*	%	*p*	**Stimulus**	**Category**
12. oGR Della Francesca	59.8%	**.002**$^{\Phi}$	57.0%	**.024**$^{\Phi}$	58.2%	**.009**	58.4%	54.6%
13. oGR Michelangelo	49.2%	.803	52.7%	.382	41.4%	**.006**	51.0%	
14. oGR Caravaggio	60.2%	**.001**$^{\Phi}$	58.6%	**.006**$^{\Phi}$	52.7%	.382	59.4%	
15. noGR van Gogh	43.0%	**.024**	55.9%	.061	45.7%	.169	49.5%	
16. Neutral line section	47.3%	.382	42.6%	**.018**	48.0%	.532	45.0%	48.0%
17. Neutral rectangle	47.3%	.382	54.7%	.134	40.2%	**.002**	51.0%	

Note: In the last two columns, the values of the mean golden rule (GR) were reported for each stimulus and each category of stimuli.

For geometric shapes, GR was chosen in 48% of trials, a percentage not significantly different from those of other proportions (p = 0.353). The sum of all responses highlighted a mean significant preference for GR (53%, $p<.001$).

Many statistically significant differences were found regarding the comparison between GR and R1.8 proportion (12 out of 17 stimuli). Indeed, 10 comparisons were in favor of GR (57.5% vs. 42.5%, $p = 0.018$), while only two were in favor of the R1.8 proportion (the painting of a woman and the painting of van Gogh, both noGR).

When compared with the R1.5 proportion, the preference for GR was reduced: It was significantly in favor of GR for three stimuli (the picture of a woman, Della Francesca's painting, both oGR stimuli, and Caravaggio's painting, a noGR stimulus) and in favor of R1.5 proportion for five stimuli. The overall preference was in line with chance: 50% for GR and 50% for R1.5 proportion. The comparison R1.5 versus R1.8 showed a not significant preference for R1.5 ($p = 0.134$).

Figure 2 shows the comparisons of preferences for GR with respect to the other two proportions, finding statistically significant differences for photographs of humans (OR = 1.19, 95% CI = 1.05–1.35, p = 0.005), sculptures (OR = 1.15, 95% CI = 1.01–1.30, p = 0.029), and paintings (OR = 1.20, 95% CI= 1.06–1.36, $p = 0.004$) but for neither virtual humanoids (OR = 1.18, 95% CI = 0.99–1.41, p = 0.057) nor geometric shapes (OR = 0.92, 95% CI = 0.77–1.10, $p = 0.353$).

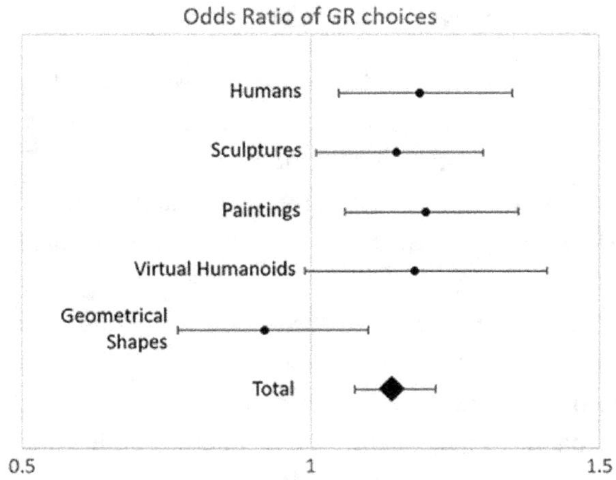

Figure 2. Forest plot of odds ratios (OR) and relevant 95% confidence intervals estimated for stimulus originally in golden rule (GR) with respect to chance (OR = 1 corresponds to no association between object's proportion and subject's choice) for the categories of the stimuli.

The total preference for GR was found related to neither age nor any cultural skills (age: $p = 0.574$, schooling time: $p = 0.860$, self-assessed art knowledge: $p = 0.173$, math knowledge: $p = 0.059$, music knowledge: $p = 0.209$). Dividing the results for the categories of stimuli, the preference for GR was found to significantly correlate with self-assessed artistic ($p = 0.035$) and mathematical ($p = 0.023$) knowledge for human photographs, whereas GR preference for sculptures was found to significantly correlate with artistic ($p = 0.015$) and musical ($p = 0.020$) knowledge.

Materials and Methods of Experiment 2

Participants

Fifteen healthy subjects (nine female) with an average age of 28.7 ± 6 years and average schooling of 18.9 ± 2.2 years, with normal or corrected-to-normal vision, and naïve to the purpose of the experiments were enrolled. The sample size was chosen in accordance with the data from a previous study (Di Dio *et al.*, 2007).

Stimuli and procedure

The same set of 51 stimuli from Experiment 1 was used but presented one by one during individual eye movement recording sessions in which the subject sat in front of a monitor at a viewing distance of 57 cm. The rectangular stimuli subtended a viewing angle of approximately 8.0 × 10.0 and 4.0 × 8.0 degrees (width by height), according to a previous study (McManus *et al.*, 2010).

The modification of the ratio was made without altering the visual angle for each stimulus; this allowed comparison of the same stimulus in the three experimental ratios maintaining the same visual angle.

The participants were asked to watch each stimulus and to express a verbal judgment using a 10-point Likert scale (from 1 = *strongly dislike* to 10 = *strongly like*). Before the experiment started, each subject was instructed about the task, placed in front of the screen using a head support, and trained to carry out the experiment through a pilot including a nine-point calibration procedure and a set of 10 stimuli that were different but homologous to the stimuli of the main experiment.

Setting and apparatus

The experiment was carried out on a desktop computer. Stimuli had a 300 dpi resolution and were presented at 85 Hz on a 17-inch CRT monitor, with a screen resolution of 1,024 × 768 pixels. Eye movements were recorded in binocular vision from the dominant eye via an EyeLink 1000 eye-tracker (SR Research), sampling at 1,000 Hz. The appearance of the stimulus on the screen was triggered by the fixation of a cross outside the stimulus area. Subjects' vocal responses scoring the aesthetic judgments were digitally recorded by a Shure microphone, a pre-amplifier, an E-MU sound card, and an ASIO driver, which was interfaced to the eye tracker by Eye Link software.

Data processing

Eye movement data were processed via EyeLink Data Viewer software (SR Research). Eye blinks were automatically signaled by the software

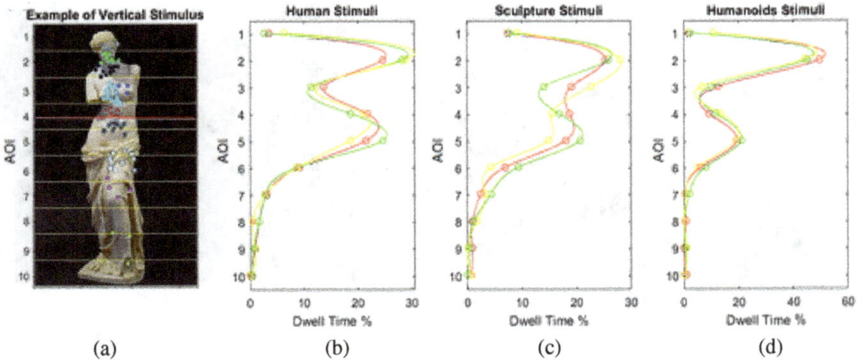

Figure 3. Graphical representation of how fixations were distributed among the vertical anthropomorphic stimuli. In the first panel (a), the Venus of Milo, as an example, is in its original golden rule (GR) proportion with a red line marking where the alteration of ratio was done. Colored points represent groups of fixations spatially closed. In the other panels, we reported the percentage of dwell time spent in each area of interest (AOI) normalized on the total dwell time for all the stimuli belonging to human figures (b), anthropomorphic sculptures (c), and virtual humanoids (d) category, comparing the three experimental proportions: GR (red), R1.5 (green), and R1.8 (yellow).

and manually discarded during the offline visual inspection carried out by the experimenter over each single trial. Fixation positions were measured and referred to areas of interest (AOI) for each stimulus trial according to 10 horizontal (for portrait stimuli, as in Figure 3) or vertical (for landscape stimuli, as in Figure 4) AOI of equal size. In particular, for each trial, the specific AOI in which the modification of ratio occurred was designated as AOIm. The dwell time per AOI (AOI dwell time) and the dwell time per trial (i.e., the total stimulus dwell time, TOT dwell time) were computed, along with the percentage ratio of time spent fixating the AOI over the time spent fixating the whole stimulus (% AOI/TOT dwell time).

Statistical analysis

The statistical analyses were performed using IBM SPSS v.23 statistical software. Data results were not normally distributed (Shapiro–Wilk test), and hence non-parametric statistics were used. Friedman's analysis was used to detect differences among the three conditions (GR, R1.5, and

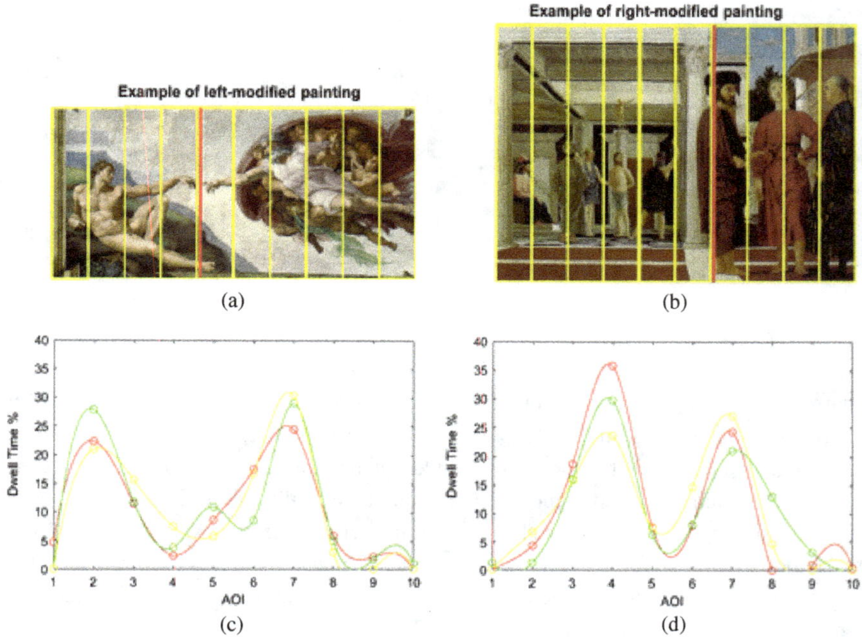

Figure 4. Example of horizontal stimuli in which ratio alteration (performed in correspondence of red vertical lines) occurred in the left (panel a) or right (panel b) side. Panels (c) and (d) show the distribution of dwell times for the paintings on the left in each interest area (defined by the yellow vertical lines in panels a and b), comparing the three experimental proportions: GR (red), R1.5 (green), and R1.8 (yellow).

R1.8), and Wilcoxon test for the paired sample was used for *post hoc* analysis between conditions. Spearman's rank correlation coefficient was computed to verify possible comparisons among these parameters and the given aesthetic judgment. The alpha level of significance was set at 0.05, but Bonferroni's correction was applied for *post hoc* multiple comparisons, reducing the alpha level at 0.025 for *post hoc* tests and at 0.017 for correlations.

Results of Experiment 2

The aesthetic judgments given in Experiment 2 were analogous to those found in Experiment 1. The judgments assigned to the GR stimulus resulted slightly but significantly higher in GR (6.71 ± 1.36) than in R1.5 (6.45 ± 1.45, $p = 0.046$), whereas those of R1.8 showed higher variability (6.34 ± 1.61, $p = 0.056$). In general, the score assigned to GR was higher than the mean score assigned to the other two proportions ($p = 0.003$), in particular for human figures (6.87 ± 1.62 vs. 6.54 ± 1.55, $p = 0.015$). Figure 3 shows an example of fixation distribution on a vertical stimulus in terms of average percentages of AOI dwell time for GR, R1.5, and R1.8.

As shown in Figure 3, the face (AOI2) was the most observed, with another peak around the navel. This double-peak pattern was evident for human photographs, sculptures, and virtual humanoids. For the photographs of humans and sculptures, when the stimulus was altered, the gaze moved accordingly: If the height of the navel was lowered, as in R1.5, the eyes moved downward and time spent in the lower AOI4 increased; when the height of the navel was lifted up, as in R1.8, the eyes moved upward, and time spent in AOI3 increased. This pattern was not observed for virtual humanoids, for which the face was observed for more than half of the total dwell time.

Friedman's test highlighted significant differences among GR, R1.5, and R1.8 regarding dwell time spent to explore the AOIm (chi-squared = 6.845, $p = 0.033$) and regarding the percentage AOIm dwell time/TOT dwell time (chi-squared = 6.345, $p = 0.042$) but not in the TOT dwell time (chi-squared = 0.808, $p = 0.668$). *Post hoc* analyses with Wilcoxon test showed a statistical difference for AOIm dwell time only regarding the comparison between R1.8 and R1.5 ($p = 0.009$). Furthermore, the percentage of AOIm compared to the total was different between GR and R1.5 ($p = 0.009$) as well as between R1.5 and R1.8 ($p = 0.008$).

Table 3 reports the correlations between the eye movement parameters and the reported esthetic judgments, resulting in a significant correlation to AOIm time and AOIm/TOT dwell time percentages only for GR stimuli.

Table 3 reports the correlations among the eye movement parameters and the reported aesthetic judgments. The aesthetic judgment significantly correlated to AOIm time and AOIm/TOT dwell time percentage only for GR stimuli.

Table 3. Spearman's rank correlation coefficient (*R*) and relevant *p*-values (in bold if statistically significant) computed between aesthetic judgment and temporal parameters of eye movements (Experiment 2).

		R1.5			GR			R1.8		
Eye movement parameters		**AOIm**	**TOT**	**AOIm/ TOT (%)**	**AOIm**	**TOT**	**AOIm/ TOT (%)**	**AOIm**	**TOT**	**AOIm/ TOT (%)**
Judgment correlation	*R*	−0.108	−0.018	−0.076	−0.175	−0.112	−0.175	−0.059	−0.069	−0.049
	p	0.085	0.774	0.225	**0.005**	0.074	**<0.001**	0.350	0.275	0.436

Note: AOIm dwell time stands for the average value of fixation time spent in exploring the area of interest in which the stimulus ratio has been modified; TOT dwell time refers to the total fixation time spent exploring the stimulus; AOIm/TOT (%) is the ratio between the average fixation time spent exploring the AOIm with respect to the total dwell time expressed in percentage.

Finally, the %AOI/TOT dwell time grouped according to the category of the stimulus showed statistically significant differences ($p<0.005$) only for painting stimuli among the three proportions.

Figure 4 shows these results, with a trend similar to those of Figure 3 but along the horizontal axis. Again, the pattern followed the modification of the ratio between the two parts of the figure. In the left-modified stimuli (as the Creation of Adam), significant differences were found for AOI3 (chi-squared = 13.273, $p = 0.001$) with higher percentage dwell time for R1.8, as well as AOI5 for R1.5 (chi-squared = 12.991, $p = 0.002$). In The Flagellation of Christ, the AOIm was on the right: The AOI7 received more fixation time when the division was put in its middle as in R1.8, than in R1.5, for which the division moved in AOI6.

Discussion

The main purpose of this study was to establish whether the presence of the GR may favor the perception of beauty in aesthetic judgments and which factors might influence it, clarifying the main controversies found in previous studies (Benjafield & Adams-Webber, 1976; Berlyne, 1970; Boselie, 1984; Davis & Jahnke, 1991; Di Dio *et al.*, 2007; Fechner, 1865; Green, 1995; Witmer, 1893; Zeising, 1855).

First of all, we found a slight but significant preference for GR, but its entity depended on the category of the stimulus. This preference was statistically significant for human photographs, in both the experiments, and for sculptures and paintings in Experiment 1 (being only close to a significant threshold for virtual humanoid figures).

Conversely, we did not find a significant preference for GR in geometric stimuli. These findings are in line with some previous results (Di Dio *et al.*, 2007, 2011) but in contrast with other studies finding a preference for GR also for geometric figures (Fechner, 1865; Russell, 2000). There are many possible explanations for the high variability among studies related to the possible preference for golden rectangles. First of all, it could be related to different methodologies related to how central tendency was computed: by means of mean, median, or modal values, with the GR that could emerge as mean value even if it was not the most frequent choice, as also previously suggested (Green, 1995). Indeed, in contrast to previous research on lines and geometric shapes' aesthetic preference, recent studies used more ecological experimental paradigms including natural and anthropomorphic figures, highlighting that geometric simple stimuli could be poorly attractive for our culture (Di Dio *et al.*, 2007, 2011). Furthermore, with respect to the study of Fechner (1865) and successive ones, subjects are now more often exposed to rectangular shapes different from GR, such as those of television screens, monitors, or smartphones having proportions closer to 1.8 (16:9 = 1.78, or 1980:1020 pixels = 1.78) or also paper sheets closer to 1.5 (A4: 1.41).

The GR is often present in many natural stimuli (Iosa *et al.*, 2018). Indeed, subjects might prefer stimuli recognized as original, as in our study occurred for human photographs, with GR preferred for stimuli originally in GR (oGR vs. noGR), despite participants not knowing which was the original stimuli.

The golden section was conceivably identified by anatomical Greek investigations in the 5th century BCE (as in the canon of Protagoras) with the idea to reproduce the harmony of the human body in the anthropomorphic sculptures and then in other artworks (Haines & Davies, 1904; Iosa *et al.*, 2018). It should be considered that anthropomorphic stimuli have their own symmetry given by the right–left relationship along the horizontal axis, whereas the golden section defines the vertical harmony of a standing human being. Hence, the GR preference could occur more frequently on the vertical than horizontal axis. Despite our studies not investigating this aspect, it should be noted that the rectangles of Fechner (1890) had the longer segment along the vertical axis, whereas in our and other studies (McManus, 1980; Russell, 2000), the longer dimension of rectangles and that of the bisected line was the horizontal one.

Despite this, Renaissance artists provided a horizontal division of their paintings according to the golden section, such as in The Flagellation of Christ and The Creation of Adam (de Campos *et al.*, 2015), with these stimuli that can be mentally divided into two parts and visually explored horizontally. A slight preference for GR was found in our paintings (54.6%), quite independent if the original stimuli were or not in golden proportion. Wölfflin (1994) questioned the preference for GR when observing an abstract rectangle because it should be related to a mental comparison of the lengths of basis, height, and their sum, again claiming that GR may "present an average measure conforming to man".

This hypothesis put in relationship the GR with "objective beauty" because it was related to a harmonic property intrinsic to the stimuli, but there is also an alternative hypothesis, reported in the so-called constructal theory, referring to "perceptual beauty" suggesting that the preference for GR was related to the hypothesis that visual system scans the world approximatively in golden proportion (Bejan, 2009). Di Dio *et al.* (2007) also identified the "subjective beauty" driven by one's own emotional experiences and related to the activation of the amygdala, whereas the objective beauty related to the intrinsic properties of the stimulus is based on a joint activation of sets of cortical neurons of lateral occipital cortex, parietal cortex, and anterior insula. All these processes could be not mutually exclusive and cohabit, forming our sense of beauty.

This integrated approach works even better considering GR as an affordance, and hence a property intrinsic to the stimulus that is recognized by our visual system and favoring the visual process. In fact, Gibson proposed the theory of affordances, suggesting that subjects noted the features of a stimulus that may constitute functional relations between the stimulus and the perceiver (Gibson, 2014; Lobo *et al.*, 2018). The abstraction of geometric figures and even of virtual humanoids might hence have mitigated this preference, in accordance with the ecological approach suggested by Gibson for the visual system (Gibson, 2014). The cultural level of subjects did not seem to influence the general preference for GR. This result could be read in conjunction with previous findings that the aesthetic judgment of beauty was mainly independent of sociocultural factors, intelligence, personality, and age (Eysenck & Tunstall, 1968). However, in our study, culture was assessed in terms of scholarly level, but different cultures were not analyzed, and further studies could investigate the preferences for GR comparing subjects of Western versus Eastern Countries.

When we analyzed single categories of stimuli, the preference for GR in human photographs and sculptures was related to art knowledge. These two types of stimuli have in common the same criterion for GR: the ratio between the distance of the navel from the ground and that between the navel and the top of the head. Virtual humanoids had the same proportion, but the judgments about them seemed to be not affected by cultural level. Probably, this judgment mainly depended on the realism of their faces, as shown by the long time spent by observers looking at the virtual faces in Experiment 2. Furthermore, being human bodies on average in GR, humans might have reflected this proportion in the concept of beauty, and it led to the adoption of this parameter also in some classical artistic masterpieces, starting from anthropomorphic sculptures (Hernández-Castro, 2007).

Human beings may associate, more or less consciously, the concept of beauty with their own body proportions (Burrell, 1932), as is the case with the preference of symmetric figures that could be related to the symmetry of the human body (Evans *et al.*, 2012; Huang *et al.*, 2018; Mühlenbeck *et al.*, 2016). As well as symmetry being considered as an affordance in accordance with the ecological approach suggested by Gibson for vision (2014), the GR that can be considered as a symmetry of higher order (Liu & Sumpter, 2018) could also be another visual natural affordance, preferred by humans and put in artworks by artists.

In our study, the proposed paintings have the longest dimension horizontal; indeed, the golden ideal division was along the horizontal axis. A slight preference (54.6%) was found for GR, quite independently if the stimulus was or was not originally in GR. With respect to the geometric figures, participants could perceive the harmony of a scene ideally divided into two parts according to the GR, even if this scene is represented with the horizontal length longer than the vertical. Considering that for symmetry there is a perceptual sensitivity to the orientation of the stimulus (1905; Fisher & Bornstein, 1982; Mach, 1883), and that there are no studies that directly investigated the relationship between GR and the orientation of the stimulus, future studies should investigate whether there is a dependence on the orientation with which it is presented the stimuli and the possible preference for GR.

Despite our study lacking this direct comparison, the bimodal distribution of dwell time found in Experiment 2 for human figures for which the golden section regards the vertical axis is similarly replicated for the paintings in which it regards the horizontal axis. In general, the results of Experiment 2 were found in accordance with those of Experiment 1 and seem to confirm the hypothesis of GR as an affordance. In Experiment 2, we found no significant differences between the three proportions related to the total exploration time but also some significant differences with respect to specific areas of interest. Further, analyzing the dwell times, significant differences among GR, R1.5, and R1.8 were found regarding the exploration of the AOIm, that is, the specific AOI where the image modification was performed. A significant negative correlation was found between the AOIm dwell time and the aesthetic judgment only for GR stimuli. This could indicate that the subject does not need to observe that area for too long to judge it as beautiful if it is in golden proportion.

Despite previous studies showing that the aesthetic appreciation of beauty is not based on an "immediate" response and may even require more time to discriminate the main characteristics of the stimuli (Bar *et al.*, 2006; Evans *et al.*, 2012; Huang *et al.*, 2018; Liu & Sumpter, 2018; Mühlenbeck *et al.*, 2016), the presence of GR may have reduced this time because it was quickly recognized and preferred by the observers. So, hypothesizing GR as an affordance, it may be implicitly recognized, facilitating the processing of visual information relating to a stimulus, making aesthetic experience more fluent, in accordance with the "processing fluency theory" for which that fluency could contribute to the pleasantness of the experience associated with the perception of beauty (Witmer, 1893).

It is noteworthy that the GR divides the human stature in a region roughly close to the body's center of mass and could even play an important role in human motor behavior (Iosa *et al.*, 2018). It should be noted that the exposition to GR does not regard only the visual system: GR was recently found as a harmonic feature of voluntary and involuntary rhythmic human movements, such as cardiac (Yetkin *et al.*, 2013), respiratory (Iared *et al.*, 2016), and walking rhythms (Iosa *et al.*, 2013; Iosa *et al.*, 2016; Iosa *et al.*, 2019; Serrao *et al.*, 2017). In fact, walking performed with GR between gait phases allows humans to harmonize locomotor acts (Iosa *et al.*, 2016; Iosa *et al.*, 2019), thus reducing energy expenditure (Serrao *et al.*, 2017). Furthermore, it has been demonstrated that perception of internal gait rhythm may be modulated (De Bartolo, Belluscio, *et al.*, 2020a) and that the harmonic structure of music can ameliorate harmony of walking in patients with Parkinson's disease (De Bartolo, Morone, *et al.*, 2020b), especially if acoustic stimulation presents GR harmony (Belluscio *et al.*, 2021), confirming the possibility to consider GR as an affordance to exploit by the observer/listener/acting subject.

However, we must bracket this proposal with some caveats. The main limitation of Experiment 1 was the online administration of the questionnaire that did not allow for controlling some variables such as the eyes' positions with respect to the screen, the response time of the subjects, or their attention level. For example, the questionnaire was formed by 51 items, and it might have led to a progressive physiological decline in the participants' attention due to fatigue.

However, the online administration allowed for enrolling a wide sample of subjects ($N = 256$) that could have attenuated the above-mentioned limits. Experiment 2 was performed in more controlled conditions, and its results confirmed those obtained in Experiment 1. Subjects were required to judge the most liked stimulus, without any investigations related to the judgment of naturalness, so caution is needed in discussing the role of naturalness of stimuli and its relationship to the preference for stimuli originally or not in GR.

Finally, in our study, we did not manipulate canonical orientation of stimuli, so we included only vertical stimuli for anthropomorphic figures and horizontal for images of artworks. Further studies should investigate the possible differences in the GR preferences for vertical versus horizontal stimuli, and these results could contribute to understanding the relationship between GR and visual scan process.

Conclusions

The results of this study showed a general preference for GR, although not for all types of stimuli; the second one showed that eye fixations were concentrated in the areas related to GR and, when stimuli are in that proportion, a higher aesthetic judgment was correlated with a lower dwell time. The GR has some features analogous to symmetry: both are geometric characteristics widely present in nature (Iosa *et al.*, 2018), including physiological rhythms (Belluscio *et al.*, 2021; Iared *et al.*, 2016; Iosa *et al.*, 2013; Iosa *et al.*, 2016; Iosa *et al.*, 2019; Serrao *et al.*, 2017; Yetkin *et al.*, 2013), both may facilitate the visual processing, hence both could be considered as ecological affordances (Bar *et al.*, 2006; Belluscio *et al.*, 2021; Burrell, 1932; Evans *et al.*, 2012; Gibson, 2014; Huang *et al.*, 2018; Iared *et al.*, 2016; Iosa *et al.*, 2013; Iosa *et al.*, 2016; Iosa *et al.*, 2019; Liu & Sumpter, 2018; Lobo *et al.*, 2018; Serrao *et al.*, 2017; Yetkin *et al.*, 2013), and may be related to the perception of beauty, as artists have well known since ancient time.

References

Angier, R. P. (1903). *The Aesthetics of Unequal Division*, Doctoral Dissertation, Harvard University.

Bar, M., Neta, M., & Linz, H. (2006). Very first impressions. *Emotion*, 6(2), 269–278. https://doi.org/10.1037/1528-3542.6.2.269.

Bastioni, M., Re, S., & Misra, S. (2008, January). Ideas and methods for modeling 3D human figures: The principal algorithms used by MakeHuman and their implementation in a new approach to parametric modeling. In *Proceedings of the 1st Bangalore Annual Compute Conference*, p. 10. ACM. https://doi.org/10.1145/1341771.1341782.

Bejan, A. (2009). The golden ratio predicted: Vision, cognition and locomotion as a single design in nature. *International Journal of Design and Nature and Ecodynamics*, 4(2), 97–104. https://doi.org/10.2495/DNE-V4-N2-97-104.

Belluscio, V., Iosa, M., Vannozzi, G., Paravati, S., & Peppe, A. (2021). Auditory Cue based on the Golden ratio can improve gait patterns in people with Parkinson's disease. *Sensors*, 21(3), 911. https://doi.org/10.3390/s21030911.

Benjafield, J. & Adams-Webber, J. (1976). The golden section hypothesis. *British Journal of Psychology*, 67(1), 11–15. https://doi.org/10.1111/j.2044-8295.1976.tb01492.x.

Berlyne, D. E. (1970). The golden section and hedonic judgments of rectangles: A cross-cultural study. *Sciences de l'art/Scientific Aesthetics*, 7(1), 1–6. https://doi.org/10.2307/1422021.

Boselie, F. (1984). The aesthetic attractivity of the golden section. *Psychological Research Psychologische Forschung*, 45(4), 367–375. https://doi.org/10.1007/BF00309712.

Bundgaard, P. F. (2015). Feeling, meaning, and intentionality — A critique of the neuroaesthetics of beauty. *Phenomenology and the Cognitive Sciences*, 14(4), 781–801. https://doi.org/10.1007/s11097-014-9351-5.

Bundgaard, P. F., Heath, J., & Østergaard, S. (2017). Aesthetic perception, attention, and non-genericity: How artists exploit the automatisms of perception to construct meaning in vision. *Cognitive Semiotics*, 10(2), 91–120. https://doi.org/10.1515/cogsem-2017-0011.

Burrell, P. S. (1932). Man the measure of all things: Socrates versus Protagoras (I). *Philosophy*, 7(25), 27–41. https://www.jstor.org/stable/3747050.

Chassy, P., Lindell, T. A., Jones, J. A., & Paramei, G. V. (2015). A relationship between visual complexity and aesthetic appraisal of car front images: An eye-tracker study. *Perception*, 44(8–9), 1085–1097. https://doi.org/10.1177/0301006615596882.

Chenier, T. & Winkielman, P. (2018). The origins of aesthetic pleasure: Processing fluency and affect in judgment, body, and the brain. In *Neuroaesthetics*, pp. 275–289. London, England: Routledge.

Danikas, D. & Panagopoulos, G. (2004). The golden ratio and proportions of beauty. *Plastic and Reconstructive Surgery*, 114(4), 1009. https://doi.org/10.1097/01.PRS.0000138702.13724.26.

Davis, F. C. (1933). Aesthetic proportion. *The American Journal of Psychology*, 45(2), 298–302. https://doi.org/10.2307/1423158.

Davis, S. T. & Jahnke, J. C. (1991). Unity and the golden section: Rules for aesthetic choice? *The American Journal of Psychology*, 104(2), 257–277. https://doi.org/10.2307/1423158.

Davis, T. A. & Altevogt, R. (1979). Golden mean of the human body. *Fibonacci Quarterly*, 17(1979), 340–344.

De Bartolo, D., Belluscio, V., Vannozzi, G., Morone, G., Antonucci, G., Giordani, G., Santucci, S., Resta, F., Marinozzi, F., Bini, F., Paolucci, S., & Iosa, M. (2020a). Sensorized assessment of dynamic locomotor imagery in people with stroke and healthy subjects. *Sensors*, 20(16), 4545. https://doi.org/10.3390/s20164545.

De Bartolo, D., Morone, G., Giordani, G., Antonucci, G., Russo, V., Fusco, A., Marinozzi, F., Bini, F., Spitoni, G. F., Paolucci, S., & Iosa, M. (2020b). Effect of different music genres on gait patterns in Parkinson's disease. *Neurological Sciences*, 41(3), 575–582. https://doi.org/10.1007/s10072-019-04127-4.

de Campos, D., Malysz, T., Bonatto-Costa, J. A., & Oxley da Rocha, A. (2015). More than a neuroanatomical representation in the creation of Adam by Michelangelo Buonarroti, a representation of the Golden ratio. *Clinical Anatomy*, 28(6), 702–705. https://doi.org/10.1002/ca.22580.

Di Dio, C., Canessa, N., Cappa, S. F., & Rizzolatti, G. (2011). Specificity of esthetic experience for artworks: An fMRI study. *Frontiers in Human Neuroscience*, 5, 139. https://doi.org/10.3389/fnhum.2011.00139.

Di Dio, C., Macaluso, E., & Rizzolatti, G. (2007). The golden beauty: Brain response to classical and renaissance sculptures. *PLoS One*, 2(11), e1201. https://doi.org/10.1371/journal.pone.0001201.

Evans, D. W., Orr, P. T., Lazar, S. M., Breton, D., Gerard, J., Ledbetter, D. H., Janosco, K., Dotts, J., & Batchelder, H. (2012). Human preferences for symmetry: Subjective experience, cognitive conflict and cortical brain activity. *PLoS One*, 7(6), e38966. https://doi.org/10.1371/journal.pone.0038966.

Eysenck, H. J. & Tunstall, O. (1968). La personnalité et l'esthétique des formes simples [Personality and aesthetics of simple forms]. *Sciences de l' Art/ Scientific Aesthetics*, 5, 3–9.

Fechner, G. T. (1865). Über die Frage des goldenen Schnitts [About the question of the golden section]. *Archiv fur diezeichnenden Kunste*, 11, 100–112.

Fechner, G. T. (1890). *Wissenschaftliche briefe von Gustav Theodor Fechner u. W. Preyer* [*Scientific letters from Gustav Theodor Fechner and W. Preyer*], 1st edn. Verlag v. L. Voss.

Fink, B. & Penton-Voak, I. (2002). Evolutionary psychology of facial attractiveness. *Current Directions in Psychological Science*, 11(5), 154–158. https://doi.org/10.1111/1467-8721.00190.

Fisher, C. B. & Bornstein, M. H. (1982). Identification of symmetry: Effects of stimulus orientation and head position. *Perception & Psychophysics*, 32(5), 443–448. https://doi.org/10.3758/BF03202774.

Gamwell, L. (2015). *Mathematics and Art: A Cultural History*. Princeton, NJ: Princeton University Press.

Gibson, J. J. (2014). *The Ecological Approach to Visual Perception.* Classic Edition. New York, NY: Psychology Press. https://doi.org/10.4324/9781315740218.

Godkewitsch, M. (1974). The 'golden section': An artifact of stimulus range and measure of preference. *The American Journal of Psychology*, 43(1), 269–277. https://doi.org/10.2307/1422021.

Green, C. D. (1995). All that glitters: A review of psychological research on the aesthetics of the golden section. Perception, 24(8), 937–968. https://doi.org/10.1068/p240937.

Haines, T. H. & Davies, A. E. (1904). The psychology of aesthetic reaction to rectangular. *Psychological Review*, 11, 249–281. https://doi.org/10.1037/h0076096.

Hernández-Castro, F. (2007). *Study of principles and methods of proportion* (Doctoral dissertation, Universität Duisburg-Essen, Fakultät für Geisteswissenschaften» Institut für Kunst und Kunstwissenschaft» Institut für Kunst-und Designwissenschaften [IKUD]).

Huang, Y., Xue, X., Spelke, E., Huang, L., Zheng, W., & Peng, K. (2018). The aesthetic preference for symmetry dissociates from early-emerging attention to symmetry. *Scientific Reports*, 8(1), 6263. https://doi.org/10.1038/s41598-018-24558-x.

Iared, V. G., de Oliveira, H. T., & Payne, P. G. (2016). The aesthetic experience of nature and hermeneutic phenomenology. *The Journal of Environmental Education*, 47(3), 191–201. https://doi.org/10.1080/00958964.2015.1063472.

Iosa, M. (2018). Il numero meraviglioso: la sezione aurea. Bellezza della matematica, armonia dell'universo, musica della natura [*The Astonishing Number: The Golden Ratio. Beauty of Mathematics, Harmony of The Universe, Music of Nature*], First Edition. Edited by Tangram Scientific Edition.

Iosa, M., De Bartolo, D., Morone, G., Boffi, T., Mammucari, E., Vannozzi, G., Bini, F., Marinozzi, F., Antonucci, G., & Paolucci, S. (2019). Gait phase proportions in different locomotion tasks: The pivot role of golden ratio. *Neuroscience Letters*, 699, 127–133. https://doi.org/10.1016/j.neulet.2019.01.052.

Iosa, M., Fusco, A., Marchetti, F., Morone, G., Caltagirone, C., Paolucci, S., & Peppe, A. (2013). The golden ratio of gait harmony: Repetitive proportions of repetitive gait phases. *BioMed Research Internationa*l, 2013, 918642. https://doi.org/10.1155/2013/918642.

Iosa, M., Morone, G., Bini, F., Fusco, A., Paolucci, S., & Marinozzi, F. (2016). The connection between anthropometry and gait harmony unveiled through the lens of the golden ratio. *Neuroscience Letters*, 612, 138–144. https://doi.org/10.1016/j.neulet.2015.12.023.

Iosa, M., Morone, G., & Paolucci, S. (2018). Phi in physiology, psychology and biomechanics: The golden ratio between myth and science. *Biosystems*, 165, 31–39. https://doi.org/10.1016/j.biosystems.2018.01.001.

Kong, F., Zhang, Y., Tian, Y., Fan, C., & Zhou, Z. (2015). Self-relevant beauty evaluation: Evidence from an event-related potentials study. *Journal of Integrative Neuroscience*, 14(01), 85–95.

Kumar, M. & Garg, N. (2010). Aesthetic principles and cognitive emotion appraisals: How much of the beauty lies in the eye of the beholder? *Journal of Consumer Psychology*, 20(4), 485–494. https://doi.org/10.1016/j.jcps.2010.06.015.

Kwart, D. G., Foulsham, T., & Kingstone, A. (2012). Age and beauty are in the eye of the beholder. *Perception*, 41(8), 925–938. https://doi.org/10.1068/p7136.

Liu, Y. & Sumpter, D. J. (2018). Is the golden ratio a universal constant for self-replication? *PLoS One*, 13(7), e0200601. https://doi.org/10.1371/journal.pone.0200601.

Lobo, L., Heras-Escribano, M., & Travieso, D. (2018). The history and philosophy of ecological psychology. *Frontiers in Psychology*, 9, 2228. https://doi.org/10.3389/fpsyg.2018.02228.

Mach, E. (1883/1912] 1988). In R. Wahsner & H.-H. Borzeszkowski (eds.), *Die Mechanik in ihrer Entwicklung historisch-kritisch dargestellt* [*The Mechanics in Their Development Presented Historically and Critically*], 7th Edition. Akademie-Verlag. See McCormack (tr.) [1893/1960] 1974.

Mach, E. (1905/1926] 2002). *Erkenntnis und Irrtum: Skizzen zur Psychologie der Forschung* [*Knowledge and Error: Sketches for the Psychology of Research*], Reprint of the 5th Edition. In M. Eberhardt (ed.), Parerga. See McCormack and Foulkes (trs.) 1976.

McManus, I. C. (1980). The aesthetics of simple figures. *British Journal of Psychology*, 71(4), 505–524. https://doi.org/10.1111/j.2044-8295.1980.tb01763.x.

McManus, I. C., Cook, R., & Hunt, A. (2010). Beyond the golden section and normative aesthetics: Why do individuals differ so much in their aesthetic preferences for rectangles? *Psychology of Aesthetics, Creativity, and the Arts*, 4(2), 113–126. https://doi.org/10.1037/a0017316.

Mühlenbeck, C., Liebal, K., Pritsch, C., & Jacobsen, T. (2016). Differences in the visual perception of symmetric patterns in orangutans (Pongo pygmaeus abelii) and two human cultural groups: A comparative eye-tracking study. *Frontiers in Psychology*, 7, 408. https://doi.org/10.3389/fpsyg.2016.00408.

Nienstedt Jr., C. W. & Ross, S. (1951). Preferences for rectangular proportions in college students and the aged. *The Pedagogical Seminary and Journal of Genetic Psychology*, 78(2), 153–158. https://doi.org/10.1080/08856559.1951.10533573.

Piehl, J. (1976). The 'golden section': An artifact of stimulus range and demand characteristics. *Perceptual and Motor Skills*, 43(1), 47–50. https://doi.org/10.2466/pms.1976.43.1.47.

Ramachandran, V. S. & Hirstein, W. (1999). The science of art: A neurological theory of aesthetic experience. *Journal of Consciousness Studies*, 6(6–7), 15–51.

Reber, R. (2012). Processing fluency, aesthetic pleasure, and culturally shared taste. In A. P. Shimamura & S. E. Palmer (eds.), *Aesthetic Science: Connecting Minds, Brains, and Experience*, pp. 223–249. Oxford University Press.

Regmi, P. R., Waithaka, E., Paudyal, A., & van Teijlingen, E. (2016). Guide to the design and application of online questionnaire surveys. *Nepal Journal of Epidemiology*, 6(4), 640–644. https://doi.org/10.3126/nje.v6i4.17258.

Russell, P. A. (2000). Testing the aesthetic significance of the golden-section rectangle. *Perception*, 29(12), 1413–1422. https://doi.org/10.1068/p3037.

Serrao, M., Chini, G., Iosa, M., & Draicchio, F. (2017). Harmony as a convergence attractor that minimizes the energy expenditure and variability in physiological gait and the loss of harmony in cerebellar ataxia. *Clinical biomechanics*, 48, 15–23. https://doi.org/10.1016/j.clinbiomech.2017.07.001.

Thompson, G. G. (1946). The effect of chronological age on aesthetic preferences for rectangles of different proportions. *Journal of Experimental Psychology*, 36(1), 50–58. https://doi.org/10.1037/h0054675.

Wever, E. G. & Zener, K. E. (1928). The method of absolute judgment in psychophysics. *Psychological Review*, 35(6), 466–493. https://doi.org/10.1037/h0075311.

Witmer, L. (1893). *Zur experimentellen Aesthetik einfacher räumlicher Formverhältnisse* [*The Experimental Aesthetics of Simple Spatial Form Relationships*]. Philosophische Studen, 9, 96144.

Wölfflin, H. (1994). Prolegomena zu einer Psychologie der Architektur, 1886 [Prolegomena to a psychology of architecture, 1886]. In H. F. Mallgrave & E. Ikonomou (eds.), *Empathy, Form, and Space: Problems in German Aesthetics*, 1873–1893, pp. 149–190. Getty Center for the History of Art and the Humanities.

Yetkin, G., Sivri, N., Yalta, K., & Yetkin, E. (2013). Golden ratio is beating in our heart. *International Journal of Cardiology*, 168(5), 4926–4927. https://doi.org/10.1016/j.ijcard.2013.07.090.

Zeising, A. (1855). *Aesthetische Forschungen* [*Aesthetic Research*]. Frankfurt, Germany: Medinger.

Zeki, S. (1999a). *Inner Vision: An Exploration of Art and the Brain*. Oxford, England and New York, NY: Oxford University Press.

Zeki, S. (1999b). Art and the brain. *Journal of Consciousness Studies*, 6, 76–96.

Zhang, Y., Luo, N., Hong, F. F., Yang, C. H., Xie, Y. F., Wu, J. Y., Wang, G. X., Zhao, P. Q., Chen, J. W., & Aashiq, K. (2020). Attractiveness-related recognition bias captures the memory of the beholder. *Journal of Integrative Neuroscience*, 19(4), 629–639. https://doi.org/10.31083/j.jin.2020.04.166.

Zhang, Z. & Deng, Z. (2012). Gender, facial attractiveness, and early and late event-related potential components. *Journal of Integrative Neuroscience*, 11(4), 477–487. https://doi.org/10.1142/S0219635212500306.

https://doi.org/10.1142/9789811298639_0014

Chapter 14

Neuroaesthetics and the Iconography in Photography*

Hennric Jokeit[†] and Daniel Blochwitz[‡]

[†]*Department of Neuropsychology, University of Zurich and Swiss Epilepsy Centre, Zürich, Switzerland*
[‡]*Photography Curator, Zürich, Switzerland*

Abstract

Can neurosciences explain art? No, but it can help us to understand why some images are more memorable and, thus, more successful than others. This chapter aims to identify certain factors that may influence the artistic success of photographic images. These factors are discussed within the contexts of basic neuropsychological concepts, visual perception, and visual memory. A new computational and neuroscientifically based model, the *predictive coding theory*, provides a powerful framework for integrating social and individual factors that influence aesthetic experience and activity. A case study of Dorothea Lange's iconic photograph *Migrant Mother* demonstrates the importance of identifiable factors that influence and determine a photograph's potential success. We are convinced that a future systemic approach will enable the

*This chapter was originally published in the special collection on neuroaesthetics in *PsyCh Journal*. Jokeit, H. & Blochwitz, D. (2020), Neuro-aesthetics and the iconography in photography. *PsyCh Journal*, 9(4), 444–457. https://doi.org/10.1002/pchj.379.

complementary integration of neuroscientific, perceptual, cognitive, emotional, and sociopsychological insights through the framework of *predictive coding theory* with socioscientific, art-theoretical, and art-historical as well as neuro- and behavioral-economical models.

Keywords: art; memory; neuroscience; photography; predictive coding theory; visual system

In this chapter, we combine concepts from the neurosciences with knowledge of the social production and appropriation conditions of artistic photography to identify factors that are likely to influence the artistic success of photographic images. In line with Ernst Gombrich (1995), we understand artistic success to be a publicly recognized formal and content-related contribution of an image to the further development of the artistic medium itself as well as its public and academic perception.

When the first photographs found an enthusiastic audience some 180 years ago, it was an event with far-reaching consequences for modern bourgeois society and its art. The Impressionism that developed during the 1860s can also be interpreted as a reaction to the loss of painting as the foremost medium of public and private pictorial memory. As soon as it had outgrown its infancy, photography almost completely usurped the role of the predominant pictorial medium. The roll film, introduced at the end of the 19th century, and its external development were the prerequisites for the incipient capitalist mass production and exploitation of photography. But the new medium and its technological advances also carried the spark of further liberation and radical emancipation. Professionalized photographic practices were established in journalistic, artistic, scientific, medical-diagnostic, and documentary fields. Hobby photography soon became an integral part of bourgeois leisure activity and everyday culture, later followed by proletarian hobbyist practices when access to photographic tools and materials became more affordable and readily available.

For more than a century, a vast industry has controlled the development and production of most available cameras, their lenses, chemistry, films, and papers, and for the past quarter century following the decline of analog photographic technologies, sensors, smartphones, screens, imaging software, and digital printers. Technically, photography today seems to be effortless, yet there is still the wonder at certain images and their obvious and invisible effects.

Social Framing

What makes artistic photography successful? Images that have found their way into our collective memory, such as those by Gertrude Käsebier, Lewis W. Hine, Alfred Stieglitz, Edward Weston, Dorothea Lange, August Sander, Ansel Adams, Lee Miller, Henri Cartier-Bresson, Gordon Parks, or Diane Arbus, for example, are now found in important museums, galleries, and private collections. At the same time, they are more frequently reproduced than other images in books, magazines, and online media. Digitally, many of these works are now available free of charge in high resolution, such as Dorothea Lange's (1936) *Migrant Mother*. This, however, also marks the beginning of the difficulty of a retrospective analysis of the causes, conditions, and effects of their success. The question as to whether the images were institutionally recognized because they have special qualities or whether they have received special attention because they were institutionally represented refers to the problem of *ex post facto* explanations for popularity and public artistic success.

The success of a picture may partly depend on visual design features but cannot be separated from its subjective and social significance and its effects on appropriating actions as practices of repeated viewing, remembering, collecting, evaluating, and sharing. Thus, object-related (objectivistic) ways of looking at things appear to us to be of limited suitability to the complexity of art. These initially tried to derive from certain visual characteristics and associated neurophysiological and psychological response patterns, species-invariant preference characteristics of artistic objects that were interpreted as criteria of beauty or aesthetics (Conway & Rehding, 2013; Shimamura, 2015; Zeki, 1999).

The fact that one quickly loses oneself in dead ends in thinking about and researching art and its related categories such as beauty, when searching for object-immanent qualities of art, was formulated early on by David Hume (1757) and Immanuel Kant (1790/1983) and later by representatives of a Marxist aesthetic (Hirdina, 1976; Lukács, 1963). What people from different epochs and cultures have judged to be beautiful, artistic, and worthy of preservation is simply too diverse and disparate, as ethnological collections impressively demonstrate. Every art theory, even one that refers to psychology or neuroscience, must be thought of in terms of social and political history, anchored in its particular time and space (Lukács, 1963). The individual engagement with art is described by those consciousness-related (subjectivistic) aesthetics as a very specific kind of aesthetic appropriation behavior. Art is thus always understood conceptually as a relationship and process between subjects and objects in visual arts or between subjects in performing arts. This concept of a process-related relationship is a highly complex, dynamic, interactive, and an ontologically hybrid one. The mutual relations exist between the art object, the subject, and society and are not only receiving but also shaping and communicating (Hirdina, 1976; Lukács, 1963).

The entanglement of object, subject, and society then takes place in individual appropriating action. Neuroaesthetics should take into account both the individually innate and acquired as well as the social conditions of aesthetic appropriation behavior. This activity of appropriation in turn defines conditions of visual perception that are at the center of neuroscientific approaches to the explanation of aesthetic phenomena. Perception is not a passive process to be thought of as linearly progressing from the sensory periphery but as a selective, informed, and purposeful productive activity for the maintenance and future improvement of homeostatic processes concerning biological, psychological, and social needs.

Cognitive and Affective Vision

The visual system is considered to be a model system for the investigation of perceptual processes because it represents the basic principles of information processing in the brain in an experimentally accessible way. No other sensory modality processes such large amounts of information. Thus, spatial position, form, color, speed, and identity are continuously made available so that the body immediately can respond physically and precisely to its visible environment (Pöppel, 1989). Certain visual stimulus classes such as fearful faces, snakes, spiders, or vomit trigger autonomous nervous system reactions even after exposure durations of less than 40 ms (Dolan, 2002; Vuilleumier, 2015); that is, long before a visual-semantic analysis, which takes at least 200 ms, signals the harmlessness of the stimulus, for example, because it is simply a photograph. At this point, blood pressure and the pulse are already increased by an inhibition of the vagus nerve and only fall again after several seconds. The very high speed of affective evaluation in parallel, but significantly slower, cognitive analyses produces a non-simultaneity that creates a predominance of affective stimulus processing at temporally early stages and thus also affectively tones further cognitive processes that neutralize or reevaluate the initial affective evaluation only in the later stages.

Relatively new findings have shown that an extreme compression of the visual signals to be processed begins already at the level of retinal ganglion cells (Gollisch & Meister, 2010). In about 24 distinct channels, retinal ganglion cells' signals reach the so-called *subcortical structures*, whose importance has long been underestimated (Bourne & Morrone, 2017; Podvigin, Jokeit, Pöppel, Chizh, & Kiselyeva, 1992). Early subcortical processing structures such as pulvinar and corpus geniculatum laterale form important integration and switching spots for the modulation of the selectivity of visual processing, known as *visual attention*. Rapid affective assessment, in which the amygdala probably plays a key role, focuses visual attention on emotionally significant information that can trigger either rewarding or aversive responses (Dolan, 2002; Vuilleumier, 2015). The control of visual attention, which manifests itself in easily measurable eye movements, is systematically adapted to the perceptual context through probabilistic reward and avoidance learning. Affective modulation of visual perception therefore occurs at basal and higher processing levels, including direct influences of amygdala, basal ganglia, and adrenergic projections.

From the beginning, information from the foveal and extrafoveal areas of the retina is processed separately (Kaplan & Shapley, 1986). The characteristics contained therein, such as form, movement, color, and luminance, are spatially distributed and represented in the extrastriatric visual cortex. The understanding of these specific processing pathways is complicated by the functional and anatomical distinction between a ventral stream that reveals the identity and meaning of visual objects (What is being processed?) and a dorsal stream that holds egocentric and allocentric spatial information (Where in my space is the object located?) available to them (Ungerleider & Haxby, 1994).

In addition to this, (a) cortical topographic separation of the processing of content and localization characteristics, (b) basal perceptual separation of an exactly resolving parvocellular and a rapidly operating magnocellular visual pathway, and (c) two complementary modes of information processing in the brain are to be distinguished: an ascending, bottom-up mode and a descending, top-down mode (Awh, Belopolsky, & Theeuwes, 2012). The bottom-up mode describes essentially ontogenetically fixed processing pathways in which physical stimuli of sensory cells are converted into neuronal signals that are further processed in a cascade of process steps. This standard model is used to explain why a certain color and luminance distribution of a visual stimulus can trigger a specific affective sensation. The objectivistic neuroaesthetic rationale would be that certain physical features activate specific neuron populations that evoke an aesthetic sensation (Chatterjee & Vartanian, 2014; Ramachandran & Freeman, 2001; Ramachandran & Hirstein, 1999).

Theoretically more complex is the so-called *top-down mode* of visual processing, which is central to the *predictive coding theory* (PCT) to be described later. In the top-down model of perception, first formulated by Hermann von Helmholtz (1860/1962), the process of perception is controlled by explicit prior knowledge and is based on active selection of information at very early stages of subcortical processing. The linking of visual attention and memory processes plays a central role here (Pelowski, Markey, Forster, Gerger, & Leder, 2017). Visual perception is therefore not a passive event that happens to the viewer but still an unconscious affective evaluation taking place in less than 40 ms that influences the further process of perception. At the same time, a multidimensional representation of typical meaningful characteristics emerges, including their topology, which top-down and recursively makes perception conscious as experience. This is accompanied by processes of lasting memory formation.

Pictorial Memory

Memory systems can be described with respect to their function and mode of action on sociological, neuropsychological, and neurophysiological levels. Memory enables an organism proactively to adapt to past experience by modifying current and future behavior. The sociological distinction of a society's collective memory as well as an intermediate community (family, friends) and an individual memory interlinks the cultural collective memory with the individual memory (Marcel & Mucchielli, 2008). Whereas collective memory appears as an object of communicative cultural techniques of memory, individual memory has a personal external, objectified side in the sense of potentially presentable analogous and digital objects (collections, catalogs) and an internal level of memory that is partially accessible to consciousness and thus partially communicable, enabling active retrieval as well as passive recognition.

In addition, there is an inherited, genetically fixed species memory in which unconditional reflexes and reflex-like reactions to stimuli such as spiders, snakes, fearful faces, and vomit are stored. In the history of art, there are numerous examples that operate with genetically fixed fear reactions. Probably the most prominent examples are Rubens (1618) and Caravaggio's (1596) "Head of Medusa". Surrounded by snakes to which even non-human primates react with fear due to species' memory, Rubens and Caravaggio showed the face of Medusa marked by fright and shock. Emotions in the face of conspecifics trigger mimicry-like processes of imitation in the viewer, which enable emotional contagion. It can be assumed that species memory and early postnatal learning intertwine here (Preston & de Waal, 2002). Individual visual or pictorial memory describes a complex neuropsychic memory system that is limited to the visual modality but is not isolated from linguistic processes. For aspects of time and content, it proves to be as heterogeneous as language-bound memory but is anthropogenetically much older. The pictorial memory, which plays a key role in the spatial orientation and identification of predators and prey to secure survival, is not subject to the capacity limits of sequentially operating language-bound memory.

Along the time axis, a sensory memory in the form of an ultrashort time memory (sensory register) is postulated to be 200–400 ms, which allows for stable further processing of the already compressed information from the retinal image (Atkinson & Shiffrin, 1968). The predominantly cortically represented visual working memory, on the other hand, allows topographic and semantic information of visual elements to be provided and executed in the seconds range for analysis and mental manipulation but can easily be disturbed by interfering stimuli. The probability of a transfer of information into long-term memory mostly depends on the duration and intensity of the processing in working memory. Long-term memory, which is now the focus of consideration, can be divided into a visual anterograde episodic memory and a visual semantic memory (Squire, 2004; Tulving, 2002).

Taxonomically, memories of pictorial information that we have seen only once can be assigned to anterograde episodic memory. This should be distinguished from semantic visual memory, in which we almost ineradibly remember visual content, because we have seen it countless times, perhaps every day, and usually have no memory of when we saw the object for the first time. A semantic memory is thus seen as a biographically decontextualized memory that no longer provides information about the place and time of its encoding. However, other semantically contextual information, such as the photographer's name or the epoch in which the image was created, is usually retrievable. This should be contrasted with an autobiographical episodic memory that stores particular events that can be located biographically (e.g., a first rendezvous).

When retrieved from long-term memory, two modes are distinguished: remembering and recognition. Memorability and recognizability are indisputably necessary conditions for the success of a picture. Memorability means that the potential to generate a memory image in front of an inner mental eye and to describe it verbally exists. Recognizability, on the other hand, means that we cannot necessarily actively remember the visual information but can recognize it from a selection. The complexity of the content and form of the visual content does not influence recognition performance in memory experiments (Madigan, 1983).

Which conditions lead to a visual event being inscribed in the individual memory? This is (a) the duration and intensity with which topographic and visual semantic analyses and feature comparisons are made in the visual working memory, (b) the repeated activity of the working memory through a renewed confrontation with the object or imagining the object, and (c) affective valence (positive and negative) and intensity (arousal) also play essential roles (Mastria, Ferrari, & Codispoti, 2017). Emotions strengthen the formation of memory; their absence impedes it. Amygdalar activation, always associated with emotions, adrenergically stimulates hippocampal activity, which is a bottleneck structure for the formation of memory (Phelps, 2004). The emotional valence of the image (d) has to be distinguished from its personal relevance. The subjective relevance is probably also mediated by the intensity of emotions that catalyze the formation of lasting memories. (e) "Novelty" is the door opener for an activation of processes that lead to encoding into memory. However, repetition of information without activation of working memory processes does not reliably contribute to the formation of memory.

According to the top-down model, the visual memory of our environment essentially shapes the process of perception, which is shaped by probabilistic experience (e.g., light comes from above in nature), visual semantic knowledge (e.g., mushrooms preferably grow on the forest floor) supplemented by possible episodic events (e.g., mushroom poisoning). According to the PCT, only prediction errors are processed because ultimately only these have a high value for behavioral adaptation and, on the other hand, can be processed very economically. Regardless of situation and object, memory representation of our visual environment would thus be the key to understanding perceptual (aesthetic) behavior.

Predictive Coding

For some years now, a new, explanatory computational approach has dominated the cognitive sciences and cognitive neurosciences, which is also important for the further development of artificial intelligence, the PCT (Adams, Shipp, & Friston, 2013; Clark, 2013; Henson & Gagnepain, 2010;

Kirov, 2016; McNally, Johansen, & Blair, 2011; Pessiglione, Seymour, Flandin, Dolan, & Frith, 2006; Rao & Ballard, 1999; Rauss, Schwartz, & Pourtois, 2011). This concept describes perception as an active process of foresighted coding and not as a passive "information sponge." The consequences of this approach for neuroaesthetic models are obvious and have already been taken up in initial theoretical discussions (Kesner, 2014; Van de Cruys & Wagemans, 2011).

The PCT is a change of perspective in an attempt to answer one of the central basic questions of neuroscience: How can an immense amount of information be economically processed by a comparatively small organ, the brain, which operates very slowly compared to microprocessors? Their main assumption is that the brain's only possible task is to make predictions (Clark, 2013). Real-world deviations from the predictions are evaluated as prediction errors and continuously optimized in the prediction model. The theory thus promises to elegantly explain the essential relationships between perception, memory, and action in mathematical terms. In this context, Andy Clark (2013), referring to Ramish Jain (1996), formulated that "perception itself is a kind of controlled hallucination." As soon as we have a functional representation of the outside world, only the necessary corrections of the deviations from our model are sparingly coded.

Our perception depends essentially on what seems useful to our brain. Otherwise, phenomena such as change or attentional blindness (Simon & Chabris, 1999) can hardly be explained. We expect steps to have a fixed height, so we take steady steps because we have an internal model of the stairs and do not need to measure each step laboriously. We expect the bus to stop at the stop known to us and increase muscle tone in time to compensate for the forces of braking acceleration. These examples illustrate how current sensory signals from the sensorimotor and equilibrium systems are combined with semantic and procedural knowledge to smoothly compensate for acceleration forces or to adapt the stride length when climbing stairs. In fact, our economical 20-W brain would have far too little capacity and computing power to passively record and evaluate the permanent multisensory information stream. Only a much lower-cost form of coding is compatible with the limits and comparatively slower neuronal processing of our brain. Predictive coding is based on representation models of sensory perception. Within a few 100 ms, we are able to ascribe meaning in the sense of visual semantics to a novel scene. Perception always seems to be modeling the next future in dense spatiotemporal dynamics. Form and content are not to be newly represented but merely the difference to a preexisting, action-relevant representation.

von Helmholtz (1860/1962) was the first to suggest that the brain does not represent sensory images, *per se*, but rather the causes of the emergence of these images. Since these causes are not directly perceptible, they must be derived from sensory impressions. After all, we perceive the world more or less as it is and not as the retinal image suggests, namely, with different sizes depending on distance and in a color that depends more on the lighting conditions than on the color of the object. von Helmholz concluded from this that the brain, based on its knowledge, reconstructs the meaning of the environment and does not reproduce its physical properties. The brain therefore does not perceive passively but generates a model that makes sense for sensorimotor action. Because similar retinal images can have different causes, experience knowledge about the environment and an understanding of the interactions between different causes must flow into the internal modeling. However, this is not a matter of completeness but of relevance for the execution of action.

The Art of Predictive Coding

The analytical perceptualism of Gombrich (1960, 1982) can be regarded as the art-theoretical precursor of a neuroaesthetic interpretation of predictive coding. For Gombrich, seeing consisted of constantly reviewing and correcting processes of perception, which in both the production and reception of art raise verifiable questions in a positivistic sense. Perception thus generates hypotheses whose tests reduce differences and also generate new expectations. Rudolf Arnheim (1954) arrived at similar conclusions in his gestalt psychological pictorial theory. According to Arnheim, seeing is by no means merely a mechanical recording of sensory impressions; instead, it is a creative understanding of reality. Unpredictability and surprise are thus important factors that challenge visual perception but also evoke positive emotions in their successful dissolution. Hans-Georg Gadamer (1960) also stressed the activity of the perceiving subject in relation to the aesthetic object. He described the nature of this exchange as continuous and dynamic, leading beyond the moment to a more complex understanding.

Van de Cruys and Wagemans (2011) were probably the first to formulate the potential of predictive coding for neuroscientific models of aesthetic behavior. They formulated that unpredictability is essential to the art experience, and surprise and irritation are part of the rewarding art experience. Thus, successful works of art exhibit qualities that create a balance in the viewer between predictability and surprise, between familiarity and novelty — an explanation that is especially comprehensible for laymen for the intense enjoyment of music. A successful composition reproducibly creates a balance of perceptual tension through unpredictability and relaxation from predictability, which is experienced effectively as a reward. The great potential of the PCT lies in the fact that it not only offers an interpretative framework for the direct confrontation with an art object but probably also explains why art production and reception are anthropogenetic constants. The examination of art creates an opportunity for play, training, expansion, and development of our internal representations, driven by a desire for function. It allows qualitative leaps in the development of representations that are physically and semantically normatively interpreted. A successfully completed absurd but coherent, new representation may lead to an activation of rewarding brain activity just like the successful solving of tasks does.

Cubism, which followed Impressionism at the beginning of the 20th century, was the next radical questioning of protophysical forms of representation through the geometrization of forms that Picasso found in African tribal art. In photography, after World War I, innovative paths were taken with the school of "New Vision" to break up ordinary models of representation here as well (e.g., Laszlo Moholy-Nagy, Franz Roh, and Alexander Rodchenko).

The PCT allows an integration of anthropogenecally preformed representations of stimulus classes with strong valence (aversion and lust), with which variations are operated again and again in visual art and photography. Acquired cultural references such as picture books, cartoons, films, and games dominate image-based individual representations today, which then form the starting material for predictive coding. Exposition, news, effect, repetition, and individual and collective attributions of meaning form a constantly adapting and fluid basic representation. We have identified the interfaces of visual attention, visual memory processes, and episodic and semantic long-term memory in the preceding sections. In particular, photography in the field of tension of a more natural reproduction of a subject (portrait and landscape), which deviates from what has been seen and expected so far, could be compatible with assumptions of the PCT. In the case study that follows, the peculiarities and uniqueness of the evolution of an image into an icon are traced with regard to the factors already discussed.

Figure 1. *Migrant Mother* (1936). A photograph by Dorothea Lange, of Florence Owens Thomspon and some of her children. https://commons.wikimedia.org/wiki/File:Lange-MigrantMother02.jpg.

Case Study: Lange's (1936) *Migrant Mother*

When studying well-known photographs, such as Lange's (1936) *Migrant Mother*, in an attempt to determine what factors may have led to their evolution to iconic status in the history of photography, some people might be tempted to read a certain predictive quality into our conclusions and assume them to be potential templates for reverse engineering "successful" pictures. Therefore, we would like to preface this section of our findings with a brief disclaimer: There is no universal formula for creating memorable photographs, let alone the crafting of an icon! The necessary elements, conditions, and filters are too many and too varied to predict with any amount of certainty what it would take to produce an iconic photograph from scratch and to anchor it as such in our collective memory. The constituent parts regarding the creation and reception of such images are spread across a range of pivotal contributors with differing and often competing interests that function much like *checks and balances*. However, the discriminating and policing nature of these interests is by default imperfect, biased, and uneven but at the same time prevents any one branch from dominating a definition as to what may constitute a "successful" image. That process is highly complex, multifaceted, and unstable. In the end, every iconic photograph that has entered our collective memory is in itself unique and, as such, unrepeatable.

This is not to say that one is unable to define particular image traits or narrow down certain favorable conditions that may help some photographs achieve a more elevated status within our visual culture. What the creation and evolution of an iconic image demand, in general terms, are the vision, timing, and skill of a photographer and the subsequent access to (favorable) approval, distribution, and reception channels. Here, the growing consensus about an image's singular and profound nature can enter a self-perpetuating amplification, resonating with the world enduringly. Using the example of Lange's (1936) *Migrant Mother*, we outline how this particular photograph became one of the outstanding icons of 20th-century photography and how it remains part of our global collective memory to this day.

Migrant Mother

One of the best known and most widely reproduced images in the history of photography is without a doubt Lange's (1936) *Destitute Pea Pickers in California. Mother of Seven Children. Age Thirty-Two. Nipomo, California*, better known under its short title *Migrant Mother*. It has become an iconic image that belongs to the world, speaking as much about what makes certain photographs powerful representations of the human condition as it does about the near-perfect renditions of chance encounters.

It was March 1936; Dorothea Lange (1895–1965) was done photographing at "the end of a cold, miserable winter" and on her way home with "the box containing all those rolls and packs of exposed film ready to mail back to Washington" on the car seat next to her (Lange, 1960; p. 262). When she spotted a sign for "Pea-Pickers Camp," she drove past it for many miles before turning around. Upon arrival at the camp, she parked her car, took her Graflex camera, scanned the camp for a good motif, and then slowly approached a 32-year-old woman named Florence Owens Thompson, who was stranded there with her seven children in a tent after "the pea crop at Nipomo had frozen." Without much interaction, Lange took a series of six images, working herself closer and closer toward Thompson. The last of the six exposures was what we know today as *Migrant Mother*.

What led to and eventually constituted the iconic status of Lange's *Migrant Mother* was a combination of factors that are both quite typical as well as extraordinary for images destined for such a fate. Rather *typical* is the classic portrayal of a woman who conforms to the conventions and our expectations of pictured motherhood — an archetype behind which the real person, Florence Owens Thompson, was eclipsed. The unsolicited pose and subsequent framing evoke religious iconography, with a strong, symmetric composition and appealing formal aspects. Furthermore, a wide cultural dissemination of the image in the years following its creation as well as an art-historical lineage that taps into our collective image memory could be called quite *typical* for an image that would achieve such an immediate and lasting popularity.

An *extraordinary* aspect and the reason for the wide dissemination of the image that in return helped *Migrant Mother* into its iconic existence was the establishment of the *Resettlement Administration* (RA) in 1935, renamed as the *Farm Security Agency* (FSA) in 1937, with its objective as a governmental patron to assign Lange (and many other influential U.S. photographers) to document the effects of the Great Depression. Without the FSA, *Migrant Mother* would simply not be a part of our collective memory today. We would not be able to identify the picture nor identify with (the subject in) the picture. Moreover, the predetermined nature of the copyright of all the photographs produced for and amassed by the FSA placed it in the public domain, making it easily accessible to anyone interested.

In other words, aside from the assignment to photograph the destitution brought upon the population by the Great Depression, the very fact that the negative of *Migrant Mother* was publicly owned aided its favorable reception by distributing reproductions far and wide across the country and the globe. As an image, it stood out, giving the Great Depression a face with which a majority of people could empathize. This portrait was initially popularized through the news media, but over time it also achieved art historical canonization through publications and seminal exhibitions. Furthermore, the photograph was visually so arresting and eventually so widely known that early prints, handmade by Lange herself before she passed the negative to the FSA (held since at the Library of Congress), also have developed a strong market, with these limited numbers of early prints being traded through galleries and auction houses at six-figure dollar values these days. *Migrant Mother* is the rare occasion of a truly collectively owned photograph depicting a shared experience during the Great Depression while holding universal interest over time that also made it subject to scholarly attention and commercial exploitation.

Formal Aspects

Photographs are able to reach us from the vast depth of our visual culture when they fulfill certain formal criteria, when they speak to us in a visual code that triggers reactions rooted in evolutionarily vested needs or desires, or when they strike a formal balance that we perceive as pleasing or interesting, such as a perfect circle, the golden ratio, symmetry, the human face, or vibrant colors, to name just some of the most obvious. There is a certain formal perfection and order when found in nature that attracts our attention and draws us in like moths to the light.

Lange's photograph of the pea picker and mother of seven children, Florence Owens Thompson (1903–1983), was not a quick single capture of the subject by the artist but rather a slow, 10-min approach, moving her large format camera ever closer toward the family. Lange photographed the Thompson family a total of six times during the process, with only the last image being the one that achieved international reknown and iconic status.

Faces in pictures draw our attention, even in their most vague form. So, it is safe to say that portraits and images of people tend to be the most popular form of photography, particularly if they are placed within environments or contexts that are either very familiar to the viewer or if they are depicted in rather atypical situations or even exotic settings. In other words, they either let us easily relate or they spark our curiosity through their difference and novelty. *Migrant Mother* did both: People recognized their own daily struggles and worries brought on by the Great Depression as well as let them see the precarious situation of someone else who seems even worse off.

The image that has become the most well-known among those six exposures showing the *Migrant Mother* had been taken close-up. Here, Thompson is framed tightly within the 4 × 5 in. negative, leaving out much of the context that was visible in the previous five images, which placed her in a squalid tent on barren grounds with very few belongings scattered around. An empty plate in front of the mother and her children looks almost like a staged signifier but was there from the very first exposures. It is outside the frame in the now famous final image. Rather than letting our gaze wander over the family's belongings that define their precarious situation, in that last close-up portrait, our focus is directed instead toward the details of Thompson's weary features as well as her gesture and posture. The narrow focal range also emphasizes the conditions and revealing textures of Thompson's skin, hair, and clothing as well as those of her children. Everything that is signified through more context in the previous five exposures is still present in this final portrait but in the minutiae.

Only three of her children are in the picture; none of them look toward the camera. The two older children have turned away entirely — mirroring each other to the left and right of their mother and thus framing her — while the baby lies barely noticeable in Thompson's left arm. In many of her most expressive photographs, Lange puts particular emphasis on her subjects' hands and hand gestures. In this image, too, Thompson's hand becomes a central focus, as she touches her lower face gently with it in a rather pensive pose while resting her elbow on her lap. She gazes past the camera, almost as if unaware of the photographer and her substantial apparatus.

Lange makes the most of the unsolicited pose, emphasizing a triangular field of interest as well as strong diagonal lines. At the same time, the composition animates our gaze around the image — and then returns back to her face. Thompson is at the very center of the camera's attention, the children obscured yet present to imply motherhood. What might seem a simple enough composition, at first glance, reveals a certain complexity upon further study. While answering to the conventions of portraiture, Lange also did not shy away from showing us destitution in rarely seen detail. Therefore, the image can hold our attention over an extended period, offering ever more information upon close observation.

Obviously, the entire scene seems to imply artworks within a religious iconography, including, for example, Raphael's *Madonna del Prato* (1506) with the Virgin Mary and the infants Jesus and John the Baptist, or Michelangelo's *Pietá* (1498–1499), and even the sculpture *Ugolino and His Sons* (1865–1867) by Jean-Baptiste Carpeaux. Yet, it is the iconography of the Madonna, or motherhood in general, that seems most relevant to *Migrant Mother* because it has become one of the most deeply ingrained parts of our collective cultural memory, triggering a recall of historic versions of such images instantaneously, even if we cannot always name any specific image that conforms to that effect.

Contribution

From the start, the image was destined for popular reception, given that Roy E. Stryker, head of the FSA, already had called Lange's *Migrant*

Mother "*the* picture" of the FSA. It became the signature image of the FSA and, as such, captured the attention of people in the United States and around the globe, and in the process shaped documentary photography and paved the career for a number of now well-known photographers. Lange recognized early on that economic migration was one of the fundamental ills of modern capitalist societies, that the search for a place to live and make a living kept a population in limbo, strained society, and jeopardized democracy. Uprooted people cannot reach their potential and contribute to the best of their abilities because their focus is solely on survival. Lange sought and delivered pictures of what she recognized as an important problem at the time and for the future, even before she was hired for the RA, the later FSA.

Lange's photographs, along with those of her fellow FSA colleagues, were meant from the very beginning as documentation of the ill effects of the economic depression in the 1930s, particularly what the collapse of sharecropping agriculture meant for the rural areas across the United States. It aimed to make visible what had rarely before been subject to the focused and joint attention of photographers. However, because the hiring of photographers to this end was in the public interest through a governmental body rather than the typical private or corporate interests, the assignments reflected the lives and struggles of a much broader spectrum of U.S. citizens. In a way, this has turned *Migrant Mother* into an American self-portrait: It became a testament and an icon of the United States during the Great Depression.

What may have helped with this public reception of the image and its symbolic dimension — one has to insert somewhat cynically — was the misconception that the *Migrant Mother* Thompson was, as had been commonly assumed at the time, a White mother rather than of Cherokee descent. "That detail," wrote *The New York Times* (11/28/2018), citing Museum of Modert Art (MoMA) photography curator Sarah Meister, raises the compelling question of whether *Migrant Mother* would have resonated so widely if viewers knew the subjects were Native American. "We have never been a race-blind country," Ms. Meister elaborates (Estrin 2018). "I wish that I can say that the response would have been the same if everyone had been aware that she was Cherokee, but I don't think that you can." Yet, once the image became an icon, it transcended mere representation and thus any specific identity.

This was also expressed by Paula Rabinowitz (1994), when she reiterates, "the power of images - a power that includes their ability to exceed the original impulse of the creation" (p. 86), and then specifically pointing out the following:

> "The troubling story of Lange's *Migrant Mother*, told and retold, [which] offers with acute poignancy an example of discourse as repository of meaning — the photograph as much as its checkered history includes a woman and her children, a photographer, a government bureau, popular magazines, museums, scholars, and a changing public — an image and tale composed, revised, circulated, and reissued in various venues until whatever reality its subject first possessed has been drained away and the image becomes icon". (p. 86)

Her image has been included in most books and survey exhibitions on the history of photography, notably the famous postwar exhibition *Family of Man* (1955), curated by Edward Steichen for the MoMA that traveled the world for years and was seen by millions of people, let alone being subject to countless international group and solo exhibitions and publications. The first major publication of her Depression-era photographs was her landmark photo book *An American Exodus: A Record of Human Erosion* in 1939. A more recent and well-received exhibition *Dorothea Lange: The Politics of Seeing*, in which special attention was also given to *Migrant Mother* and its history, traveled from Oakland (The Oakland Museum of California), Nashville (The First Art Museum), Paris (Jeu de Paume), and London (Barbican Centre). The success of this exhibition demonstrates the continued interest in Lange's work and *Migrant Mother* in particular. In addition, the MoMA will follow up on it and will dedicate an exhibition to Lange's work in 2020, *Words & Pictures*, and Meister (2019), photography curator at the MoMA, recently released a book that focuses solely on the iconic photograph: *Dorothea Lange*: *Migrant Mother*.

Distribution

Migrant Mother was initially distributed mainly through news media, picking up images that illustrated stories about the Great Depression and poverty, and later via a number of first exhibitions and publications. But the main step toward its iconic status was the commodification of our relationship with the image, which is tied up in the ability of potentially endless reproduction as well as market-driven efforts to curb that supply.

Migrant Mother is a curious case because it demystifies the widely shared belief and accepted rule that the rarified print is a prerequisite for its lucrative and appreciative market while also supporting this convention. *Migrant Mother* is probably still the most reproduced image in the history of photography. Potentially everyone has access to a print straight from the original negative, provided by the Library of Congress. This still holds true, only these days one receives a high-resolution digital file — but still made from the original negative. This freed the image from the tyranny of the material world and may have led to a new wave of dissemination. This means that literally thousands, if not millions, of straight reproductions from the original negatives are in circulation today. The prints made from the original negative are not very good, given that there is no artistic intent involved in the printing process other than an automated and standard exposure and chemical development. A digital file, on the other hand, allows the technically skilled printer to manipulate the available image information into a somewhat decent print quality. But it remains a print that only has the license or "authorization" from the Library of Congress as copyright holder but not from the artist. The only original here regarding the artist is the negative. Yet, our relationship is with the object, even if immaterial.

Migrant Mother, with its profound formal strength, its high-profile visibility, and its importance within 20th-century photography as well as distinguishable features of its printed reiteration made its rarer versions a desired asset. This image of a *Destitute Pea Picker in California and Mother of Seven Children*, who never got "a penny out of" being the most recognizable face of the Great Depression, can fetch high six-figure dollar amounts in today's market. The sheer ubiquity of the image and its reproductions seems to have no impeding effect on the price of its rarified early prints. On the contrary, the iconic status of the image boosts the market for vintage prints of *Migrant Mother*, despite the fact that thousands, if not millions, of additional prints have been made from the same original negative thereafter. In other words, you can find a price range from about $100

U.S. to a high of $337,500 U.S. at auction (Phillips, New York, 2019) and even higher (but undisclosed) amounts at art fairs or in galleries, depending on their history and provenance as well as the owner's "motivation to sell."

Conclusion from a Case Study

What made Dorothea Lange's *Migrant Mother* such an iconic image? What helped burn the image so deeply into each of our consciousness and memory? How did it become one of the key images in our collective memory, a picture we can reference and discuss without having to show and look at it? Is it because of certain compositional trademarks that aid recognition and trigger interest? Is it the fact that the subject depicted is a human being with a striking face and pose? Or is it the way it taps into and further informs our visual knowledge as well as art-historical lineage, referencing the nearly universal icons of motherhood, religious imagery, and the human condition? Or is it because it has been reproduced more than any other image of that era? Or is it the historical importance that has been attached to it by scholars, curators, publishers, and picture editors of newspapers and illustrated magazines? Or is it even the monetary value early rare versions of the print have achieved through collectors' competition? Obviously, the answer is "all of the above." An image that attracts our attention because of its visual features and the human interest contained within, its art historical significance and derivation, its excessive reproduction and dissemination, its encapsulation of myth, and its public ownership and parallel import in the art market is bound to reach a level of recognition and consensus that establishes it high up on the Mount Olympus of our visual culture. The term *migrant mother* has become synonymous with Lange's photograph. The image eclipses the words.

Synopsis

Our attempt to integrate neuroscientific concepts to explain the iconicity (Paul, 2009) of photographs refers to the great importance of historically fateful, objective social constellations (the Great Depression), biographically unpredictable events (assignment by the FSA and meeting with

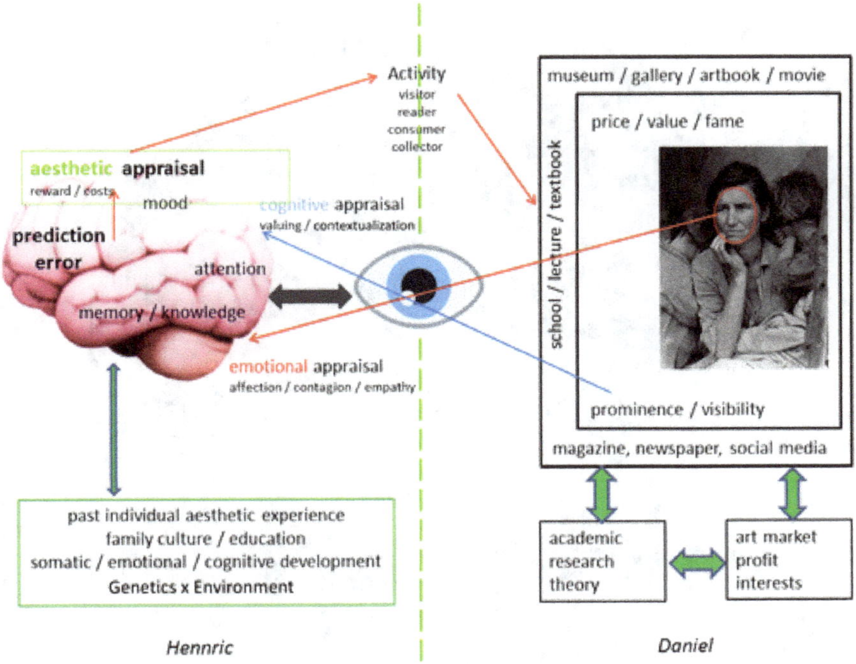

Figure 2. Description of interactions between the photographic image, contextual information, vision, and aesthetic appraisal based on a "prediction error".

Florence Owens Thompson); and finally the art object itself (*Migrant Mother*). Photography, on the other hand, exhibits conceptual (iconography), formal (composition), and physical qualities (gray value distribution), which together are inviting to an aesthetic appropriation. It also follows that the perception of a photographic image cannot be separated from its pre-evaluation and filtering by social conditions, such as location (museum), purpose (education), status (fame), and visibility (publicity). These factors have both a direct and an indirect influence on the viewer's cognitive and affective evaluation of the art object (see Figure 2). The strength of the affective reaction (amygdala), the duration and intensity of the cognitive and emotional confrontation with the image (working memory), and its perceptual and conceptual novelty (hippocampus) lead to the formation of a retrievable memory of the image (long-term memory) (see Figure 3).

Individual memory, in turn, contributes cumulatively to an image-related collective memory, which on one hand supports an artistic

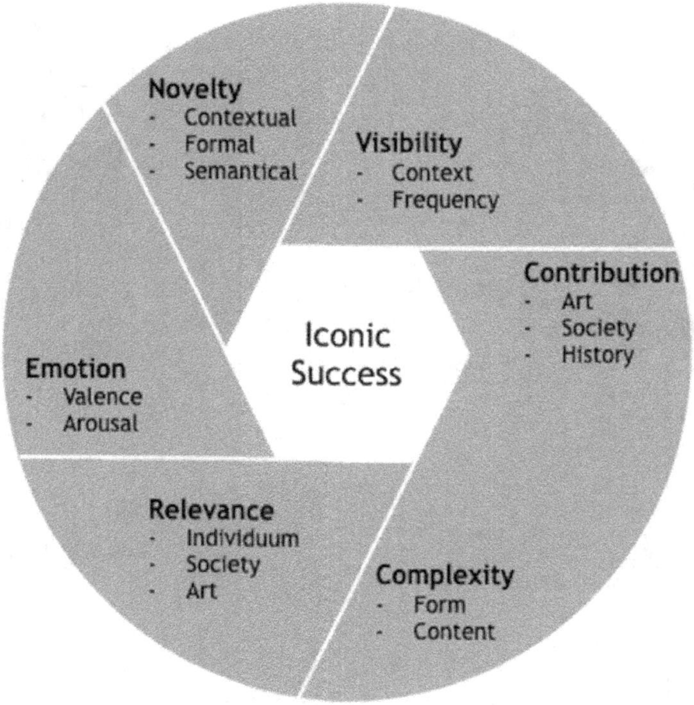

Figure 3. Factors that might contribute to the artistic success of a photograph.

recognition process and on the other hand, like a meme, actively and passively contributes to the spread and iconization of the image. New media (e.g., Pinterest and Instagram) function as an individual, collective, and public form of communicative pictorial memory, capable of triggering recognition processes globally and in real time, which, in the case of Lange's *Migrant Mother* in the pre-Internet epoch, still took years.

With the PCT, a core process of aesthetic behavior can be described, which represents the interface to social aesthetic appropriation behavior but also presupposes subject-related inner modeling of potentially perceptive changes in the environment. Particularly affect-charged news (amygdala and hippocampus) and subjective relevance (frontoparietal brain structures) require an immediate adaptation of the inner model according to the PCT. The fact that art itself is depicted as a representative model of a theoretical execution of action as a representation of the second order may enable the pleasurable effort and reward (nucleus accumbens) of a successful aesthetic appropriation. The predictive coding could then be precisely the mechanism that generates oscillatory resonance phenomena that either individually or collectively generate an artistic experience of appropriation (Assmann, 2012). The process of selecting images with potential iconicity is a collective process of consensus-based selection and authentication determined through the judgments and approval of opinion leaders, art experts, and institutional gatekeepers. The author of the image, the image itself, or a single viewer cannot initiate this process of canonized iconization. It is a collaborative process that integrates experience and opinions while also shaping them.

From a systems theory perspective, the feedback and resonances take place between the neuroscientific, psychological, and sociological levels of subject–object relations. This means that a systemically isolated visual neuroaesthetic will only be able to describe feature levels that do not correspond to the complexity of a contemporary concept of art. The PCT seems to be an excellent tool to model neuropsychological resonance phenomena of aesthetic appropriation. In the individual, it assumes an elaborate and open art model that enters into dynamic interaction and resonance within a fluid social framework of meaning of the art object (Figure 2). Negatively formulated, a purely objectivist art model will remain blind to the essential developments of modernism and postmodernism, such as a readymade (Marcel Duchamp), cubist art (Pablo Picasso), or conceptual art (Sol Lewitt).

We are convinced that a future systemic approach will allow the complementary integration of neuroscientific, perceptual, cognitive, emotional, and sociopsychological findings via the interface of the PCT with social-scientific, art-theoretical, and art-historical and also behavioral-economic models. The producibility of iconization may be desired by participants in the art market, but fortunately, it stands in the way of the surprise and novelty sensitivity of the PCT. The attempts by book publishers and the music and film industries to reproduce success with imitative models are rarely successful, as they serve only short-term trends as retorts. They do not replace the authentic voice or hand of the artist. Further, the creativity, the degree of innovation, and the skills of the artistic creation, which tend to move in their individual development toward increased complexity and abstraction, are to be considered in an extended PCT. The questioning, ignoring, negating, and changing of existing formal and semantic model views or representations and the planned recourse to these also within the framework of feedback processes appear to be essential aspects of creative work. Art production and reception are thus two sides of a specific field of action of the PCT, which explains that artistship, no matter in which medium, presupposes a receptive confrontation with artistically significant positions.

References

Adams, R. A., Shipp, S., & Friston, K. J. (2013). Predictions not commands: Active inference in the motor system. *Brain Structure and Function*, 218(3), 611–643. https://doi.org/10.1007/s00429-012-0475-5.

Arnheim, R. (1954). *Art and Visual Perception: A Psychology of the Creative Eye*. Berkeley, CA: University of California Press.

Assmann, A. (2012). Resonanz und Einschlag: Zur Affektlogik von Bildern im kulturellen Gedächtnis. *Nova Acta Leopoldina*, 113(386), 23–36.

Atkinson, R. C. & Shiffrin, R. M. (1968). Human memory: A proposed system and its control processes. In *Psychology of Learning and Motivation*, Vol. 2, pp. 89–195. New York, NY: Academic Press. https://doi.org/10.1016/S0079-7421(08)60422-3.

Awh, E., Belopolsky, A. V., & Theeuwes, J. (2012). Top-down versus bottom-up attentional control: A failed theoretical dichotomy. *Trends in Cognitive Sciences*, 16(8), 437–443. https://doi.org/10.1016/j.tics.2012.06.010.

Bourne, J. A. & Morrone, M. C. (2017). Plasticity of visual pathways and function in the developing brain: Is the pulvinar a crucial player? *Frontiers in Systems Neuroscience*, 11(3), 3. https://doi.org/10.3389/fnsys.2017.00003.

Caravaggio, M. M. (1596). *Medusa*. https://en.wikipedia.org/wiki/Medusa_(Caravaggio).

Carpeaux, J.-B. (1827–1875). *Ugolino and His Sons (1865–1867)*. New York, NY: Metropolitan Museum of Art.

Chatterjee, A. & Vartanian, O. (2014). Neuroaesthetics. *Trends in Cognitive Sciences*, 18(7), 370–375. https://doi.org/10.1016/j.tics.2014.03.003.

Clark, A. (2013). Whatever next? Predictive brains, situated agents, and the future of cognitive science. *Behavioral and Brain Sciences*, 36(3), 181–204. https://doi.org/10.1017/S0140525X12002427.

Conway, B. R. & Rehding, A. (2013). Neuroaesthetics and the trouble with beauty. *PLoS Biology*, 11(3), e1001504. https://doi.org/10.1371/journal.pbio.1001504.

Dolan, R. J. (2002). Emotion, cognition, and behavior. *Science*, 298, 1191–1194. https://doi.org/10.1126/science.1076358.

Estrin, J. (2018). Unraveling the Mysteries of Dorothea Lange's 'Migrant Mother'. *The New York Times*, 28 November. https://www.nytimes.com/2018/11/28/lens/dorothea-lange-migrant-mother.html.

Gadamer, H. G. (1960). *Wahrheit und Methode* [*Truth and Method*]. Tübingen: JCB Mohr. https://doi.org/10.1080/00062278.1973.10596981.

Gollisch, T. & Meister, M. (2010). Eye smarter than scientists believed: Neural computations in circuits of the retina. *Neuron*, 65, 150–164. https://doi.org/10.1016/j.neuron.2009.12.009.

Gombrich, E. H. (1960). *Art and Illusion: A Study in the Psychology of Pictorial Representation*. New York, NY: Pantheon.

Gombrich, E. H. (1982). *The Image and the Eye: Further Studies in the Psychology of Pictorial Representation*, pp. 40–62. Oxford: Phaidon.

Gombrich, E. H. (1995). *The Story of Art*, Vol. 12. London: Phaidon.

Helmholtz, H. (1860/1962). *Handbuch der physiologischen Optik*. Leipzig: Leopold Voss.

Henson, R. N. & Gagnepain, P. (2010). Predictive, interactive multiple memory systems. *Hippocampus*, 20(11), 1315–1326. https://doi.org/10.1002/hipo.20857.

Hirdina, K. (1976). Zur Ästhetik des Bauhausfuntionalismus. *Wissenschaftliche Zeitschrift der Hochschule für Architektur und Bauwesen Weimar*. Issue 5/6, 522.

Hume, D. (1757). *Four Dissertations: Of the Standard of Taste*, First Edition. London, England: A. Millar, in the Strand.

Jain, R. (1996). Merging virtual and real. *IEEE Multimedia*, 3(4), 3–4. https://doi.org/10.1109/MMUL.1996.556532.

Kant, I. (1790/1983). *Kritik der Urteilskraft*. Darmstadt: Wissenschaftliche Buchgesellschaft.

Kaplan, E. & Shapley, R. M. (1986). The primate retina contains two types of ganglion cells, with high and low contrast sensitivity. *Proceedings of the National Academy of Sciences of the United States of America*, 83(8), 2755–2757. https://doi.org/10.1073/pnas.83.8.2755.

Kesner, L. (2014). The predictive mind and the experience of visual art work. *Frontiers in Psychology*, 5, 1417. https://doi.org/10.3389/fpsyg.2014.01417.

Kirov, R. (2016). Brain oscillations and predictive coding in the context of different conscious states and sleep-wake cycle: Implications for decision making and psychopathology. *Frontiers in Psychology*, 7, 1768. https://doi.org/10.3389/fpsyg.2016.00922.

Lange, D. (1960). The assignment i'll never forget. *Popular Photography*, 46(2), 42,126. Reprinted in (1980). *Newhall, Photography: Essays and Images*, pp. 262–265. London, England: Routledge.

Lange, D. & Dorothea, L. (1936). *Migrant Mother, Nipomo, California (Destitute Pea Pickers in California. Mother of Seven Children. Age Thirty-Two. Nipomo, California)*. Washington, DC: Library of Congress Prints and Photographs Division.

Lukács, G. (1963). *Ästhetik, Werke Bd.* 11/12. Berlin, Germany: Neuwied.

Madigan, S. (1983). Picture memory. In Yuille, J. C. (ed.), *Imagery, Memory and Cognition: Essays in Honor of Allan Paivio*, p. 65. Hillsdale, NJ: Lawrence Erlbaum.

Mastria, S., Ferrari, V., & Codispoti, M. (2017). Emotional picture perception: Repetition effects in free-viewing and during an explicit categorization task. *Frontiers in Psychology*, 8, 1001. https://doi.org/10.3389/fpsyg.2017.01001.

Marcel, J. C. & Mucchielli, L. (2008). Maurice Halbwachs's mémoire collective. In *Media and Cultural Memory/Medien und Kulturelle Erinnerung*, pp. 141–149. Berlin, Germany: Walter de Gruyter. https://doi.org/10.13140/2.1.1296.4807.

McNally, G. P., Johansen, J. P., & Blair, H. T. (2011). Placing prediction into the fear circuit. *Trends in Neurosciences*, 34(6), 283–292. https://doi.org/10.1016/j.tins.2011.03.005.

Meister, S. H. (2019). *Dorothea Lange: Migrant Mother [Photograph]*. New York, NY: The Museum of Modern Art.

Michelangelo di Lodovico Buonarroti Simoni (1475–1564). *Pietá* (1498–1499). Vatican, Italy: St. Peter's Basilica.

Paul, G. (ed.) (2009). *Das Jahrhundert der Bilder: 1900 bis 1949*. Göttingen, Germany: Vandenhoeck & Ruprecht.

Pelowski, M., Markey, P. S., Forster, M., Gerger, G., & Leder, H. (2017). Move me, astonish me… delight my eyes and brain: The Vienna integrated model of top-down and bottom-up processes in art perception (VIMAP) and corresponding affective, evaluative, and neurophysiological correlates. *Physics of Life Reviews*, 21, 80–125. https://doi.org/10.1016/j.plrev.2017.02.003.

Pessiglione, M., Seymour, B., Flandin, G., Dolan, R. J., & Frith, C. D. (2006). Dopamine-dependent prediction errors underpin reward-seeking behaviour in humans. *Nature*, 442(7106), 1042–1045. https://doi.org/10.1038/nature05051.

Phelps, E. A. (2004). Human emotion and memory: Interactions of the amygdala and hippocampal complex. *Current Opinion in Neurobiology*, 14(2), 198–202. https://doi.org/10.1016/j.conb.2004.03.015.

Podvigin, N. F., Jokeit, H., Pöppel, E., Chizh, A. N., & Kiselyeva, N. B. (1992). Stimulus-dependent oscillatory activity in the lateral geniculate body of the cat. *Naturwissenschaften*, 79(9), 428–431. https://doi.org/10.1007/BF01138580.

Pöppel, E. (1989). Taxonomy of the subjective: An evolutionary perspective. In J. W. Brown (ed.), *Neuropsychology of Visual Perception*, pp. 219–232. New York, NY: Routledge. https://doi.org/10.4324/9781315441849-11.

Preston, S. D. & De Waal, F. B. (2002). Empathy: Its ultimate and proximate bases. *Behavioral and Brain Sciences*, 25(1), 1–20. https://doi.org/10.1017/S0140525X02000018.

Rabinowitz, P. (1994) They must be represented. *The Politics of Documentary*, 1994, 86.

Ramachandran, V. S. & Hirstein, W. (1999). The science of art: A neurological theory of aesthetic experience. *Journal of Consciousness Studies*, 6(6–7), 15–51.

Ramachandran, V. S. & Freeman, A. (2001). Sharpening up 'the science of art'. *Journal of Consciousness Studies*, 8(1), 9–30. https://doi.org/10.1007/s00897980270a.

Rao, R. P. & Ballard, D. H. (1999). Predictive coding in the visual cortex: A functional interpretation of some extra-classical receptive-field effects. *Nature Neuroscience*, 2(1), 79–87. https://doi.org/10.1038/4580.

Raffaello Sanzio da Urbino (1483–1520). *Madonna del Prato* (*Madonna of the Meadow*) (1505–1506). Vienna, Austria: Kunsthistorisches Museum.

Rauss, K., Schwartz, S., & Pourtois, G. (2011). Top-down effects on early visual processing in humans: A predictive coding framework. *Neuroscience & Biobehavioral Reviews*, 35(5), 1237–1253. https://doi.org/10.1016/j.neubiorev.2010.12.011.

Rubens, P. P. (1618). Head of Medusa. *Wikipedia*. https://en.wikipedia.org/wiki/Medusa_(Rubens).

Shimamura, A. (2015). *Experiencing Art in the Brain of the Beholder*. Oxford: Oxford University Press.

Squire, L. R. (2004). Memory systems of the brain: A brief history and current perspective. *Neurobiology of Learning and Memory*, 82(3), 171–177. https://doi.org/10.1016/j.nlm.2004.06.005.

Tulving, E. (2002). Episodic memory: From mind to brain. *Annual Review of Psychology*, 53(1), 1–25. https://doi.org/10.1016/s0035-3787(04)70940-6.

Ungerleider, L. G. & Haxby, J. V. (1994). What'and 'where'in the human brain. *Current Opinion in Neurobiology*, 4(2), 157–165. https://doi.org/10.1016/0959-4388(94)90066-3.

Van de Cruys, S. & Wagemans, J. (2011). Putting reward in art: A tentative prediction error account of visual art. *Perception*, 2(9), 1035–1062. https://doi.org/10.1068/i0466aap.

Vuilleumier, P. (2015). Affective and motivational control of vision. *Current Opinion in Neurology*, 28(1), 29–35. https://doi.org/10.1097/WCO.0000000000000159.

Zeki, S. (1999). *Inner Vision. An Exploration of Art and the Brain*. Oxford: Oxford University Press.

https://doi.org/10.1142/9789811298639_0015

Chapter 15

The "Third Abstraction" of the Chinese Artist LaoZhu: Neural and Behavioral Indicators of Aesthetic Appreciation*

Yan Bao[†,‡], Taoxi Yang[†,‡], Jinfan Zhang[‡], Jiyuan Zhang[†], Xiaoxiong Lin[†,§], Marco Paolini[¶], Ernst Pöppel[†,‡] and Sarita Silveira[‡]

[†]*School of Psychological and Cognitive Sciences, and Beijing Key Laboratory of Behaviour and Mental Health, Peking University, Beijing, China*

[‡]*Institute for Medical Psychology, Ludwig Maximilian University, Munich, Germany*

[§]*Graduate School of Systemic Neurosciences, Ludwig Maximilian University, Munich, Germany*

[¶]*Institute for Clinical Radiology, Ludwig Maximilian University, Munich, Germany*

*This chapter was originally published in *PsyCh Journal*. Bao, Y., Yang, T., Zhang, J., Zhang, J., Lin, X., Paolini, M., Pöppel, E., & Silveira, S. (2017). The "third abstraction" of the Chinese artist LaoZhu: Neural and behavioral indicators of aesthetic appreciation. *PsyCh Journal,* 6, 110–119. https://doi.org/10.1002/pchj.167.

Abstract

The eminent Chinese artist LaoZhu has created a homogeneous set of abstract pictures that are referred to as the "third abstraction." By definition, these pictures are meant to be representations of the artist's personal involvement and as such to create an internal point of view in the observer on an implicit level of processing. Aiming at investigating whether the artist's choice of a specific color is experienced in a specific way by the recipient, we assessed both explicit and implicit (i.e., neurocognitive) correlates in naive viewers of LaoZhu's pieces. The behavioral results reveal a preference of the original red paintings over color-changed counterparts in green or black. Paradoxically and inconsistent with predictions, we found higher levels of neural activation in several brain regions (predominantly in the frontal and parietal cortices) for the color-changed compared to the original red conditions. These observations add empirically to the complementarity of early visual pathways and higher-order cognition as well as of explicit and implicit information processing during aesthetic appreciation. We discuss our findings in light of processing effort and top-down control of the color-changed paintings. With regard to the third abstraction as defined by LaoZhu, in particular to the distinction between an external and internal point of view when viewing abstract art, our results contribute to an understanding of "abstraction and empathy" as a fundamental part of aesthetic appreciation.

Keywords: abstract art; aesthetics; color; emotion; empathy; functional magnetic resonance imaging; knowledge systems; visual art

Single case studies are usually associated with research on a single subject, in particular patients with brain injuries or other pathologies, or a group of subjects with specific characteristics; such single case studies with patients or healthy subjects have often provided unique insights into neural and psychological processes (e.g., Bao *et al.*, 2015; Luria, 1968; Pöppel, Brinkmann, von Cramon, & Singer, 1978; Silveira, Bao, *et al.*, 2015; Strasburger & Waldvogel, 2015; Wang *et al.*, 2016; Zaytseva *et al.*, 2014; Zhou *et al.*, 2016) and have often opened new fields of research, for instance, with respect to memory research, the famous case H.M. (Scoville & Milner, 1957). Here we expand this experimental paradigm by adopting a new strategy in relating a single person with his creative achievements in a complementary way. The artwork of the eminent Chinese artist LaoZhu on the "third abstraction" (Sun, 2011) provides a unique opportunity to follow this new line; some of his artwork is characterized by a surprising homogeneity (see his pictures from the series *Zhu Mo* in Figure S1 of the Supporting Information). This unique constellation of the art and artist invites us to treat the artwork and the artist together as a "single case." Considering that paintings are the artist's expressions of inner states, the use of this strategy provides insights into common responses in several subjects and the success in transferring artistic content and meaning.

It has been suggested by LaoZhu himself that, historically, abstract art can be divided into three periods: The artworks of Wassily Kandinsky (1912/2009), Kazimir Malevich, and Piet Mondrian represent the first phase, and the cubism of Pablo Picasso and the abstract expressionism of Jackson Pollock stands for the second phase. In these first two phases, specific shapes or colors in the pictures are emphasized, forms or unique compositions with defined optical elements are dominant, and emotions may be expressed. For these two phases, one experiences or views (or even inspects) pictures preferentially with an external point of view, whereas pictures of the "third abstraction" create an internal point of view in which the personal involvement of the artist and the identification with his work is experienced by the viewer without explicit reflection but rather on an implicit level of sensory and *Cognitive Processing* (Pöppel & Bao, 2011). Art of the third abstraction as it is created and also understood by LaoZhu is strongly influenced by the typical brushstrokes from traditional Chinese art reflecting the spiritual state of the calligrapher. It captures the open-minded viewer who may feel and enjoy becoming a part of the picture; such an experience of being embedded in a picture has also been suggested when viewing Chinese landscapes, and it has been argued that such an identification with a piece of art supports personal and cultural identity (Bao *et al.*, 2016).

What may be some critical factors that contribute to the differentiation of external or internal points of view? Such criteria may be found in "what" is represented in the picture and also "how" the "what" is represented. We consider this distinction between the content (the what) and the way the content is represented (the how) to be crucial for a deeper understanding of an external or internal point of view and aesthetic appreciation; this conceptual (even logical) difference has been emphasized for cognitive processes in general (Bao & Pöppel, 2012; Pöppel, 1989). Without much reflection, one might assume that the content of a picture has to be dominant with respect to this question; however, a closer look at the pre-semantic level of processing appears also to be useful. In fact, substantial evidence has already been gained about aesthetic appreciation on this operational level, which may contribute also to a better understanding of these different perspectives. Not surprisingly, visual features of an artwork — such as symmetry — matter in aesthetic appreciation. Spectral analyses of different classes of visual stimuli have indicated that specific visual features of an artwork correspond to the sensory and perceptual processing in the visual stream (Graham & Redies, 2010); higher-order statistics for natural images and their usefulness for understanding visual processing in general has been demonstrated previously by Zetzsche & Krieger (2001), which may be relevant also for artwork. But simpler stimulus characteristics are also important; by swapping the colors in the tri-color compositions of Piet Mondrian's original abstract paintings, Locher, Overbeeke, and Stappers (2005) showed that manipulating the colored regions influenced the viewer's perceived weight of the areas and the judgment of the balance centers of the compositions. Thus, aesthetic appreciation of a painting may have its source in extracting basic elements of stimuli already in the early visual pathway (Cinzia & Vittorio, 2009; Zhou *et al.*, 2016). The sensory information, derived from early and intermediate visual processing, includes color, tone (dark–light range), texture, symmetry, and contrast (Berlyne, 1971; Cupchik & Winston, 1996). But aesthetic appreciation of pictures is also controlled by top-down mechanisms, often on an implicit level (Pöppel & Bao, 2011), such as prior experiences (Silveira *et al.*, 2012), expectations (Silveira, Fehse, Vedder, Elvers, & Hennig-Fast, 2015; Silveira et al., 2015), or the historical, cultural, and economical background in general (Jacobsen & Wolsdorff, 2007; Ritterfeld, 2002).

With respect to neural mechanisms, important observations have been made by Thakral, Moo, and Slotnick (2012), who showed that when subjects viewed paintings of van Gogh, the motion experience was associated with activity in motion-processing regions of the brain; the experience of pleasantness was associated with neural activations in the anterior prefrontal cortex, which implies that aesthetic experiences depend both on sensory processing and conceptual processing. In another study using functional magnetic resonance imaging (fMRI), Cupchik, Vartanian, Crawley, and Mikulis (2009) observed that representational paintings containing ill-defined forms elicited higher activation of the left superior parietal lobe than "hard-edge" paintings (containing well-defined forms), especially when subjects were requested to approach the paintings in a subjective and engaged manner rather than a detached and objective manner.

Compared to representational art, abstract artwork does not consist of clearly and distinctly characterized visual objects. While the "what" diminishes in significance in most abstract artwork, the "how" rises in strength, presumably causing a perceptual bias toward basic visual elements (Aviv, 2014). Interestingly, the distinction between content (or what-functions) and logistical (or how-functions) on the neural level (Bao & Pöppel, 2012; Pöppel, 1989) finds its analog in the historic development of pictorial art in the last 100 years. In our view, abstract art is well suited for investigating aesthetic appreciation. Particular techniques of an artist, like the manipulation of texture, color, shape, or composition, in abstract pictures can stimulate different aesthetic experiences. This study was designed to address the modulation of aesthetic appreciation by basic visual features, focusing on LaoZhu's abstract art. The specific choice of the color red for the brushstrokes that are meant to represent LaoZhu's inner states was investigated in its function to transfer aesthetic appeal and artistic meaning. We examined aesthetic appreciations of LaoZhu's original abstract paintings and color manipulations in naive viewers, both on behavioral and neural levels. According to LaoZhu's theory on the third abstraction, his original choice of color should lead to higher aesthetic appeal than color-changed conditions. This is assumed to be accompanied by higher neural activation in reward-related brain regions for the original color condition. A further exploratory question (which, however, can only be reflected on a speculative level) was whether the choice of a specific color as a basic sensory feature of LaoZhu's pictures contributes to the creation of an internal, as compared to an external, point of view, thus, providing perhaps a scientific perspective to the third abstraction.

Method

Subjects

Sixteen healthy right-handed German subjects (mean age = 24.45 years, *SD* = 4.51; seven females) took part in the experiment after giving informed written consent. All had normal or corrected-to-normal visual acuity and color vision and had no history of neurological disease. None of them was an art specialist. This study was approved by the ethical committee of Ludwig-Maximilian University (LMU) in Munich, in agreement with the Declaration of Helsinki.

Materials

Twenty-one pictures were selected from LaoZhu's painting series *Zhu Mo* with the artist's permission. All pictures were prepared in bitmap file format, and the image dimensions were 700 × 700 pixels. These paintings are referred to as the "original paintings." Then we constructed the RGB image and swapped the red and green channels or removed color; the 42 paintings (21 green and 21 black) that were created through this process are referred to as the "color-changed paintings." Scramble versions for each original and color-changed painting were created in MATLAB by dividing the picture matrix into non-overlapping blocks with the size specified by S (10 × 10 pixels) and randomly shuffling these blocks. Scramble paintings served as a control condition. The above process resulted in 126 paintings that varied along the dimensions of color (red, green, or black) and their corresponding scrambled versions (Figure 1(a)).

Procedure

All stimuli were projected onto a translucent screen through a video projector that subjects could view inside the scanner via a head coil-compatible mirror system. Subjects participating in the imaging study were instructed to carefully watch the paintings without giving a response. To ensure sufficient attention to these paintings, subjects were informed of the requirements of filling out a questionnaire after the functional

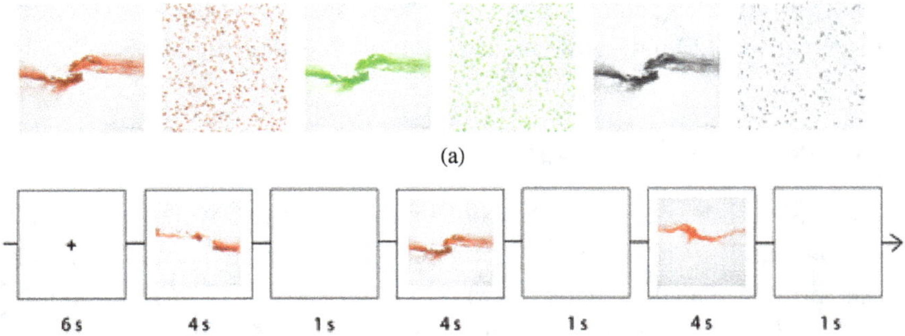

(a)

(b)

Figure 1. (a) *Examples of the stimulus material used in the experiment. From left, original red, scrambled red, green, scrambled green, black, and scrambled black.* (b) Experimental paradigm. Stimulus presentation was 4,000 ms followed by a 1,000-ms blank screen. The inter-block interval was 6,000 ms.

imaging session. The decision not to collect behavioral responses in the scanner was made to facilitate maximal concentration on viewing the pictures. Paintings were presented in a block design. There were six conditions, each including seven blocks and twenty-one paintings; this way, each block included three paintings. The order of stimuli and blocks was pseudo-randomized. The duration of each block was 15 s. Each painting was presented for 4 s, followed by a 1-s blank screen. The inter-block interval was 6 s (Figure 1(b)). After the scanning session, each subject viewed the same paintings presented in the same sequence as in the scanner; stimulus presentation was self-paced. The subjects were asked "How much do you like this painting?" to which they responded using a 5-point scale ranging from 1 (*no feeling*) to 5 (*I like it very much*).

Data acquisition

Brain-imaging data were obtained with a 3T MRI scanner with a standard head coil at the University Hospital of LMU, Munich. For BOLD signals, T2*-weighted echo planar imaging was used (repetition time/echo time = 2,500/30 ms; flip angle = 90°; acquisition matrix = 80 × 80; slice thickness = 3 mm, no gap between slices). One run consisting of 358 functional volumes was acquired for each subject. A high-resolution T1-weighted scan of each participant's brain anatomy was acquired with a 240 × 240 matrix, field-of-view = 220 mm, covering the brain with a slice thickness of 1 mm.

Data preprocessing and analysis

The behavioral data on liking or not liking the pictures were subjected to a one-way analysis of variance (ANOVA) with color (red, green, or black) as the variable. The data analysis was performed using SPSS for Windows (version 21.0). The MRI data were analyzed with Statistical Parametric Mapping V12 (SPM12; http://www.fil.ion.ucl.ac.uk/spm). The preprocessing procedure included the removal of the first eight volumes to allow for T1-equilibration effects, 3-D motion correction by applying an affine rigid registration as part of the realignment procedure built in SPM12, and spatial smoothing with an 8-mm Gaussian kernel. In further preprocessing analysis, the mean functional image was co-registered with the anatomical image and normalized to the Montreal Neurological Institute (MNI) template provided in SPM12. Individual events were modeled by a hemodynamic response function. By using the general linear model, we obtained parameter estimates for each condition and each subject and then acquired statistical parametric maps of the t-statistic resulting from linear contrasts between the three experimental conditions (original red, green, or black) compared with their corresponding control conditions (scrambled red, scrambled green, or scrambled black). For the group analysis, these individual contrast images were entered in a second-level analysis treating subjects as a random effect. The average BOLD response across the brain in viewing original red paintings was compared to that in the other two, the color-changing conditions, with paired t-tests. For these t-tests, significant voxels initially passed a voxel-wise statistical threshold of $p \leq 0.01$, and a cluster-level threshold was obtained at the family-wise-error (FWE)-corrected statistical significance level of $p < 0.05$.

Results

Behavioral results

The responses to the questions showed that the experimental manipulation by changing the colors had some important consequences. Average preference ratings for the original red paintings and their color-changed counterparts (green or black) were 4.04 (*SE* = 0.11), 3.53 (*SE* = 0.09), and 3.63 (*SE* = 0.13). The one-way ANOVA demonstrated that there was a main effect of color, $F(1, 15) = 19.648$, $p < 0.001$, $\eta_p^2 = 496$, indicating a significant influence of changed colors on the aesthetic appreciation of the paintings. Bonferroni-adjusted *post hoc* tests revealed that red was more preferable than green or black ($p = 0.002$ and $p = 0.011$, respectively), whereas the difference between green and black ($p = 0.561$) was not significant. This suggests that the subjects did not discriminate in their affective responses to the two color-changed conditions.

fMRI results

The comparison of the BOLD responses in the viewing of green paintings versus original red paintings revealed greater activations to green in several brain regions. The main clusters of an increased green-related activation were located in the left precuneus, the inferior frontal gyrus, and the inferior parietal lobules. These results are summarized in Table 1 and Figure 2. The reverse comparison revealed no significant activation.

There was also a greater activation in response to black paintings compared to the original red paintings in several brain areas, particularly

Table 1. Location of brain regions that respond to comparison of color-changed conditions vs. original red condition.

	MNI coordinates				
Brain regions	*x*	*y*	*z*	**Z-scores**	**Number of voxels**
Green–red					
L precuneus	−9	−64	35	4.68	1,092
L inferior frontal gyrus	−36	4	27	4.00	449
L inferior parietal lobule	−60	−46	38	3.94	687
Black–red					
L supplementary motor areas	−6	2	62	4.38	1,011
L medial frontal gyrus	−33	11	44	3.21	

Note: None of the reverse comparisons (red vs. green, or red vs. black) reached significance. Regions are designated using the MNI coordinates. MNI = Montreal Neurological Institute; L = left hemisphere.

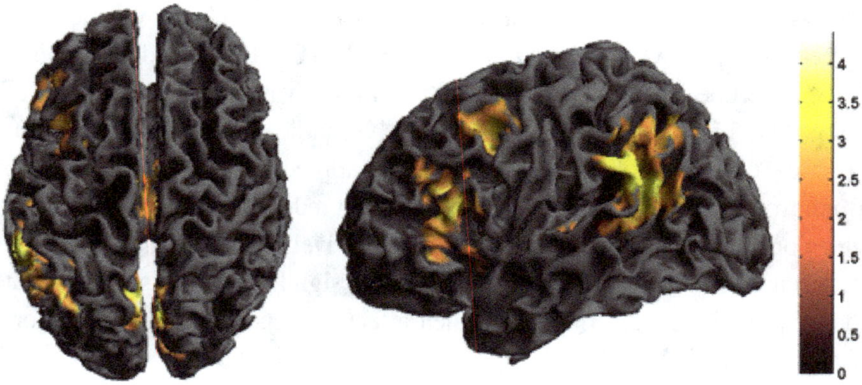

Figure 2. The green-red contrast revealed activation in the left precuneus, inferior frontal gyrus, and inferior parietal lobule.

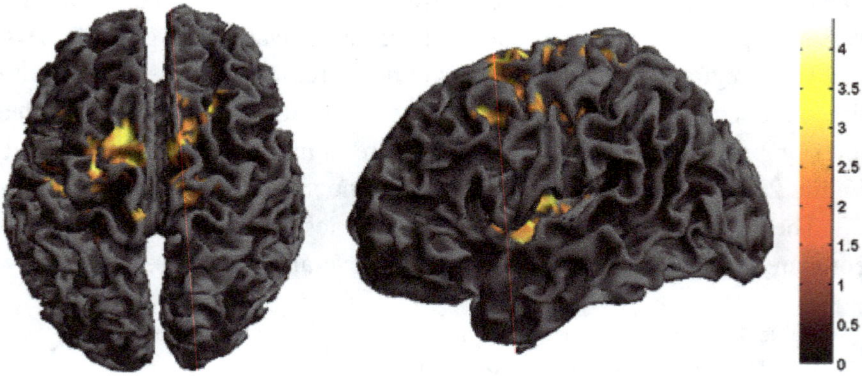

Figure 3. The comparison of black and red conditions resulted in activations in supplementary motor areas and medial frontal gyrus.

in the supplementary motor area and the medial frontal gyrus. The lower parts of Table 1 and Figure 3 show a summary of these results. There was no significant activation for the reverse comparison.

Discussion

In this study, we report the influence of color on the aesthetic appreciation of paintings by the Chinese artist LaoZhu representing the third abstraction. Both behavioral and brain-scanning methods were employed to identify potential neural correlates. The behavioral results showed that the original paintings were preferred when compared with their color-changed counterparts; surprisingly and contrary to expectation, neural activities in several brain regions which are known to be involved in aesthetic appreciation were greater when subjects looked at the color-changed paintings compared to viewing the original red paintings. Despite this discrepancy, our findings support the hypothesis that color information itself (or certain regions of the electromagnetic spectrum) can generate differential effects on neural activities when viewing abstract paintings. This may not be too surprising since specific pathways within the visual system are devoted to the processing of colors (e.g., Gegenfurtner, 2003). This functional segmentation may already be implemented at the retinal level, as experiments on color induction would suggest (Pöppel, 1986; Zhou *et al.*, 2016). Any visual experience requires an initial analysis of optical features in the early visual pathway, and this applies also to aesthetic appreciation when viewing artwork (i.e., specific stimulus properties of perceptual objects have to be extracted). Our results add to the body of evidence stressing the importance of such primary features on the aesthetic appreciation of paintings (for a review, see also Palmer, Schloss, & Sammartino, 2013). For example, by displacing an object within the original paintings, Vartanian and Goel (2004) showed that the originals were preferred compared to the changed paintings, and this comparison of original and changed pictures revealed significant activations in several relevant brain areas. Furthermore, the effects of symmetry and complexity of the stimulus configurations on aesthetic judgments have been shown to be robust in several studies (Jacobsen, Schubotz, Höfel, & Cramon, 2006; Tinio & Leder, 2009).

As indicated, our observations that color-changed artworks evoked higher neural activations are not in line with previous findings that showed higher aesthetic preferences together with higher neural activity as judged from neural processing in the reward circuitry and those neural areas assumed to be involved in the aesthetic assessment of stimuli (Lacey *et al.*, 2011; Montague, Harvey, & Kirk, 2014). In those studies, the less preferred artworks led to less reward, less emotional valence, and less attentional orienting, all of which correlated with reduced neural activities in different brain areas. One possible explanation of our contradictory result might be that the red paintings of LaoZhu are experienced as being more natural or harmonious; the subjective experience of viewing the original red paintings may require less processing effort on the neural level; such a different processing effort has been suggested already in a previous study when beautiful and non-beautiful environments being stored in episodic memory were compared (Vedder *et al.*, 2015). The color-changed versions of the pictures may result in an experiential inconsistency or a violation of a good gestalt; such a subjective inconsistency may well be implemented on an implicit level (Pöppel & Bao, 2011). The higher neural activation at the cortical level whenever a psychological mismatch, as suggested here, takes place may reflect an enhanced top-down influence. Indeed, unusual or unnatural stimuli may lead to more neural activity; for example, Kirk (2008) found that objects viewed in unexpected or abnormal figure–ground combinations led to significantly more activations compared to normal contextual conditions. More support for this notion comes also from studies when viewing abnormally colored objects (Zeki & Marini, 1998), novel stimuli (Cardillo, Watson, Schmidt, Kranjec, & Chatterjee, 2012), or irregular temporal patterns (Zeki, Hulme, Roulston, & Atiyah, 2008). Furthermore, activations were mainly lateralized to the left hemisphere in this study, which may relate to the complex, effortful processing of color-changed paintings.

In detail, the green–red contrast revealed neural activations in the precuneus, the inferior frontal gyrus, and the inferior parietal lobule. These areas are often observed as being activated in the aesthetic evaluation of artworks (Boccia *et al.*, 2016; Vartanian & Skov, 2014). Specifically, activation of the inferior frontal gyrus may be involved in the decoding of stimuli (Kirk, 2008), whereas the precuneus, when related to vision and imagery, has been involved in voluntarily guided attention and manipulation of mental images (Cavanna & Trimble, 2006). As part of the associative cortices, the precuneus is in general related to rather abstract higher cognitive functions, in particular to an integration of external and self-generated information, which is why it has previously been linked to (1) the integration of external information in a self-referential context (Northoff *et al.*, 2006), (2) the integration of egocentric and allocentric spatial relations (Cavanna & Trimble, 2006), and (3) the differentiation of first-person and third-person perspective (Vogeley *et al.*, 2001). Thus, the differential involvement of the precuneus might indicate that color modulates the strength of creating an internal point of view when observing LaoZhu's artwork.

The direct comparison of black and red conditions resulted in activations in the supplementary motor area and the medial frontal gyrus; this observation can be related to the commonly activated brain structures when viewing artworks (Boccia *et al.*, 2016; Vartanian & Skov, 2014). The medial frontal gyrus is likely to be engaged in higher mental functions involving executive control (Talati & Hirsch, 2005). Thus, taken together, the paradoxical result that the processing of green and black paintings compared to the original red paintings required more neural effort in the participating networks may not be at all paradoxical.

In summary, on this more technical level of analysis, the behavioral results indicate that the original red paintings of LaoZhu are preferred when compared with their color-changed counterparts, and on the neural level, higher activations for the color-changed conditions have been uncovered. These observations suggest that, on the one hand, lower-level visual properties modulate the aesthetic appreciation of abstract art, but, on the other hand, higher-level neural mechanisms are also involved. This latter point invites speculation about some further aspects of our study, such as reflecting the concept of "abstraction and empathy" (Worringer, 1908/2007; i.e., the internal point of view in comparison to the external point of view) when one is exposed to pictures of the third abstraction by LaoZhu.

One question that is related to this antinomy of perspectives is why red in the pictures of LaoZhu is preferred on the subjective level than green or black. We suggest that the color red is more deeply rooted in the neural systems of humans than the other colors. Such a hypothesis is motivated, for instance, by a single case study with a patient who had suffered a bilateral occipital lobe infarction (Pöppel *et al.*, 1978). This patient with an extremely restricted visual field had suffered a complete loss of color perception after a stroke; however, he could still differentiate between light and darkness. After some time, color perception returned to his visual world, and it was the color red that was perceived again at first. Gradually, the other colors on the perceptual level came back too; however, surprisingly green and blue could not be discriminated initially.

This return of seeing colors by the patient was unrelated to retinal mechanisms that had remained intact after the stroke. This specific temporal sequence of seeing colors again after their loss emphasizes the higher importance of red, and that, indeed, red is more deeply rooted in the neural machinery of the visual system. Thus, we speculate that LaoZhu was drawn to express the third abstraction in the color red for reasons related to an ease of processing; the artist himself claims no alternative seemed suitable. This selection of the right color does not have to take place on an explicit level of rational conjecture; it happens as a self-evident process on an implicit level without necessary reflection (Pöppel & Bao, 2011). As the artist identifies himself with his own artwork, the viewer is similarly affected by the choice of "the right color." An internal view is automatically created, and one may feel being part of the abstract scene when viewing a picture.

In fact, the viewer sees a scene and not a meaningless picture, and this makes the third abstraction of LaoZhu very different from the earlier phases of abstract art. Most of the pictures of the series studied in our experiment are characterized by a visual duality stretching along the horizontal meridian. Whatever the accidental priming of a viewer might be in a given moment, one is mentally forced (or perhaps seduced) to impose meaning onto the picture either on a concrete or a conceptual level. One person may see a landscape that might be reminiscent of a landscape picture in Eastern art (Bao *et al.*, 2016); such "internal landscapes" may create a feeling of belongingness, again supporting and even verifying the internal point of view. Another situational priming may result in seeing red clouds in the evening sky, or one may see blood spread over a surface. On a more conceptual level, the duality of many pictures initiates a thought process about the ambivalence of life, being separated from the other and longing for togetherness.

All of these internal scenes arise automatically and without mental effort, and they touch on a fundamental principle of human knowledge systems. Other than what is often implied, we do not possess just one knowledge system, which is traditionally referred to as being rational, explicit, and semantic (Descartes, 1637/1990): We possess two additional knowledge systems (Pöppel & Bao, 2011). In addition to explicit knowledge, we are in fact dominated by an implicit knowledge system that is, for instance, activated when viewing the art of the third abstraction by LaoZhu. And we have to add a third knowledge system, which is the visual (or sensory in general) experience itself as has been stressed in particular by Zeki (1999). Seeing is knowing. Thus, when we are exposed to the third abstraction, all three knowledge systems are activated; automatically and without mental effort, both the visual and implicit knowledge systems are entrained. Then, when we reflect on what we see, or what we have seen, and when we visit our "inner museum" (i.e., the pictures that are stored in our episodic memory), we activate the explicit knowledge systems. The question remains as to whether our knowledge systems are independent of each other, or whether they are bound together in a complementary way by a common principle. It has been argued that on a conceptual level, two such unifying principles can be localized (Pöppel & Bao, 2011). In a nutshell, all knowledge obeys the mimetic principle (i.e., that knowledge is always the representation of something as something, whatever the mode of neural representation may be). Furthermore, on a formal level, the knowledge systems obey the aesthetic principle (i.e., in whatever way knowledge is represented in our neural systems or on the behavioral level, it has to please our aesthetic sense).

References

Aviv, V. (2014). What does the brain tell us about abstract art? *Frontiers in Human Neuroscience*, 8, 85. https://doi.org/10.3389/fnhum.2014.00085.

Bao, Y. & Pöppel, E. (2012). Anthropological universals and cultural specifics: Conceptual and methodological challenges in cultural neuroscience. *Neuroscience and Biobehavioral Reviews*, 36, 2143–2146. https://doi.org/10.1016/j.neubiorev.2012.06.008.

Bao, Y., Pöppel, E., Wang, L., Lin, X., Yang, T., Avram, M., Blautzik, J., Paolini, M., Silveira, S., Vedder, A., Zaytseva, Y., & Zhou, B. (2015). Synchronization as a biological, psychological and social mechanism to create common time: A theoretical frame and a single case study. *PsyCh Journal*, 4, 243–254. https://doi.org/10.1002/pchj.119.

Bao, Y., Yang, T., Lin, X., Fang, Y., Wang, Y., Pöppel, E., & Lei, Q. (2016). Aesthetic preferences for Eastern and Western traditional visual art: Identity matters. *Frontiers in Psychology*, 7, 1596. https://doi.org/10.3389/fpsyg.2016.01596.

Berlyne, D. E. (1971). *Aesthetics and Psychobiology*. New York, NY: Appleton-Century-Crofts.

Boccia, M., Barbetti, S., Piccardi, L., Guariglia, C., Ferlazzo, F., Giannini, A. M., & Zaidel, D. W. (2016). Where does brain neural activation in aesthetic responses to visual art occur? Meta-analytic evidence from neuroimaging studies. *Neuroscience and Biobehavioral Reviews*, 60, 65–71. https://doi.org/10.1016/j.neubiorev.2015.09.009.

Cardillo, E. R., Watson, C. E., Schmidt, G. L., Kranjec, A., & Chatterjee, A. (2012). From novel to familiar: Tuning the brain for metaphors. *NeuroImage*, 59, 3212–3221. https://doi.org/10.1016/j.*NeuroImage*.2011.11.079.

Cavanna, A. E. & Trimble, M. R. (2006). The precuneus: A review of its functional anatomy and behavioural correlates. *Brain*, 129, 564–583. https://doi.org/10.1093/brain/awl004.

Cinzia, D. D., & Vittorio, G. (2009). Neuroaesthetics: A review. *Current Opinion in Neurobiology*, 19, 682–687. https://doi.org/10.1016/j.conb.2009.09.001.

Cupchik, G. C., Vartanian, O., Crawley, A., & Mikulis, D. J. (2009). Viewing artworks: Contributions of cognitive control and perceptual facilitation to aesthetic experience. *Brain and Cognition*, 70, 84–91. https://doi.org/10.1016/j.bandc.2009.01.003.

Cupchik, G. C. & Winston, A. S. (1996). Confluence and divergence in empirical aesthetics, philosophy, and mainstream psychology. In M. P. Friedman & E. C. Carterette (eds.), *Handbook of Perception and Cognition: Cognitive Ecology*, 2nd Edition, pp. 61–85. San Diego, CA: Academic Press. https://doi.org/10.1016/b978-012161966-4/50005-0.

Descartes, R. (1990). *Discours de la méthode* [*A Discourse on Method*]. Hamburg, Germany: Felix Meiner (Original Work Published 1637).

Gegenfurtner, K. R. (2003). Cortical mechanisms of colour vision. *Nature Reviews Neuroscience*, 4, 563–572. https://doi.org/10.1038/nrn1138.

Graham, D. J. & Redies, C. (2010). Statistical regularities in art: Relations with visual coding and perception. *Vision Research*, 50, 1503–1509. https://doi.org/10.1016/j.visres.2010.05.002.

Jacobsen, T., Schubotz, R. I., Höfel, L., & Cramon, D. Y. V. (2006). Brain correlates of aesthetic judgment of beauty. *NeuroImage*, 29, 276–285. https://doi.org/10.1016/j.NeuroImage.2005.07.010.

Jacobsen, T. & Wolsdorff, C. (2007). Does history affect aesthetic preference? Kandinsky's teaching of colour-form correspondence, empirical aesthetics, and the Bauhaus. *Design Journal*, 10, 16–27. https://doi.org/10.2752/146069207789271902.

Kandinsky, W. (2009). *Über das Geistige in der Kunst* [*On the Spiritual in Art*], 3rd Edition. Bern, Switzerland: Benteli Verlag (Original Work Published 1912).

Kirk, U. (2008). The neural basis of object-context relationships on aesthetic judgment. *PLoS One*, 3, e3754. https://doi.org/10.1371/journal.pone.0003754.

Lacey, S., Hagtvedt, H., Patrick, V. M., Anderson, A., Stilla, R., Deshpande, G., Hu, X., Sato, J. R., & Sathian, K. (2011). Art for reward's sake: Visual art recruits the ventral striatum. *NeuroImage*, 55, 420–433. https://doi.org/10.1016/j.NeuroImage.2010.11.027.

Locher, P., Overbeeke, K., & Stappers, P. J. (2005). Spatial balance of color triads in the abstract art of Piet Mondrian. *Perception*, 34, 169–189. https://doi.org/10.1068/p5033.

Luria, A. R. (1968). *The Mind of a Mnemonist: A Little Book about a Vast Memory*. New York, NY: Basic Books.

Montague, P. R., Harvey, A. H., & Kirk, U. (2014). Using fMRI to study valuation and choice. In T. D. Papageorgiou, G. I. Cristopoulous, & S. M. Smirnakis (eds.), *Advanced Brain Neuroimaging Topics in Health and Disease: Methods and Applications*, pp. 605–618. Rijeka, Croatia: Intech. https://doi.org/10.5772/58257.

Northoff, G., Heinzel, A., de Greck, M., Bermpohl, F., Dobrowolny, H., & Panksepp, J. (2006). Self-referential processing in our brain—A meta-analysis of imaging studies on the self. *NeuroImage*, 31, 440–457. https://doi.org/10.1016/j.*NeuroImage*.2005.12.002.

Palmer, S. E., Schloss, K. B., & Sammartino, J. (2013). Visual aesthetics and human preference. *Annual Review of Psychology*, 64, 77–107. https://doi.org/10.1146/annurev-psych-120710-100504.

Pöppel, E. (1986). Long-range colour-generating interactions across the retina. *Nature*, 320, 523–525. https://doi.org/10.1038/320523a0.

Pöppel, E. (1989). Taxonomy of the subjective: An evolutionary perspective. In J. W. Brown (ed.), *Neuropsychology of Visual Perception*, pp. 219–232. Hillsdale, NJ: Lawrence Erlbaum.

Pöppel, E. & Bao, Y. (2011). Three modes of knowledge as basis for intercultural cognition and communication: A theoretical perspective. In S. Han & E. Pöppel (eds.), *Culture and Neural Frames of Cognition and Communication*, pp. 215–231. Heidelberg, Germany: Springer-Verlag. https://doi.org/10.1007/978-3-642-15423-2_14.

Pöppel, E., Brinkmann, R., von Cramon, D., & Singer, W. (1978). Association and dissociation of visual functions in a case of bilateral occipital lobe infarction. *Archiv für Psychiatrie und Nervenkrankheiten*, 225, 1–22. https://doi.org/10.1007/BF00367348.

Ritterfeld, U. (2002). Social heuristics in interior design preferences. *Journal of Environmental Psychology*, 22, 369–386. https://doi.org/10.1006/jevp.2002.0276.

Scoville, W. B. & Milner, B. (1957). Loss of recent memory after bilateral hippocampal lesions. *Journal of Neurology, Neurosurgery, and Psychiatry*, 20, 11–21. https://doi.org/10.1136/jnnp.20.1.11.

Silveira, S., Bao, Y., Wang, L., Pöppel, E., Avram, M., Simmank, F., Zaytseva, Y., & Blautzik, J. (2015). Does a bishop pray when he prays? And does his brain distinguish between different religions? *PsyCh Journal*, 4, 199–207. https://doi.org/10.1002/pchj.116.

Silveira, S., Fehse, K., Vedder, A., Elvers, K., & Hennig-Fast, K. (2015). Is it the picture or is it the frame? An fMRI study on the neurobiology of framing effects. *Frontiers in Human Neuroscience*, 9, 528. https://doi.org/10.3389/fnhum.2015.00528.

Silveira, S., Graupmann, V., Frey, D., Blautzik, J., Meindl, T., Reiser, M., Chen, C., Wang, Y., Bao, Y., Pöppel, E., & Gutyrchik, E. (2012). Matching reality in the arts: Self-referential neural processing of naturalistic compared to surrealistic images. Perception, 41, 569–576. https://doi.org/10.1068/p7191.

Silveira, S., Gutyrchik, E., Wetherell, G., Bao, Y., Pöppel, E., Blautzik, J., Reiser, M., Frey, D., & Graupmann, V. (2014). Ceci n'est pas la mort: Evidence for the recruitment of self-reference from surrealistic art under mortality salience. *European Journal of Social Psychology*, 45, 255–266. https://doi.org/10.1002/ejsp.2076.

Strasburger, H. & Waldvogel, B. (2015). Sight and blindness in the same person: Gating in the visual system. *PsyCh Journal*, 4, 178–185. https://doi.org/10.1002/pchj.109.

Sun, D. (2011). Laozhu: The third abstract. *Leap*, 12. Retrieved from http://www.leapleapleap.com/2012/02/zhu-qingsheng-and-tan-ping/.

Talati, A. & Hirsch, J. (2005). Functional specialization within the medial frontal gyrus for perceptual go/no-go decisions based on "what," "when," and "where" related information: An fMRI study. *Journal of Cognitive Neuroscience*, 17, 981–993. https://doi.org/10.1162/0898929054475226.

Thakral, P. P., Moo, L. R., & Slotnick, S. D. (2012). A neural mechanism for aesthetic experience. *NeuroReport*, 23, 310–313. https://doi.org/10.1097/WNR.0b013e328351759f.

Tinio, P. P. & Leder, H. (2009). Just how stable are stable aesthetic features? Symmetry, complexity, and the jaws of massive familiarization. *Acta Psychologica*, 130, 241–250. https://doi.org/10.1016/j.actpsy.2009.01.001.

Vartanian, O. & Goel, V. (2004). Neuroanatomical correlates of aesthetic preference for paintings. *NeuroReport*, 15, 893–897. https://doi.org/10.1097/00001756-200404090-00032.

Vartanian, O. & Skov, M. (2014). Neural correlates of viewing paintings: Evidence from a quantitative meta-analysis of functional magnetic resonance imaging data. *Brain and Cognition*, 87, 52–56. https://doi.org/10.1016/j.bandc.2014.03.004.

Vedder, A., Smigielski, L., Gutyrchik, E., Bao, Y., Blautzik, J., Pöppel, E., Zaytseva, Y., & Russell, E. (2015). Neurofunctional correlates of environmental cognition: An fmri study with images from episodic memory. *PLoS One*, 10, e0122470. https://doi.org/10.1371/journal.pone.0122470.

Vogeley, K., Bussfeld, P., Newen, A., Herrmann, S., Happé, F., Falkai, P., Maier, W., Shah, N. J., Fink, G. R., & Zilles, K. (2001). Mind reading: Neural mechanisms of theory of mind and self-perspective. *NeuroImage*, 14, 170–181. https://doi.org/10.1006/nimg.2001.0789.

Wang, L., Bao, Y., Zhang, J., Lin, X., Yang, L., Pöppel, E., & Zhou, B. (2016). Scanning the world in three seconds: Mismatch negativity as an indicator of temporal segmentation. *PsyCh Journal*, 5, 170–176. https://doi.org/10.1002/pchj.144.

Worringer, W. (2007). *Abstraktion und Einfühlung: Ein Beitrag zur Stilpsychologie* [*Abstraction and Empathy: A Contribution to the Psychology of Style*]. Munich, Germany: Wilhelm Fink Verlag (Original Work Published 1908).

Zaytseva, Y., Gutyrchik, E., Bao, Y., Pöppel, E., Han, S., Northoff, G., Welker, L., Meindl, T., & Blautzik, J. (2014). Self processing in the brain: A paradigmatic fMRI case study with a professional singer. *Brain and Cognition*, 87, 104–108. https://doi.org/10.1016/j.bandc.2014.03.012.

Zeki, S. (1999). *Inner Vision: An Exploration of Art and the Brain.* Oxford, UK: Oxford University Press.

Zeki, S., Hulme, O. J., Roulston, B., & Atiyah, M. (2008). The encoding of temporally irregular and regular visual patterns in the human brain. *PLoS One*, 3, e2180. https://doi.org/10.1371/journal.pone.0002180.

Zeki, S. & Marini, L. (1998). Three cortical stages of colour processing in the human brain. *Brain*, 121, 1669–1685. https://doi.org/10.1093/brain/121.9.1669.

Zetzsche, C. & Krieger, G. (2001). Nonlinear mechanisms and higher-order statistics in biological vision and electronic image processing: Review and perspectives. *Journal of Electronic Imaging*, 10, 56–99. https://doi.org/10.1117/1.1333056.

Zhou, B., Pöppel, E., Wang, L., Yang, T., Zaytseva, Y., & Bao, Y. (2016). Seeing without knowing: Operational principles along the early visual pathway. *PsyCh Journal*, 5, 145–160. https://doi.org/10.1002/pchj.141.

https://doi.org/10.1142/9789811298639_0016

Chapter 16

Enchanted in the Multiverse of LaoZhu: A Reflection

Yan Bao and Ernst Pöppel

School of Psychological and Cognitive Sciences, Peking University, Beijing, China
Institute of Medical Psychology, Ludwig Maximilian University, Munich, Germany

"Habent sua fata libelli": This classical Latin sentence says that books have their own fate. This applies also to pictures. In the moment an artist opens the atelier and invites the visitors to look at the pictures, they start a new life, their ownership is now shared by others. This happened to the two of us when we visited LaoZhu in his atelier. He looked at one of us (EP) who looked in this moment at a picture of the *Zhu Mo* series (Figure 1). LaoZhu spontaneously said that the picture is now part of you. Indeed, in this moment it became a part of one of us (EP) but also a part of the other one of us (YB), although in a different way, when she looked at the same picture. Different bridges were created between our seeing eyes and our inner eyes, the representation in our mind. Even LaoZhu may have created in this moment a new bridge to his pictures because of our physical presence and our shared ownership. Although every one of us saw the picture in a different way possibly shaped by our different cultural

Figure 1. (top) & (bottom): Pictures of LaoZhu from the series "Zhu Mo" ("red ink").

imprinting (Chinese and German), by prior experiences, implicit expectations, hidden prejudices, we were connected with each other by the one picture, the artist and we visitors. Although every one of us was (and always is) a "victim" of a very personal process of abstraction when seeing a picture, unity was created in this moment by a common reference, namely, this one picture. The founder of aesthetics in modern times, Alexander Baumgarten (1750/58), asked this question in his classic book *Aesthetica*: *quid enim est abstractio, si iactura non est?* — "what else means abstraction than a loss?" But how much have we gained by the abstract bridges to the one picture in spite of the necessary theoretical loss due to abstraction; we three found ourselves together in this moment at one location in an abstract space, enchanted in the multiverse of LaoZhu.

What are typical features that characterize individual perspectives such that everyone sees a different picture in his or her inner museum, although everybody has the same picture in front of the eyes? These features are emphasized differently in every person, some features may be irrelevant for some, but in each case they are abstractions because the focus on one feature neglects necessarily others. To mention such specific abstracted features may be even annoying to the artist, but pictures have their own fate. What is seen in the picture by others is beyond the artist. The more is beyond an explicit or implicit intention of the artist, the more it touches personal experiences, and the richer is each picture.

What then are some typical features of the pictures of the Zhu Mo series (e.g., Figure 1), each one representing different trajectories of abstractions? What immediately strikes the eye is the fact that these pictures are caught within a frame. The structure of the pictures implies that they go beyond the frame. They comprise a much larger area in the visual field as they physically represent. The pictures are much larger than they are. When looking at a picture one is invited to imagine what could be beyond the frame. Thus, all pictures are enigmatic; they all share a different and individual secret. This extension beyond the physical frame by itself creates attention and curiosity.

This spatial aspect is complemented by a temporal aspect. The pictures are highly dynamic but paradoxically represent "frozen time". Thus, they invite to create a story in the mind; the present moment of the pictures in a complementary way extends into the past and into the future. Here the concept of the "third abstraction" finds its realization; LaoZhu relates his art to calligraphy which is characterized by a highly dynamic process capturing in the moment what has been and what might be. Complementarity as a characteristic of a creative process finds its experiential confirmation in his art.

With respect to spatial organization, most pictures are characterized by a dual structure along the horizontal meridian. This left–right orientation is complemented by different vertical composition patterns. What is most striking in the pictures is a tendency toward the center, but it is an "empty center". This creates a unique tension and emotional involvement. The empty center provokes questions like the following: "What is going to happen? What might be hidden? Is there a way to follow? Do I have to fight being endangered? Am I safe being protected?" One question follows the other without leading to an answer.

Although superficially the pictures look similar in their structural composition, they are extremely different. Each picture is unique, it has its own harmony, and it creates a distinct position in the inner museum of the receptive mind. This uniqueness can, however, only be discovered if one is willing to open not only the eyes in front of the head but the "inner eye". If one walks along many pictures hanging on the wall with a superficial attitude and a mind in the mode of inspection, one will only see that they are red and very similar. What a waste of time by dedicating too little time.

What can one see in this multiverse? Whatever the accidental priming of a viewer might be in a given moment, one is mentally forced (or perhaps seduced) to impose meaning onto the picture either on a concrete or a conceptual level in spite of their enigmatic nature. One person may see a three-dimensional landscape which might be reminiscent of a landscape picture in Eastern art; such "internal landscapes" may create a feeling of belongingness by triggering a "floating view". Another situational priming may result in seeing red clouds in the evening sky and a feeling of longing may arise. One may see blood spread over a surface remembering a cruel event. Another one or the same person in another frame of mind may associate an erotic event. On a more conceptual level, the duality of many pictures initiates a thought process about the ambivalence of life, being separated from each other, and feeling lonely. And one may think of a sentence of the Greek philosopher Anaximander with whom Western philosophy started and which is perhaps one of the most important characteristics of the art of LaoZhu, that the origin of everything is the undefined, the unlimited, the "apeiron".

https://doi.org/10.1142/9789811298639_0017

Chapter 17

The Impact of Action Observation and Motor Imagery on the Aesthetic Preference of Chinese Calligraphy*

Mingcheng He, Xianyou He, Zixin Zheng, Jiamin Deng and Chenjing Wu

Key Laboratory of Brain, Cognition and Education Sciences (South China Normal University), Ministry of Education, Guangzhou, China
School of Psychology, South China Normal University, Guangzhou, China
Center for the Studies of Psychological Application, South China Normal University, Guangzhou, China
Guangdong Key Laboratory of Mental Health and Cognitive Science, South China Normal University, Guangzhou, China

*This chapter was originally published in the special collection on neuroaesthetics in *PsyCh Journal*. He, M., He, X., Zheng, Z., Deng, J., & Wu, C. (2022). The impact of action observation and motor imagery on the aesthetic preference of Chinese calligraphy. *PsyCh Journal*, 11(5), 645–655. https://doi.org/10.1002/pchj.488.

Abstract

The creation of artwork requires motor activity. However, few empirical studies have directly explored the relationship between the experience of an artist's action and aesthetic experience. This study aimed to examine the effect of observing and imagery of the artist's action on the participants' aesthetic preferences. In Experiments 1 and 2, we took hard-pen and brush-pen Chinese calligraphy images as the stimulus, respectively, to explore the influence of action observation on the aesthetic preference by manipulating the artists' actions. The results of both Experiments 1 and 2 show that when participants observed the artists' actions, they tended to report a higher preference for calligraphy images compared with the control condition. In Experiments 3 and 4, we used instructions to manipulate the motor imagery tasks and investigated the effect of imaging the artist's action on the participants' aesthetic preferences. The results showed that both kinesthetic imagery and visual imagery increased the participants' preferences. In general, our study shows that both action observation and motor imagery contribute to participants' aesthetic preferences. The results are discussed in terms of how artists' actions possibly influence the aesthetic preference of Chinese calligraphy.

Keywords: action observation; artist's action; Chinese calligraphy; aesthetic preference; motor imagery

Introduction

In the visual arts, traditional aesthetic studies have mainly focused on the influence of perceptual features, such as contrast, color, orientation, proportion, and grouping (Leder *et al.*, 2012; Taylor *et al.*, 2012). However, the creation of an artwork requires motor activity (Bullot, 2009; Ticini *et al.*, 2014). We sometimes watch live creations by artists, including their actions. For example, as an important symbolic art in China, observers can often appreciate calligraphy through the on-site creation of calligraphers. Therefore, some views have noted that the experience of an artist's actions may also play a crucial role in our aesthetic experience (Freedberg & Gallese, 2007; Leder *et al.*, 2012; Taylor *et al.*, 2012). Previous studies have shown that observers could indirectly perceive artists' actions through clues, such as the brushstrokes on the canvas or the trace of sculpture (Sbriscia-Fioretti *et al.*, 2013; Taylor *et al.*, 2012), whereas few empirical studies have directly explored the relationship between the experience of artists' action and aesthetic experience. Here, this study mainly aims to explore whether two common ways to experience the artists' actions, which encompass action observation and motor imagery, impact the observer's aesthetic experience.

"Action observation" refers to the observation of action performed by others (Jeannerod, 2001). Imaging studies have shown that mirror neurons in the brain are activated when performing the action and observing others performing the action (Rizzolatti & Sinigaglia, 2010; Shaw & Czekóová, 2013), which indicated that observation of action induced a simulation of the corresponding action in the observer's brain (Avenanti *et al.*, 2013). This allows us to experience the other's action in an "embodied" way and to infer the internal state of others, including understanding the intentions and emotions behind the action (Gardner *et al.*, 2017; Hamilton, 2013; Molnar-Szakacs, 2011). For example, observing others' gestures influenced the observers' reasoning (Alibali *et al.*, 2014) because they could effectively perceive their surroundings from the others' perspective (Frischen *et al.*, 2009). The psycho-historical framework, which was proposed by Bullot and Reber (2013), suggests that artistic understanding of the work is based on knowledge of the art-historical contexts. Art-historical contexts encompass artists' actions which leave causal information in each work of art. Perception of the artists' actions led the observers to engage in the design stance of art and this engagement triggers the search for what caused the features perceived in an artwork. Such a search for causal information in artworks is a requisite for artistic understanding. Therefore, we further infer that observing artists' actions might instruct observers in an "embodied" way to engage in the design stance of art and eventually gain artistic understanding and promote aesthetic appreciation.

Moreover, previous studies have shown that observers tend to approach beauty or to avoid ugliness in an embodied way during aesthetic perception and judgments (Kornysheva *et al.*, 2010; Zhang *et al.*, 2016). Lebreton *et al.* (2012) found that observers had a stronger desire to simulate the observed action when the action-directed goal was attractive (e.g., toy). Based on daily experience, we will observe not only the live creations of art experts but also those of art novices. So, we are interested in whether observing artists' actions at different levels (e.g., expert or novice) also would influence the observer's desire for action simulation and aesthetic experience. We infer that the observer would increase their desire for action simulation when they observe the action-directed goal was attractive (e.g., expert's work) and reduce their inner simulation desire when the action-directed goal was unattractive (e.g., novice's work). In short, we hypothesized that if action observation is causally involved in the affective response, it would contribute to the participant's aesthetic experience and also prime similar actions for the observers (Ticini *et al.*, 2014). Meanwhile, the beauty degree of the action-directed goal (e.g., beauty or ugliness calligraphy) may impact their desire for action simulation, which would ultimately be reflected in the participants' button-pressing.

In addition to direct observation, the experience of the artist's action could be acquired through motor imagery. Motor imagery is a type of mental simulation that involves imagining the body performing an action in the absence of motor output (Jeannerod, 2001). Krishna *et al.* (2014) found that active motor imagery boosted the consumers' physiological responses (e.g., salivating). Keesman *et al.* (2016) manipulated motor imagery through instructions and found that the participants who had guidelines to imagine the experience of eating food had increased salivation compared to those who only saw the food. Steinmetz *et al.* (2018) required the participants to imagine themselves in a very cold/warm environment and report their current feelings. The results suggested that the participants in the cold/warm condition preferred more warming/cold activities. These studies showed that people who actively imaged a given state developed more extra feelings paralleled to the real experience or shifted the people's preferences for relevant activities. In other words, we could mentally simulate an experience by imagining it in detail and evoke the same consequences as actually experiencing it. Further, we expect that such benefits would transfer to aesthetic appreciation and speculate that actively imagining the actions needed to create an artwork might additionally improve the spectator's perceptual communications with the artwork. In particular, some forms of art also place great emphasis on the impact of the imagination of the artists' actions on the aesthetic experience. For example, traditional Chinese calligraphy critics believe that imagining the creative action of calligraphy involves the process of artistic re-creation;

it also promotes our aesthetic experience. However, this theoretical view has not been supported by empirical studies.

Moreover, active motor imagery contains kinesthetic imagery (KI) and visual imagery (VI), which establish the subjective "distance" among the self and his own imaginal experience and distinguish KI from VI (Jeannerod, 2001). KI requires people to feel the movement and to perceive muscle contractions and stretching mentally from a first-person (internal) perspective, which corresponds to the representation of a movement as if the individual takes part in the action himself, whereas VI requires self-visualization of a movement from a third-person (external) perspective, which corresponds to the representation of the movement as if the subject is a spectator and that somebody (himself or another person) performs the action (Guillot *et al.*, 2009). According to the aesthetic fluency theory proposed by Reber *et al.* (2004), positive aesthetic experiences are linked to processing ease. So, we suggest that KI tasks are easier to perform because of their closer subjective distance than VI. Therefore, we further hypothesize that performing a KI task might be more helpful for participants to make positive aesthetic judgments than VI.

This Study

We test whether two common ways to experience an artist's action, including action observation and motor imagery, contribute to the participant's aesthetic experience. In Experiments 1 and 2, we used hard-pen and brush-pen Chinese calligraphy images as the stimulus, respectively, to test whether observing artists' actions increased participants' preferences for calligraphy images and how robust this effect was. Chinese calligraphy was used as the stimuli for the following reasons: On the one hand, observers can often appreciate calligraphy through the on-site creation of calligraphers. We are interested in whether this observation has positive implications. On the other hand, the creative action of other forms of art was new to the participants. However, Chinese calligraphy is an art form based on Chinese characters and both follow almost the same writing rules. Therefore, compared with the other forms of art, Chinese calligraphy ensured that the participants performed the motor imagery tasks. In other words, Chinese calligraphy is a suitable artistic medium to examine the influence of action observation and motor imagery on observers' aesthetic preferences. Similarly, In Experiments 3 and 4, we tested whether actively simulating a visceral state (e.g., imaging the action needed to create calligraphy) influenced the participants' preference and behavior, and how robust this effect was.

Experiment 1

Method

Participants

Thirty right-handed, native Mandarin-speaking Chinese undergraduate students from South China Normal University (8 males, 22 females, M_{age} = 20.2 years, $SD = 1.67$ years) participated in the experiment. Participants provided written consent before the study and were paid for their participation. All studies were approved for testing by the Research Ethics Committee of South China Normal University.

Materials

The stimuli consisted of 128 images of hard-pen calligraphy. To control the complexity of the stimuli, the number of character strokes ranged from two to four ($M = 3.63$, $SD = 0.59$). To control the meaning preference of characters, another 32 participants (6 males, 26 females, ages ranging from 17 to 29 years) participated in a seven-point (values from 1–7) scale rating questionnaire about the meaning preference of 32 Chinese characters. The results showed that the average meaning ratings for these Chinese characters were 4.2 ($SD = 0.27$). Then, a calligraphy expert and a calligraphy novice were invited to write these 32 characters. Finally, a total of 64 hard-pen calligraphy image pairs were selected as the stimulus set. For each pair, two images were photographed. One image was with the writing hand to create the condition of action observation while the other image was without the writing hand to create the condition of not observing the action. We believe that this treatment not only allows the observers to watch the artists' actions but also ensures the integrity of the artwork. Thus, four photographs were made for each pair of objects: Two artistic levels (expert's work/novice's work) × two observation conditions (with action observation/without action observation). All images were featured in high resolution (534 × 300 pixels) for the rating task.

Procedure

The experiment began with instructions followed by practice. Each trial of the rating task started with a fixation cross displayed for 1.5 s, which was immediately followed by an image of hard-pen calligraphy presented for 2 s. Participants were asked to watch each image closely and to provide ratings on a 9-point scale as quickly as possible. The scale ranged from 1 (*I do not like it at all*) to 9 (*I like it very much*). The keyboard reaction time was also recorded. To control the influence of physical distance on reaction time, the participants were asked to press 1–2–3–4 keys with their left hand and 6–7–8–9 keys with their right hand, respectively. The images were presented in random order by the E-prime 2.0 software system (as seen in Figure 1).

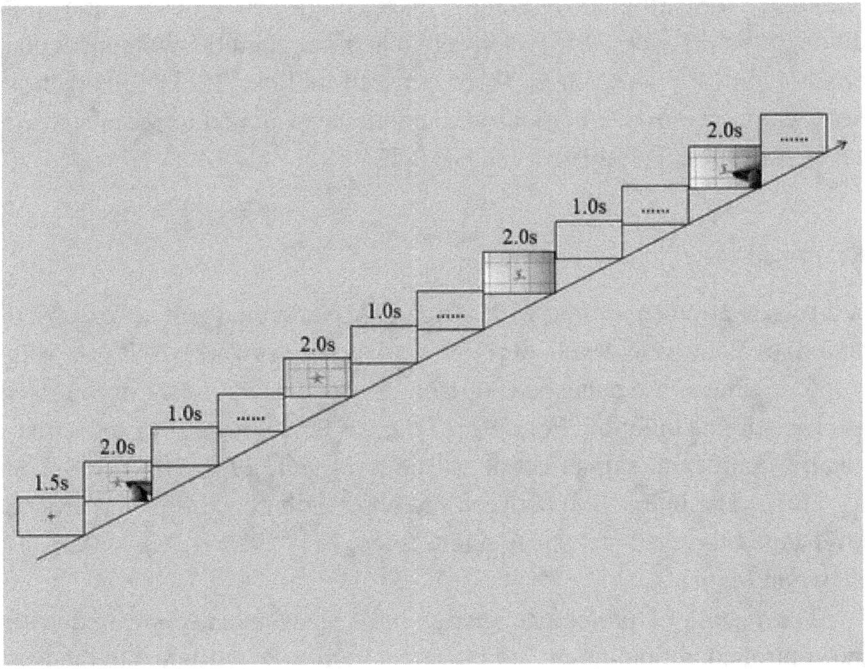

Figure 1. Experimental design, procedure, and examples of stimuli in Experiment 1.

Table 1. Results of preference ratings in Experiment 1 ($M \pm SD$).

	Observation condition	
Artistic level	**With action observation**	**Without action observation**
Expert's work	6.48 ± 0.83	6.30 ± 0.87
Novice's work	3.01 ± 0.76	2.73 ± 0.78

Results and discussion

Preference ratings

A 2 (observation condition: with action observation/without action observation) × 2 (artistic level: expert's work/novice's work) ANOVA with the preference ratings as the dependent variable revealed that there was a significant main effect of artistic level, $F(1, 29) = 170.17$, $p < .001$, $\eta_p^2 = 0.85$, and a significant main effect of observation condition, $F(1, 29) = 27.22$, $p < .001$, $\eta_p^2 = 0.48$, meaning that participants expressed a greater preference for hard-pen calligraphy images in the action observation condition ($M = 5.01$, $SD = 0.17$) relative to that of the condition without action observation ($M = 4.72$, $SD = 0.16$) (as seen in Table 1). The interaction between observation conditions and artistic levels failed to reach statistical significance, $F(1, 29) = 1.50$, $p = 0.23$.

Response time

A 2 (observation condition: with action observation/without action observation) × 2 (artistic level: expert's work/novice's work) ANOVA with response time as the dependent variable revealed a significant main effect of observation condition, $F(1, 29) = 5.59$, $p = 0.025$, $\eta_p^2 = 0.16$, and a marginally significant main effect of artistic level, $F(1, 29) = 3.34$, $p = 0.078$, $\eta_p^2 = 0.10$. The interaction between the observation condition and artistic level failed to reach statistical significance, $F(1, 29) = 1.0$, $p = 0.33$, as shown in Figure 2.

The results of preference ratings have shown that, compared with the control condition, observing the artists' actions contributed to the participants' aesthetic preference, and our previous hypothesis was verified.

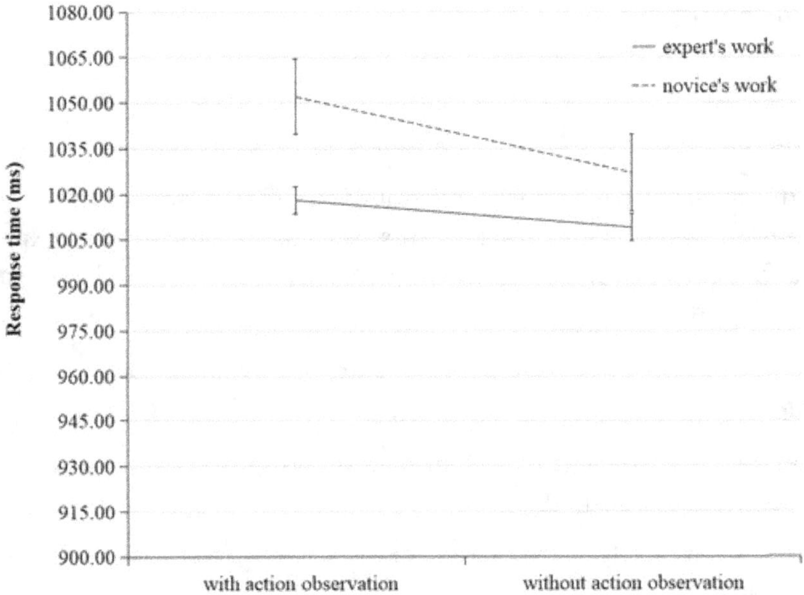

Figure 2. Mean response times for Experiment 1 as a function of the observation condition and artwork type. Error bars are the SE of the mean.

Meanwhile, the results of response time indicated that the observation condition affected the participants' button-pressing responses. In other words, observing the artists' actions induced the participants' action simulation, and this simulation influenced the participants' button-pressing speed. However, we did not find that observing the expert's work significantly positively influenced the response speed in the condition with action observation. One possible reason was that we ignored the influence of the participants' own writing experiences. And the modern hard-pen calligraphy is the main form of daily writing. Participants with more beautiful handwriting may be less prone to simulating the actions of less beautiful calligraphy, which may eventually interfere with their response speed. Although as a representative art form in China, most people have little access to brush-pen calligraphy. Therefore, to exclude the influence of the participants' writing experience, we tried to replicate the procedure of Experiment 1 in Experiment 2 by using brush-pen calligraphy images as stimuli to continue to explore the relationship between action observation and aesthetic experience.

Experiment 2

Method

Participants

Thirty-five undergraduate students from South China Normal University (11 males, 24 females, $M_{age} = 20.6$ years, $SD = 0.32$) participated in Experiment 2. Participants provided written consent before the study and were paid for their participation.

Materials

Similar to Experiment 1, 128 images of brush-pen calligraphy were created for Experiment 2.

Procedure

The design and procedure for Experiment 2 were identical to that for Experiment 1.

Results and discussion

Preference ratings

A 2 (observation condition: with action observation/without action observation) × 2 (artistic level: expert's/novice's work) ANOVA with the preference ratings as the dependent variable revealed that there was a significant main effect of artistic level, $F(1, 34) = 383.29$, $p < 0.001$, $\eta_p^2 = 0.92$, and a significant main effect of observation condition, $F(1, 34) = 25.32$, $p < 0.001$, $\eta_p^2 = 0.43$, meaning that participants had a higher preference for brush-pen calligraphy images in the condition with action observation ($M = 5.12$, $SD = 0.12$) relative to the condition without action observation ($M = 4.91$, $SD = 0.11$) (as seen in Table 2). The interaction between observation conditions and artistic levels did not reach statistical significance, $F(1, 34) = 2.55$, $p = 0.12$.

Table 2. Results of preference ratings in Experiment 2 ($M \pm SD$).

	Observation condition	
Artistic level	**With action observation**	**Without action observation**
Expert's work	6.18 ±1.14	5.95 ±1.16
Novice's work	3.84 ±0.94	3.49 ±0.84

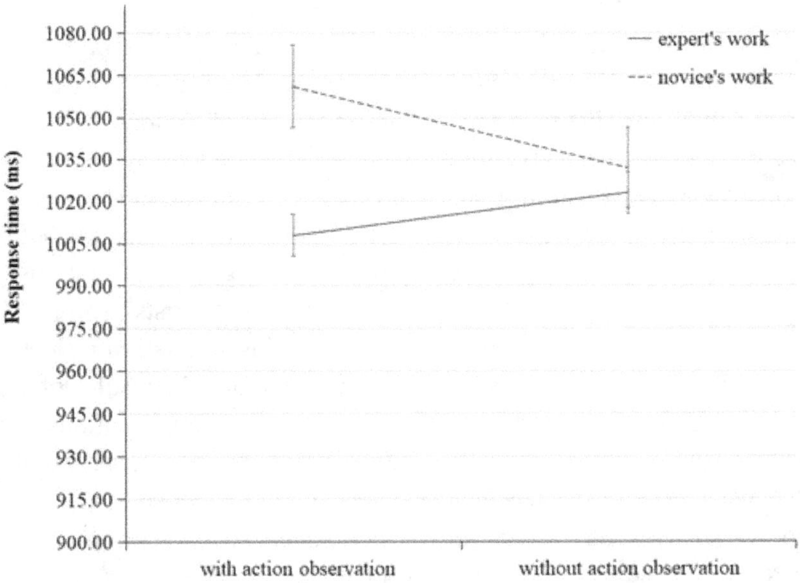

Figure 3. Mean response times for Experiment 2 as a function of the observation condition and artwork type. Error bars are the SE of the mean.

Response time

A 2 (observation conditions: with action observation/without action observation)×2 (artistic levels: expert's/novice's work) ANOVA with response time as the dependent variable revealed that the main effects of artistic level and observation condition failed to reach statistical significance, $F(1, 34) = 4.02$, $p = 0.053$; $F(1, 34) = 0.94$, $p = 0.34$. There was a significant interaction between observation conditions and artistic levels, $F(1, 34) = 10.51$, $p = 0.003$, $\eta_p^2 = 0.24$, as shown in Figure 3. Simple effect analysis showed that, for the expert's work, participants in the action observation condition responded faster (M = 1,008 ms, SD = 186.0) than those who were not (M = 1,023 ms, SD = 194.1), $F(1, 34) = 4.6$, $p = 0.039$, $\eta_p^2 = 0.12$. However, for the novice's work, participants in the action observation condition responded slower (M = 1,061 ms, SD = 193.9) compared to those who were not (M = 1,032 ms, SD = 203.6), $F(1, 34) = 6.16$, $p = 0.018$, $\eta_p^2 = 0.15$.

We carried out a direct replication of Experiment 1 in Experiment 2. Similarly, the results of preference ratings indicated that participants tended to express higher ratings with the action observation relative to the control condition. That is to say, observing the artists' actions contributed to the participants' aesthetic preferences. So, our previous hypothesis was verified once again. The results of response time indicated that both the observation condition and artistic levels affected the participants' button-pressing responses. In other words, after excluding the influence of the participants' own writing experiences, observing the artists' actions may induce the participants' action simulation, and this simulation influenced the participants' button-pressing speed. In general, the results of both Experiment 1 and 2 suggest that observing the artists' live creations could have a positive impact on our aesthetic experience. In Experiments 3 and 4, we further tested whether the alternative experience of the artists' actions, such as actively imagining the action needed to create calligraphy, affected the participants' aesthetic experience.

Experiment 3

In Experiment 3, participants were required either to imagine themselves performing the creating actions of calligraphy or to imagine others performing the same artists' actions to compare the effects of the two forms of imagery on the aesthetic preference of Chinese calligraphy. As mentioned in the Introduction, if our inference is reasonable, we could find that, compared with the control condition, active motor imagery may be more likely to enhance the participants' experience of the artist's action.

Method

Participants

Thirty-fiveundergraduate students from South China Normal University (9 male, 26 female, M_{age} = 21.4 years, SD = 2.01) participated in Experiment 3. Participants provided written consent before the study and were paid for their participation.

Materials

A total of 36 pairs of hard-pen calligraphy images (both expert's and novice's work without the writing hand) were selected from Experiment 1 to serve as the stimuli set.

Procedure

All calligraphy images were further divided into three groups. Each group (including 12 pairs of calligraphy images) was assigned to one of the three blocks: KI, VI, and control condition. Each calligraphy image was randomly presented on the screen for 2 s. Afterward, a tone was presented that lasted for 2 s. According to instructions, participants were required to close their eyes and perform imagery tasks. In the KI block, they imagined themselves performing the actions to create the presented calligraphy images. In the VI block, participants imagined the calligraphy creator of the presented calligraphy images performing the action of creation. In the control condition block, participants closed their eyes. Six seconds later, the participants heard another tone that meant they should open their eyes quickly and report the degree of their preference (1 = *I do not like it at all*, 9 = *I like it very much*). The pilot study showed that it lasted about 6 s for participants to imagine the complete action of writing a character. In addition, after participants imagined the action of creating each calligraphy image, they needed to rate the intensity/vividness of the kinesthetic/visual imagery by using a 9-point scale from 1 (*no sensation/no image*) to 9 (*sensation as intense as actually writing/image as clear as seeing*

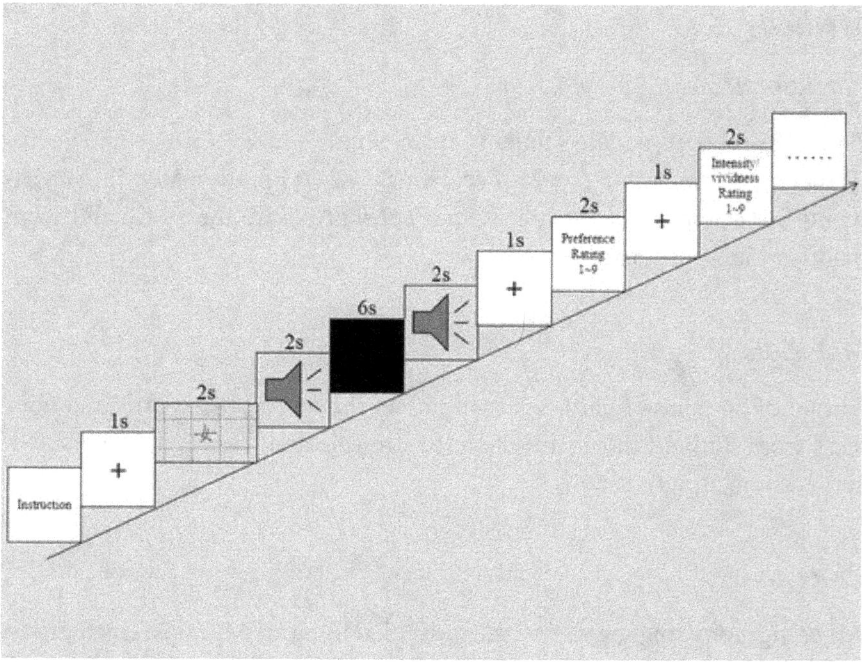

Figure 4. Experimental design, procedure, and examples of stimuli in Experiment 3.

others write). The order of imagery conditions was counterbalanced across participants. To reduce the fatigue effect, participants rested for 5 min between two blocks, as shown in Figure 4.

Imagery task instructions

In the kinesthetic/visual imagery condition, participants were instructed to close their eyes and imagine performing the action to create each calligraphy image as follows: "We want you to focus on the calligraphy image as much as possible, and then imagine yourself creating this calligraphy image/imagine the calligraphy creator creating this image, and close your eyes for 6 s, from the first stroke to the end of the last stroke." In the control condition, participants were instructed to do as follows: "We want you to focus on the calligraphy images as much as possible and close your eyes for 6 s."

Results and discussion

A 3 (imagery condition: control condition/VI/KI) × 2 (artistic level: novice's work/expert's work) repeated-measures ANOVA with preference ratings as the dependent variables revealed that both the main effect of imagery condition and artistic level reached statistical significance, $F(2, 68) = 13.56$, $p<0.001$, $\eta_p^2 = 0.29$; $F(1, 34) = 84.71$, $p<0.001$, $\eta_p^2 = 0.71$. The interaction between imagery condition and artistic level reached statistical significance, $F(2, 68) = 7.54$, $p = 0.001$, $\eta_p^2 = 0.18$. Simple effect analysis revealed that, for experts' work, the differences among imagery conditions reached statistical significance, $F(2, 68) = 6.92$, $p = 0.004$, $\eta_p^2 = 0.28$; the preference ratings in the KI condition were higher than those of the control condition ($p = 0.003$) and were also higher than those in the VI condition but without a significant difference ($p = 0.57$). The preference ratings in the VI condition were higher than those in the control condition ($p = 0.045$). For the novice's work, the differences among imagery conditions reached statistical significance, $F(2, 68) = 17.22$, $p < 0.001$, $\eta_p^2 = 0.56$; the preference ratings in the KI condition were higher than those in the control condition but without significance ($p = 0.31$); also, ratings in the VI condition were higher than those in the control ($p<0.001$) and KI ($p = 0.003$) conditions, as shown in Table 3 and Figure 5.

Correlations between the participants' preference ratings for the expert's work and vividness ($r = 0.40$, $p = 0.018$) and intensity ($r = 0.48$, $p = 0.004$) were significant, indicating that both KI and VI significantly predicted the preference ratings of hard-pen calligraphy images. Similarly, the relationship between the participants' preference ratings for novice's work and vividness ($r = 0.38$, $p = 0.025$) and intensity ($r = 0.42$, $p = 0.012$) was significant, also indicating that both the KI and VI significantly predicted the preference ratings of hard-pen calligraphy images. In sum, the

Table 3. Results of preference ratings in Experiment 3 ($M \pm SD$).

	Imagery condition		
Artistic level	**Control condition**	**Visual imagery (VI)**	**Kinesthetic imagery (KI)**
Expert's work	6.52 ± 0.20	6.26 ± 0.18	6.88 ± 0.17
Novice's work	3.12 ± 0.16	4.35 ± 0.20	3.62 ± 0.22

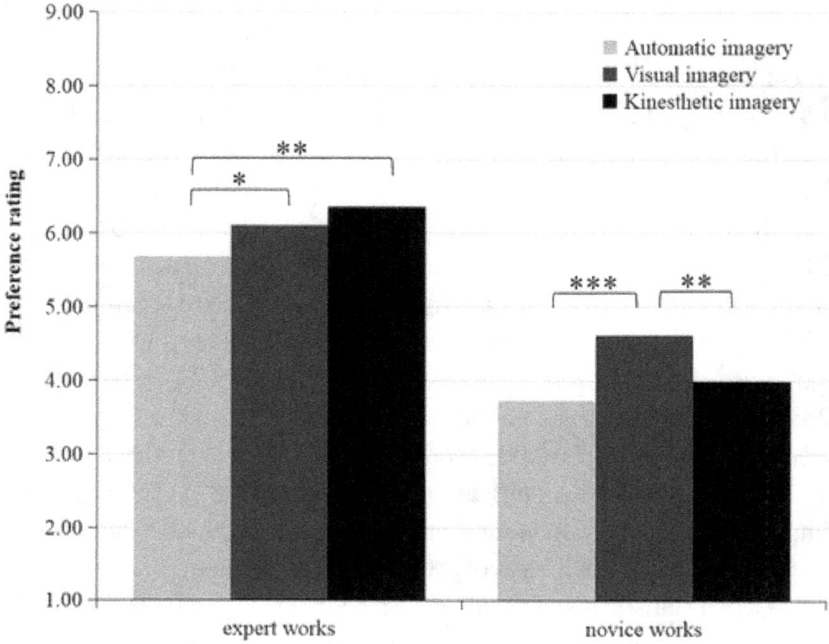

Figure 5. Means of the preference ratings in Experiment 3. The *, **, and *** indicate significant differences at $p<0.05$, $p < 0.01$, and $p<0.001$, respectively.

results of Experiment 3 show that, compared with the control condition, motor imagery (KI or VI) of artists' actions contributed to the participants' preference ratings. Correlation analysis also indicated that the more vividly the participants imagined the action of creation, the higher the participants' aesthetic ratings were.

Experiment 4

Similarly, we replicate the procedure of Experiment 3 in Experiment 4 by using hard-pen calligraphy images as stimuli.

Method

Participants

Thirty-three undergraduate students from South China Normal University (9 males, 24 females, $M_{age} = 20.9$ years, $SD = 2.14$) participated in Experiment 4.

Participants provided written consent before the study and were paid for their participation.

Materials

A total of 36 pairs of brush-pen calligraphy images (without writing hand) were selected from Experiment 2 to serve as the stimuli set.

Procedure

The design and procedure for Experiment 4 were identical to Experiment 3.

Results and discussion

A 3 (imagery condition: control condition/VI/KI) × 2 (artistic level: novice's work/expert's work) repeated-measures ANOVA with preference ratings as the dependent variables revealed that both the main effects of imagery condition and artistic level reached statistical significance, $F(2, 64) = 9.25$, $p < 0.001$, $\eta_p^2 = 0.22$, $F(1, 32) = 242.05$, $p < 0.001$, $\eta_p^2 = 0.88$, as seen in Table 3. The interaction between imagery condition and artistic level reached statistical significance, $F(2, 64) = 31.78$, $p < 0.001$, $\eta_p^2 = 0.49$. Simple effect analysis revealed that for the expert's work, the imagery condition reached statistical significance, $F(2, 64) = 9.96$, $p < 0.001$, $\eta_p^2 = 0.39$, and the preference ratings in the KI condition were significantly higher than those in the control condition ($p = 0.045$) and the VI condition ($p < 0.001$). However, the difference between the control condition and VI was not significant ($p = 0.24$). For the novice's work, the imagery condition reached statistical significance, $F(2, 64) = 26.94$, $p < 0.001$, $\eta_p^2 = 0.64$; the preference ratings in the KI condition were higher than that those in the control condition ($p < 0.041$), and the preference ratings for VI were higher than those in the control condition ($p < 0.001$) and in the KI condition ($p < 0.001$), as shown in Table 4 and Figure 6.

Correlations between the participants' preference ratings for the expert's work and vividness ($r = 0.48$, $p = 0.005$) and intensity ($r = 0.74$, $p < 0.001$) were significant, indicating that both KI and VI significantly predicted the preference ratings of brush-pen calligraphy images. Similarly,

Table 4. Results of preference ratings in Experiment 4 ($M \pm SD$).

	Imagery condition		
Artistic level	**Control condition**	**Visual imagery (VI)**	**Kinesthetic imagery (KI)**
Expert's work	5.74 ± 1.34	6.20 ± 1.45	6.43 ± 1.21
Novice's work	3.81 ± 1.50	4.69 ± 1.35	4.06 ± 1.49

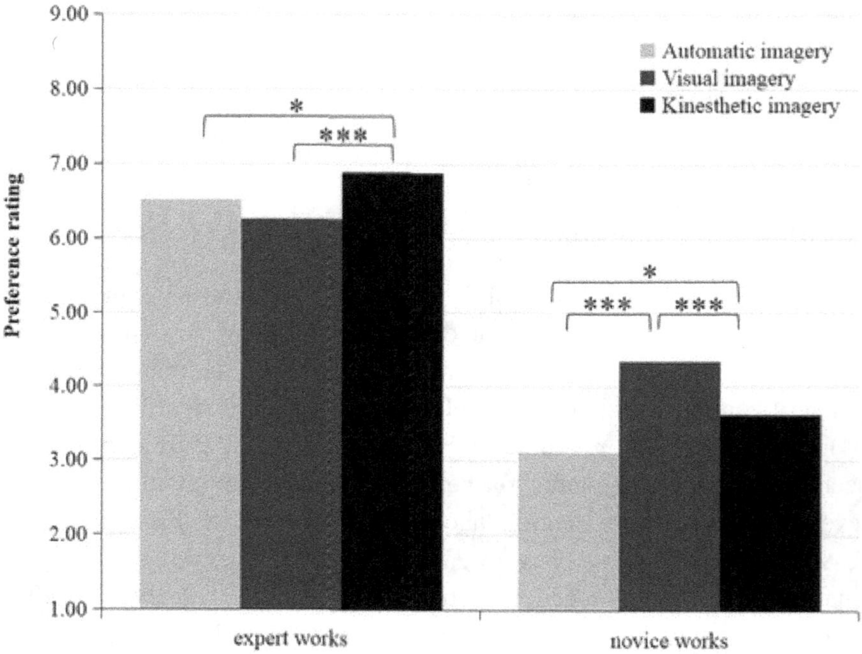

Figure 6. Means of the preference ratings in Experiment 4. The * and *** indicate significant differences at $p<0.05$ and $p<0.001$, respectively.

the relation between the participants' preference ratings for the novice's work and vividness ($r = 0.34$, $p = 0.05$) and intensity ($r = 0.64$, $p < 0.001$) was significant, also indicating that both the KI and VI significantly predicted the preference ratings of brush-pen calligraphy images. In sum, the results of Experiment 4 show that, compared with the control condition, motor imagery (KI or VI) of artists' actions contributed to the participants' preference ratings. Correlation analysis also indicated that the more vividly the participants imagined the action of creation, the higher the participants' aesthetic ratings were.

Experiment 4 was a direct replication of Experiment 3, but used brush-pen calligraphy, and succeeded in reproducing the results of Experiment 3. The results of Experiment 4 show that the motor imagery of the action of the creation of brush-pen calligraphy could also increase the participants' preference ratings, and our hypothesis was verified again. Therefore, the results of both Experiments 3 and 4 show that imagining the action of creation could indeed affect participants' aesthetic preferences.

General Discussion

For the first time, this study used Chinese calligraphy as the object of aesthetic judgments and investigated the impact of two guided ways of experiencing the artist's action, including action observation and motor imagery, on the participants' aesthetic preferences. It was found that both action observation and motor imagery affected the participants' aesthetic preference for Chinese calligraphy images.

In Experiments 1 and 2, we manipulated the conditions of action observation through the presence or absence of the artist's actions to explore the impact of action observation on the aesthetic preference of calligraphy images. The results of both Experiments 1 and 2 showed that compared with the condition in which no action was observed, observing the artist's action of creating calligraphy images induced participants to simulate the same writing action, and this action simulation increased the participants' aesthetic preference. Notably, one may argue that the higher ratings of the calligraphy image might be simply due to the extra gestures observed. For participants, the gesture may enlarge the size of the stimulus and thus may attract their prior attention, resulting in higher ratings and faster response. However, the following reasons could explain the above concerns. In Experiment 1, if the size of the stimulus impacted the participants' preference, they should respond faster to both the expert's and novice's work with action observation conditions. However, we found that the mean difference in response time between the action observation and the control condition in the expert's work failed to reach statistical significance ($p = 0.37$). On the contrary, the participants expressed slower button-pressing speed to the novice's work with action observation than without action observation. Similarly, in Experiment 2, we did not find that the participants' responses were any faster to either the expert's or novice's work with action observation conditions. Therefore, these results exclude the effects of the size of the stimulus. Further, we used the brush-pen calligraphy images as the stimulus to exclude the influence of the participants' writing experience. We found that the participants expressed faster button-pressing speed to the expert's work with action observation than without action observation; conversely, they expressed slower button-pressing speed to the novice's work with action observation than without action observation. These results indicated that the action-directed goal which was attractive increased the participants' desire for action simulation and promoted their button-pressing speed, while the action-directed goal which was unattractive reduced the participants' desire for action simulation and interfered with their button-pressing. In short, both the results of preference ratings and response time in Experiment 1 and Experiment 2 suggested that observing the action of creation affected the participants' aesthetic ratings of calligraphy images.

Given these results, how does observing the action of creation affect the participants' aesthetic ratings on calligraphy images? We speculate that observing the action of creation may initiate action simulation in the participants' minds. Previous studies suggested that when we are watching someone performing an action, our brains may simulate the action we observe (Jeannerod, 1994, 1995). Imaging studies have also shown that action observation in humans involves an internal motor simulation of the observed action (Calvo-Merino *et al.*, 2004; Tipper, 2010). This covert action simulation allows us to better understand the intentions and emotions behind the action. Similarly, in this study, this internal simulation makes it possible for the observers to "copy" the psychological state of the calligraphy creator at the time of creation and to experience the creator's action in a "personal" way. Then, observers could better appreciate the purpose, intention, and emotions of the artist's action. As a result, they may think highly of the work. Our interpretation is similar to the hypothesis that emphasizes the role of action in aesthetics experience (Cross, 2015;

Freedberg & Gallese, 2007). These studies argued that the body's response to the action in paintings is crucial to the aesthetic experience. The main focus of these researchers is the action in the static visual art content, such as sculpture traces (Di Dio *et al.*, 2011) and brushstrokes (Leder *et al.*, 2012; McLean *et al.*, 2015; Umilta *et al.*, 2012). Unlike previous empirical research, our study further explored the idea that the direct observation of the creator's action could also affect the observer's aesthetic experience. The results of Experiment 1 and Experiment 2 were also basically consistent with those reported by Ticini *et al.* (2014). After visual-motor training, participants were required to connect different actions (including a precision grip, forced grip, and flat grip of the brush pen) with the styles of painting. Compared with the control condition (flat hand), when the correct grip of the brush pen was observed, the participants tended to have higher preference ratings for subsequent paintings with consistent styles. This indicated that observing the action of creation initiated the participants' simulation of the corresponding actions, and this simulation would increase the participants' aesthetic ratings of subsequent paintings with consistent styles (Ticini *et al.*, 2014). Overall, the results of Experiment 1 and Experiment 2 further showed that not only the experience of the action in the content of the artwork could enhance participants' emotional response to the artwork, but the direct observation of the artist's action of creation also had a similar effect.

In Experiment 3 and Experiment 4, we manipulated motor imagery tasks through the instructions provided to participants to investigate the influence of the different imagery conditions on the aesthetic preference of calligraphy images, and the results were largely in line with our expectations. From the results of Experiment 3 and Experiment 4, we found that, compared with the control condition, both KI and VI of the artist's action could improve the participants' preference ratings of calligraphy images. We speculated that this was the case because, compared with the sensations accompanying the control condition, participants could experience psychological events for each stroke of the character, even the rhythm of writing, with both KI and VI regardless of imagining others (VI) or imagining themselves (KI). This psychological experience imitated the real experience or provided a desirable sensation (Pavlik & Nordin-Bates, 2016). When the participants actively engaged in the motor imagery of the calligrapher's action of creation, the participants simulated the action through motor imagery, and this simulation "reproduced" the creative state for calligraphy, which improved the participants' understanding of the calligrapher's creation goals, intentions, and emotions and further enhanced their aesthetic experience. In general, to better understand the aesthetic experience of art, we must consider the artist behind the artwork and their action (Tinio, 2013; Williams *et al.*, 2018).

Also, the theory of visual imagination proposed by Walton (1990) provided another possible explanation. Walton believed that the marks on the perceptible surface of a work of fine art function to trigger imaginative events that are partially constitutive of the perceptual content of that work. For example, oil painting is a planar perceptual stimulus of a two-dimensional pattern and observers could perceive a painting as realistic three-dimensional scenes by the imagery (Seely, 2006). In other words, by imagination, observers' perceptual communications with artworks could be increased. Additionally, artistic levels could affect the positive influence of observing the action of creating calligraphy on the participants' experience. In Experiment 3 and Experiment 4, we found that there was a certain difference between imagining others and imagining oneself creating the novice's work, which may be related to oneself being more compatible with first-person motor imagery than with third-person imagery. Humans tend to hold positive views about themselves; in other words, they evaluate themselves more positively than they evaluate others (Chen *et al.*, 2014) and attribute good things to themselves (Mezulis *et al.*, 2004). For example, Xia *et al.* (2021) found that the participants responded faster to positive traits than to negative traits in the self-name cue conditions. This reasonably explained why in Experiment 3, when imagining the artist's action of the novice's work, participants tended to provide higher preference ratings from the third-person perspective (VI) than from the first-person perspective (KI). In general, however, experiencing the action of creation by imagining oneself or others could contribute to participants' aesthetic preferences.

Our research also has some limitations. First, we chose still images as a condition of action observation instead of short films as Avanzino *et al.* (2015), Kirsch *et al.* (2016), and Mierowsky *et al.* (2019) did, which were more likely to provide the actual situation of action observation. Second, there is no objective indicator to measure whether the participants effectively performed imagination tasks, which may affect the results. Third, we speculated that the mechanism that affected calligraphy aesthetics was the result of action simulation, whether it was elicited by action observation or motor imagery. However, the most direct evidence should originate from the activation of the brain. Imaging methods are needed to supplement our study in the future. Fourth, this study only examined the impact of the experience of the artists' actions on calligraphy aesthetics. Does this effect also exist in other art forms, especially those that emphasize action elements, such as "action painting," an art form showing how the act of creating is inseparable from the final product (Taylor *et al.*, 2012)? Since this art form particularly emphasizes the creative action, the action experience (action observation and motor imagery) may be better able to enhance the observer's understanding of the emotions the artist wishes to convey. Finally, the self-reported aesthetic ratings were subjective and may be affected by many other factors, such as personal experience and expectations (Cross & Ticini, 2012). Meanwhile, both the corrugator supercilii and zygomaticus major muscles could be engaged during the observation of affect-evoking stimuli (Kirsch *et al.*, 2016). Therefore, in future research, we can also record these muscles by facial electromyography to explore the relationship between the experience of the artist's action and aesthetic preference.

To conclude, our study indicates that the aesthetic experience is related to the art-making process of the artist, revealing that both action observation and motor imagery could enhance a person's aesthetic experience.

References

Alibali, M. W., Boncoddo, R., & Hostetter, A. B. (2014). Gesture in reasoning: An embodied perspective. In L. Shapiro (ed.), *The Routledge Handbook of Embodied Cognition*, pp. 150–159. London, England: Routledge.

Avanzino, L., Lagravinese, G., Bisio, A., Perasso, L., Ruggeri, P., & Bove, M. (2015). Action observation: Mirroring across our spontaneous movement tempo. *Scientific Reports*, 5, 10325. https://doi.org/10.1038/srep10325.

Avenanti, A., Candidi, M., & Urgesi, C. (2013). Vicarious motor activation during action perception: Beyond correlational evidence. *Frontiers in Human Neuroscience*, 7, 185. https://doi.org/10.3389/fnhum.2013.00185.

Bullot, N. (2009). Material anamnesis and the prompting of aesthetic worlds. Journal of Consciousness Studies, 16(1), 85–109. https://doi.org/10.1007/978-1-4020-9485-9_26.

Bullot, N. J. & Reber, R. (2013). The artful mind meets art history: Toward a psycho- historical framework for the science of art appreciation. *Behavioral and Brain Sciences*, 36(2), 123–145. https://doi.org/10.1017/S0140525X12001641.

Calvo-Merino, B., Glaser, D. E., Grèzes, J., Passingham, R. E., & Haggard, P. (2004). Action observation and acquired motor skills: An fMRI study with expert dancers. *Cerebral Cortex*, 15(8), 1243–1249. https://doi.org/10.1093/cercor/bhi007.

Chen, Y., Zhong, Y., Zhou, H., Zhang, S., Tan, Q., & Fan, W. (2014). Evidence for implicit self-positivity bias: An event-related brain potential study. *Experimental Brain Research*, 232(3), 985–994. https://doi.org/10.1007/s00221-013-3810-z.

Cross E. S. & Ticini L. F. (2012). Neuroaesthetics and beyond: new horizons in applying the science of the brain to the art of dance. *Phenomenology and the Cognitive Sciences*, 11(1), 5–16. https://doi.org/10.1007/s11097-010-9190-y.

Cross, E. S. (2015). Beautiful embodiment: The shaping of aesthetic preference by personal experience. In J. P. Huston, M. Nadal, F. Mora, L. F. Agnati, & C. J. Cela-Conde (eds.), *Art, Aesthetics and the Brain*, pp. 189–208. Oxford University Press. https://doi.org/10.1093/acprof:oso/9780199670000.003.0010.

di Dio, C., Canessa, N., Cappa, S. F., & Rizzolatti, G. (2011). Specificity of esthetic experience for artworks: An fMRI study. *Frontiers in Human Neuroscience*, 5, 139. https://doi.org/10.3389/fnhum.2011.00139.

Freedberg, D. & Gallese, V. (2007). Motion, emotion and empathy in esthetic experience. *Trends in Cognitive Sciences*, 11(5), 197–203. https://doi.org/10.1016/j.tics.2007.02.003.

Frischen, A., Loach, D., & Tipper, S. P. (2009). Seeing the world through another person's eyes: Simulating selective attention via action observation. *Cognition*, 111(2), 212–218. https://doi.org/10.1016/j.cognition.2009.02.003.

Gardner, T., Aglinskas, A., & Cross, E. S. (2017). Probing the action observation network response to varying levels of action familiarity. *NeuroImage*, 156, 174–189. https://doi.org/10.1016/j.NeuroImage.2017.04.060.

Guillot, A., Collet, C., Nguyen, V. A., Malouin, F., Richards, C., & Doyon, J. (2009). Brain activity during visual versus kinesthetic imagery: An fMRI study. *Human Brain Mapping*, 30(7), 2157–2172. https://doi.org/10.1002/hbm.20658.

Hamilton, A. F. (2013). The mirror neuron system contributes to social responding. *Cortex*, 49(10), 2957–2959. https://doi.org/10.1016/j.cortex.2013.08.012.

Jeannerod, M. (1994). The representing brain: Neural correlates of motor intention and imagery. *Behavioral and Brain Sciences*, 17(2), 187–202. https://doi.org/10.1017/S0140525X00034026.

Jeannerod, M. (1995). Mental imagery in the motor context. *Neuropsychologia*, 33(11), 1419–1432. https://doi.org/10.1016/0028-3932(95)00073-C.

Jeannerod, M. (2001). Neural simulation of action: A unifying mechanism for motor cognition. *NeuroImage*, 14(1), 103–109. https://doi.org/10.1006/nimg.2001.0832.

Keesman, M., Aarts, H., Vermeent, S., Häfner, M., & Papies, E. K. (2016). Consumption simulations induce salivation to food cues. *PLoS One*, 11(11), e0165449. https://doi.org/10.1371/journal.pone.0165449.

Kirsch, L. P., Snagg, A., Heerey, E., & Cross, E. S. (2016). The impact of experience on affective responses during action observation. *PLoS One*, 11(5), e0154681. https://doi.org/10.1371/journal.pone.0154681.

Kornysheva, K., von Cramon, D. Y., Jacobsen, T., & Schubotz, R. I. (2010). Tuning-in to the beat: Aesthetic appreciation of musical rhythms correlated with a premotor activity boost. *Human Brain Mapping*, 31, 48–64. https://doi.org/10.1002/hbm.20844.

Krishna, A., Morrin, M., & Sayin, E. (2014). Smellizing cookies and salivating: A focus on olfactory imagery. *Journal of Consumer Research*, 41(1), 18–34. https://doi.org/10.1086/674664.

Lebreton, M., Kawa, S., Forgeot d'Arc, B., Daunizeau, J., & Pessiglione, M. (2012). Your goal is mine: Unraveling mimetic desires in the human brain. *Journal of Neuroscience*, 32(21), 7146–7157. https://doi.org/10.1523/JNEUROSCI.4821-11.2012.

Leder, H., Bar, S., & Topolinski, S. (2012). Covert painting simulations influence aesthetic appreciation of artworks. *Psychological Science*, 23(12), 1479–1481. https://doi.org/10.1177/0956797612452866.

McLean, C., Want, S. C., & Dyson, B. J. (2015). The role of similarity, sound and awareness in the appreciation of visual artwork via motor simulation. *Cognition*, 137, 174–181. https://doi.org/10.1016/j.cognition.2015.01.002.

Mezulis, A. H., Abramson, L. Y., Hyde, J. S., & Hankin, B. L. (2004). Is there a universal positivity bias in attributions? A meta-analytic review of individual, developmental, and cultural differences in the self-serving attributional bias. *Psychological Bulletin*, 130, 711–747. https://doi.org/10.1037/0033-2909.130.5.711.

Mierowsky, R., Marcus, N., & Ayres, P. (2019). Using mimicking gestures to improve observational learning from instructional videos. *Educational Psychology*, 40, 1–20. https://doi.org/10.1080/01443410.2019.1650896.

Molnar-Szakacs, I. (2011). From actions to empathy and morality-a neural perspective. *Journal of Economic Behavior & Organization*, 77, 76–85. https://doi.org/10.1016/j.jebo.2010.02.019.

Pavlik, K. & Nordin-Bates, S. (2016). Imagery in dance: A literature review. *Journal of Dance Medicine & Science*, 20(2), 51–63. https://doi.org/10.12678/1089-313X.20.2.51.

Reber, R., Schwarz, N., & Winkielman, P. (2004). Processing fluency and aesthetic pleasure: Is beauty in the perceiver's processing experience? *Personality and Social Psychology Review*, 8, 364–382. https://doi.org/10.1207/s15327957pspr0804_3.

Rizzolatti, G. & Sinigaglia, C. (2010). The functional role of the parieto-frontal mirror circuit: Interpretations and misinterpretations. *Nature Review Neuroscience*, 11, 264–274. https://doi.org/10.1038/nrn2805.

Sbriscia-Fioretti, B., Berchio, C., Freedberg, D., Gallese, V., & Umiltà, M. A. (2013). ERP modulation during observation of abstract paintings by Franz Kline. *PLoS One*, 8(10), e75241. https://doi.org/10.1371/journal.pone.0075241.

Seely, W. P. (2006). Naturalizing aesthetics: Art and the cognitive neuroscience of vision. *Journal of Visual Art Practice*, 5(3), 195–213. https://doi.org/10.1386/jvap.5.3.195_1.

Shaw, D. J. & Czekóová, K. (2013). Exploring the development of the Mirror neuron system: Finding the right paradigm. *Developmental Neuropsychology*, 38(4), 256–271. https://doi.org/10.1080/87565641.2013.783832.

Steinmetz, J., Tausen, B. M., & Risen, J. L. (2018). Mental simulation of visceral states affects preferences and behavior. *Personality and Social Psychology Bulletin*, 44(3), 406–417. https://doi.org/10.1177/0146167217741315.

Taylor, J. E. T., Witt, J. K., & Grimaldi, P. J. (2012). Uncovering the connection between artist and audience: Viewing painted brushstrokes evokes corresponding action representations in the observer. *Cognition*, 125(1), 26–36. https://doi.org/10.1016/j.cognition.2012.06.012.

Ticini, L. F., Rachman, L., Pelletier, J., & Dubal, S. (2014). Enhancing aesthetic appreciation by priming canvases with actions that match the artist's painting style. *Frontiers in Human Neuroscience*, 8, 391. https://doi.org/10.3389/fnhum.2014.00391.

Tinio, P. P. (2013). From artistic creation to aesthetic reception: The mirror model of art. *Psychology of Aesthetics, Creativity, and the Arts*, 7(3), 265–275. https://doi.org/10.1037/a0030872.

Tipper, S. P. (2010). From observation to action simulation: The role of attention, eye-gaze, emotion, and body state. *Quarterly Journal of Experimental Psychology*, 63(11), 2081–2105. https://doi.org/10.1080/17470211003624002.

Umilta, M. A., Berchio, C., Sestito, M., Freedberg, D., & Gallese, V. (2012). Abstract art and cortical motor activation: An EEG study. *Frontiers in Human Neuroscience*, 6, 311. https://doi.org/10.3389/fnhum.2012.00311.

Walton, K. (1990). *Mimesis as Make-Believe*. Cambridge, MA: Harvard University Press.

Williams, L., McSorley, E., & McCloy, R. (2018). The relationship between aesthetic and drawing preferences. *Psychology of Aesthetics, Creativity, and the Arts*, 12(3), 259–271. https://doi.org/10.1037/aca0000188.

Xia, R., Shao, H., Cui, L., Zhang, P., & Li, S. (2021). Evidence for self-positivity bias in a subliminal self-cue: An event-related potential study. *Neuroscience Letters*, 744, 135625. https://doi.org/10.1016/j.neulet.2021.135625.

Zhang, W., Lai, S., He, X., Zhao, X., & Lai, S. (2016). Neural correlates for aesthetic appraisal of pictograph and its referent: An fmri study. *Behavioural Brain Research*, 305, 229–238. https://doi.org/10.1016/j.bbr.2016.02.029.

https://doi.org/10.1142/9789811298639_0018

Chapter 18

Factor Structure of Audiences' Physical Experience While Watching Dance*

Maja S. Vukadinović[†] and Slobodan Marković[‡]

[†]*Novi Sad Business School, Novi Sad, Serbia*
[‡]*University of Belgrade — Faculty of Philosophy, Belgrade, Serbia*

Abstract

The aim of this study is to explore the factor structure of audiences' physical experience and their related bodily sensations when watching dance choreographies. This study also includes the process of developing an instrument for measuring the observers' physical experience of dance choreographies and their related kinesthetic responses to watching dance. In the main study, participants rated their physical experiences and kinesthetic responses when watching 16 different dance choreographies of various dance forms, including contemporary dance, jazz, tango, and hip-hop. Three factors of the observers' physical experiences when watching dance were identified: Action Tendency, Arousal, and

*This chapter was originally published in the special collection on neuroaesthetics in *PsyCh Journal*. Vukadinović, M. S. & Marković, S. (2022). Factor structure of audiences' physical experience while watching dance. *PsyCh Journal*, 11(5), 660–672. https://doi.org/10.1002/pchj.513.

Relaxation. Furthermore, the results show that the structure of the observers' kinesthetic responses when watching dance consisted of three factors: Focus, Excitement, and Embodied Anticipation. The spectators' physical experiences and bodily sensations in response to an observed dance are the result of their engagement in the dance, the pleasure they draw from the dance, the emotions provoked in them, and their admiration for the performance. These findings are in line with the concept of kinesthetic empathy, which can be understood as the experience of sharing a dancer's movements.

Keywords: body sensations; dance choreographies; kinesthesia; kinesthetic empathy; observer's physical experience

Introduction

Dance is a complex artistic, cultural, social, psychological, and physical phenomenon that provokes the most diverse reactions and responses in the observer. The aesthetic experience of a dance audience may vary from admiration for virtuosity to amazement and fascination (Marković, 2017; Vukadinović, 2019). Observing dance provokes different emotions, such as joyfulness, cheerfulness, or even sadness. It also leads to physical experience and bodily sensations, such as holding one's breath or getting goosebumps (Brownlow *et al.*, 1997; Reasons & Reynolds, 2010; Sawada *et al.*, 2003).

In this chapter, "aesthetic experience" is defined as an exceptional state of mind that is characterized by a strong focus on a certain object that engages and fascinates the subject, whereas all other objects and actions in the environment are excluded from consciousness (cf. Beardsley, 1982; Csikszentmihalyi, 1990; Cupchik, 1974; Kubovy, 1999; Koestler, 1970; Marković, 2010; Ognjenović, 2003; Polovina & Marković, 2006; Telegan & Atkinson, 1974). Further explorations by Marković (2017) that were related to aesthetic experience have indicated that there is a motivational, semantic, and affective aspect of aesthetic experience.

Previous studies have mainly focused on the affective aspect and have shown that aesthetic experience is often associated with pleasant feelings and emotions (Leder *et al.*, 2004; Leder & Nadal, 2014; Winkielman & Cacioppo, 2001), but it can also be followed by unpleasant ones (Cooper & Silvia, 2009; Rawlings, 2003; Silvia & Brown, 2007; Wagner *et al.*, 2014). Furthermore, Marković (2010, 2012) points out that aesthetic feeling is characterized by the observer's fascination, amazement, and profound relationship with the object (whether the object of experience is a dance, painting, film, sculpture, music piece, etc.).

However, little is known about the physical experience as the fourth aspect of aesthetic experience. Observed in general and regardless of an artistic discipline, there are few studies dealing with the physical aspect of aesthetic experience (Benedek & Kaernbach, 2011; Cervellin & Lippi, 2011; Cova & Deonna, 2014; Menninghaus *et al.*, 2015; Miceli & Castelfranchi, 2003; Niedenthal, 2007; Schnall *et al.*, 2010; Schubert *et al.*, 2016; Tan & Frijda, 1999; Vingerhoets & Bylsma, 2016; Wassiliwizky *et al.*, 2017; Wassiliwizky *et al.*, 2015; Zickfeld, 2015). Furthermore, there are even fewer studies in the domain of dance (Brownlow *et al.*, 1997; Foster, 2008; Reasons & Reynolds, 2010; Sawada *et al.*, 2003; Strukus, 2011). Thus, the main focus of this study is to explore different aspects of audiences' experiences related to the body (including sensory, motor, and emotional experiences) when watching dance.

Two key concepts related to observing dance are explored in this chapter: physical experience and bodily sensations. The physical experience of observing dance, according to different studies (Foster, 2008; Jola *et al.*, 2011; Martin, 1939; Reasons & Reynolds, 2010; Strukus, 2011), is connected with the effects watching dance has on the observer's body as well as with the description of feelings that result from observing the dance movements of others. Spectators of dance often report that they feel cheerful, touched, amazed, proud, and so on (Foster, 2008; Jola *et al.*, 2011; Reasons & Reynolds, 2010; Strukus, 2011).

On the other hand, bodily sensations include a wide range of different somatovisceral and motoric responses as a part of the physical experience, such as, for example, holding one's breath or getting goosebumps (Foster, 2008; Jola *et al.*, 2011; Reasons & Reynolds, 2010; Strukus, 2011). Previous studies have reported that bodily sensations are related to kinesthesia as an integral part of physical experience (Batson, 2008; Foster, 2007, 2008, 2011; Montero, 2006, 2012). Moreover, kinesthesia, which refers to the sensitivity of body position and muscle tension, may be defined as the awareness of the position and movement of the parts of one's own body mediated by receptors in the muscles, joints, tendons, ligaments, and skin that are stimulated by body tensions (cf. Batson, 2008; Foster, 2008, 2011; Montero, 2006, 2012; Sherringtone, 1907). Furthermore, kinesthesia (or proprioception) is considered to be integral to perception (Berthoz, 2000; Reynolds, 2007), as well as to distinct multicomponent sensory modality (Jola *et al.*, 2011). Thus, bodily sensations (i.e., kinesthetic responses) associated with observing dance could include different somatovisceral and motoric responses, such as tapping the feet, shaking, and clenching the fists. Based on the abovementioned discussion, in this chapter, the terms "bodily sensations" and "kinesthetic responses" are used as synonyms. Furthermore, concerning the relationship between these two key concepts, bodily sensations (i.e., kinesthetic responses) are understood in this chapter as a somatic part of physical experience that includes both bodily sensations and emotional experience. In other words, it could be said that physical experience includes bodily sensations and emotional experience.

Kinesthetic Empathy

Previous studies related to the bodily sensations (i.e., kinesthetic responses) in a spectator's body (Foster, 2007, 2008, 2011; Hagendoorn, 2004; Jola *et al.*, 2011; Reasons & Reynolds, 2010, Reynolds & Reasons, 2012; Strukus, 2011) have shown that the physical experience of watching dance is often associated with "shared kinesthetic experience" (Lipps, 1906) and "kinesthetic empathy" (Martin, 1939). Lipps (1906) assumes that shared kinesthetic experience could be described as the viewer's physical and kinesthetic response when observing a painting. The concept of "shared kinesthetic experience" was adopted and upgraded into "kinesthetic empathy" by dance critic John Martin (1939). Underlying concepts of kinesthetic empathy are of particular interest for the studies of dance.

In dance research, the concept of kinesthetic empathy is related to the physical responses of some spectators when watching dance. Furthermore, it may be understood as the feeling of sharing another person's movement, or vicariously experiencing another person's movement simply by watching (cf. Foster, 2007, 2008, 2011; Hagendoorn, 2004; Jola *et al.*, 2011; Reasons & Reynolds, 2010; Reynolds & Reasons, 2012; Strukus, 2011).

However, kinesthetic empathy is a complex phenomenon as the experience of kinesthetic empathy depends on stimuli, characteristics of the perceiver, and various cultural and historical factors (Barker, 2006; Foster, 2007; Reasons & Reynolds, 2010; Strukus, 2011), as well as on contextual differences (Strukus, 2011). All of these have to be taken into account simultaneously. Regarding the nature of the perceiver, Jola *et al.* (2011) found that motor simulation, which is related to kinesthetic experience, is modified by both visual expertise and the empathic abilities of the dance observer. Moreover, Reasons & Reynolds (2010) reported that different experiences in dance training influenced the kinesthetic experience of dance. They have shown that observers who are not trained in dance often describe their admiration in terms of their own inability to do the movements, while trained dancers often derive enjoyment through "inner mimicry". In other words, their admiration stems from bridging the gap between them and the dancer by imagining themselves dancing.

Some studies have suggested that kinesthetic empathy, as a special way of engaging in dance, may give pleasure to spectators through inner mimicry as well as through admiration of virtuosity, which is often related to effortlessness, grace, and the flow of executing movements (Reasons & Reynolds, 2010). Furthermore, kinesthetic empathy may represent an important motivational factor in the reasons why people are inspired to watch dance. Reasons & Reynolds (2010) suggested that observers are often motivated by hopes, expectations, and pleasures derived from a sense of closeness to the dancers, from the awareness of their effort, and from seeking an escape from reality. On the other hand, Jola *et al.* (2011) identified similar responses from the observers when watching dance, such as a "desire to move," "admiration of virtuosity," "connection to the dancer," "embodied response," "evaluation of quality," and so forth.

Research Perspective

Until now, the aesthetic experience of dance has been researched from different perspectives in previous studies, such as the somaesthetic (cf. Arnold, 2005; Fenemor, 2003; Turner, 2008), neuroaesthetic (cf. Brown *et al.*, 2006; Calvo-Merino *et al.*, 2005, 2006, 2009; Cross *et al.*, 2006), and cognitive perspectives (cf. Glass, 2005; Stevens & McKechnie, 2005). In this study, we intended to look into the phenomenon of audiences' physical experiences when watching dance by applying a factor analysis in our approach. By using this approach, we attempted to explore the factor structure of the audiences' physical experiences and their related bodily sensations when watching dance. Earlier studies have used this approach to explore the factors of aesthetic experience of paintings (Polovina & Marković, 2006). Moreover, in their earlier study that applied factor analysis to explore the structure of aesthetic experience of dancers and their observers, Vukadinović & Marković (2012) defined three factors in the audience's aesthetic experience (dynamism, exceptionality, and affective evaluation). The results have shown that an audience's aesthetic experience stems from the overall context. This includes the scenography, lighting, music, physical characteristics of the dancers, the dancers' interpretation of choreography, and its staging (c.f. Vukadinović, 2019; Vukadinović & Marković, 2012, 2017). Considering the structure of an audience's aesthetic experience, as well as the kinesthetic nature of dance as a form of art, one question remained open: What happens with the body of the observer while they are watching dance performances? That is, what is the physical experience of the audience when watching dance?

Exploring the factor structure of an audience's physical experience and its related bodily sensations may contribute to a better understanding of aesthetic experience. Based on the abovementioned concept of kinesthetic empathy and the aspects of the experience related to the body, this study aims to investigate the observers' physical experiences and bodily sensations when watching dance performances. It can be hypothesized that the structure of observers' physical experiences while watching dance will include motivational and affective dimensions. Concerning the exploration of the bodily sensations, that is, kinesthetic responses that participants have while watching dance, it can be hypothesized that the observers' structure of kinesthetic responses will also be multidimensional, including embodiment, stimulation, and direction.

Purpose of the Study

Following the contextual background described in the Introduction, the main purpose of this study is to investigate the physical experience when observing dance and its related kinesthetic responses (i.e., body sensations) by using exploratory factor analysis. Physical experience and bodily sensations are operationalized through a series of descriptors. The primary aim of this study is to explore the factor structure of the ratings on a set of descriptors of the observers' physical experience and bodily sensations when watching dance performances and to determine whether the descriptors of the observers' physical experience and related bodily sensations are grouped into two, three, or more factors. In the following paragraphs, the selection of the descriptors of physical experience (i.e., scales on which the stimuli were rated) is elaborated upon.

Preliminary Study

The aim of the preliminary study was to select the descriptors of observers' physical experience when watching dance (i.e., scales for stimuli ratings) that would be used in the main study.

Method

Participants

Two groups participated in the experiment. The first group consisted of non-dancers who were students of Novi Sad Business School — Higher Education Institution for Applied Studies (n = 54), between 20 and 26 years of age (M = 21.83 years; SD = 1.28 years). There were 48.1% men and 51.9% women participating in the study. The second group consisted of professional dancers with 8 years of experience or more who performed different forms of dance (n = 33). They were between 25 and 33 years of age (M = 28.69 years; SD = 2.83 years) and 36.4% were men and 63.6%

were women. These two groups of participants were included because previous studies have shown that there is a significant difference in perceiving and appreciating dance between professional dancers and non-dancers (Calvo-Merino *et al.*, 2005, 2006, 2009; Cross *et al.*, 2006; Jola *et al.*, 2011; Jola *et al.*, 2012; Reasons & Reynolds, 2010).

Stimuli

Ten audiovisual recordings of different forms of dance, including classical ballet, modern ballet, contemporary dance, flamenco, folklore, and tango, were presented to the two groups of participants through 10 audiovisual recordings via a video projector. The main criterion for the selection of dance video recordings was to cover a wide range of different styles of dance. These recordings of different dance forms were chosen randomly from the Internet (Appendix A contains the list of websites).

Procedure

The tasks given to the participants, both non-dancers and professional dancers, were the same. The first task for both groups was to write down as many descriptors as possible that could describe their physical experiences and bodily sensations related to watching different dance forms. The instructions given to the participants were as follows: "Write down as many adjectives or descriptions as possible that best illustrate your physical experience and bodily sensations while watching a dance." The second task was to note down as many descriptors as possible related to their physical experience and bodily sensations while watching each dance from the recordings presented. The instructions given to the participants were as follows: "After each presented choreography on the video recording, please write down as many adjectives or descriptions as possible that best illustrate your physical experience and bodily sensations while watching this particular dance."

Results

As a result of two tasks given to both non-dancers and professional dancers, four groups of descriptors were obtained: The first and second groups consisted of the descriptors produced by the non-dancers, whereas the third and fourth groups consisted of the descriptors produced by the dancers. First, the descriptors that were not adequate for describing the physical experience when watching dance were eliminated from each group (e.g., *morbid, selfish, and rational*). The common category included the descriptors that were synonymous or similar to each other (e.g., *I feel calm* and *I feel relaxed*, or *I feel alive* and *I feel vivacious*). Second, the dancers' and non-dancers' lists of descriptors were formed. On the dancers' list, there were 86 descriptors while on the non-dancers' list, there were 78. The frequency of the occurrence of each particular descriptor was calculated in each list. The descriptors were compared, and those with the highest frequencies (those appearing at least 20 times) were singled out. When the list of the most frequent descriptors produced by the dancers was compared with the list of the most frequent descriptors produced by the non-dancers, it could be seen that the only difference was that the dancers' list included the following descriptions: *I can feel my body flutter, I feel vibrations*.

Furthermore, one common list of descriptors was formed for dancers and non-dancers. On that common list, there were 48 descriptors that had a frequency of occurrence of minimum 20 times. Further analyses of the list have shown that two categories of descriptors could be singled out. One is related to the observers' physical experience and the other to the bodily sensations felt by the observers. Finally, two common lists of descriptors were formed.

The first common list specified the 31 most frequent descriptors of the observers' physical experiences when watching dance: *I feel like moving, I feel like dancing, I feel like squealing, I feel like getting up, I feel like singing, I feel like clapping, I feel like whistling, I feel pleasant, I am excited, I feel longing, I feel warmth, I feel blissful, I feel calm, I am touched, I am surprised, I am fulfilled, I am cheerful, I feel admiration, I feel vivacious, I am amazed, I can feel my body flutter, I am fascinated, I feel vibrations, I feel pride, I feel wonder, I am inspired, I am happy, I feel impassioned, I feel effused, I am astonished*, and *I am delighted.*

The second common list contained the 17 most frequent bodily sensations felt by the observers (i.e., kinesthetic responses) when watching dance: *I tap my feet, I hold my breath, I get goosebumps, My heart beats faster*, *My knees buckle*, *I get teary-eyed*, *I feel chills*, *My muscles clench*, *I feel vibrations in my body*, *I have butterflies in my stomach*, *I cannot look away*, *I smile*, *I cannot blink*, *I inhale slowly and deeply*, *I shake*, *I clench my fists*, and *I mimic the movements while I sit*.

These descriptors, given in the form of a statement, served as the basis for the construction of an instrument for measuring the physical experience of watching dance choreographies.

The Main Study

Method

Participants and procedure

Thirty-eight students from Novi Sad Business School — Higher Education Institution for Applied Studies, aged between 18 and 23 years ($M = 19.68$ years, $SD = 1.27$ years), participated in the study. There were 10 (26.3%) male and 28 (73.3%) female students. The participants did not have any formal experience, either in dance training or in any other form of professional physical training. Moreover, the participants' experience in watching dance performances was controlled for, that is, they were "novices." This means that they had not watched more than one dance theater performance in the past 5 years (Jola *et al.*, 2011). The students participated voluntarily and they did not receive any course credit or monetary compensation for their participation. Variables such as the sexual orientation of the participants, familiarity with dance styles (i.e., the level of viewing experience), as well as the attractiveness of the dancers were not controlled for.

After the participants had given their consent to participate in the study, they answered a set of questions related to their sociodemographic characteristics (age and gender) and their dance practice and experience in watching dance performances. After that, they rated the choreographies. Sixteen choreographies were presented to the participants via an LCD projector on a screen, in a randomized order. The stimuli were observed from a distance of around 4 m and the dimensions of the screen projections were $h = 1.25\,m \times w = 2.21\,m$.

Participation in the study was voluntary, anonymous, and without any financial compensation. The study was conducted in accordance with the Helsinki Declaration.

Stimuli

The stimulus set consisted of 16 dance choreographies of different types of dance, including modern, contemporary, jazz, lyrical jazz, tango, and hip-hop dance choreographies. The choreographies had been made by choreographers who were more than three-time nominees for *Primetime Emmy Awards for Outstanding Choreography*. The choreographies were originally performed in the American television dance competition show *So You Think You Can Dance* that aired on Fox between 2008 and 2016. Stimuli were video recordings of 16 original dance performances. The videos included the original music selected by the choreographers. Choreographies were performed by a couple (two female, two male, or male and female dancers) or a group of eight dancers (of both genders). The average duration of all choreographies was around 1 min 50 s. The video recordings of the choreographies were taken from the Internet and adapted for research purposes (the websites are listed in Appendix B).

All choreographies presented in the stimuli were accompanied with music. Although the performance of a choreography is almost always closely connected to music (Carrol & Moore, 2012), it has been noted that music can be a confounding factor when studying the aesthetic experience of dance. On one hand, the way in which the cognitive system combines music and dance into a unique aesthetic experience has not been explored enough yet (Christensen & Calvo-Merino, 2013). On the other hand, it has been noticed in previous studies (Thaut *et al.*, 2014; Thaut *et al.*, 2015) that music stimulates blood to pump into the muscles of legs or arms. This may be the reason why people tap their feet or move their fingers to the rhythm. Furthermore, rhythm may cause changes in the heart rate and in the respiratory system so that the person can synchronize with the music (Cervellin & Lippi, 2011; Reinhardt, 1999; Tormodsdatter Færøvik, 2017). In addition, it is worth mentioning that one study (Tsay, 2013) has dealt with the relative influence of vision versus audition and showed that visual information outweighs the auditory (cf. Woolhouse & Lai, 2014). Although we had in mind this problem related to music as a possible confounding variable, in this study, dance and music were given in their original merged form. Namely, we intended to investigate the observers' physical experience of dance as a complex and usually multimedial artistry that includes both vision and sound (i.e., music).

Instrument

The observers' physical experiences when watching dance were measured by 31 seven-point scales specified in the preliminary study (see Table 1).

The observers' bodily sensations (i.e., kinesthetic responses) when watching dance were measured by 17 dichotomous (*yes/no*) scales specified in the preliminary study (please see Table 2).

The scales were presented to the participants in a randomized order. The participants observed the audiovisual recordings of choreographies, and having finished watching each of them, they immediately made their ratings on seven-point scale (1 = *least intense feeling provoked within them*; 7 = *the strongest intensity of the feeling provoked in them*)

Table 1. Descriptors of the observers' physical experiences.

I feel like moving	I feel blissful	I feel vibrations
I feel like dancing	I feel calm	I feel pride
I feel like squealing	I am touched	I feel wonder
I feel like getting up	I am surprised	I am inspired
I feel like singing	I am fulfilled	I am happy
I feel like clapping	I am cheerful	I feel impassioned
I feel like whistling	I feel admiration	I feel effused
I feel pleasant	I feel vivacious	I am astonished
I am excited	I am amazed	I am delighted
I feel longing	I can feel my body flutter	
I feel warmth	I am fascinated	

Table 2. Descriptors of the observers' body sensations.

I tap my feet	I feel chills	I cannot blink
I hold my breath	My muscles clench	I inhale slowly and deeply
I get goosebumps	I feel vibrations in my body	I shake
My heart beats faster	I have butterflies in my stomach	I clench my fists
My knees buckle	I cannot look away	I mimic the movements while I sit
I get teary-eyed	I smile	

containing 31 descriptors of the observers' physical experience when watching dance, as well as on dichotomous scale (1 = *they perceived a bodily sensation*; 0 = *they did not perceive any bodily sensations*) containing 17 descriptors of the observers' bodily sensations (i.e., kinesthetic responses). The time given to them for rating each choreography was not limited. Correlations between the descriptors of physical experience, as well as bodily sensations (i.e., kinesthetic responses), are shown in the Supporting Information (please see Appendix S1).

Data analysis

A matrix for analysis was created using the string-out method proposed by Osgood and collaborators (Osgood *et al.*, 1975; see also Vukadinović & Marković, 2012). According to this method, the 3-D matrix (Participants × Stimuli × Scales) was organized into a 2-D matrix by arranging the single stimuli matrices one under the other. In other words, matrices for 16 dance choreographies were arranged in a single string-out matrix.

Data analysis was performed using statistical software SPSS for Windows Version 25.0. To explore the factor structure of participants' ratings on the scales of physical experience and kinesthetic responses when watching dance, an exploratory factor analysis (EFA) was conducted. In order to decide how many factors would be kept, a parallel analysis using syntax was also performed. For this purpose, an online syntax was used (https://oconnor-psych.ok.ubc.ca/nfactors/parallel.sps).

Results

The observers' physical experiences when watching dance

In performing EFA, the maximum likelihood method was used to extract the factors from the data matrix. Bartlett's test of sphericity, which tests the overall significance of all the correlations within the correlation matrix, was significant, χ^2 (465) = 17,426.229, $p<0.001$, indicating that it was appropriate to use the factor analytic model on this set of data. The Kaiser–Mayer–Olkin (KMO) measure of sampling adequacy indicated that the strength of the relationship among variables was high (KMO = 0.95). Four factors with eigenvalues greater than 1 were extracted. In order to decide how many factors would be kept, a parallel analysis using syntax was performed. It has been indicated that only three factors have a mean (random data eigenvalues obtained in parallel syntax) that is lower than the initial eigenvalues obtained in SPSS (see Table 3). These analyses indicated that the three factors gave the most interpretable solution.

Altogether, the three factors have a common explained variance of 63.13%. A Promax rotation was performed as the factors were expected to be correlated. The highest factor loadings on each factor are displayed in Table 4. Only descriptors with factor loadings of 0.70 and above are shown.

With regards to the contents of the obtained factors (i.e., items with the highest factor loadings), the factors are named as follows: Factor 1 — Action

Table 3. Parallel analysis with SPSS and syntax.

	Initial eigenvalues			**Extraction sums of squared loadings**			**Random data eigenvalues**
	Eigenvalue	**% variance**	**% Cumulative variance**	**Eigenvalue**	**% variance**	**% Cumulative variance**	**Means**
Factor 1	15.270	49.259	49.259	14.907	48.086	48.086	1.498
Factor 2	3.625	11.695	60.954	3.279	10.578	58.664	1.386
Factor 3	1.801	5.809	66.763	1.387	4.473	63.136	1.341
Factor 4	1.118	3.608	70.371	.758	2.445	65.582	1.303

Note: The results from Parallel analysis using syntax are shown in the last column.

Table 4. Four extracted factors presented with factor loadings based on a maximum likelihood procedure with Promax rotation for 31 descriptors of the observers' physical experience when watching dance.

	F1 Action tendency	**F2 Arousal**	**F3 Relaxation**
Descriptors of the observers' physical experiences when watching dance			
I feel like dancing	**0.891**		
I feel like moving	**0.792**		
I feel vibrations	**0.789**		
I feel vivacious	**0.780**		
I feel like getting up	0.778		
I can feel my body flutter	0.759		
I feel like singing	0.724		
I feel like clapping			
I feel like squealing			
I feel impassioned		**0.829**	
I feel pride		**0.815**	
I am cheerful		**0.814**	
I feel effused		**0.804**	
I am inspired		0.802	
I am happy		0.785	
I feel like whistling		0.704	
I feel wonder		0.700	
I feel longing			
I am touched			**0.830**
I feel warmth			**0.818**
I feel calm			**0.792**
I feel blissful			**0.756**
I feel pleasant			0.756
I am amazed			
I feel admiration			

(*Continued*)

Table 4. (*Continued*)

	F1 Action tendency	F2 Arousal	F3 Relaxation
I am delighted			
I am fascinated			
I am fulfilled			
I am excited			
I am astonished			
I am surprised			

Note: Bolded values are significant at *p* level, 0.001.

Table 5. Descriptive statistics for factors of scale measuring physical experience.

Scale 1: Physical experience					
	Factors	*M*	*SD*	*MIC*	α
1	Action Tendency	3.32	2.07	0.671	0.891
2	Arousal	2.85	1.88	0.651	0.881
3	Relaxation	2.79	1.75	0.614	0.864

Note: MIC, mean inter-item correlation.

Tendency, Factor 2 — Arousal, and Factor 3 — Relaxation. Action Tendency and Relaxation showed low inter-correlation between them ($r = 0.381$). Other inter-correlations between Action Tendency and Arousal ($r = 0.550$) and between Relaxation and Arousal ($r = 0.660$) were found to be medium strong.

To resume and indicate our final solution, the results of EFA have shown that the structure of the observers' physical experience when watching dance consisted of three factors: Action Tendency (*I feel like dancing, I feel like moving, I feel vibrations* and *I feel vivacious*), Arousal (*I feel impassioned, I feel pride, I am cheerful* and *I feel effused*), and Relaxation (*I am touched, I feel warmth, I feel calm* and *I feel blissful*). For each factor, only the first four items with higher loadings ($r > 0.75$) were kept. The descriptive statistic for each factor (*M*, *SD*, MIC, and Cronbach's alpha) after the item selection is shown in Table 5.

Furthermore, the correlations between factors after the item selection were calculated. The correlation between Action Tendency and Relaxation is low and significant ($r = 0.167$, $p < 0.001$) while those between Action Tendency and Arousal ($r = 0.527$, $p < 0.001$) and between Arousal and Relaxation ($r = 0.555$, $p < 0.001$) are significant and medium strong.

The correlations between factors of the scale that measures the observers' physical experiences when watching dance and factors of the scale measuring the observers' bodily sensations are displayed in the following section.

The spectators' bodily sensations (i.e., kinesthetic responses) when observing dance

In performing EFA, the maximum likelihood method was used to extract the factors from variable data. Bartlett's test of sphericity, which tests the overall significance of all the correlations within the correlation matrix, was significant, χ^2 (136) = 3,439.522, $p < 0.001$, indicating that it was appropriate to use the factor analytic model on this set of data. The KMO measure of sampling adequacy indicated that the strength of the relationship among variables was high (KMO = 0.84). Four factors with eigenvalues greater than 1 were extracted. Furthermore, to decide how many factors should be retained, a parallel analysis using syntax was also performed. According to the results of the analyses conducted, it is indicated that the three factors gave the most interpretable solution (please see Table 6).

Table 6. Parallel analysis with SPSS and syntax.

	Initial eigenvalues			**Extraction sums of squared loadings**			**Random data eigenvalues**
	Eigenvalue	**% Variance**	**% Cumulative variance**	**Eigenvalue**	**% Variance**	**% Cumulative variance**	**Means**
Factor 1	5.145	30.264	30.264	4.565	26.885	26.885	1.299
Factor 2	1.972	11.598	41.862	1.261	7.420	34.275	1.238
Factor 3	1.716	10.114	51.976	1.426	8.386	42.661	1.192
Factor 4	1.106	6.504	58.489	.634	3.727	46.389	1.150

Note: The results from Parallel analysis using syntax are shown in the last column.

Altogether, the three factors have a common explained variance of 42.66%. A Promax rotation was performed as the factors were expected to be correlated. The highest factor loadings on each factor are displayed in Table 7. Only the descriptors of kinesthetic responses with factor loadings above 0.50 are shown.

In relation to the content of the obtained factors (i.e., items with the highest factor loadings), the factors are named as follows: Factor 1 — Focus, Factor 2 — Excitement, and Factor 3 — Embodied Anticipation. The inter-correlation between factors Excitement and Embodied

Table 7. The three factors extracted and presented with factor loadings based on a maximum likelihood procedure with Promax rotation for 17 descriptions of kinesthetic responses (i.e., body sensations) when watching dance.

	F1 focus	**F2 excitement**	**F3 embodied anticipation**
The observers' kinesthetic responses when watching dance			
I hold my breath	**0.815**		
I get goosebumps	**0.660**		
I cannot look away	**0.615**		
I cannot blink	**0.600**		
I inhale slowly and deeply	0.568		
I feel chills	0.535		
My heart beats faster		**0.667**	
I have butterflies in my stomach		**0.638**	
I feel vibrations in my body		**0.626**	
I mimic the movements while I sit		**0.608**	
I tap my feet		0.505	
I smile			
I clench my fists			
My knees buckle			**0.724**
I get teary-eyed			**0.691**
I shake			**0.654**
My muscles clench			0.530

Note: Bolded values are significant at *p* level, 0.001.

Table 8. Descriptive statistics for factors of scale measuring physical experience.

	Scale 2: Bodily sensations (i.e., kinesthetic responses)				
Factors		***M***	***SD***	***MIC***	**α**
1	Focus	1.86	1.54	0.432	0.754
2	Excitement	1.08	1.26	0.408	0.727
3	Embodied Anticipation	0.28	0.71	0.476	0.730

Note: MIC, mean inter-item correlation.

Anticipation is low (r = 0.208). Other inter-correlations between Excitement and Focus (r = 0.413) and between Focus and Embodied Anticipation (r = 0.327) are medium strong.

To resume, the results of the exploratory factor analysis have shown that the structure of the observers' kinesthetic responses when watching dance consists of three factors: Focus (*I hold my breath, I get goosebumps, I cannot look away*, and *I cannot blink*), Excitement (*My heart beats faster, I have butterflies in my stomach, I feel vibrations in my body*, and *I mimic the movements while I sit*) and Embodied Anticipation (*My knees buckle, I get teary-eyed*, and *I shake*). For each factor, only items with higher loadings ($r > 0.60$) were kept. The descriptive statistic for each factor (*M*, *SD*, MIC, and Cronbach's alpha) after the item selection is shown in Table 8.

Next, the correlations between factors after the item selection were calculated. The correlations between Focus and Excitement (r = 0.421, $p < 0.001$), as well as between Focus and Embodied Anticipation (r = 0.358, $p < 0.001$) are significant and medium strong. Furthermore, the correlation between Excitement and Embodied Anticipation (r = 0.309, $p < 0.001$) is also significant and medium strong.

The correlation between the factors of two scales, that is, the one measuring the observers' physical experiences and the other measuring their bodily sensations (kinesthetic responses) when watching dance, shows that all factors are correlated. Correlations are significantly low or medium strong (see Table 9).

Excitement has a positive and medium strong correlation with Action Tendency, which means that the higher the Excitement is, the higher is the

Table 9. Correlation between factors of the observers' physical experience and bodily sensations when watching dance.

Physical experience	Bodily sensations (kinesthetic responses)		
	Focus	**Excitement**	**Embodied anticipation**
Action Tendency	0.350**	0.577**	0.167**
Arousal	0.574**	0.242**	0.128**
Relaxation	0.506**	0.085*	0.219**

*$p<0.01$.
**$p<0.001$.

Action Tendency. Moreover, Focus has a positive and medium strong correlation with Arousal as well as with Relaxation. That means that the higher the Focus is, the higher are Arousal and Relaxation.

Discussion

The purpose of this study was to explore the factor structure of the observers' physical experiences of dance and their bodily sensations when watching dance choreographies. The results show that the structure of the observers' physical experiences when watching dance choreographies consists of three factors: Action Tendency, Arousal, and Relaxation.

Factors of physical experience when observing dance choreographies.

The factor of Action Tendency includes descriptors such as *I feel like dancing*, *I feel like moving*, *I feel vibrations*, and *I feel vivacious*. This finding is in line with previous studies reporting that one of the most frequent experiences of observers watching dance is their desire to move (Christensen & Calvo-Merino, 2013; Hagendoorn, 2004; Jola *et al.*, 2011; McFee, 1992; Montero, 2016; Reasons & Reynolds, 2010; Strukus, 2011). The fact that the observers have such an active engagement with dance is closely related to the concept of kinesthetic experience. In other words, kinesthetic empathy is understood as the feeling of sharing another person's movement (Martin, 1939; Strukus, 2011). The conceptualization of kinesthesia as "the muscular connection with our deepest feelings" (Foster, 2008, p. 52) suggests that the observers' kinesthetic engagement when watching dance has the function "either to awake or to enliven feelings" (Foster, 2008, p. 53). Thus, it could be assumed that the observers' desire to move and their tendency for action represent some of the crucial responses to dance. In short, the factor of Action Tendency is more closely related to the effect of dance as kinesthetic art, that is, "art of the muscular sense" (Arnheim, 1966, p. 21). This is an effect that may be experienced in the entire body. Furthermore, the results of this study indicate that the scale measuring physical experience is correlated with the scale measuring bodily sensations. Between the factors of Action Tendency (physical experience) and Excitement (bodily sensations, i.e., kinesthetic responses) there is a positive and medium strong correlation. This implies that Action Tendency as a part of physical experience when watching dance is closely related with kinesthetic responses (i.e., Excitement — described by sensations such as *My heart beats faster*, *I have butterflies in my stomach*, *I feel vibrations in my body*, and *I mimic the movements while I sit*).

The factor of Arousal, which includes descriptors such as *I feel impassioned*, *I feel pride*, *I am cheerful*, and *I feel effused*, can be associated with the feelings that frequently accompany the observers' physical experiences when watching dance. Joyfulness, happiness, and cheerfulness are reported as the observers' most frequent and most recognizable emotional experiences when watching dance (Brownlow *et al.*, 1997; Dittrich *et al.*, 1996; Jola *et al.*, 2011; Reasons & Reynolds, 2010; Sawada *et al.*, 2003). Furthermore, the factor of Arousal, which forms the structure of the observers' physical experience when watching dance, may be strengthened with a suggestion given by Niedenthal (2007). He proposed that the aspects of experience related to the body include not just sensory and motor experience but emotional experience as well.

The factor of Relaxation is the factor that includes descriptors such as *I am touched*, *I feel warmth*, *I feel calm*, and *I feel blissful*. It seems that these descriptors are very close to what could be comprehended as the observers' pleasure-based responses when watching dance. In previous studies, it has been reported that when engaged in the observation of dance, many spectators are motivated by the pleasures that they find within the dance choreographies themselves, for example, "It gladdens the heart," "It calms me down," "It relaxes me," and "It makes me happy." The audiences' emotional experiences relate not only to the specific quality of the dance movement but also to the observers' "interpretative abilities in which the individual spectators engage with that dance movement" (Reasons & Reynolds, 2010, p. 66). Therefore, it may be assumed that the Relaxation factor of the observers' physical responses when watching dance may be related to the viewers' motivation, that is, to what the observers are looking for in a dance performance and how they experience it (Jola *et al.*, 2011).

This factor structure suggests that the observers' physical experience of dance choreographies is a multidimensional phenomenon and its structure could be related to different aspects of the observers' physical experience. In other words, Action Tendency is mostly related to the concept of kinesthetic empathy, Arousal with emotional experience, and Relaxation with pleasure-based motivation.

Factors of kinesthetic responses when observing dance choreographies.

In the context of the observers' physical experience when watching dance, there are a number of somatovisceral and sensory-motor kinesthetic responses that form a part of their physical experience (e.g., tapping one's feet, shaking, getting goosebumps, and changing the tempo of breathing). The results of this study have also shown that a wide range of kinesthetic responses when observing dance choreographies may as well be articulated through a factor structure that consists of three factors: Focus, Excitement, and Embodied Anticipation.

The factor of Focus, which includes bodily sensations such as *I hold my breath*, *I get goosebumps*, *I cannot look away*, and *I cannot blink*, may be related to the "wow" factor previously described by Reason & Reynolds (2010, p. 58). They connect the "wow" factor with an audience's amazement, fascination, and admiration of virtuosity. Moreover, they propose that these sensations are the most prominent audience's kinesthetic responses when observing dance. In addition to this, the results of this study have shown that Focus has a positive and medium strong correlation with Arousal as well as with Relaxation, which are the factors of the scale measuring observers' physical experience when watching dance. That means that the higher the Focus is, the higher are Arousal (i.e., emotional experience) and Relaxation (pleasure-based motivation). It can be recognized that kinesthetic responses referring to the Focus may also be related to the components of aesthetic experience, such as "admiration of virtuosity" and "fascination with the unusual" described by Kubovy (1999) and Marković (2010). Kinesthetic responses related to the Focus may be understood as a way for the observers to engage in dance. They are a means of participating with the dancers in their movement. This assumption suggests a strong link between the Focus and aesthetic experience, which has previously been defined as an exceptional state of mind that is characterized by a strong focus on a certain object that engages and fascinates a subject, whereby all other objects and actions in the environment are excluded from consciousness (cf. Beardsley, 1982; Csikszentmihalyi, 1990; Cupchik & Winston, 1996; Koestler, 1970; Kubovy, 1999; Marković, 2010; Ognjenović, 2003; Polovina & Marković, 2006; Telegan & Atkinson, 1974). However, the link between kinesthetic responses and the dimensions of aesthetic experience (Vukadinović & Marković, 2012) would require further empirical testing.

The factor of Excitement includes bodily sensations such as *My heart beats faster*, *I have butterflies in my stomach*, *I feel vibrations in my body*, and *I mimic the movements while I sit*. Kinesthetic responses referring to Excitement seem to be closely related to what has previously been described as "inner mimicry" or "imagining movement" (Foster, 2008; Jola *et al.*, 2011; Martin, 1939; Reasons & Reynolds, 2010; Strukus, 2011). Those kinesthetic responses may indicate the spectators' excitement that is drawn from the pleasure of imagining themselves doing the movement. Furthermore, the results of this study regarding the correlation between factors of two scales, one measuring the observers' physical experience and the other measuring the observers' bodily sensations (i.e., kinesthetic responses), showed that Excitement has positive and medium strong correlation with Action Tendency, which means that the higher the Excitement, the higher the Action Tendency. Thus, these results may be interpreted with the fact that bodily sensations related to Excitement may represent a means by which viewers attempt to connect with a dancer (cf. Foster, 2008; Jola *et al.*, 2011; Martin, 1939; Strukus, 2011). However, as Strukus (2011) suggested, seeing somebody else's movements would not necessarily include imagining movement nor result in "mirroring." Previous research reported the differences in the spectators' kinesthetic responses related to imagining themselves performing the movements, depending on the observers' level of experience in watching dance, their dance training, and their being professional dancers themselves (Calvo-Merino *et al.*, 2005, 2006, 2009; Christensen & Calvo-Merino, 2013; Cross *et al.*, 2006; Jola *et al.*, 2011; Jola *et al.*, 2012). Since in this study some of the students who participated had no experience in dance training or in watching dance, the results should be taken with reservations, concerning the generalization related to all the different "types" of dance observers. The question of the relationship between the observers' factors of kinesthetic responses and their previous experience with dance remains open for further empirical studies. Furthermore, in future studies, the question of the gender of the dancer and the gender of the observer (Strukus, 2011) in the context of the observers' factors of kinesthetic responses should be empirically tested and addressed.

The factor of Embodied Anticipation is related to bodily sensations such as *My knees buckle*, *I get teary-eyed*, and *I shake*. Reason and Reynolds (2010) propose that there is a connection between empathy and Embodied Anticipation, reporting that kinesthetic responses involve embodied sensations referring to anticipation connected with the development of choreography. They suggest that there is a relationship between breathing, body rhythms, and emotions while perceiving dynamics through dance observation. According to these authors (2010, p. 66), these kinds of kinesthetic responses are related to the effects of suspense and involve the "automatic/instinctive anticipation" of movement, such as falling or jumping, which changes the observers' breathing pattern and muscular tension. Previous studies (Vukadinović & Marković, 2017) have also explained the influence of the characteristics of dance choreography on the observers' aesthetic experience. Furthermore, the characteristics of a dance choreography that are listed as dynamism structure, progression, and temporality (Foster, 2008; Hagendoorn, 2004) are suggested to be crucial in shaping the observers' anticipation (cf. Berthoz, 2000; Reasons & Reynolds, 2010). Further research would be required to answer the question of how the objective characteristics of dance choreographies (regarding the effect of suspense) influence the factors of the observers' kinesthetic responses when watching dance.

Limitations

Even though this study provided a better insight into the phenomenon of the observers' physical experience and their related bodily sensations when watching dance, there are three aspects of limitations which should be mentioned. There are limitations concerning the sample, which included a small and unbalanced number of male and female participants. Also, there are limitations regarding the instrument used in the study because there were no other variables for its validation. Also, the limitation related to the instrument is related to marginal Cronbach's α obtained as a reliability measure for the scale of the observers' bodily sensations, which implies that this result should be taken with caution.

Furthermore, there are limitations regarding the stimuli used in the main study where the variable of dance type (modern, contemporary, lyrical jazz, tango, and hip-hop) was not controlled for. Moreover, there is a lack of control over how the objective characteristics of dance choreographies, such as dance technique, elegance, and dynamics (Vukadinović, 2019), influence the observers' physical experience of watched performances. Another limitation related to the stimuli is that it was presented as a video recording and not a live performance. Previous studies have shown that there is a significant effect on the observers' experience in general when they watch live performances compared to video recordings (Jola & Grosbras, 2013; Vukadinović & Marković, 2017). There is also a high probability that these results may be shaped by the influence of music, which could be a confounding variable in perceiving dance. Since in the previous studies (Cervellin & Lippi, 2011; Reinhardt, 1999; Thaut *et al.*, 2014; Thaut *et al.*, 2015; Tormodsdatter Færøvik, 2017) it has been shown that listening to music involves not only the auditory areas of the brain but also motor and sensory cortex among others, there is a possibility that some bodily sensations are the result of the influence of music and rhythm. For example, there is evidence that "auditory stimulation primes the motor system in a state of readiness to move, because rhythm provides anticipatory time cues for the brain to plan ahead and be ready" (Thaut *et al.*, 2015, p. 2).

In future studies, when exploring the structure of the observers' physical experience and the structure of the observers' kinesthetic responses while watching dance, it would be important to take into account the possible influence of music. It would also be significant to assure control of the music/rhythm variable, by, for example, comparing the observers' physical experiences and bodily sensations while only listening to music and while watching dance choreographed to that particular music.

Conclusion

One of the benefits of this study is the construction of an instrument for measuring the observers' experiences when watching dance. While the previous studies dealing with physical experience and bodily sensations when watching dance did not quantify responses (Batson, 2008; Foster, 2008; Jola *et al.*, 2011; Montero, 2006, 2012; Reasons & Reynolds, 2010; Strukus, 2011), the instrument constructed in the present study measures and quantifies kinesthetic responses when watching dance. Using a factor analysis to research the structure of audiences' physical experiences and their related bodily sensations when watching dance, the instrument constructed in this study does not just enable a more precise quantification of kinesthetic empathy, but it also provides a better insight into this phenomenon.

This instrument could be used in future studies to address the differences in the observers' physical experiences when watching dance performances depending on the audience's previous experience in dance training. Also, it could be applied to the research into the differences between the audiences' physical experiences when watching live dance compared to recorded dance. Studying particular dance types (e.g., classical ballet, contemporary dance, jazz dance, street styles of dance [e.g., hip-hop and break dance], folk, and flamenco), or specific characteristics of choreographies (e.g., complexity, tempo, and dance technique), or specific characteristics of individual dancers (e.g., staging or attractiveness of the dancers' face) in relation to the observers' physical experiences and related bodily sensations when watching dance would also be fruitful. Likewise, with this instrument, the relationship between the aesthetic experience of dance choreographies and the observers' kinesthetic responses that accompany the aesthetic experience could also be studied. In addition, using this instrument for exploring observers' physical experiences and bodily sensations in other artistic disciplines (e.g., painting, music, literature, theater, and film) may be very useful in order to compare spectators' body reactions. It would be especially challenging to see if there are some similarities among observers in response to different disciplines of art.

Generally, it can be concluded that observers' physical experiences of dance choreographies are complex, multidimensional phenomena that have a structure consisting of Action Tendency, Arousal, and Relaxation. In addition to this factor structure, we explored the factor structure of a bodily sensation — kinesthetic responses — which accompany the observers' physical experience when watching dance. Compared to earlier studies, our scale measuring bodily sensation is very closely related to what is most frequently assumed to be "kinesthetic experience" (Jola *et al.*, 2011; Martin, 1939; Reasons & Reynolds, 2010; Strukus, 2011). Factors such as Focus, Excitement, and Embodied Anticipation form the structure of the observers' kinesthetic responses (bodily sensations) when they engage in observing dance.

Appendix A

The list of dances downloaded from the Internet for the preliminary study:

1. Folk: "Čačak kolo" http://www.youtube.com/watch?v=eisq2319P7U
2. Contemporary Dance: "Philia" http://www.youtube.com/watch?v=ysG2dyV3Vt4
3. Flamenco: "Farruca" http://www.youtube.com/watch?v=y5triYK92IA
4. Giselle "Duet" http://www.youtube.com/watch?v=WyShMJxep7c
5. "Kolo — Great Final of the Balkan Odyssey" http://www.youtube.com/watch?v=YFOU87LAnfM

6. Swan Lake "Odette Variation" http://www.youtube.com/watch?v=d3jAD4Dr7BI
7. Flamenco: "Solea por Bulerias" http://www.youtube.com/watch?v=plTTJFP7YY0
8. "Duet — Srodne duše" http://www.youtube.com/watch?v=YWCwMGnGbQU&feature=PlayList&p=9CF4C2BA82377382&
9. "Tango Fire — Verano Portenas" http://www.youtube.com/watch?v=cFeAwZslAHY
10. Contemporary dance: "Why" http://www.youtube.com/watch?v=qaKJNs3cSLcž

Appendix B

The list of dance choreographies originally performed within the American televised dance competition show *So You Think You Can Dance* that aired on Fox between 2008 and 2016. Downloaded from the Internet and used as stimuli in this research:

1. "How it ends" — De VotchKa https://www.youtube.com/watch?v=9YbEtfJryXA
2. "Tore my Heart" — Oona & Dave Tweedie https://www.youtube.com/watch?v=I6tp8xByIAE
3. "Wave" — Back https://www.youtube.com/watch?v=SnoQlthExKA
4. "The Gulag Orkestar Beirut" https://www.youtube.com/watch?v=lecijwOXX1g
5. "So Broken" (Live version) — Björk https://www.youtube.com/watch?v=0UO4dLY3vBI
6. "Medicine" — Daughter https://www.youtube.com/watch?v=LXvCEfq39ws
7. "Game on" — District 78 https://www.youtube.com/watch?v=DdsTQoN4WcQ
8. "Mirror" — Alexandre Desplat https://www.youtube.com/watch?v=DctLJHiJgOI
9. "Run the World (Girls)" — Beyonce https://www.youtube.com/watch?v=qdxxN0sUsUI

10. "Outta your Mind" — Lil Jon & *LM*AO* https://www.youtube.com/watch?v=mhyWzC7df-0
11. "Raise your Weapon" — Deadmouse https://www.youtube.com/watch?v=VkICjZyjVBw
12. "Get Low" — Dilan Francis & DJ Snake https://www.youtube.com/watch?v=neexFho8Z0I
13. "Libertango" — Bond Quartet https://www.youtube.com/watch?v=k4A97bZ0mZ4
14. "Hello Good Morning (Remix)" — Diddi Dirty Monay feat. Nicky Minay & Rick Ross https://www.youtube.com/watch?v=tCb_UOakEQI
15. "My Chick Bad" — Ludacris & Nicky Minay https://www.youtube.com/watch?v=XrT5ca9EbTw
16. "Brotsjór" — Olafur Arnolds https://www.youtube.com/watch?v=Skb_urlQ4Zg

References

Arnheim, R. (1966). *Umetnost i vizuelno opažanje*. Beograd: Univerzitet umetnosti u Beogradu.

Arnold, P. J. (2005). Somaesthetics, education and the art of dance. *Journal of Aesthetic Education*, 39(1), 48–64.

Barker, M. (2006). I have seen the future and it is not here yet; or on being ambitious for audience research. *The Communication Review*, 9(2), 123–141. https://doi.org/10.1080/10714420600663310.

Batson, G. (2008). Proprioception. The international association for dance medicine and science. Resource paper. Retrieved on 01 Aug, 2016: http://c.ymcdn.com/sites/www.iadms.org/resource/resmgr/imported/info/proprioception.pdf.

Beardsley, M. (1982). *The Aesthetic Point of View*. London, England: Cornell University Press.

Bertoz, A. (2000). *The Brain's Sense of Movement*. London, England: Harvard University Press.

Benedek, M. & Kaernbach, C. (2011). Psychological correlates and emotional specifcity of human piloerection. *Biological Psychology*, 86, 320–329. https://doi.org/10.1016/j.biopsycho.2010.12.012.

Brown, S., Martinez, M. J., & Parsons, L. M. (2006). The neural basis of human dance. *Cerebral Cortex*, 16(8), 1157–1167. https://doi.org/10.1093/cercor/bhj057.

Brownlow, S., Dixon, A. R., Egbert, C. A., & Radcliffe, R. D. (1997). Perception of movement and dancer characteristics from point-light dysplay of dance. *The Psychological Record*, 47, 411–421.

Calvo-Merino, B., Ehrenberg, S., Leung, D., & Haggard, P. (2009). Experts see it all: Configural effects in action observation. *Psychological Research*, 74(4), 400–406. https://doi.org/10.1007/s00426-009-0262-y.

Calvo-Merino, B., Glaser, D. E., Grezes, J., Passingham, R. E., & Haggard, P. (2005). Action observation and acquired motor skills: An fMRI study with expert dancers. *Cerebral Cortex*, 15(8), 1243–1249. https://doi.org/10.1093/cercor/bhi007.

Calvo-Merino, B., Grezes, J., Glaser, D. E., Passingham, R. E., & Haggard, P. (2006). Seeing or doing? Influence of motor familiarity in action observation. *Current Biology*, 16(19), 1905–1910. https://doi.org/10.1016/j.cub.2006.07.065.

Carrol, N. & Moore, M. (2012). Moving in concert: Dance and music. In P. Goldie & E. Schellekens (eds.), *The Aesthetic Mind: Philosophy and Psychology*, pp. 1–7. Oxford University Press.

Cervellin, G. & Lippi, G. (2011). From music-beat to heart-beat: A journey in the complex interactions between music, brain and heart. *European Journal of Internal Medicine*, 22(4), 371–374. https://doi.org/10.1016/j.ejim.2011.02.019.

Christensen, J. F. & Calvo-Merino, B. (2013). Dance as a subject for empirical aesthetics. *Psychology of Aesthetics, Creativity, and the Arts*, 7(1), 76–88. https://doi.org/10.1037/a0031827.

Cooper, J. M. & Silvia, P. J. (2009). Opposing art: Rejection as an active tendency of hostile aesthetic emotions. *Empirical Studies of the Arts*, 27, 111–128. https://doi.org/10.2190/EM.27.1.f.

Cova, F. & Deonna, J. A. (2014). Being moved. *Philosophical Studies*, 169, 447–466. https://doi.org/10.1007/s11098-013-0192-9.

Cross, E. S., Hamilton, A. F., & Grafton, S. T. (2006). Building a motor simulation de novo: Observation of dance by dancers. *NeuroImage*, 31(3), 1257–1267. https://doi.org/10.1016/j.NeuroImage.2006.01.033.

Csikszentmihalyi, M. (1990). *Flow: The Psychology of Optimal Experience*. New York, NY: Harper & Row.

Cupchik, G. C. (1974). An experimental investigation of perceptual and stylistic dimensions paintings suggested by art history. In D. E. Berlyne (ed.), *Studies in New Experimental Aesthetics*, pp. 235–257. Hemisphere Publishing Corporation.

Cupchik, G. C. & Winston, A. C. (1996). Confluence and divergence in empirical aesthetics, philosophy, and main stream psychology. In M. P. Friedman & E. C. Carterette (eds.), *Handbook of Perception & Cognition: Cognitive Ecology*, pp. 62–85. San Diego, CA: Academic Press.

Dittrich, W. H., Troscianko, T., Lea, S. E. G., & Morgan, D. (1996). Perception of emotion from dynamic point-light displays represented in dance. *Perception*, 25, 727–738. https://doi.org/10.1068/p250727.

Fenemor, A. (2003). On being moved by performance. *Performance Research*, 8(4), 107–114. https://doi.org/10.1080/13528165.2003.10871975.

Foster, S. (2007). Kinesthetic empathies and politics of compassion. In J. Reinelt & J. Roach (eds.), *Critical Theory and Performance*, pp. 245–258. University of Michigan Press.

Foster, S. (2008). Movement's contagion: The kinesthetic impact of performance. In C. Davis Tracy (ed.), *The Cambridge Companion to Performance Studies*, pp. 46–59. Cambridge University Press.

Foster, S. (2011). *Choreographing Empathy: Kinesthesia in Performance*. London, England: Routledge.

Glass, R. (2005). Observer response to contemporary dance. In R. Grove, C. Stevens, & S. McKechnie (eds.), *Thinking in Four Dimensions: Creativity and Cognition in Contemporary Dance*, pp. 107–121. Victoria, Australia: Melbourne University Press.

Hagendoorn, I. (2004). Some speculative hypotheses about the nature and perception of dance and choreography. *Journal of Consciousness Studies*, 11(3–4), 79–110.

Jola, C., Abedian-Amiri, A., Kuppuswamy, A., & Grosbras, M. H. (2012). Motor simulation without motor expertise: Enhanced corticospinal excitability in visually experienced dance spectators. *PLoS One*, 7(3), e33343. https://doi.org/10.1371/journal.pone.0033343.

Jola, C., Ehrenberg, S., & Reynolds, D. (2011). The experience of watching dance: Phenomenological-neuroscience duets. *Phenomenology and the Cognitive Sciences*, 11(1), 17–37. https://doi.org/10.1007/s11097-010-9191-x.

Jola, C. & Grosbras, M. H. (2013). In the here and now. Enhanced motor cortico spinal excitability in novices when watching live compared to video recorded dance. *Cognitive Neuroscience*, 4(2), 90–98. https://doi.org/10.1080/17588928.2013.776035.

Koestler, A. (1970). *The Act of Creation*. London, England: Pan Books.

Kubovy, M. (1999). On the pleasures of the mind. In D. Kahneman, E. Diener, & N. Schwartz (eds.), *Well-Being: The Foundation of Hedonic Psychology*, pp. 134–154. New York, NY: Russell Sage.

Leder, H., Belke, B., Oeberst, A., & Augustin, D. (2004). A model of aesthetic appreciation and aesthetic judgments. *British Journal of Psychology*, 95, 489–508. https://doi.org/10.1348/0007126042369811.

Leder, H. & Nadal, M. (2014). Ten years of a model of aesthetic appreciation and aesthetic judgments: The aesthetic episode — Developments and challenges in empirical aesthetics. *British Journal of Psychology*, 105, 443–464. https://doi.org/10.1111/bjop.12084.

Lipps, T. (1906). *Asthetik: Psychologie des Schönen und der Kunst*, 2nd Edition. Hamburg, Germany: Leopold Voss.

Marković, S. (2010). Aesthetic experience and the emotional content of paintings. *Psihologija*, 43(1), 47–64. https://doi.org/10.2298/PSI1001047M.

Marković, S. (2012). Components of aesthetic experience: Aesthetic fascination, aesthetic appraisal, and aesthetic emotion. *Perception*, 3(1), 1–17. https://doi.org/10.1068/i0450aap.

Marković, S. (2017). *Domeni estetske preferencije. Od estetskog stimulusa do estetskog doživljaja. Filozofski fakulte*t, Univerzitet u Beogradu.

Martin, J. (1939). *Introduction to the Dance*. New York, NY: Dance Horizons.

McFee, G. (1992). *Understanding Dance*. London, England/New York, NY: Routledge.

Menninghaus, W., Wagner, V., Hanich, J., Wassiliwizky, E., Kuehnast, M., & Jacobsen, T. (2015). Towards a psychological construct of being moved. *PLoS One*, 10, e0128451.

Miceli, M. & Castelfranchi, C. (2003). Crying: Discussing its basic reasons and uses. *New Ideas in Psychology*, 21, 247–273.

Montero, B. (2006). Proprioception as an aesthetic sense. *The Journal of Aesthetics and Art Criticism*, 64(2), 231–242. https://doi.org/10.1111/j.0021-8529.2006.00244.x.

Montero, B. (2012). Practice makes perfect: The effect of dance training on the aesthetic judge. *Phenomenology and the Cognitive Science*, 11(1), 59–68. https://doi.org/10.1007/s11097-011-9236-9.

Montero, B. (2016). *Though in Action. Expertise and the Conscious Mind.* Oxford: Oxford University Press.

Niedenthal, P. M. (2007). Embodying emotion. *Science*, 316(5827), 1002–1005. https://doi.org/10.1126/science.1136930.

Ognjenović, P. (2003). *Psihološka teorija umetnosti* [*The Psychological Theory of Art*]. Gutenbergova galaksija.

Osgood, C., May, W., & Miron, M. (1975). *Cross — Cultural Universals of Affective Meaning*. Urbana, IL: University of Illinois Press.

Polovina, M., Marković, S., & Estetski doživljaj umetničkih slika (2006). *Psihologija*, 39(1), 39–55. https://doi.org/10.2298/PSI0601039P.

Rawlings, D. (2003). Personality correlates of liking for "unpleasant" paintings and photographs. *Personality and Individual Differences*, 34, 395–410. https://doi.org/10.1016/S0191-8869(02)00062-4.

Reasons, M. & Reynolds, D. (2010). Kinesthesia, empathy and related pleasures: An inquiry into audience experience of watching dance. *Dance Research Journal*, 42(2), 49–75. https://doi.org/10.1017/S0149767700001030.

Reinhardt, U. (1999). Investigations into synchronization of heart rate and musical rhythm in a relaxation therapy in patients with cancer pain. *Forschende Komplementarmedizin*, 6(3), 135–141.

Reynolds, D. (2007). *Rhythmic subjects: Uses of energy in the dances of Mary Wigman, Martha Graham and Merce Cunningham*. Hampshire, England: Dance Books.

Reynolds, D. & Reasons, M. (eds.) (2012). *Kinesthetic Empathy in Creative and Cultural Practices*. Chicago, IL: University of Chicago Press.

Sawada, M., Suda, K., & Ishii, M. (2003). Expression of emotion in dance: Relationship between arm movement characteristics and emotion. *Perceptual and Motor Skill*, 97, 697–708. https://doi.org/10.2466/pms.2003.97.3.697.

Schnall, S., Roper, J., & Fessler, D. M. (2010). Elevation leads to altruistic behaviour. *Psychological Science*, 21, 315–320. https://doi.org/10.1177/0956797609359882.

Schubert, T. W., Zickfeld, J. H., Seibt, B., & Fiske, A. P. (2016). Momento-moment changes in being moved match changes in perceived closeness, weeping, goosebumps, and warmth: Time series analyses. *Cognition and Emotion*, 32(1), 174–184. https://doi.org/10.1080/02699931.2016.1268998.

Sherringtone, C. S. (1907). On the proprioceptive system, especially on its reflex aspect. *Brain*, 29, 467–485.

Silvia, P. J. & Brown, E. M. (2007). Anger, disgust, and the negative aesthetic emotions: Expanding an appraisal model of aesthetic experience. *Psychology of Aesthetics, Creativity, and the Arts*, 1, 100–106. https://doi.org/10.1037/1931-3896.1.2.100.

Stevens, K. & McKechnie, S. (2005). Thinking in action: Thought made visible in contemporary dance. *Cognitive Processing*, 6, 243–252. https://doi.org/10.1007/s10339-005-0014-x.

Strukus, W. (2011). Mining the gap: Physically integrated performance and kinesthetic empathy. *Journal of Dramatic Theory and Criticism*, 25(2), 89–105.

Tan, E. S. & Frijda, N. H. (1999). Sentiment in film viewing. In U. C. Plantinga & G. M. Smith (eds.), *Passionate Views. Film, Cognition and Emotion*, pp. 48–64. Baltimore, MD: John Hopkins University Press.

Telegan, A. & Atkinson, G. (1974). Openness to absorbtion and self-altering experiences: A trait related to hypnotic susceptibility. *Journal of Abnormal Psychology*, 83, 268–277.

Thaut, M. H., McIntosh, G. C., & Hoemberg, V. (2015). Neurobiological foundations of neurologic music therapy: Rhythmic entrainment and the motor system. *Frontiers in Psychology*, 5, 1185. https://doi.org/10.3389/fpsyg.2014.01185.

Thaut, M. H., Trimarchi, P. D., & Parsons, L. M. (2014). Human brain basis of music rhythm perception: Common and distinct neural substrates for meter, tempo and pattern. *Brain Sciences*, 4, 428–452. https://doi.org/10.3390/brainsci4020428.

Tormodsdatter Færøvik, H. U. (2017). *Music and heart rate physiological effects from listening to music in different tempos*, Master Thesis, Universitetet i Bergen det Psykologiske Fakultet.

Tsay, C. (2013). Sight over sound in the judgment of music performance. *Proceedings of the National Academy of the Sciences USA*, 110, 14580–14585. https://doi.org/10.1073/pnas.1221454110.

Turner, B. (2008). Somaesthetics and the critique of Cartesian dualism: Body counsciousness: A philosophy of mindfulness and Somaesthetics. *Body and Society*, 14(3), 129–133. https://doi.org/10.1177/1357034X08093575.

Vingerhoets, A. J. J. M. & Bylsma, L. M. (2016). The riddle of human emotional crying: A challenge for emotion researchers. *Emotion Review*, 8, 207–217. https://doi.org/10.1177/1754073915586226.

Vukadinović, M. (2019). *Psihologija plesa i umetničke igre* [*Psychology of Dance*]. Sombor: Pedagoški fakultet, Novi Sad: Novosadski centar za istraživanje plesa i umetnost flamenka — La Sed Gitana.

Vukadinović, M. & Marković, S. (2012). Aesthetic experience of dance performances. *Psihologija*, 45(1), 23–41. https://doi.org/10.2298/PSI1201023V.

Vukadinović, M. & Marković, S. (2017). The relationship between the dancers' and the audience's aesthetic experience. *Psihologija*, 50(4), 465–481. 10.2298/PSI160222009V.

Wagner, V., Menninghaus, W., Hanich, J., & Jacobsen, T. (2014). Art schema effects on affective experience: The case of disgusting images. *Psychology of Aesthetics, Creativity, and the Arts*, 8, 120–129. https://doi.org/10.1037/a0036126.

Wassiliwizky, E., Jacobsen, T., Heinrich, J., Schneiderbauer, M., & Menninghaus, W. (2017). Tears falling on goosebumps: Co-occurrence of emotional lacrimation and emotional piloerection indicates a psychophysiological climax in emotional arousal. *Frontiers in Psychology*, 8, 41. https://doi.org/10.3389/fpsyg.2017.00041.

Wassiliwizky, E., Wagner, V., Jacobsen, T., & Menninghaus, W. (2015). Art elicited chills indicate states of being moved. *Psychology of Aesthetics, Creativity, and the Arts*, 9, 405–416. https://doi.org/10.1037/aca0000023.

Winkielman, P. & Cacioppo, J. T. (2001). Mind at ease put a smile on the face: Psychophysiological evidence that processing facilitation elicits positive effect. *Journal of Personality and Social Psychology*, 8, 1989–1000. https://doi.org/10.1037/0022-3514.81.6.989.

Woolhouse, M. H. & Lai, R. (2014). Traces across the body: Influence of music-dance synchrony on the observation of dance. *Frontiers in Human Neuroscience*, 8, 965. https://doi.org/10.3389/fnhum.2014.00965.

Zickfeld, J. H. (2015). *Heart-Warming Closeness: Being Moved Induces Communal Sharing and Increases Feelings of Warmth*. Master's Thesis, University of Oslo.

https://doi.org/10.1142/9789811298639_0019

Chapter 19

Emotional Neuroaesthetics of Color Experience: Views from Single, Paired, and Complex Color Combinations*

Naoyuki Osaka

Kyoto University, Kyoto, Japan

Abstract

Colors are critical for understanding the emotional aspect of the human artistic mind, such as that found in painting a landscape, still life, or portrait. First, we report how single colors are memorized in the brain; second, how pairs of colors harmonize in the dissociated brain under the influence of the emotional brain; third, we see how colored paintings are appreciated as beautiful or ugly in the dissociated brain areas led by the intrinsic reward system in the human brain. The orbitofrontal brain is probably one of the vital brain areas that brings us a value-based reward system that makes a unique contribution to emotional neuroaesthetics.

*This chapter was originally published in the special collection on neuroaesthetics in *PsyCh Journal*. Osaka, N. (2022). Emotional neuroaesthetics of color experience: Views from single, paired, and complex color combinations. *PsyCh Journal*, 11(5), 628–635. https://doi.org/10.1002/pchj.577.

Keywords**:** color; emotion; functional magnetic resonance imaging; harmony; neuroaesthetics; reward

How are Colors Memorized in the Brain?

Neuroaesthetics is a new field that aims to understand the brain mechanisms engaged during experimental esthetics and related experiences in the broadest sense (Pearce *et al.*, 2016; Zeki, 1999; Zeki *et al.*, 2020). Neuroaesthetics is also an element of empirical esthetics, which aims to understand the esthetic experience through the brain's reward system using different colors and painting recognition. Here, we show how it works.

Colors and shapes are the critical factors at work in understanding an object, such as a painting of a landscape, still life, or portrait. *Still Life with Apples* by Paul Cezanne is an example. He believed that the eye must comprehend and bring the fruits together, and the brain gives the apple red from different points of view. *Girl with a Pearl Earring* by Johannes Vermeer impresses by centering on the girl's gaze toward the viewer. The face, the eyes, and the draperies were painted using ochres, ultramarine, and lead white, making for a soft expression. We could recall these famous paintings in the mind and be quickly reminded of the dark red of the apples with a background of the blue cloth in the still life and the ultramarine of an oriental turban with the black wall in the painting of the girl. How are these colors memorized in our brain? One possibility stems from a description of the structure of painting as being based on verbal color memory.

Here, we provide a unique view that typical colors are verbalized, whereas atypical colors are memorized in visual form. Interestingly, colors could be memorized in verbal or visual memory, depending on the color category borders. The colors of the apple and the turban appear as the focused-on color in each painting with the background of dark blue cloth and a black wall in each. These focused-on colors can easily be memorized and maintained in the perceptual and verbal memory, making some culturally driven stories take on emergent properties, with the paired background color making for an easy clue to retrieving the memory. In the view of the Gestalt psychologist, the whole (ground) and part (figure) issue here corresponds to the background color and the focused object color as a part. Presumably, these whole/part structures would make paired colors harmonize with emotionally driven pleasant or unpleasant feelings and help us recall a painting in the visual working memory (Baddeley & Hitch, 2007).

Color is used broadly as a visual stimulus; the representation of color in the brain is possibly manipulated either verbally or visually in the dissociated brain areas. Here, we offer a short review of how colors are memorized and harmonized, followed by the process of appreciation governed by the reward system in the human brain. Using functional magnetic resonance imaging (fMRI), we introduced a two-back working memory task to study the verbal and visual components of working memory in color (Ikeda & Osaka, 2007). The verbal expression of colors used in daily life is, essentially, categorical. Therefore, distinct colors that cross the borders of hue categories could be represented verbally because of the involvement of the corresponding specific color names with significant hue differences.

In contrast, colors within the border of the single-color category are likely to be difficult to code verbally because of the involvement of slight hue differences. After successive visual scans of some colored landscape paintings based on quick, saccadic eye movements, our visual appreciation in short-term memory comes to make an evaluation based on perceptual or verbal coding. Thus, the dual coding of perceptual colors and verbal color names may affect our neuroaesthetic experience.

In Baddeley's model (Figure 1), working memory is a limited-capacity system that temporarily retains and manipulates information. The model involves an executive function as a supervisory system and three domain-specific subsystems: The phonological loop (PL) is responsible for language processing, whereas, at the same time, the visuospatial sketchpad (VSSP) manages visuospatial processing. The episodic buffer is assumed to be capable of storing information in a multi-dimensional code which provides a temporary interface between the slave systems: the PL, the VSSP, and the long-term memory (Figure 1(b)). The slave systems are assumed to be controlled by the central executive (attention controller),

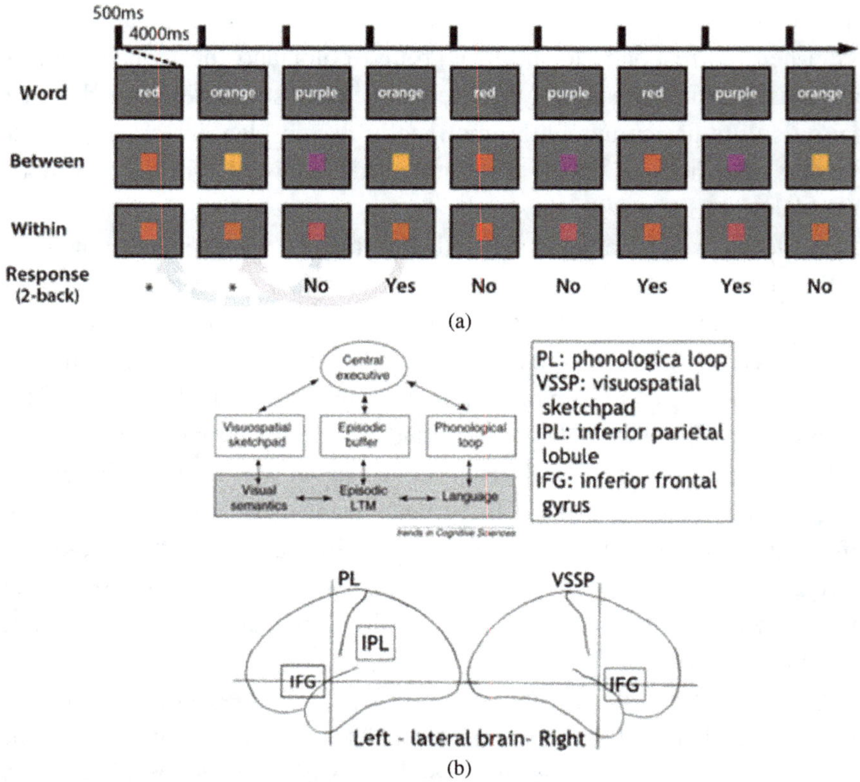

Figure 1. (a) Two-back working memory experiment under three conditions. The observer must monitor whether the current stimulus is the same as those presented in two previous trials. Word: read the word; Between: distinct colors cross the hue border; Within: colors within hue borders with slight hue change. The response arrow underneath indicates that the third trial (purple) is not the same as the first trial (red); therefore, the correct selection is no, whereas the fourth trial (orange) is the same as the second trial (orange); therefore, the correct selection is the yes sequence (reproduced with permission from Ikeda & Osaka, 2007). (b) The three-component model of working memory (Baddeley, 2000). The episodic buffer is assumed to store information in a multi-dimensional code that provides a temporary interface between the slave systems: the PL, the VSSP, and the episodic buffer. The slave systems are assumed to be controlled by the central executive (attention controller), which is responsible for binding information from the PL, the VSSP, and the episodic buffer into coherent episodes (reproduced with permission from Baddeley, 2000). Brain regions associated with assumed PL and VSSP are indicated in the left and right brains below.

which is responsible for binding information from the PL, the VSSP, and the episodic buffer into coherent episodes (Baddeley, 2000). We have assumed that the PL and VSSP each have a unique role in memorizing color names and visual colors. Our results indicated that colors across the categories, defined by basic color names, activated the left inferior frontal gyrus (IFG) and left inferior parietal lobule (IPL) corresponding to the PL as verbal working memory.

In contrast, colors within the same category strongly activated the right IFG corresponding to the VSSP operating as visual working memory (Ikeda & Osaka, 2007). The choice of colors to memorize could modulate the cognitive load balance between the PL and the VSSP. This balance is thought to be influenced by cultural differences; cultures have many color names balanced to more verbal code. This balance suggests how humans memorize colors in working memory. This dual-code property of color evokes a cognitive conflict between the two codes. For example, in the Stroop color task (Stroop, 1935), participants who were asked to report the color of the ink in which a name of a color was written were presented with words whose meaning and color are congruent (e.g., "red" written in red ink) or incongruent (e.g., "red" written in green ink). In the latter scenario, the response time was consistently longer, and the response error observed in the incongruent condition suggests cognitive conflict derived from the lexical processing of the words in the task.

Behavioral studies showed that verbal representation affects color memory (Bornstein, 1976), and the effects are maximized when the colors to be memorized cross the border of hue categories (Roberson & Davidoff, 2000). Berlin & Kay (1969) defined hue categories in terms of the 11 basic color names (i.e., white, black, red, green, yellow, blue, brown, purple, pink, orange, and gray). They argued that color terms in different languages had to meet the same criteria: red, not red–yellow, yellow–red (monolexemic), or blue, but not bluish (monomorphemic). The process generates typical basic color term systems of cultural evolution under the influence of focal colors. Interestingly enough, Russian blues reveal such cultural effects of language on color discrimination. Unlike English-speaking people, Russian-speaking people make an obligatory distinction between the border of lighter blues ("goluboy") and darker blues ("siniy"), which demonstrates that the categories in language affect performance on the simple perceptual color task (Winawer *et al.*, 2007).

Based on the current neuroimaging studies (Owen *et al.*, 2005; Wager & Smith, 2003), we initially thought colors cross different categories (Between conditions) that would activate the PL because colors have significant hue differences, as in the case of color words (Word condition; Figure 1). In contrast, colors within the same category with slight hue differences (Within condition) could not activate the PL. Under the Within condition, memory for adjacent colors should not rely on the PL (Mohr & Linden, 2005). Thus, left frontoparietal network activation is expected to increase under both the Between and Word conditions, whereas the right frontoparietal network activation would be expected to increase under the Within condition.

We used nine healthy adult participants with ages ranging 20–28 years and introduced the two-back task in which the task requires the participant to monitor and update the information they could remember (Owen *et al.*, 2005). They were required to monitor whether the current stimulus is the same as the one previously presented in two trials. We introduced three conditions here: a series of nine stimuli presented on the screen in a single block. Each stimulus was presented for 0.5 s followed by a 4-s interstimulus interval. The observer was asked to press a key with the left or right thumb as instructed.

A single block had a group of colors consisting of three stimuli, from which one was selected as the specified color (red, yellow, green, or blue). Under the Between condition, the stimuli were shown as a colored square of a specified color along with two typical colors of adjacent hue categories (e.g., red group: purple, red, and orange; Figure 1), whereas under the Within condition, the stimuli were a specified color along with two adjacent colors in the same category to minimize possible verbal labeling. Under the Word condition, we found dominant activation in the left IFG. Although previous studies have addressed the possible influence of language on color memory at the behavioral level, our brain imaging study suggests that the PL is employed when we memorize colors under the Between condition. Furthermore, we investigated the involvement of the PL and VSSP by analyzing signal changes across the three conditions.

Considerable evidence of the PL from neuroimaging studies and neurological patients has suggested that the IPL, specifically the left supramarginal gyrus, serves as a passive storage buffer.

Furthermore, frontal areas, such as the IFG, comprise the active rehearsal circuit (Smith & Jonides, 1999; Vallar *et al.*, 1997). It was assumed that the right intraparietal sulcus (IPS) is a visual storage buffer, analogous to the PL in the left, which is increasingly activated as the number of color patches increases (Todd & Marois, 2004). Moreover, the right IFG would expect to involve the right IPS as a buffer for a visual color image. The right IFG activation uniquely observed under the Within condition might plausibly be considered the recruitment of that area, instead of a verbal strategy in the left IFG, to retrieve visual colors under the Within condition. These findings suggest that both the PL and VSSP are employed depending on the property of the colors memorized.

How are Colors Harmonized in the Brain?

In the second section, we move from a single-color study to paired color-harmony studies, which bring the neuroaesthetics of color to a more complicated, emotionally driven perspective. A single color could be memorized in either verbal or perceptual code; however, there appears to be less commitment to an emotional component in a single color, whereas the two-color combination of patches makes it pleasant/unpleasant because of color harmony as in the case of the "Still life with apples" by Paul Cezanne. Neural correlates of color harmony are a critical issue in investigating the neuroaesthetics of emotional experience.

It is well established that people tend to gravitate toward harmonious color combinations as a source of general comfort. *Theory of Colors*, a book published by Goethe in 1810 (von Goethe, 1970, translated version), created a wheel depicting the principal components of colors, namely, the three primary colors (yellow, red, and blue) and three secondary colors (orange, green, and violet) in which harmonious relationships exist between colors opposite (complementary to) each other on this wheel. Munsell (1905) developed a color order system based on three perceptual properties — hue, value (lightness), and chroma (saturation) — and argued that all colors could be accommodated in a slightly distorted spherical solid. Based on this system, Moon and Spencer (1944) proposed identity, similarity, and contrast as three principles for color harmony, whereas Judd and Wyszecki (1975) proposed four principles of color theory, including order, familiarity, similarity, and unambiguity. These principles suggest that harmony is created when neighboring colors share similarities or are complementary in hue when selected harmonically from a hue circle.

These psychological studies have sought to validate and predict harmony scores based on these three properties (Ou & Luo, 2006). This study used 54 colors selected from CIELAB color space (from 1,431 color pairs) for the visual assessment of harmony and disharmony. These color samples covered 11 basic color terms proposed by Berlin and Kay (1969). The CIELAB system has three perceptually even-interval scales, which include lightness (L*), red–green (a*), and blue–yellow (b*). Additionally, chroma (C*) and hue (H*) calculated from the length and angles of vectors are used on the a*b* plane. They suggested that the most likely colors to harmonize with any given color are of similar hue and chroma but with a higher value (lightness). When two colors are seen in neighboring areas, they produce a pleasing effect; although individual preferences may exist, they produce color harmony (Judd & Wyszecki, 1975; Figure 2). The critical issue here is the definition of color harmony. The definition of harmony and pleasantness holds that the former is an esthetic measure, whereas the latter is probably a measure of emotion. The definition becomes clear if pleasantness and harmony originated from the different networks in the human brain. How are pleasantness and harmony to be dissociated in the brain?

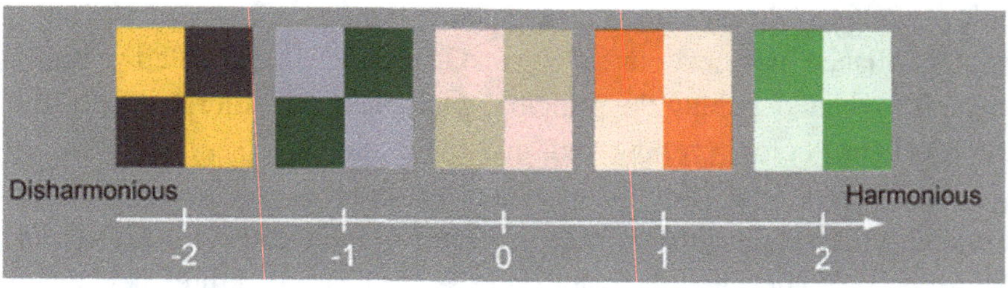

Figure 2. Example of two-color combinations based on chroma (reproduced with permission from Ikeda & Osaka, 2007).

Emotional experience comes from the orbitofrontal cortex (OFC) and the basal ganglia, including the caudate and nucleus accumbens (NACC) of the limbic system (Blood *et al.*, 1999; Brown *et al.*, 2011). However, emotional neuroaesthetics has not yet been established in the human brain. Observing paired colors with a different hue, defined by hue and lightness, creates pleasantness from harmonious combinations, whereas negative responses can emerge from disharmonious combinations.

We studied brain areas activated by harmonious or disharmonious color pairs using fMRI compared to different paired stimuli (Ikeda & Osaka, 2015). We observed that the left medial OFC (mOFC) and left amygdala had been activated when observers watched harmonious and disharmonious stimuli, respectively. These results suggest that color disharmony may depend on stimulus properties and more automatic neural processes mediated by the amygdala, whereas color harmony is revealed by the conscious esthetic value of the mOFC. Interestingly, the OFC is assumed to be one of the brain's reward centers along with the caudate and NACC in the basal ganglia.

Figure 3 shows a color palette with examples of harmonious, neutral, and disharmonious combinations. The neighboring color strongly

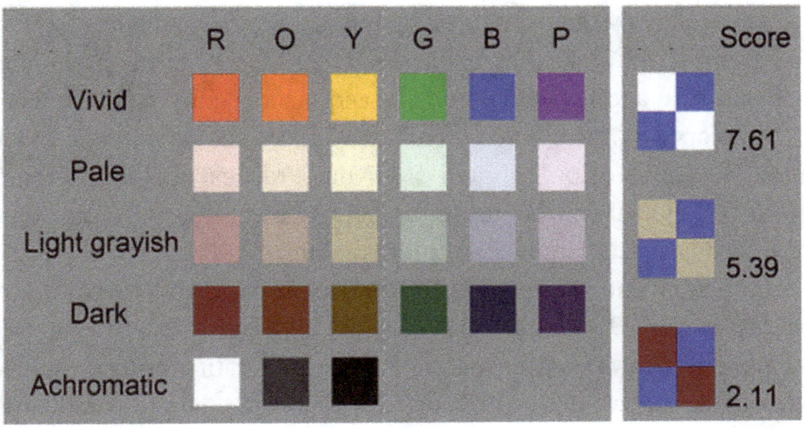

(a) (b)

Figure 3. Color patches and examples. (a) Color patches were used in the study. (b) Examples of color combination stimuli. The upper, middle, and lower patches indicate harmonious, neutral, and disharmonious, respectively. Scores indicate average color-harmony ratings from the participants (reproduced with permission from Ikeda & Osaka, 2015).

determines a pleasing effect, although the term *pleasing* may be rather loaded. However, the neural mechanisms underlying the esthetic and emotional aspects of color, specifically, harmony or disharmony, are as yet poorly understood. Evidence shows that color information is first analyzed in the retina, in the early visual cortices involving V1–V5. Then, higher visual cortices, such as the prefrontal and parietal cortices, and the brain's limbic system involving the amygdala, basal ganglia, and OFC form the reward system (Bartels & Zeki, 2000; McKeefry & Zeki, 1997). As color harmony has been reported as causing pleasing effects (Judd & Wyszecki, 1975), brain areas activated by harmonious or disharmonious color combinations should be processed in terms of sensory and affective features, particularly in their esthetic aspects. Moreover, disharmonious color combinations might provoke an opposite, negative emotion due to the possible contribution of the amygdala.

Previous investigations have suggested that several factors could influence color harmonies, such as various colors, shapes, sizes, and relative colors in a combination (Hård & Sivik, 2001). Our study examined the association between perceptual properties and color-harmony scores, with 351 color pairs presented against a gray background (Ikeda & Osaka, 2015). Different color combination pairs allowed us to assess the relationships between two colors and their lightness, chroma, and hue.

We employed 18 healthy adults aged 19–30 years in the preliminary psychophysical and fMRI experiments and made a color palette containing 27 colors (six hues across four tones and three achromatic colors; Figure 3). These colors, chosen from the Practical Color Coordinate System (developed by the Japan Color Research Institute), have been determined to make harmonious color combinations.

The participants were asked to rate the stimulus on a 9-point scale (1 = *disharmony*, 9 = *harmony*) during the behavioral experiment, using nine numerical keys on a keyboard. Participants' rating criteria included "harmony of the two-color combination"; following the presentation of a black fixation cross (for 1 s) in the MRI scanner, participants were instructed to rate the stimulus on a 3-point scale (1 = *disharmony*, 3 = *harmony*) within 2.5 s by pressing three buttons on the response box in their right hand (Figure 4).

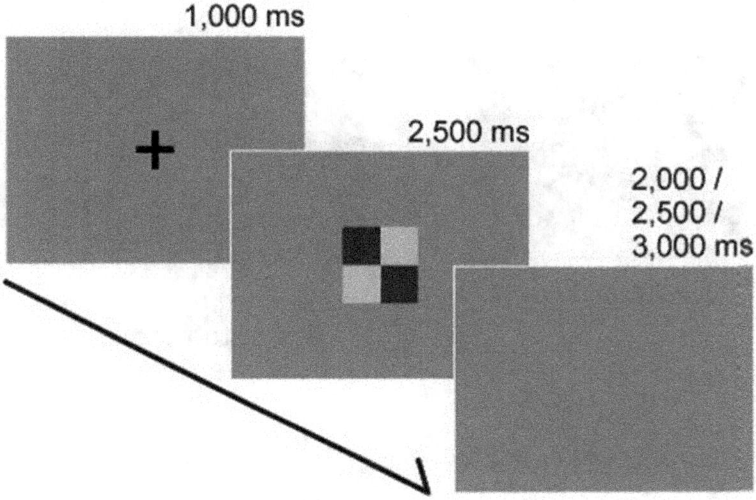

Figure 4. Trial sequence (see text for details) (reproduced with permission from Ikeda & Osaka, 2015).

Figure 5 indicates activated areas during the presentation of harmonious and disharmonious stimuli compared to other stimuli as a baseline (Ikeda & Osaka, 2015). During the presentation of the harmonious stimuli, the mOFC and the bilateral rostral anterior cingulate cortex (rACC) were significantly active, whereas the left amygdala and right posterior insula were significantly active during the presentation of disharmonious stimuli.

Observing harmonious stimulus combinations activating the mOFC is consistent with previous studies showing that this region is activated when observers are immersed in an esthetically pleasing experience. For example, in a musical context, paralimbic areas, including the medial prefrontal and OFC, are activated by harmonious sounds (Blood *et al.*, 1999). In a visuospatial context, esthetically pleasing paintings activate the mOFC, which is associated with increased esthetic preferences (Kawabata & Zeki, 2004). Ishizu & Zeki (2011) demonstrated that the mOFC is activated during experiences of musical and visual beauty. Furthermore, the mOFC is also activated when an attractive face is presented (O'Doherty *et al.*, 2003; Osaka & Minamoto, 2011), whereas the

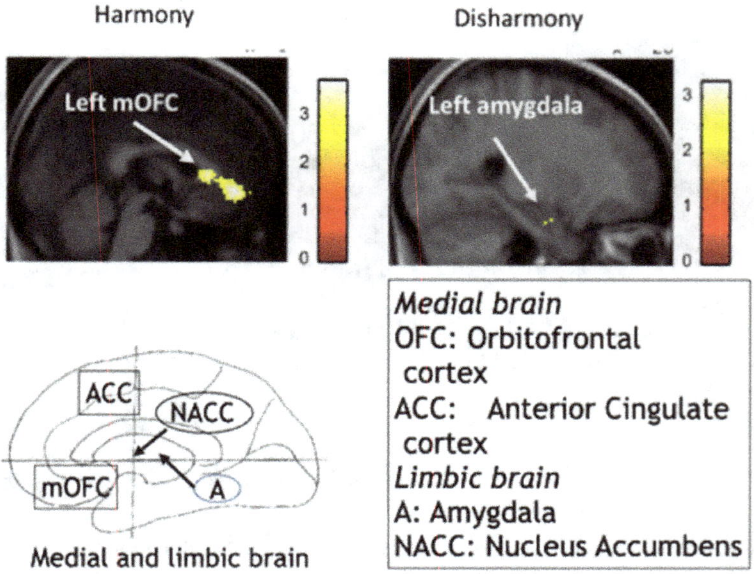

Figure 5. Activated areas under harmony (left mOFC) and disharmony (left amygdala) in sagittal view of the human brain (partly reproduced with permission from Ikeda & Osaka, 2015). Brain regions associated with assumed harmony and disharmony are indicated below.

amygdala is activated when a sad face is presented (Osaka, 2022). Therefore, mOFC activation is not restricted to experiences of color harmony but is reflective of various esthetic experiences regardless of modality. Cultural differences are also critical. In Japanese Noh theater, leading performers put on symbolic masks depicting basic emotions, such as sadness, happiness, anger, disgust, fear, and surprise (Osaka, 2022). These emotions are communicated elegantly through a performer's face, figure, and behavior. Among these basic emotions, negative emotions, such as fear and disgust, have been shown to activate the amygdala of the human brain (Osaka, 2022).

Based on the contrast results of harmony compared with the other stimuli, the cluster including the mOFC extended to the rACC. The ACC is part of the limbic lobe and functionally divided into rostral and caudal portions (Bush *et al.*, 2000) and is part of a circuit involved in attention that serves to regulate both emotional (rACC) and *Cognitive Processing*.

The rACC is also indicated as being responsive toward positive, esthetic appraisals (Brown *et al.*, 2011). Furthermore, Vartanian and Goel (2004) demonstrated that activation within the rACC correlates with individual preference ratings.

Longer reaction times in the neutral condition than in the harmonious and disharmonious conditions indicate that initial responses help determine whether the combination is harmonious or disharmonious. We also observed that disharmonious combinations activated the left amygdala and right posterior insula. The amygdala plays an essential role in evaluating the biological significance of affective visual stimuli. For instance, the amygdala is significantly activated when emotionally negative stimuli are presented (Davis & Whalen, 2001), and activation in the amygdala might involve the automatic processing of affective visual stimuli (Pessoa & Adolphs, 2010). The insula is also involved in processing negative emotions, particularly fear, sadness, and disgust (Phillips *et al.*, 1998). In summary, activation within the amygdala and insula in response to disharmonious combinations might have critical biological implications.

In this report, we discussed the neural basis of visual esthetic judgment by introducing fMRI. The evidence showed that esthetic experiences or judgments enhance the ACC, OFC, and caudate nucleus, including the NACC. These regions partially overlap with the emotional neural circuit and the reward system related to happiness.

How Do Beautiful Paintings Activate the Brain's Reward System?

In the final section, we move the discussion forward and examine how beautiful paintings relate to the brain's reward system. We noted in the first section that a single color has two ways of being memorized, followed by the two-color combination generating emotional color harmony in the second section. We propose the hypothesis that "beautiful paintings" (and, in some cases, "ugly paintings") possibly have connections with the OFC in the frontal brain and subcortical reward system, including the caudate (as a part of the striate nucleus), and the NACC in the basal ganglia. The striatal reward area has multiple aspects of cognition, including action

(a)

(b)

Figure 6. Examples of landscape paintings: (a) most beautiful and (b) most ugly (reproduced with permission from Ikeda & Osaka, 2008).

planning, decision-making, motivation, or even a happy mind, including implicit laughter (Osaka & Osaka, 2005).

Before the experiment with an event-related fMRI, 14 healthy subjects gave a 10-point scale rating (1 = *ugliest*; 5–6 = *neutral*, 10 = *most beautiful*) twice for each of 358 Japanese painting stimuli presented in random order. As a result, 20 images, each judged to be the ugliest, neutral, and most beautiful, were selected for each of the three painting categories of landscape, portrait, and still life according to the rating values (beautiful, neutral, and ugly). For example, Figure 6 shows the paintings with the highest and lowest ratings in the landscape category.

Interestingly, beautiful showed activation in the left parahippocampal gyrus, left IFG, and ACC, whereas ugly showed activation in the right amygdala, right OFC, and medial prefrontal cortex. The parahippocampal gyrus seen in beautiful landscapes is a region in the brain that includes a parahippocampal place area (PPA) that responds well to landscapes. As these areas were not activated in the ugly landscape, it is presumed that the PPA is not considered beautiful in cooperation with the IFG and the ACC. On the contrary, the amygdala, OFC, and medial prefrontal cortex were activated by the ugly painting, whereas the activation of OFC, known as the region of encoding value, is linked to the reward system (Kringelbach, 2005).

The OFC is differentially engaged during the recognition of beautiful and ugly, regardless of the painting category (Kawabata & Zeki, 2004). Another study, however, showed activation with the beauty of music and painting in the medial OFC (Ishizu & Zeki, 2011). Regardless of the difference, it is clarified that the activity reflects the intrinsic reward value of the stimulus.

The effect of reward on looking at a "beautiful painting" is suggested by the fact that the reward system acts according to the beauty of the painting, but as the term "reward" has various meanings, it is necessary to examine its relationship with beauty in detail.

Rewards can be subdivided into physiological rewards, learning-based rewards, and intrinsic rewards. Physiological rewards refer to food, drink, sexual partners, and other essentials for supporting life. Learning-based rewards are those that, like money in human society, are not directly useful for sustaining life. However, they are linked to food, drink, and other things, encouraging life, and acting as drivers. Intrinsic rewards are not, *per se*, necessary for survival, for example, listening to soothing music or looking at a beautiful painting. Social rewards, for example, those found in honorable and altruistic behavior, are somewhat more complicated but may be involved in intrinsic rewards.

Along with rewards, aversive stimuli are also critical for survival. People act for rewards while, at the same time, taking action to escape aversive stimuli. Escape from aversive stimuli brings safety to life; therefore, elimination or avoidance of aversive stimuli can also be considered a reward in a sense.

Three types of rewards usually activate the OFC, NACC, and the related reward system in the brain. Furthermore, the reward system is also activated by removing the aversive stimuli, as mentioned before. Activation of the OFC and medial prefrontal cortex is thought to lead to pleasure and removal of aversive stimuli is thought to act in search of them to promote the pleasure associated with the activation. It can be considered that the activity of these brain areas is a common currency that reflects the reward value (Rolls & Grabenhorst, 2008).

It may be accompanied by activity in different brain regions, and different types of rewards may be related to the work of different networks in the brain.

If visual beauty is associated with reward, it may be appropriate to explain it with the emotional component of pleasure in reward. It should also be noted that facial attractiveness assessments include clues to social value judgment. Many studies have shown that gender differences are a factor when assessing the faces of the opposite sex and are said to be linked to social and biological advantages, such as marriage (Cloutier *et al.*, 2008). It is uncertain whether a painting can have a reward value comparable to food or faces.

The main issue of this section is reward expectation, which is an internal state of mind related to reward. From the viewpoint of behavioral economics, reward expectations can be seen as one of the internal drivers of economic behavior. It can be said that neuroeconomics considers the origin of economic activity to be centered on the brain's reward system. Behind the desire for something is the expectation of reward, which requires making choices.

The question here is how a beautiful painting is connected to the reward system. Now, again, think about what a reward is. A reward is a strong motive for guiding behavior and is one of the fundamental qualities of humans.

Overall, the learning-based reward activates the NACC, whereas any reward, including experiencing "beautiful paintings," activates the caudate nucleus and OFC.

All rewards would be more substantial and unique to humans, and they are expected to bring a sense of satisfaction and pleasure. Turner & Poeppel (1988) first referred to the critical role of a reward system in the brain; they argued that the self-reward system is the primary motivating agent of the brain. What is more, surprisingly, it has become clear that the most common function of the reward system in the brain is behind it. Further investigation is expected to see beauty connected with the brain's reward system in terms of value-based and culture-based neuroaesthetics.

Acknowledgments

I would like to thank Professor Ernst Pöppel for his invitation and encouragement to contribute a short report on emotional neuroaesthetics. Also, thanks to Dr. Takashi Ikeda for his cooperation. Parts of the texts or experiments reported here have been published in previous studies (Ikeda & Osaka, 2007, 2008, 2015). These studies were approved by the Brain Activity Imaging Center (ATR, Kyoto, Japan) and by the Committee of Medical Ethics, Graduate School of Medicine, Kyoto University. All participants gave written informed consent before participation.

References

Baddeley, A. (2000). The episodic buffer: A new component of working memory? *Trends in Cognitive Sciences*, 4, 417–422. https://doi.org/10.1016/s1364-6613(00)01538-2.

Baddeley, A. & Hitch, G. (2007). Working memory; past, present…and future. In N. Osaka, R. Logie, & M. D'Esposito (eds.), *The Cognitive Neuroscience of Working Memory*, pp. 1–20. Oxford University Press.

Bartels, A. & Zeki, S. (2000). The architecture of the colour centre in the human visual brain: new results and a review. *European Journal of Neuroscience*, 12, 172–183. https://doi.org/10.1046/J.1460-9568.2000.00905.x.

Berlin, B. & Kay, P. (1969). *Basic Color Terms: Their Universality and Evolution*. Berkeley, CA: University of California Press.

Blood, A. J., Zatorre, R. J., Bermudez, P., & Evans, A. C. (1999). Emotional responses to pleasant and unpleasant music correlate with activity in paralimbic brain regions. *Nature Neuroscience*, 2, 382–387. https://doi.org/10.1038/7299.

Bornstein, M. H. (1976). Name codes and color memory. *American Journal of Psychology*, 89, 269–2792. https://doi.org/10.2307/1421410.

Brown, S., Gao, X., Tisdelle, I., Eickhoff, S. B., & Liotti, M. (2011). Naturalizing aesthetics: brain areas for aesthetic appraisal across sensory modalities. *NeuroImage*, 58, 250–258. https://doi.org/10.1016/j.NeuroImage.2011.06.012.

Bush, G., Luu, P., & Posner, M. I. (2000). Cognitive and emotional influences in the anterior cingulate cortex. *Trends in Cognitive Sciences*, 4, 215–222. https://doi.org/10.1016/S1364-6613(00)01483-2.

Cloutier, J., Heatherton, T. F., Whalen, P. J., & Kelly, W. M. (2008). Are attractive people rewarding? Sex differences in the neural substrates of facial attractiveness. *Journal of Cognitive Neuroscience*, 20, 941–951. https://doi.org/10.1162/jocn.2008.20062.

Davis, M. & Whalen, P. J. (2001). The amygdala: vigilance and emotion. *Molecular Psychiatry*, 6, 13–34. https://doi.org/10.1038/sj.mp.4000812.

Hård, A. & Sivik, L. (2001). A theory of colors in combination — a descriptive model related to the NCS color-order system. *Color Research and Application*, 26, 4–28. https://doi.org/10.1002/1520-6378(200102)26:1<4::AID-COL3> 3.0.CO;2-T.

Ikeda, T. & Osaka, N. (2008). Kinō no fashisuto OP shikaku — biteki handan: Shinkei undōgakuteki ofurōdo [Neural basis of visual-aesthetic judgment: A neuroaesthetic approach]. *Japanese Psychological Review*, 51, 318–329.

Ikeda, T. & Osaka, N. (2007). How are colors memorized in working memory? A functional magnetic resonance imaging study. *Neuroreport*, 18, 111–114. https://doi.org/10.1097/WNR.0b013e328010ff3f.

Ikeda, T. & Osaka, N. (2015). Color harmony represented by activity in the medial orbitofrontal cortex and amygdala. *Frontiers in Human Neuroscience*, 9, 382. https://doi.org/10.3389/fnhum.2015.00382.

Ishizu, T. & Zeki, S. (2011). Toward a brain-based theory of beauty. *PLoS One*, 6, e21852. https://doi.org/10.1371/journal.pone.0021852.

Judd, D. B. & Wyszecki, G. (1975). *Color in Business, Science, and Industry*, 3rd Edition. New York, NY: Wiley.

Kawabata, H. & Zeki, S. (2004). Neural correlates of beauty. *Journal of Neurophysiology*, 91, 1699–1705. https://doi.org/10.1152/jn.00696.2003.

McKeefry, D. J. & Zeki, S. (1997). The position and topography of the human colour centres as revealed by functional magnetic imaging. *Brain*, 120, 2229–2242. https://doi.org/10.1093/brain/120.12.2229.

Mohr, H. M. & Linden, D. E. J. (2005). Separation of the systems for color and spatial manipulation in working memory revealed by a dual-task procedure. *Journal of Cognitive Neuroscience*, 17, 355–366. https://doi.org/10.1162/0898929053124929.

Moon, P. & Spencer, D. E. (1944). Geometric formulation of classical color harmony. *Journal of the Optical Society of America*, 34, 93–103. https://doi.org/10.1364/JOSA.34.000046.

Munsell, A. H. (1905). *A Color Notation.* Boston, MA: G. H. Ellis Company.

O'Doherty, J., Winston, J., Critchley, H., Perrett, D., Burt, D. M., & Dolan, R. J. (2003). Beauty in a smile: the role of medial orbitofrontal cortex in facial attractiveness. *Neuropsychologia*, 41, 14–155. https://doi.org/10.1016/s0028-3932(02)00145-8.

Osaka, N. (2022). Cultural neuroaesthetics of delicate sadness induced by Noh mask. In A. Chatterjee & E. Cardillo (eds.), *Brain, Beauty, & Art: Essays Bringing Neuroaesthetics in Focus*, p. 122126. Oxford University Press. https://doi.org/10.1093/oso/9780197513620.993.0025.

Osaka, N. & Osaka, M. (2005). Striatal reward areas activated by implicit laughter induced by mimic words in humans: A functional magnetic resonance imaging study. *Neuroreport*, 16(15), 1621–1624. https://doi.org/10.1097/01.wnr.0000181581.18636.a7.

Osaka, N. & Minamoto, T. (2011). Happiness associated with Noh-masks modulates the mesolimbic reward system: An fMRI study based on neuroaesthetics. *Psychologia*, 54, 234–240. https://doi.org/10.2117/psysoc.2011.234.

Ou, L. C. & Luo, M. R. (2006). A colour harmony model for two-colour combinations. *Color Research & Application*, 31, 191–204. https://doi.org/10.1002/col.20208.

Owen, A. M., McMillan, K. M., Laird, A. R., & Bullmore, E. (2005). N-back working memory paradigm: a meta-analysis of normative functional neuroimaging studies. *Human Brain Mapping*, 25, 46–59. https://doi.org/10.1002/hbm.20131.

Pearce, M. T., Zaidel, D. W., Vartanian, O., Skov, M., Leder, H., Chatterjee, A., & Nadal, M. (2016). Neuroaesthetics: The cognitive neuroscience of aesthetic experience. *Perspectives on Psychological Science*, 11, 265–279. https://doi.org/10.1177/1745691615621274.

Pessoa, L. & Adolphs, R. (2010). Emotion processing and the amygdala: from a "low road" to "many roads" of evaluating biological significance. *Nature Review Neuroscience*, 11, 773–783. https://doi.org/10.1038/nrn2920.

Phillips, M., Young, A. W., Scott, S. K., Calder, A. J., Andrew, C., Giampletro, V., Williams, S. C., Bullmore, E. T., Brammer, M., & Gray, J. A. (1998). Neural responses to facial and vocal expressions of fear and disgust. *Proceedings of the Royal Society B: Biological Sciences*, 265, 1809–1817. https://doi.org/10.1098/rspb.1998.0506.

Roberson, D. & Davidoff, J. (2000). The categorical perception of colors and facial expressions: the effect of verbal interference. *Memory & Cognition*, 28, 977–986. https://doi.org/10.3758/BF03209345.

Smith, E. E. & Jonides, J. (1999). Storage and executive processes in the frontal lobes. *Science*, 283, 1657–1661. https://doi.org/10.1126/science.283.5408.1657.

Todd, J. J. & Marois, R. (2004). Capacity limit of visual short-term memory in human posterior parietal cortex. *Nature*, 428, 751–754. https://doi.org/10.1038/nature02466.

Turner, F. & Pöppel, E. (1988). Metered poetry, the brain, and time. In I. Rentschler, B. Herzberger, & D. Epstein (eds.), *Beauty and the Brain: Biological Aspects of Aesthetics*, pp. 71–90. Basel, Switzerland: Birkhaeuser Verlag.

Vallar, G., Di Betta, A. M., & Silveri, M. C. (1997). The phonological short-term store system: patterns of impairment and neural correlates. *Neuropsychologia*, 35, 795–812. https://doi.org/10.1016/s0028-3932(96)00127-3.

Vartanian, O. & Goel, V. (2004). Neuroanatomical correlates of aesthetic preference for paintings. *Neuroreport*, 15, 893–897. https://doi.org/10.1097/00001756-200404090-00032.

von Goethe, J. W. (1970). *Theory of Colours*, Translated by C. L. Eastlake, Trans. MIT Press.

Winawer, J., Witthoft, N., Frank, M. C., Wu, L., Wade, A. R., & Boroditsky, L. (2007). Russian blues reveal effects of language on color discrimination. *Proceedings of the National Academy of Sciences, USA*, 104, 7780–7785. https://doi.org/10.1073/pnas.0701644104.

Wager, T. D. & Smith, E. E. (2003). Neuroimaging studies of working memory: a meta-analysis. *Cognitive, Affective, & Behavioral Neuroscience*, 3, 255–274. https://doi.org/10.3758/cabn.3.4.255pmid:15040547.

Zeki, S. (1999). *Inner Vision: An Exploration of Art and the Brain*. Oxford: Oxford University Press.

Zeki, S., Bao, Y., & Pöppel, E. (2020). Neuroaesthetics: The art, science, and brain triptych. *PsyCh Journal*, 9, 427–428. https://doi.org/10:1002/pchj.383.

https://doi.org/10.1142/9789811298639_0020

Chapter 20

Perceived Cognitive Challenge Predicts Eye Movements While Viewing Contemporary Paintings*

Joanna Ganczarek[†], Karolina Pietras[†] and Roman Rosiek[‡]

[†]*Institute of Psychology, Pedagogical University of Cracow, Kraków, Poland*

[‡]*Institute of Physics, Pedagogical University of Cracow, Kraków, Poland*

Abstract

Contemporary art is often challenging for the viewer, especially when it violates classic rules of representation. Also, viewers usually have little knowledge about this type of art, making its reception even more difficult. Our main research question was how the cognitive challenge associated with contemporary art affects eye movement. In particular, we aimed to assess the impact on eye movements of (a) object-related cognitive challenge in terms of image properties (syntactic and semantic violations) and (b) subject-related cognitive challenge (composite subjective estimate of image inconsistency, ambiguity, and complexity).

*This chapter was originally published in the special collection on neuroaesthetics in *PsyCh Journal*. Ganczarek, J., Pietras, K., & Rosiek, R. (2020). Perceived cognitive challenge predicts eye movements while viewing contemporary paintings. *PsyCh Journal*, 9(4), 490–506. https://doi.org/10.1002/pchj.365.

The eye movements of expert and naive participants were recorded while they freely viewed digital copies of contemporary paintings (four groups of five paintings each, differing in the presence of semantic and syntactic violations). We found that neither violations nor art expertise alone predicted eye movements, although perceived, subjectively experienced cognitive challenge did. In particular, subject-related cognitive challenge was associated with an increase in visual exploration (longer and more numerous fixations, bigger area of exploration, and longer viewing time). The roles of object-related and subject-related indicators of cognitive challenge in perception of contemporary art are discussed.

***Keywords*:** challenging art; contemporary art; expertise; eye movements; paintings; syntactic and semantic violations

Art is often challenging. This may be particularly true for those works of contemporary art that are puzzling, demanding, or even shocking, deliberately hindering understanding (An & Cerasi, 2017). In addition, many viewers know little about contemporary artists and their works. Therefore, some contemporary artworks are markedly challenging either due to the artist's conscious intention to create uncertainty or due to a lack of well-structured knowledge about current artistic trends that could serve as a mental schema guiding viewers' perceptions.

In this chapter, we use the term "cognitive challenge" to describe the difficulty faced by viewers of contemporary art. Cognitive challenge can have both subject-related components, that is, how viewers perceive their own difficulty in dealing with artworks (referred to as "subject-related cognitive challenge"), and object-related components, that is, artwork properties (referred to as "object-related cognitive challenge" or "violations"). This view is in line with many models of aesthetic experience in which the outcome of an aesthetic experience is based on the interaction between characteristics of viewers and artwork properties (for reviews, see Leder & Nadal, 2014; Pelowski, Markey, Lauring, & Leder, 2016).

Research in empirical aesthetics has uncovered some of the cognitive mechanisms underlying the experience of confusion in the face of a demanding artwork. The question of how cognitive challenge affects the perception of artworks has been classically addressed by fluency theory (Reber, Schwarz, & Winkielman, 2004). According to this theory, violations of classical rules of composition or pictorial representation evoke effortful processing, lower artworks' aesthetic appeal, and consequently lead to a decrease in visual exploration. Other authors (Jakesch & Leder, 2009; Jakesch, Leder, & Forster, 2013) agree that violations can cause non-fluent processing, although rather than leading to a lower aesthetic appeal, they might actually provoke increased appreciation and liking. In other words, violations in art can be pleasing despite their effortful processing. Perception of complex artworks is rewarding because of an actually lived or merely anticipated "Aha!" moment when viewers gain insight into these puzzling works (Muth & Carbon, 2013; Muth, Hesslinger, & Carbon, 2015). This mechanism may be particularly relevant for contemporary artworks that often play on the sense of surprise viewers experience when they are confronted with violations. Therefore, we assumed that the cognitive challenge in contemporary art would be linked to liking even if it involved effortful processing.

Despite the fact that there is a growing body of research on how viewers experience challenging art (Jakesch & Leder, 2009; Jakesch *et al.*, 2013; Muth & Carbon, 2013; Muth *et al.*, 2015), little is known about how both subject- and object-related cognitive challenge affect eye movement. The question of brain correlates of violations in surrealist paintings has been addressed by Markey, Jakesch, and Leder (2016), although to the best of our knowledge there is no study that has investigated directly the impact of violations in contemporary art on eye movement. Eye-movement recordings afford the observation of several parameters of visual perception, thus allowing the study of the impact of cognitive challenge directly on the famously quoted "eye of the beholder" (see Table 1 for an overview). For example, long fixations are frequently treated as markers of effortful processing of image content (for a review, see Henderson, 2007), whereas increasing spatial dispersion of fixations and their number is linked to the perception of challenging pictorial features in abstract art (Pihko *et al.*, 2011). Together these parameters provide a good measure of overt attention to both content and form during the aesthetic experience (for a review, see Locher, 2006).

Table 1. Overview of research on the effects of selected art-related factors on eye movements.

Art-related factor	**Effects on eye movements**
Artwork properties	– Fine details (DiPaola, Riebe, & Enns, 2013), color (Fuchs, Ansorge, Redies, & Leder, 2011), dynamism (Massaro *et al.*, 2012), and contrast attract attention (Quiroga & Pedreira, 2011). – Increasing level of abstraction is associated with shorter, more numerous fixations, and longer scanpaths (Pihko *et al.*, 2011; but see also Uusitalo, Simola, & Kuisma, 2009). – In figurative paintings, gaze is mostly directed to recognizable objects; in abstract paintings, viewers usually look at the center (Uusitalo *et al.*, 2009). – Initial fixations fall on the center of composition and main figures (Kapoula, Bucci, Yang, & Bacci, 2010) and constitute the gist phase of viewing (Locher, Krupinski, Nodine, & Mello-Thoms, 2007). – Human faces are associated with smaller-amplitude saccades (Tatler, Wade, & Kaulard, 2007), longer fixation durations (Pihko *et al.*, 2011), and numerous fixations when agents in paintings are involved in social interactions (Villani *et al.*, 2015); also, the human body receives many fixations (Massaro *et al.*, 2012; Villani *et al.*, 2015).
Expertise	– Fixations of experts are less driven by low-level factors in paintings (Koide, Kubo, Nishida, Shibata, & Ikeda, 2015) and they focus more on formal abstract elements (Vogt, 1999; Vogt & Magnussen, 2007), particularly when viewing figurative paintings (Pihko *et al.*, 2011). – Naive viewers spend more time on recognizable objects and human figures (Vogt, 1999; Vogt & Magnussen, 2007). – Experts tend to view abstract works in a global way (Zangemeister, Sherman, & Stark, 1995) with fixations away from the center (Pihko *et al.*, 2011) and show long-to-short gazes (specific exploration followed by a diverse one; Nodine, Locher, & Krupinski, 1993). – Increase in the number of fixations associated with abstract paintings is greater for naive viewers than expert viewers (Pihko *et al.*, 2011) and the maximum fixation duration is longer for naive viewers (Locher, Krupinski, & Schaefer, 2015). – Art knowledge and frequency of attendance to art events correlate positively with the number of fixations on paintings but are not related to fixation durations or revisits (Lin & Yao, 2018).

Table 1. (*Continued*)

Art-related factor	Effects on eye movements
Subjective evaluations of artworks	– No correlations between subjective evaluations and eye movements (Heidenreich & Turano, 2010; Massaro *et al.*, 2012). – Liking and interest predict positively total fixation time for graffiti art and sculptures (Mitschke, Goller, & Leder, 2017). – Increase in fixation durations is greater when paintings are liked more (Plumhoff & Schirillo, 2009). – Art appreciation, understanding, and ambiguity positively predict viewing time. The relationship between ambiguity and viewing time depends on the context: It is positive in a museum setting but negative in a laboratory (Brieber, Nadal, Leder, & Rosenberg, 2014).

Consequently, when it comes to object-related cognitive challenge, we focus here on violations in pictorial form and content. Muth & Carbon (2016) have coined the term "SeIns" to describe the experience of semantic instability induced by perceptual or conceptual challenge. However, as the authors point out themselves (Muth, Hesslinger, & Carbon, 2018), different variants of SeIns are difficult to separate and may refer to various aspects of *Cognitive Processing*. As a result, it would make predictions about their influence on viewers' gazes rather difficult. Therefore, in order to test how violations affect eye movement, we focus on a basic level that is the structure of an artwork itself, namely its form and content. Research investigating these two aspects of artworks (Augustin, Defranceschi, Fuchs, Carbon, & Hutzler, 2011; Augustin, Leder, Hutzler, & Carbon, 2008) suggests that their effects on viewers can be studied separately and that viewers react initially to content and the formal aspects are analyzed subsequently. Given the importance of form and content, when either of them is violated, viewers may feel confused or shocked. For example, with respect to form, a painting can contain different styles, such as abstract and figurative elements, making it difficult for viewers to comprehend it as an integral whole. Meanwhile, taking into consideration the content, a painting can include a juxtaposition of elements, like an elephant sitting on the top of the Eiffel Tower, which might create confusion and a sense of surprise.

The distinction between form and content resembles the linguistic distinction between the syntax (form) and semantics (content) of language. The two concepts were applied to the domain of visual perception for the first time by Biederman, Mezzanotte, and Rabinowitz (1982) in order to describe different types of relationships between objects in real-world scenes. More recently, the terms visual syntax and semantics have also been used in studies involving event-related potentials (Võ & Wolfe, 2013) and eye-tracking (Võ & Henderson, 2009). According to the definition by Võ & Wolfe (2013), an implausible object in the right place would be classed as a semantic violation, whereas a plausible object in the wrong place would be classed as a syntactic violation. In our study, we transferred these definitions of syntactic and semantic violations from real-world scenes into contemporary art in order to operationalize at least part of the cognitive challenge that some contemporary art poses.

However, the transition from real-world scenes to artworks is not straightforward and requires one major adjustment. According to Võ & Wolfe (2013), a syntactic violation relates solely to an unusual placement of an object in a scene. Importantly, an unusual placement of an object can be defined only if there is a clear spatial context that we can refer to. Yet in paintings, objects can be floating in the air without any well-defined spatial context (e.g., painter Luc Tuymans's cups in the middle of an empty canvas; Tuymans, 2012). Moreover, "unusual location" refers to a single location, but in art, an object's form can be shattered and spread in different areas of a painting (e.g., fragmented landscapes in *Broken City* by Alejandro Ospina; Ospina, 2012). Or on the contrary, many single objects are presented in the same location, superimposed, or merged, leading to a more difficult recognition of single forms (e.g., the tangled parts of animals and objects in *Bad Harvest* by Michael Bauer; Bauer, 2008). Finally, a painting can have areas depicted in a realistic way and others where geometrization is used (e.g., contrasting styles in *My Soul Hath A Remembrance And Is Humbled In Me II* by David Brian Smith; Smith, 2011). Therefore, syntactic violation in art, unlike in real-world scenes, can encompass different aspects connected by the fact that they all refer to formal features of artworks and that these formal aspects are of great importance.

Summing up, for the sake of our study, "syntactic violations" refers to a painting's form, that is, (a) use of different artistic styles, (b) undefined forms, or (c) form that is unclear or lacking spatial context. "Semantic violations," on the other hand, refers to a painting's content, that is, a scene is considered semantically violated where (a) an atypical object appears that fails to match the global sense of a given scene or (b) the relationship between objects is unusual.

In eye-movement research, it is well established that effortful processing is associated with longer and more numerous fixations and a more intense visual exploration in general (for a review, see Henderson, 2007). Most of the literature on object–scene inconsistencies deals with semantic violations that are usually connected with longer, denser fixations and more revisits to previously fixated areas as compared with their consistent counterparts (De Graef, Christiaens, & d'Ydewalle, 1990; Henderson & Pierce, 2008; Henderson, Weeks, & Hollingworth, 1999; Underwood & Foulsham, 2006). To the best of our knowledge, Võ & Henderson (2009) were the first researchers to investigate the effects of syntactic violations on eye movements. They observed that the total amount of attention (total viewing time and total number of fixations) on either semantically or syntactically inconsistent objects was greater for the inconsistent objects than their consistent equivalents, which follows the earlier results regarding semantic violations. Therefore, we expected that the presence of these specific violations in art would influence eye movement accordingly. Moreover, we also expected that their effects might be different for naive and expert viewers. Experts are more aware of formal and content-related artistic manipulations (Cela-Conde, Marty, Munar, Nadal, & Burges, 2002). Also, they are more sensitive to composition (Nodine, Locher, & Krupinski, 1993) and outperform naive viewers in recognizing visual patterns (Kozbelt, 2001). They appreciate more original than altered versions of artworks (Hekkert & Van Wieringen, 1996), show better understanding, form their evaluations faster and more accurately (Mullennix & Robinet, 2018), and rely more on their preferences when estimating the artistic quality of objects (Haertel & Carbon, 2014). Their eye movements are also less influenced by low-level factors (Koide, Kubo, Nishida, Shibata, & Ikeda, 2015) and challenging abstract features (Pihko *et al.*, 2011; see

Table 1 for more details). Finally, they demonstrate attenuated reactions to negatively valenced artworks (Leder, Gerger, Brieber, & Schwarz, 2014). The effect of facilitation of processing in experts has also been observed in neuroimaging studies (Pang, Nadal, Müller-Paul, Rosenberg, & Klein, 2013; Solso, 2001) and is postulated by many models of aesthetic processing (Leder, Belke, Oeberst, & Augustin, 2004). As a consequence, their processing may be less effortful, particularly when stimuli contain violations. Conversely, the eye movements of a naive group might indicate increased response to cognitive challenge (i.e., more numerous and longer fixations, larger exploration area, and longer viewing times) due to a lack of expertise in coping with such stimuli. On the other hand, as experts perform more profound and complex analyses of visual stimuli (Cupchik & Gebotys, 1988), we could not exclude the possibility that their processing would be more effortful. Therefore, even though the effects of expertise on eye movement were expected, the direction of these effects was to yet be determined.

Finally, we also took into account the subject-related cognitive challenge, that is, the subjective reactions evoked by an artwork. Among various estimates of cognitive challenge studied in empirical aesthetics (Jakesch & Leder, 2009; Jakesch *et al.*, 2013; Muth & Carbon, 2013; Muth *et al.*, 2015), we chose to focus on perceived inconsistency, complexity, and ambiguity, as these measures allowed us to capture different aspects of cognitive challenge in contemporary art. Whereas "ambiguity" is often associated with multiple meanings and interpretations that could be attributed to the stimuli (Muth *et al.*, 2015), "complexity" usually refers to more formal aspects, such as the variety of elements and ways they are organized (Nadal, Munar, Marty, & Cela-Conde, 2010). In turn, "inconsistency" captures a more general category of an existing conflict. Therefore, by choosing these three characteristics as a proxy of a subject-related cognitive challenge, we were able to address both the general difficulties (inconsistency) as well as the more specific ones (i.e., either semantic [ambiguity] or syntactic [complexity] difficulties) that viewers encountered with demanding art.

Subjective ratings of aesthetic judgments have been classically considered a final output of an aesthetic experience (Leder *et al.*, 2004), with liking, among others, being a key component of aesthetic appraisals that has been studied most broadly (Leder, Gerger, Dressler, & Schabmann,

2012; Leder, Ring, & Dressler, 2013). Here, however, we proposed that they might also be regarded as input for eye movement. Despite the fact that the latter is a theoretically sound assumption (Pelowski *et al.*, 2016), analyses using subjective ratings as correlates or predictors of various eye-movement parameters are employed rather rarely. However, results of some studies suggest that total dwell time increases with increasing subjective ratings of liking or interest (Brieber, Nadal, Leder, & Rosenberg, 2014; Mitschke, Goller, & Leder, 2017; Plumhoff & Schirillo, 2009), complexity (Chassy, Lindell, Jones, & Paramei, 2015), or ambiguity (Brieber *et al.*, 2014; see Table 1 for more details). Consequently, we expected that subject-related cognitive challenge would also positively predict visual exploration not only when it came to dwell time but also other parameters, such as, for example, the duration and number of fixations. This prediction is also based on the aforementioned assumption that cognitive challenge and liking are positively linked. Indeed, if challenging artworks were never liked, they would not necessarily be intensely explored.

Finally, familiarity with stimuli might play an important role in aesthetic evaluations (Leder *et al.*, 2004) and in subjective estimations of cognitive challenge in particular (Belke, Leder, & Carbon, 2015). Also, as familiarity is known to influence eye movements (Francuz, Zaniewski, Augustynowicz, Kopi , & Jankowski, 2018), in this study, we controlled one aspect of it, namely, the effects of previous exposure to stimuli.

Summing up, we hypothesized that (1) object-related cognitive challenge (syntactic and/or semantic violations) would predict an increase in visual exploration of images, (2) subject-related cognitive challenge (subjective ratings of cognitive challenge) would predict an increase in visual exploration of images, and (3) this effect of object- and subject-related cognitive challenge on eye movement would be different in naive and expert viewers.

Method

Participants

Ninety-three participants (M_{age} = 29.2 years, SD_{age} = 9.4 years) took part in our study. The sample included 56 naive subjects (23 males) with no art-related education (mainly social science and technical science) and 37 experts (12 males) with art-related education (fine arts and history of art). The average age for the naive group was 27 years (SD = 9 years) and that for the expert group was 32 years (SD = 9 years). The higher mean of age for the expert group reflected the fact that participants in the expert group had to be either senior year students or already graduated. Participants were recruited through announcements publicized in universities and fine art academies, which were directed toward both senior year students and academic staff.

Besides art experience based on formal education, we assessed the frequency of attendance at visual art events. As expected, experts (Mdn = 4) attended art events more often than naive viewers (Mdn = 2; Mann–Whitney test, $U = 275$, $p < 0.001$, $r = 0.63$). None of the naive participants tended to attend such an event once a week, compared to 22% of experts. Just 12% of naive viewers tended to visit art galleries monthly, 30% did so quarterly, and 14% did so half-yearly, while these rates for experts were 51%, 16%, and 11%, respectively. Moreover, only naive viewers declared that they either had never taken part in a visual art event (14%) or did so once a year only (29%).

All participants had normal or corrected-to-normal vision, gave written informed consent prior to participation in our study, and were offered a gift voucher to a bookshop for their contribution. The study received ethical approval from the ethics board of the Pedagogical University of Cracow.

Stimuli

One hundred and nineteen digital reproductions of contemporary paintings from the online collections of recognized art museums and galleries were preselected by the authors according to the criteria of semantic and syntactic violations presented above. The images had to have either no violations at all, mostly semantic violations, mostly syntactic violations, or equally prominent semantic and syntactic violations. None of the paintings were solely abstract and all of them contained figurative elements. Subsequently, the images were presented in an online survey addressed to visual art experts. After being instructed and given examples, eight experts independently judged paintings on two separate dimensions (i.e., presence of semantic/syntactic violations) using a 7-point Likert-type scale (from 1 [*completely semantically/syntactically inconsistent*] to 7 [*completely semantically/syntactically consistent*]). Eventually, 20 paintings were chosen (five for each of four possible combinations of the intensity of semantic and syntactic violations) utilizing interjudge reliability (Kendall's *W* coefficient was 0.85 for semantic violations, 0.81 for syntactic violations, and 0.81 for both violations). The artworks' details are listed in the Appendix and examples of paintings used as stimuli are presented in Figure 1. All images were adjusted to the same height, that is, 1,000 pixels, whereas the width differed between images (referred to as "width" in the subsequent analyses) so the original image format was preserved. A gray frame was added around each image resulting in the total (original image plus frame) image size of 1,920 × 1,200 pixels. Images were presented on a 24-in. monitor.

Measures

Eye-tracking

Eye movements were recorded using the GazePoint GP3 eye tracker (GazePoint, Vancouver, British Columbia), which is a video-based remote eye tracker with a sampling rate of 60 Hz and accuracy of 0.5°–1°. Participants were seated 60 cm away from a centrally located monitor. Stimuli subtended 29° × 48°, respectively. Stimulus presentation was controlled with GazePoint Control software (Gazepoint, Vancouver, British Columbia).

Subjective evaluations

Participants were asked about previous exposure to a given image (*Yes* [I have seen the image before]/*No* [I have not seen the image before]) and were then asked to use a 5-point (1 [lowest] to 5 [highest]) scale to indicate their subjective evaluation of a painting's level of perceived (a)

Figure 1. Examples of artworks used as stimuli. (a) Image without syntactic or semantic violations: Matvey Levenstein, Orient at Dusk (Flowers in Glass Vase on Corner of Deck), 2012, oil on linen, 38 × 51 cm, courtesy of Matvey Levenstein and Galleria Lorcan O'Neill. (b) Image with syntactic violations only: David Brian Smith, My Soul Hath a Remembrance and is Humbled in Me II, 2011, oil on herringbone linen, 180 × 150 cm, courtesy of David Brian Smith. (c) Image with semantic violations only: Jeff Koons, Niagara, 2000, oil on canvas, 299.7 × 431.8 cm, Solomon R. Guggenheim Museum, New York. Commissioned by Deutsche Bank AG in consultation with the Solomon R. Guggenheim Foundation for the Deutsche Guggenheim, Berlin. (d) Image with syntactic and semantic violations: Kei Imazu, Journey of PUMA, 2016, oil on canvas, 194 × 194 cm, courtesy of Kei Imazu.

inconsistency ("how consistent/inconsistent an image was"), (b) complexity ("how simple/complex an image was"), and (c) ambiguity ("how unambiguous/ambiguous an image was"). These three measures were averaged and served as an estimate of perceived cognitive challenge. Moreover, participants were also asked to indicate if the perceived inconsistency related to a painting's content or form and to evaluate on a 5-point scale how much they liked an image (liking).

Procedure

The study was divided into three blocks: (1) a "free-viewing" block with eye-tracking, (2) a "subjective evaluations" block, and (3) a "demographic survey" block. Before the session, participants signed a consent form and were told that they could withdraw at any time during the study. They were tested individually in a dimly lit room. Each session lasted approximately 90 min. The session started with vision testing (acuity, contrast, and color vision). Participants' visual acuity and contrast sensitivity were assessed with the Freiburg Visual Acuity and Contrast Test (Bach, 1996), whereas their color vision was examined with a web-based color vision test available at https://www.city.ac.uk/health/research/centre-for-applied-vision-research/a-new-web-based-colour-vision-test (Barbour, Harlow, & Plant, 1994). The monitor was calibrated accordingly for both tests. Subsequently, the first block commenced with eye-tracking calibration (9 points, accuracy 0.5°–1°) followed by a validation, after which 20 images were presented in a randomized order, separated by an additional 5-point validation. The additional validation between stimuli was introduced to allow correction of the gaze position for data analysis. During this validation, participants were asked to fixate on five crosses appearing for 1 s each in the four corners of the screen and in the center of the screen. The final cross was placed centrally to ensure uniform gaze position at the beginning of each trial.

Participants were asked to freely view the paintings with no time limitation while their eye movements were recorded. They proceeded to the next trial by a verbal command to the experimenter who terminated the presentation of a current image, and started the validation followed by the next stimulus. Each trial, that is, presentation time, lasted 37.14 s on average.

After the eye-tracking session, the second block was completed during which participants evaluated previously seen artworks. This time participants again viewed each image for 3 s in a randomized order (visualization was controlled with PsychoPy v1.85.6; Peirce *et al.*, 2019). After a stimulus was removed, participants assigned ratings for a specific image on all scales presented in the same order between trials (scales ranged from 1 to 5). After all ratings were completed, participants were presented with the next stimulus. This block was also self-paced. Finally, in the third block, participants answered questions about their exposure to art and demographic characteristics.

Eye-movement analyses

The eye-movement data were exported from GazePoint Analysis software (version 4.1.0; Gazepoint, Vancouver, British Columbia). The original file included time, gaze position, fixation sequence, fixation durations, blink rate, and blink duration. The data were initially explored in order to eliminate any possible anomalies. For this reason, blink artifacts and first fixations were removed. First fixations were eliminated as they fell on the central fixation cross displayed before the onset of test images. Moreover, fixations lasting less than 90 ms and more than 2,000 ms were treated as outliers and excluded from further analyses resulting in the elimination of 1.1% of the data. Four different parameters of eye movements were computed to quantify temporal and spatial characteristics of viewers' gaze patterns and included mean fixation duration, number of fixations, area of exploration, and total viewing time. Mean duration and number of fixations were the average duration/total number of all valid fixations on a given image. The area of exploration was calculated as the number of 100 × 100-px cells where at least one valid fixation had been counted divided by the total image surface (i.e., total number of 100 × 100-px cells) and multiplied by 100 resulting in the percentage of image surface that

Table 2. Descriptive statistics for eye-movement parameters.

Eye-movement parameter	*M*	*SD*	Min.	Max.
Average fixation duration (ms)	370.04	57.22	233	589
Number of fixations	79.70	68.04	10	697
Area of exploration (%)	30.26	15.62	4.62	94.44
Viewing time (s)	37.14	33.11	3.07	373.84

received fixations. This procedure was similar to the one used by Locher (2006). Total viewing time was the average time spent looking at an image. Descriptive statistics are presented in Table 2.

In our study, we checked for the effect of previous exposure to the stimuli. On average, participants knew only 2.28% of images and 80.85% of participants (in both groups) had never seen any of the images before. The difference between the groups was highly significant with experts more familiar with the images than non-experts (experts: $M = 4.86\%$, $SD = 8.34\%$; naive viewers: $M = 0.53\%$, $SD = 2.06\%$; $t(92) = 3.73$, $p<0.001$). Consequently, all analyses were run only on unfamiliar trials.

Results

Subjective ratings of syntactic and semantic violations

In order to validate that images containing violations are indeed perceived as more challenging, we ran repeated-measures analyses of variance (ANOVAs) for the subjective ratings of inconsistency, complexity, and ambiguity (for descriptive statistics, see Table 3). Whenever Mauchly's test indicated that the assumption of sphericity was violated, we reported the Greenhouse–Geisser corrected degrees of freedom. For *post hoc* comparisons, Bonferroni adjustment was applied.

The main effects of image category (four levels: *none* [no violations], *syn* [syntactic violations alone], *sem* [semantic violations alone], and *syn + sem* [syntactic and semantic violations]) were significant for all selected measures: perceived inconsistency, $F(3{,}243) = 99.70$, $p < 0.001$, $\eta_p^2 = 0.55$; perceived ambiguity, $F(2.66, 218.46) = 110.34$, $p < 0.001$, $\eta_p^2 = 0.57$; and

Table 3. ANOVA *post hoc* Bonferroni comparisons of subjective evaluations of images with different violations.

Type of violations	*M*	*SD*	None	Syntax	Semantics
Inconsistency					
None	1.67	0.61			
Syntax	2.87	0.97	$t=-11.18$ $p<0.001$ $r=0.58$		
Semantics	3.10	0.95	$t=-13.09$ $p<0.001$ $r=0.64$	$t=-1.91$ $p=0.12$	
Syntax+Semantics	3.47	0.87	$t=-16.54$ $p<0.001$ $r=0.73$	$t=-5.36$ $p<0.001$ $r=0.33$	$t=-3.45$ $p<0.01$ $r=0.22$
Ambiguity					
None	2.34	0.78			
Syntax	3.45	0.73	$t=-12.11$ $p<0.001$ $r=0.61$		
Semantics	3.45	0.67	$t=-12.00$ $p<0.001$ $r=0.61$	$t=-0.11$ $p=1$	
Syntax+Semantics	4.07	0.59	$t=-18.77$ $p<0.001$ $r=0.77$	$t=-6.66$ $p<0.001$ $r=0.39$	$t=-6.77$ $p<0.001$ $r=0.40$
Complexity					
None	2.02	0.75			
Syntax	3.53	0.76	$t=-17.00$ $p<0.001$ $r=0.73$		
Semantics	2.79	0.68	$t=-8.88$ $p<0.001$ $r=0.49$	$t=8.00$ $p<0.001$ $r=0.45$	
Syntax+Semantics	3.95	0.55	$t=-21.66$ $p<0.001$ $r=0.81$	$t=-12.66$ $p<0.001$ $r=0.63$	$t=-4.66$ $p<0.001$ $r=0.63$

perceived complexity, $F(3{,}249) = 179.53$, $p < 0.001$, $\eta_p^2 = 0.68$. In line with our assumptions, *post hoc* comparisons indicate that all images containing violations were rated as more challenging than images without violations and images with double violations were rated as more challenging than images with single violations (for detailed statistics of the pairwise comparisons, see Table 3).

Moreover, we performed additional analyses in order to determine if the inconsistencies were correctly noted (i.e., corresponded to content in the case of semantic violations and form in the case of syntactic violations) by the participants. The analyses reveal that the four groups of paintings differ significantly when it comes to the perceived presence of formal, $\eta^2(3) = 92.58$, $p < 0.001$, $W = 0.50$, and content-related, $\eta^2(3) = 81.90$, $p < 0.001$, $W = 0.44$, inconsistencies. A Bonferroni correction was applied so all *post hoc* comparisons are reported at a 0.008 level of significance. In accordance with the evaluations by the independent judges, the formal inconsistencies were noted more often in paintings with syntactic violations and with double violations as compared to paintings with semantic violations (*syn* vs. *sem*, $T = 55$, $p < 0.001$, $r = 0.78$ and *syn+sem* vs. *sem*, $T = 23$, $p < 0.001$, $r = 0.84$) and those with no violations at all (*syn* vs. *none*, $T = 33$, $p < 0.001$, $r = 0.83$ and *syn+sem* vs. *none*, $T = 7.50$, $p < 0.001$, $r = 0.86$). The difference between paintings with syntactic violations and with double violations was not significant (*syn* vs. *syn+sem*, $T = 528$, $p = 0.06$). On the other hand, content-related inconsistencies were found more frequently in paintings with semantic violations and those with double violations than in paintings with syntactic violations (*sem* vs. *syn*, $T = 145.50$, $p < 0.001$, $r = 0.67$ and *syn+sem* vs. *syn*, $T = 110$, $p < 0.001$, $r = 0.71$) and those without any violations (*sem* vs. *none*, $T = 32$, $p < 0.001$, $r = 0.84$ and *syn+sem* vs. *none*, $T = 19$, $p < 0.001$, $r = 0.85$). The difference between paintings with semantic violations and those with double violations was not significant ($T = 632$, $p = 0.34$).

Correlations between subjective ratings

In order to explore the relationship between subjective measures and test the assumption regarding the relationship between perceived

Table 4. Means, standard deviations, and correlations with confidence intervals for subjective evaluations of images.

Variable	*M*	*SD*	1	2	3
1. Inconsistency	2.78	0.61			
2. Ambiguity	3.33	0.46	0.18		
			[−0.03, 0.38]		
3. Complexity	3.08	0.46	0.25*	0.67**	
			[0.04, 0.44]	[0.53, 0.77]	
4. Liking	3.05	0.46	0.17	0.27*	0.27*
			[−0.04, 0.37]	[0.06, 0.46]	[0.06, 0.45]

Note: Values in square brackets indicate the 95% confidence interval for each correlation. All subjective evaluations reported here are averaged for all paintings.
* $p<0.05$.
** $p<0.01$.

inconsistency, ambiguity, and complexity with appreciation, we performed Pearson correlations (for detailed data, see Table 4).

The analyses revealed that perceived image ambiguity and complexity correlated positively with liking but the level of perceived inconsistency did not. Therefore, the more complex and more ambiguous the stimuli were, the more they were liked.

Moreover, ambiguity and complexity correlated highly with one another, whereas image inconsistency correlated significantly with complexity but not with ambiguity. This pattern of correlations is in accordance with our assumptions that ambiguity and complexity relate to particular image characteristics and thus correlate very strongly, whereas inconsistency is a more generic category referring to an overall difficulty. The three measures were further combined into one variable called "subject-related cognitive challenge," which in turn also correlated positively with liking ($r = 0.31$, $p = 0.01$).

Models for eye-movement data

Models for the eye-movement data were aimed at evaluating the impact on eye movements exerted by object-related (i.e., syntactic or semantic violations) and subject-related cognitive challenge in expert and naive viewers.

All the eye-movement data were analyzed using *R* (R Core Team, 2017) and the *R* package *lme4* (Bates, Maechler, Bolker, & Walker, 2015). The structure of our experiment was suitable for the use of linear mixed-effect models, which accounts for correlated data from repeated measures and does not require unnecessary averaging (Kliegl, Wei, Dambacher, Yan, & Zhou, 2011). The use of mixed-effects models allowed us to include simultaneously crossed random effects of both participants (e.g., baseline fixation duration or tendency to assign certain ratings) and stimuli (properties of images, such as colors or composition), rendering the results much more reliable than separate ANOVAs. Moreover, mixed-effect models can manage interactions between categorical factors (e.g., syntactic and semantic violations) and continuous variables (e.g., subject-related cognitive challenge).

A separate mixed-effect model was built for each eye-movement parameter. As random effects, we included participants and stimuli as well as random slopes for the effects of subject-related cognitive challenge. We accounted for the confounding effects of low-level image features by including stimuli as random factors. Treating stimuli as random factors is known to eliminate the effects of uncontrolled image properties that are inherent in the sampling of stimuli, therefore reducing the risk of type 1 error (Judd, Westfall, & Kenny, 2012). As fixed effects, we included the following factors: formal expertise (binary variable coded as expert vs. naive; reference category: naive), syntax and semantics (object-related cognitive challenge in terms of syntactic and semantic violations; two binary variables coded as no violations vs. violations; reference category: no violation), and subject-related cognitive challenge (aggregated scores of inconsistency, complexity, and ambiguity ranging from 1 to 5). We also included image width (image width in pixels; see the Appendix for details), which afforded control of its unwanted effects. The continuous variables were centered as follows: subject-related cognitive challenge was group-mean centered (within subjects, i.e., observations coming from a single person), whereas width was grand-mean centered.

We checked for linearity, normality, and homogeneity of variance by visual inspections of plots of residuals against fitted values. Problems with normality of residuals for viewing time were corrected using a log transformation on the outcome variable. The model for number of fixations was modeled using a Poisson distribution as well as a normal distribution with a log transformation of the outcome variable. For each dependent variable, the following strategy was applied: Initially, models with the most complex structure of interactions between predictors were run and compared against simpler models. Subsequently, the best structure of random effects was evaluated by comparing models without random slopes and with random slopes for the effect of subject-related cognitive challenge within participants and within images. The models were compared with the ANOVA function. The simplest models with the relatively best fit assessed by chi-square tests were selected for further data analyses. Finally, all mediation analyses were run in accordance with recommendations by Zhang, Zyphur, and Preacher (2009). For mediation effects, Sobel's test statistics are presented.

Models for eye-movement parameters

Mean duration of fixations

Initially, several models with interaction effects between expertise and other predictors were run and compared with each other. None of the models with interaction terms showed a significantly better fit or explained more variance; therefore, for further analyses, a simpler model with the main effects only was selected.

The model revealed that the object-related cognitive challenge in terms of syntactic and semantic violations did not predict average fixation durations, $b = -1.36$, $SE = 5.21$, $t(16.32) = -0.26$, $p = 0.79$ and $b = -2.38$, $SE = 6.19$, $t(16.16) = -0.38$, $p = 0.70$, respectively. However, the subject-related cognitive challenge did. In particular, the more challenging an image, the longer the average fixations: an increase in subject-related cognitive challenge of 1 unit was associated with an increase in the average fixation duration of 3.27 ms, $b = 3.27$, $SE = 1.05$, $t(1{,}469.42) = 3.10$, $p = 0.01$.

Mediation analyses revealed that syntactic and semantic violations were significant predictors of the subject-related cognitive challenge — $b = 0.50$, $SE = 0.08$, $t(17.02) = 6.43$, $p<0.001$ and $b = 0.40$, $SE = 0.08$,

$t(17.01) = 5.23, p < 0.001$, respectively — but not of fixation duration — $b = 3.02$, $SE = 6.13$, $t(17.01) = 0.49$, $p = 0.62$ and $b = 8.79$, $SE = 6.13$, $t(17.01) = 1.43$, $p = 0.15$, respectively. Subject-related cognitive challenge, on the other hand, significantly predicted average fixation duration, $b = 2.89$, $SE = 1.11$, $t(196.87) = 2.61$, $p = 0.01$. Therefore, it seems that object-related cognitive challenge explained some part of the subjectively experienced challenge but it was the subject-related cognitive challenge that explained the average fixation duration. In other words, although object-related cognitive challenge did not predict average fixation duration either directly or indirectly, it could have triggered subjective experience of processing difficulty, which in turn predicted deeper processing reflected in longer average fixation durations.

Number of fixations

Initially, several models with interaction effects between expertise and other predictors were run and compared with each other. The model that fit best for the number of fixations included an interaction between expertise and syntactic or semantic violations as well as the interaction between expertise and subject-related cognitive challenge. It revealed that the object-related cognitive challenge in terms of syntactic and semantic violations did not predict number of fixations — $b = 7.89$, $SE = 1.09$, $t(19.49) = 1.53$, $p = 0.13$ and $b = -6.43$, $SE = 1.09$, $t(18.95) = -1.36$, $p = 0.17$, respectively — but that the subject-related cognitive challenge did, $b = 2.32$, $SE = 1.02$, $t(140.96) = 2.02$, $p = 0.04$. In other words, the more challenging an image, the more fixations it received: An increase of 1 unit in subject-related cognitive challenge was associated with an increase of 2.32 fixations. Moreover, an interaction between expertise and semantic violations was found: Semantic violations were related to a decrease in the number of fixations and this decrease was greater in the case of experts (no semantic violations $M = 67.36$ and semantic violations $M = 54.05$) than in the case of naive viewers (no semantic violations $M = 60.95$ and semantic violations $M = 54.05$). However, simple effects analysis with a correction for multiple comparisons showed that these pairwise comparisons were not significant, $t(20.70) = 2.52$, $p = 0.09$ and $t(18.99) = 1.36$, $p = 0.54$, respectively. Furthermore, an interaction between expertise and subject-related cognitive challenge was found, suggesting that the relationship between the number of fixations and subject-related cognitive challenge was stronger in the case of experts, $b = 5.68$, $t(51.28) = 3.32$, $p = 0.002$, than in naive viewers, $b = 1.75$, $t(94.53) = 2.11$, $p = 0.04$. In other words, for experts, an increase of 1 unit in subject-related cognitive challenge was associated with an increase of 5.68 fixations, whereas for naive viewers the number of fixations increased only by 1.75 fixations. However, trend analyses revealed that this effect was only marginally significant, $t(131.38) = 1.96$, $p = 0.05$; therefore, these results should be considered with caution.

Area of exploration

Initially, several models with interaction effects between expertise and other predictors were run and compared with each other. None of the models with interaction terms showed a significantly better fit or explained more variance; therefore, for further analyses, a simpler model with the main effects only was selected.

The model for the area of exploration (for details, see Eye movements analyses) revealed that the object-related cognitive challenge in terms of syntactic and semantic violations did not predict the area of exploration, $b = 2.44$, $SE = 1.37$, $t(16.25) = 1.79$, $p = 0.07$ and $b = -1.87$, $SE = 1.62$, $t(16.14) = -1.16$, $p = 0.25$, respectively. Instead, the subject-related cognitive challenge did, $b = 0.76$, $SE = 0.28$, $t(149.81) = 2.67$, $p = 0.01$. This means that an increase in subject-related cognitive challenge of 1 unit was associated with an increase in the area of exploration by 0.76%.

As mentioned earlier (see mediation analyses for average duration of fixations), both syntactic and semantic violations predicted subject-related cognitive challenge. However, only syntactic violations predicted an increase in the area of exploration by 3.53%, $b = 3.53$, $SE = 1.57$, $t(17.01) = 2.24$, $p = 0.03$. Inclusion of subject-related cognitive challenge decreased the influence of syntactic violations, $b = 3.14$, $SE = 1.58$, $t(17.20) = 1.99$, $p = 0.05$. Sobel's test suggested that the mediation effect was significant ($z = 2.42$, $p = 0.02$). Therefore, it was the subject-related cognitive challenge that explained the effects of syntactic violations on the area of exploration.

Viewing time

Initially, several models with interaction effects between expertise and other predictors were run and compared with each other. None of the models with interaction terms showed a significantly better fit or explained more variance; therefore, for further analyses, a simpler model with the main effects only was selected. The model for viewing time included a log transformation of the outcome variable. It revealed that the object-related cognitive challenge in terms of syntactic and semantic violations did not predict viewing time, $b = 1.93$, $SE = 1.04$, $t(17.78) = 1.87$, $p = 0.06$ and $b = -1.04$, $SE = 1.04$, $t(17.49) = -1.18$, $p = 0.24$, respectively. However, the subject-related cognitive challenge did, $b = 1.93$, $SE = 1.01$, $t(150.76) = 4.88$, $p < 0.001$: An increase of 1 unit in the perceived cognitive challenge was associated with an increase of 1.93 s in viewing time.

As mentioned earlier (see mediation analyses for the average duration of fixations), both syntactic and semantic violations predicted subject-related cognitive challenge. However, only syntactic violations predicted an increase in viewing time, $b = 2.82$, $SE = 1.04$, $t(16.98) = 2.70$, $p = 0.01$. Also, the subject-related cognitive challenge predicted the viewing time, $b = 1.95$, $SE = 1.01$, $t(150.81) = 4.88$, $p < 0.001$. Inclusion of subject-related cognitive challenge reduced the impact of syntactic violations, $b = 1.95$, $SE = 1.04$, $t(17.79) = 1.87$, $p = 0.06$, and the mediation effect was significant ($z = 3.85$, $p < 0.001$). Therefore, it suggests that the subject-related cognitive challenge explained the impact of syntactic violations on viewing time.

Discussion

Our study aimed to evaluate the impact of object- and subject-related cognitive challenge on eye movements during perception of contemporary art by expert and naive viewers. In line with previous research (Belke *et al.*, 2015; Jakesch & Leder, 2009; Jakesch *et al.*, 2013; Muth *et al.*, 2015), the degree of ambiguity and complexity in images was positively linked with appreciation, suggesting that despite its effortful processing, this kind of challenging art can be liked. However, this pattern was not replicated for the third component of subject-related cognitive challenge as no relation between perceived inconsistency and liking was found. This difference in the results could be partially explained by the disparate status of these variables. In accordance with the correlation pattern between the variables, it seems that inconsistency represents a higher-order concept, separate from more concrete categories, such as ambiguity (with its reference to semantics) and complexity (with reference to syntax). Consequently, ambiguity and complexity could be more natural categories for viewers to evaluate paintings, which leads to a more accurate assessment of the subject-related challenge. Of course, this hypothetical explanation calls for further investigation.

Nonetheless, given that the overall estimate of subject-related cognitive challenge correlates positively with liking, we expected that image difficulty in terms of both syntactic or semantic violations and subjectively experienced challenge would predict intense visual exploration. Our hypothesis was confirmed partially in that only the subject-related experience of cognitive challenge affected eye movement. In particular, we found that when images were perceived as challenging, viewers performed long, numerous fixations over large parts of images. On the other hand, violations or the object-related cognitive challenge alone did not significantly predict eye movements.

This implies that even though violations in art were estimated as challenging, no effect of violations on eye movement was found. The eye movements were instead connected with the perceived cognitive challenge. More precisely, as demonstrated by mediation analyses, it was the subject-related cognitive challenge in response to violations that was related to an increase in visual exploration. This result adds an interesting point to the debate on the relative input of viewer- and artwork-related features to aesthetic experience (Leder & Nadal, 2014; Pelowski *et al.*, 2016). Our data suggest that the subjective experience may be even more important for the outcome of an aesthetic experience than the stimulus properties. This is in line with the study by Forster, Leder, and Ansorge (2013) stressing the role of a perceiver's feeling of subjective fluency (rather than objective ease in processing the stimuli) in art appreciation.

Now let us look into the question of why no effect of violations on eye movement was found. First of all, our expectation that violations would predict increased visual exploration came from studies of real-world scenes. We could hypothesize that the difference in results stems from different experimental stimuli. For example, a flying pot in a kitchen or a bar of soap on a laptop screen are unexpected in everyday life; however, in art, they might actually be anticipated and therefore perceived as less challenging in comparison to violations in natural scenes. Jakesch, Goller, and Leder (2017) stress that non-fluent stimuli classified as art provoke more positive reactions (measured by facial electromyography) than non-fluent stimuli in everyday life. Thus, the context of art may attenuate reactions to difficult stimuli by, among others, providing a secure state of so-called psychical distance or detached aesthetic processing mode (Cupchik, 2002, 2013). In this mode, viewers can appreciate and deliberately search for ambiguities or semantic instabilities in art (Belke *et al.*, 2015; Muth & Carbon, 2013, 2016, 2019; Muth *et al.*, 2015) because the context of art makes the experience of challenge less threatening and possibly more enjoyable than in real life. Consequently, we can speculate that the art context and the aesthetic processing mode might modify eye movement as well. The psychical distance or detachment would free viewers from the pragmatic concerns that require them to react and deal with violations and instead make them more prone to appreciating difficulties (Cupchik, Vartanian, Crawley, & Mikulis, 2009), which might influence eye movements differently than in real-life scenes.

We could also propose that our definitions of syntactic and semantic violations differ from those used in real-world scenes, thus having two various constructs might lead to different results. However, both of these explanations fail to hold when we consider that, notwithstanding the differences in definitions, participants in our study actually perceived these violations as inconsistent, complex, and ambiguous, and, similar to the judges, noted content- and form-related inconsistencies.

The second explanation is more technical and related to the sampling of the stimuli. It is possible that the effect of violations could have been significant had the stimuli been more uniform. Here, we accounted for the confounding effects of low-level image features in the random effects structure. In future studies, it would be interesting to see if violations derived from controlled image manipulation rather than artwork sampling could predict eye movement. Clearly, when we modify artworks the ecological validity decreases and raises the question as to whether we should still investigate art or merely artificial lab-prepared stimuli.

Finally, and most importantly, it is possible that we did not detect the effects of violations on eye movement because of the strong effect of subjective evaluations. In fact, this was the case for syntactic violations that lost their impact on eye movements when a model also contained the subject-related cognitive challenge. In other words, it is possible that in art the subjective experience of cognitive challenge has more bearing than what is actually in an image. This interpretation can be supported by the finding that subject-related cognitive challenge consistently predicted all eye-movement parameters in every statistical model. Moreover, albeit only as a marginally significant effect, the influence of subjective estimations on eye movements was stronger in expert viewers, which again demonstrates the importance of subject- rather than object-related properties. In other words, the findings of this study support the role of top-down factors, such as subject-related cognitive challenge and expert knowledge in the perception of art.

This conclusion might be particularly important for contemporary art when even non-violated images can be perceived as thought-provoking, especially with free-viewing. Indeed, free-viewing might have promoted a more reflexive, unconstrained exploration possibly related to cognitive mastering.

Our results are in line with eye-movement studies where the link between cognitive challenge and eye movement is well established (for a review, see Henderson, 2007). This study provides support for a similar mechanism in art perception with one major difference. Whereas studies on perception of real-world scenes suggest a more intense visual exploration when a given task is more difficult, our results suggest that even if objectively a task is easy and without time pressure, eye movement may be influenced by subjectively experienced difficulty.

Finally, we found that expertise alone did not predict eye movements but interacted with subject-related cognitive challenge. This interaction showed that the link between subject-related cognitive challenge and the number of fixations was strongest for experts, suggesting that when viewers are equipped with expert knowledge, they are more likely to refer to their subjective evaluations than to objective violations. Naive viewers, on the other hand, might tend to link their subjective evaluations with the objective violations found in artworks. This interpretation is in agreement with the study by Koide *et al.* (2015), who found that art expertise reduced the influence of low-level features on eye movements. Also, it follows the findings by Haertel & Carbon (2014), who demonstrated that art experts trust their own preferences more compared to naive viewers who rely more on object-related factors, such as originality and ambiguity. Clearly, to allow any further conclusions this result needs to be replicated given that the trend analysis was only marginally significant.

Conclusions and Further Studies

Our study provides support for the hypothesis that cognitive challenge related to the perception of contemporary art can actually be observed in the eyes of the beholders. Interestingly, it is not so much what is in the picture but what a viewer makes of it that counts.

In future studies it would be interesting to explore the nature of this positive relationship between liking, subject-related cognitive challenge, and visual exploration. Is it the eye movement that influences subjective evaluations via afferent feedback or is it the subjective experience that guides gaze? Clearly, eye-movement recording is only one way of exploring viewers' aesthetic experiences. Therefore, it would be very interesting to use neuroimaging or measure other bodily reactions to cognitive challenge, such as postural or myographic activity. Also, it would be valuable to investigate the connection between liking and subjective cognitive challenge. For example, we can ask whether this connection is only valid for artworks. Whereas giant green apples hovering in the air might be quite unusual in everyday life, in art they might be expected, at least since the times of René Magritte and the Surrealist movement. In other words, cognitive challenge or conflict may be part of art, and of contemporary art in particular. One expects to be bewildered or confused. These expectations in turn can influence different physiological measures and one's subjective experience. For example, conflict in art could actually be rewarding if this is what we expect to find (Van de Cruys & Wagemans, 2011) or when it allows viewers to engage in cognitive mastering followed by insight (Belke *et al.*, 2015; Muth & Carbon, 2013, 2016, 2019; Muth *et al.*, 2015). Therefore, in contemporary art, cognitive challenge may be anticipated and appreciated and offer an opportunity to engage in cognitive elaboration reflected in intense visual exploration. As Muth & Carbon (2019) claim, it may be the shifting in and out of semantic instability that attracts viewers to complex and ambiguous artworks. On the other hand, we may speculate that for non-artworks or other works where the cognitive challenge is unexpected or not appreciated, the effect of cognitive challenge may be different. In real-life situations, for example, people may rely more on stable perceptual interpretations that allow habitual interaction with the environment. In such contexts, challenges may be avoided and effortful ambiguous stimuli less intensely explored. Thus, in future studies, it would be interesting to include both challenging artworks and real-life scenes for comparison.

Moreover, we found that the effect of formal art expertise was rather subtle and did not allow strong conclusions. However, based on our results, we can speculate that the impact of expertise might be greater if viewers had to estimate artistic quality or memorize image details. In other areas of research, it has been demonstrated that the role of individual differences is greater when a particular task is performed (Szymura & N cka, 2005). Last but not least, our results suggest that the role of expertise is more prominent when we consider the mediating role of subjective experience. We found that the influence of subject-related cognitive challenge on eye movements was greater in experts than in naive viewers. This may mirror the relative role of top-down and bottom-up factors in eye-movement control for expert and naive viewers, respectively. Both of these hypotheses deserve empirical scrutiny.

Obviously, our study has the common limitations of lab experiments. Self-paced free-viewing was aimed at reducing the artificiality of the experimental situation but clearly it does not completely rule it out. Also, the presentation on a computer monitor limits the impact of other important factors, such as the actual physical size, texture, or the museum context. Certainly, such a setting is different from visiting a museum but, paradoxically, we think that nowadays viewing artworks on a computer screen can be as ecologically valid as viewing them in a museum due to the growing popularity of online viewing. Nonetheless, it would be interesting to compare the effects of violations in museum and laboratory contexts, especially given the importance of physical factors in art perception, such as size, material, or distance. Moreover, the effect of subject-related cognitive challenge is relatively small, which suggests that it is only one of many factors that influence gaze behavior during aesthetic experience. In addition, gaze reflects only a part of the complex processing that underlies an encounter with art, thus adding other types of measurements may be necessary to build a more complete picture of what it means to view art.

Acknowledgments

The authors received financial support from the Pedagogical University of Cracow. We thank Patrick Markey for his suggestion to consider different artistic styles as part of the syntactic violations. Also, we thank Tomasz Smoleń and Joanna Fryt for their valuable help with the preparation of this manuscript.

Appendix

List of paintings without semantic or syntactic violations:

1. Lee Chen Dao, *Lounge Bar*, 2006, 1,446 × 1,000 px
2. John Currin, *The Gardeners*, 2001, 1,439 × 1,000 px
3. Matvey Levenstein, *Orient at Dusk*, 2012, 1,338 × 1,000 px
4. Bartek Materka, *Bez tytułu. Ona śpi* [Untitled. She sleeps], 2012, 1,625 × 1,000 px
5. Shi Zhiying, *Mars*, 2010, 1,025 × 1,000 px

List of paintings with syntactic violations:

1. David Brian, *My Soul Hath a Remembrance and is Humbled in Me II*, 2011, 835 × 1,000 px
2. David Brian, *Goodwill and the Unknown Man*, 2012, 839 × 1,000 px

3. Massimiliano Carollo, *Tauromachie 002*, 2016, 1,717 × 1,000 px
4. Annalisa Fulvi, *Come topi* [Like mice], 2014, 1,852 × 1,000 px
5. Erik van Leishout, *Untitled*, 2014, 1,819 × 1,000 px

List of paintings with semantic violations:

1. Jacub Gagnon, *Storming the Castle*, 2009, 1,362 × 1,000 px
2. Ewa Juszkiewicz, *Portrait of Carol Rama*, 2014, 796 × 1,000 px
3. Jeff Koons, *Niagara*, 2000, 1,408 × 1,000 px
4. Scott Reeder, *Bombs (At Dinner)*, 2011, 1,259 × 1,000 px
5. Chris Woods, *Houdini*, 1998, 670 × 1,000 px

List of paintings with semantic and syntactic violations:

1. Massimiliano Carollo, *Neonati* [The Newborn], 2014, 627 × 1,000 px
2. Seonna Hong, *The Spill*, 2012, 992 × 1,000 px
3. Kei Imazu, *Journey of PUMA*, 2016, 991 × 1,000 px
4. Morgan Mandalay, *The Departed*, 2016, 653 × 1,000 px
5. Joram Roukes, *Mourning*, 2012, 824 × 1,000 px

References

An, K. & Cerasi, J. (2017). *Who's Afraid of Contemporary Art*? New York, NY: Thames & Hudson.

Augustin, M. D., Defranceschi, B., Fuchs, H. K., Carbon, C. C., & Hutzler, F. (2011). The neural time course of art perception: An ERP study on the processing of style versus content in art. *Neuropsychologia*, 49, 2071–2081. https://doi.org/10.1016/j.Neuropsychologia.2011.03.038.

Augustin, M. D., Leder, H., Hutzler, F., & Carbon, C. C. (2008). Style follows content: On the microgenesis of art perception. *Acta Psychologica*, 128, 127–138. https://doi.org/10.1016/j.actpsy.2007.11.006.

Bach, M. (1996). The Freiburg Visual Acuity Test: Automatic measurement of visual acuity. *Optometry and Vision Science*, 73, 49–53. https://doi.org/10.1097/00006324-199601000-00008.

Barbour, J. L., Harlow, A. J., & Plant, G. T. (1994). Insights into the different exploits of colour in the visual cortex. *Proceedings of the Royal Society of*

London, Series B: Biological Science, 258, 327–334. https://doi.org/10.1098/rspb.1994.0181.

Bates, D., Maechler, M., Bolker, B., & Walker, S. (2015). Fitting linear mixed-effects models using lme4. *Journal of Statistical Software*, 67, 1–48. https://doi.org/10.18637/jss.v067.i01.

Bauer, M. (2008). *Bad Harvest* [Painting]. London, England: SaatchiGallery. https://www.saatchigallery.com/artists/artpages/michael_bauer_bad_harvest.htm.

Belke, B., Leder, H., & Carbon, C. C. (2015). When challenging art gets liked: Evidences for a dual preference formation process for fluent and non-fluent portraits. *PLoS One*, 10, 1–34. https://doi.org/10.1371/journal.pone.0131796.

Biederman, I., Mezzanotte, R. J., & Rabinowitz, J. C. (1982). Scene perception: Detecting and judging objects undergoing relational violations. *Cognitive Psychology*, 14, 143–177. https://doi.org/10.1016/0010-0285(82)90007-X.

Brieber, D., Nadal, M., Leder, H., & Rosenberg, R. (2014). Art in time and space: Context modulates the relation between art experience and viewing time. *PLoS One*, 9, e99019. https://doi.org/10.1371/journal.pone.0099019.

Cela-Conde, C. J., Marty, G., Munar, E., Nadal, M., & Burges, L. (2002). The "style scheme" grounds perception of paintings. *Perceptual and Motor Skills*, 95, 91–100. https://doi.org/10.2466/pms.2002.95.1.91.

Chassy, P., Lindell, T. A., Jones, J. A., & Paramei, G. V. (2015). A relationship between visual complexity and aesthetic appraisal of car front images: An eye-tracker study. *Perception*, 44, 1085–1097. https://doi.org/10.1177/0301006615596882.

Cupchik, G. C. (2002). The evolution of psychical distance as an aesthetic concept. *Culture & Psychology*, 8, 155–187. https://doi.org/10.1177/1354067X02008002437.

Cupchik, G. C. (2013). I am, therefore I think, act, and express both in life and in art. In T. Roald & J. Lang (Eds.), *Art and Identity: Essays on the Aesthetic Creation of Mind*, pp. 67–92. Amsterdam, the Netherlands: Rodopi.

Cupchik, G. C. & Gebotys, R. J. (1988). The search for meaning in art: Interpretive styles and judgments of quality. *Visual Arts Research*, 14, 38–50.

Cupchik, G. C., Vartanian, O., Crawley, A., & Mikulis, D. J. (2009). Viewing artworks: Contributions of cognitive control and perceptual facilitation to aesthetic experience. *Brain & Cognition*, 70, 84–91. https://doi.org/10.1016/j.bandc.2009.01.003.

De Graef, P., Christiaens, D., & d'Ydewalle, G. (1990). Perceptual effects of scene context on object identification. *Psychological Research*, 52, 317–329. https://doi.org/10.1007/BF00868064.

DiPaola, S., Riebe, C., & Enns, J. T. (2013). Following the masters: Portrait viewing and appreciation is guided by selective detail. *Perception*, 42, 608–630. https://doi.org/10.1068/p7463.

Forster, M., Leder, H., & Ansorge, U. (2013). It felt fluent, and I liked it: Subjective feeling of fluency rather than objective fluency determines liking. *Emotion*, 13, 280–289. https://doi.org/10.1037/a0030115.

Francuz, P., Zaniewski, I., Augustynowicz, P., Kopiś, N., & Jankowski, T. (2018). Eye movement correlates of expertise in visual arts. *Frontiers in Human Neuroscience*, 12, 87. https://doi.org/10.3389/fnhum.2018.00087.

Fuchs, I., Ansorge, U., Redies, C., & Leder, H. (2011). Salience in paintings: Bottom-up influences on eye fixations. *Cognitive Computation*, 3, 25–36. https://doi.org/10.1007/s12559-010-9062-3.

Haertel, M. & Carbon, C. C. (2014). Is this a "Fettecke" or just a "greasy corner"? About the capability of laypersons to differentiate between art and non-art via object's originality. *i-Perception*, 5, 602–610. https://doi.org/10.1068/i0664.

Heidenreich, S. M. & Turano, K. A. (2010). Where does one look when viewing artwork in a museum? *Empirical Studies of the Arts*, 29, 51–72. https://doi.org/10.2190/em.29.1.d.

Hekkert, P. & Van Wieringen, P. C. W. (1996). The impact of level of expertise on the evaluation of original and altered versions of post-impressionistic paintings. *Acta Psychologica*, 94, 117–131. https://doi.org/10.1016/0001-6918(95)00055-0.

Henderson, J. M. (2007). Regarding scenes. *Current Directions in Psychological Science*, 16, 219–222. https://doi.org/10.1111/j.1467-8721.2007.00507.x.

Henderson, J. M. & Pierce, G. L. (2008). Eye movements during scene viewing: Evidence for mixed control of fixation durations. *Psychonomic Bulletin & Review*, 15, 566–573. https://doi.org/10.3758/PBR.15.3.566.

Henderson, J. M., Weeks, P. A., Jr., & Hollingworth, A. (1999). The effects of semantic consistency on eye movements during complex scene viewing. *Journal of Experimental Psychology: Human Perception and Performance*, 25, 210–228. https://doi.org/10.1037/0096-1523.25.1.210.

Imazu, K. (2016). *Journey of PUMA* [Painting]. http://www.imazukei.com/works-のコピー.html1.

Jakesch, M., Goller, J., & Leder, H. (2017). Positive fEMG patterns with ambiguity in paintings. *Frontiers in Psychology*, 8, Article 785, 1–7. https://doi.org/10.3389/fpsyg.2017.00785.

Jakesch, M. & Leder, H. (2009). Finding meaning in art: Preferred levels of ambiguity in art appreciation. *Quarterly Journal of Experimental Psychology:*

Human Experimental Psychology, 62, 2105–2112. https://doi.org/10.1080/17470210903038974.

Jakesch, M., Leder, H., & Forster, M. (2013). Image ambiguity and fluency. *PLoS One*, 8, e74084. https://doi.org/10.1371/journal.pone.0074084.

Judd, C. M., Westfall, J., & Kenny, D. A. (2012). Treating stimuli as a random factor in social psychology: A new and comprehensive solution to a pervasive but largely ignored problem. *Journal of Personality and Social Psychology*, 103, 54–69. https://doi.org/10.1037/a0028347.

Kapoula, Z., Bucci, M. P., Yang, Q., & Bacci, F. (2010). Perception of space in Piero della Francesca's annunciation: An eye-movement and art-historical study. *Leonardo*, 43, 153–158. https://doi.org/10.1162/leon.2010.43.2.153.

Kliegl, R., Wei, P., Dambacher, M., Yan, M., & Zhou, X. (2011). Experimental effects and individual differences in linear mixed models: Estimating the relationship between spatial, object, and attraction effects in visual attention. *Frontiers in Psychology*, 1, 238. https://doi.org/10.3389/fpsyg.2010.00238.

Koide, N., Kubo, T., Nishida, S., Shibata, T., & Ikeda, K. (2015). Art expertise reduces influence of visual salience on fixation in viewing abstract-paintings. *PLoS One*, 10, e0117696. https://doi.org/10.1371/journal.pone.0117696.

Koons, J. (2000). *Niagara* [Painting]. New York, NY: Solomon R. Guggenheim Museum. https://www.guggenheim.org/artwork/10734.

Kozbelt, A. (2001). Artists as experts in visual cognition. *Visual Cognition*, 8, 705–723. https://doi.org/10.1080/13506280042000090.

Leder, H., Belke, B., Oeberst, A., & Augustin, D. (2004). A model of aesthetic appreciation and aesthetic judgments. *British Journal of Psychology*, 95, 489–508. https://doi.org/10.1348/0007126042369811.

Leder, H., Gerger, G., Brieber, D., & Schwarz, N. (2014). What makes an art expert? Emotion and evaluation in art appreciation. *Cognition and Emotion*, 28, 1137–1147. https://doi.org/10.1080/02699931.2013.870132.

Leder, H., Gerger, G., Dressler, S. G., & Schabmann, A. (2012). How art is appreciated. *Psychology of Aesthetics, Creativity, and the Arts*, 6, 2–10. https://doi.org/10.1037/a0026396.

Leder, H. & Nadal, M. (2014). Ten years of a model of aesthetic appreciation and aesthetic judgments: The aesthetic episode–developments and challenges in empirical aesthetics. *British Journal of Psychology*, 105, 443–464. https://doi.org/10.1111/bjop.12084.

Leder, H., Ring, A., & Dressler, S. G. (2013). See me, feel me! Aesthetic evaluations of art portraits. *Psychology of Aesthetic, Creativity and the Arts*, 7, 358–369. https://doi.org/10.1037/a0033311.

Levenstein, M. (2012). *Orient at Dusk (Flowers in Glass Vase on Corner of Deck)* [Painting]. Rome, Italy: GalleriaLorcan O'Neill. http://www.lorcanoneill.com/site/elencoLavoriDettaglio.php?idArtista=8&pageno=10.

Lin, F. & Yao, M. (2018). The impact of accompanying text on visual processing and hedonic evaluation of art. *Empirical Studies of the Arts*, 36, 180–198. https://doi.org/10.1177/0276237417719637.

Locher, P. (2006). The usefulness of eye movement recordings to subject an aesthetic episode with visual art to empirical scrutiny. *Psychology Science*, 48, 106.

Locher, P., Krupinski, E., Nodine, C., & Mello-Thoms, C. (2007). Visual interest in pictorial art during an aesthetic experience. *Spatial Vision*, 21, 55–77. https://doi.org/10.1163/156856807782753868.

Locher, P., Krupinski, E., & Schaefer, A. (2015). Art and authenticity: Behavioral and eye-movement analyses. *Psychology of Aesthetics, Creativity, and the Arts*, 9, 356–367. https://doi.org/10.1037/aca0000026.

Markey, P., Jakesch, M., & Leder, H. (2016, August). Art and ambiguity — Semantic and syntactic processing of art: Perception of paintings and associated distinct neurophysiological brain responses. Paper presented at the 24th *Conference of the International Association Of Empirical Aesthetics*, Vienna, Austria.

Massaro, D., Savazzi, F., Di Dio, C., Freedberg, D., Gallese, V., Gilli, G., & Marchetti, A. (2012). When art moves the eyes: A behavioral and eye-tracking study. *PLoS One*, 7, e37285. https://doi.org/10.1371/journal.pone.0037285.

Mitschke, V., Goller, J., & Leder, H. (2017). Exploring everyday encounters with street art using a multimethod design. *Psychology of Aesthetics, Creativity, and the Arts*, 11, 276. https://doi.org/10.1037/aca0000131.

Mullennix, J. W. & Robinet, J. (2018). Art expertise and the processing of titled abstract art. *Perception*, 47, 359–378. https://doi.org/10.1177/0301006617752314.

Muth, C. & Carbon, C. C. (2013). The aesthetic aha: On the pleasure of having insights into gestalt. *Acta Psychologica*, 144, 25–30. https://doi.org/10.1016/j.actpsy.2013.05.001.

Muth, C. & Carbon, C. C. (2016). SeIns: Semantic instability in art. *Art & Perception*, 4, 145–184. https://doi.org/10.1163/22134913-00002049.

Muth, C. & Carbon, C. C. (2019). When art is not mastered but creates insights. Shifting in and out of semantic instability. *Art & Perception*, 7, 123–136. https://doi.org/10.1163/22134913-20191118.

Muth, C., Hesslinger, V. M., & Carbon, C. C. (2015). The appeal of challenge in the perception of art: How ambiguity, solvability of ambiguity, and the opportunity for insight affect appreciation. *Psychology of Aesthetics, Creativity, and the Arts*, 9, 206. https://doi.org/10.1037/a0038814.

Muth, C., Hesslinger, V. M., & Carbon, C. C. (2018). Variants of semantic instability (SeIns) in the arts: A classification study based on experiential reports. *Psychology of Aesthetics, Creativity, and the Arts*, 12, 11. https://doi.org/10.1037/aca0000113.

Nadal, M., Munar, E., Marty, G., & Cela-Conde, C. J. (2010). Visual complexity and beauty appreciation: Explaining the divergence of results. *Empirical Studies of the Arts*, 28, 173–191. https://doi.org/10.2190/EM.28.2.d.

Nodine, C. F., Locher, P. J., & Krupinski, E. A. (1993). The role of formal art training on perception and aesthetic judgment of art compositions. *Leonardo*, 26, 219. https://doi.org/10.2307/1575815.

Ospina, A. (2012). *Broken City* [Painting]. London, England: Saatchi Gallery. https://www.saatchigallery.com/artists/artpages/alejandro_ospina_brken_city_2.htm.

Pang, C. Y., Nadal, M., Müller-Paul, J. S., Rosenberg, R., & Klein, C. (2013). Electrophysiological correlates of looking at paintings and its association with art expertise. *Biological Psychology*, 93, 246–254. https://doi.org/10.1016/j.biopsycho.2012.10.013.

Peirce, J. W., Gray, J. R., Simpson, S., MacAskill, M. R., Höchenberger, R., Sogo, H., Kastman, E., & Lindeløv, J. K. (2019). PsychoPy2: experiments in behavior made easy. *Behavior Research Methods*, 51, 195–203. https://doi.org/10.3758/s13428-018-01193-y.

Pelowski, M., Markey, P. S., Lauring, J. O., & Leder, H. (2016). Visualizing the impact of art: An update and comparison of current psychological models of art experience. *Frontiers in Human Neuroscience*, 10, 160. https://doi.org/10.3389/fnhum.2016.00160.

Pihko, E., Virtanen, A., Saarinen, V. M., Pannasch, S., Hirvenkari, L., Tossavainen, T., Haapala, A., & Hari, R. (2011). Experiencing art: The influence of expertise and painting abstraction level. *Frontiers in Human Neuroscience*, 5, 94. https://doi.org/10.3389/fnhum.2011.00094.

Plumhoff, J. E. & Schirillo, J. A. (2009). Mondrian, eye movements, and the oblique effect. *Perception*, 38, 719–731. https://doi.org/10.1068/p6160.

Quiroga, R. Q. & Pedreira, C. (2011). How do we see art: An eye-tracker study. *Frontiers in Human Neuroscience*, 5, 98. https://doi.org/10.3389/fnhum.2011.00098.

R. Core Team, View (2017). R: A language and environment for statistical computing. Vienna, Austria: R Foundation for Statistical Computing. https://www.R-project.org/.

Reber, R., Schwarz, N., & Winkielman, P. (2004). Processing fluency and aesthetic pleasure: Is beauty in the perceiver's processing experience? *Personality and Social Psychology Review*, 8, 364–382. https://doi.org/10.1207/s15327957pspr0804_3.

Smith, D. B. (2011). *My Soul Hath A Remembrance and Is Humbled In Me II* [Painting]. London, England: Saatchi Gallery. https://www.saatchigallery.com/artists/artpages/david_brian_smith_my_soul_hath.htm.

Solso, R. L. (2001). Brain activities in a skilled versus a novice artist: An fMRI study. *Leonardo*, 34, 31–34. https://doi.org/10.1162/002409401300052479.

Szymura, B. & Nęcka, E. (2005). Three superfactors of personality and three aspects of attention. In A. Eliasz, S. E. Hampson, & B. Raad (eds.), *Advances in Personality Psychology*, pp. 65–79. Hove, England: Psychology Press.

Tatler, B. W., Wade, N. J., & Kaulard, K. (2007). Examining art: Dissociating pattern and perceptual influences on oculomotor behaviour. *Spatial Vision*, 21, 165–184. https://doi.org/10.1163/156856808782713807.

Tuymans, L. (2012). *Teacups I* [Painting]. New York , NY: David Zwirner Gallery.

Underwood, G. & Foulsham, T. (2006). Visual saliency and semantic incongruency influence eye movements when inspecting pictures. *Quarterly Journal of Experimental Psychology*, 59, 1931–1949. https://doi.org/10.1080/17470210500416342.

Uusitalo, L., Simola, J., & Kuisma, J. (2009). Perception of abstract and representative visual art. In *Proceedings of AIMAC, 10th Conference of the International Association of Arts and Cultural Management*, pp. 1–12.

Van de Cruys, S., & Wagemans, J. (2011). Putting reward in art: a tentative prediction error account of visual art. *i-Perception*, 2, 1035–1062. https://doi.org/10.1068/i0466aap.

Villani, D., Morganti, F., Cipresso, P., Ruggi, S., Riva, G., & Gilli, G. (2015). Visual exploration patterns of human figures in action: An eye tracker study with art paintings. *Frontiers in Psychology*, 6, 1636. https://doi.org/10.3389/fpsyg.2015.01636.

Võ, M. L. H. & Henderson, J. M. (2009). Does gravity matter? Effects of semantic and syntactic inconsistencies on the allocation of attention during scene perception. *Journal of Vision*, 9, 24. https://doi.org/10.1167/9.3.24.

Võ, M. L. H. & Wolfe, J. M. (2013). Differential electrophysiological signatures of semantic and syntactic scene processing. *Psychological Science*, 24, 1816–1823. https://doi.org/10.1177/0956797613476955.

Vogt, S. (1999). Looking at paintings: Patterns of eye movements in artistically naïve and sophisticated subjects. *Leonardo*, 32, 325. https://doi.org/10.1162/002409499553325

Vogt, S. & Magnussen, S. (2007). Expertise in pictorial perception: Eye-movement patterns and visual memory in artists and laymen. *Perception*, 36, 91–100. https://doi.org/10.1068/p5262.

Zangemeister, W. H., Sherman, K., & Stark, L. (1995). Evidence for a global scan-path strategy in viewing abstract compared with realistic images. *Neuropsychologia*, 33, 1009–1025. https://doi.org/10.1016/0028-3932(95)00014-T.

Zhang, Z., Zyphur, M. J., & Preacher, K. J. (2009). Testing multilevel mediation using hierarchical linear models: Problems and solutions. *Organizational Research Methods*, 12, 695–719. https://doi.org/10.1177/1094428108327450.

Part III

Faces in Art

https://doi.org/10.1142/9789811298639_0021

Chapter 21

Symmetric in the Striate but Asymmetric in the Extra-Striate Cortex When Processing Three-Quarter Faces: Neural Underpinnings for Aesthetic Appreciations*

Yifan Zeng[†,‡,§], Xuanyu Wang[†,¶], Sarita Silveira[‡,||] Jana von Trott zu Solz,[‡] Fabian Simmank[‡], Yuliya Zaytseva[‡,**,††], Yan Bao[†,‡,‡‡] and Marco Paolini[§§]

[†]*School of Psychological and Cognitive Sciences, Peking University, Beijing, China*
[‡]*Institute of Medical Psychology, Ludwig Maximilian University of Munich, Germany*
[§]*Department of General and Experimental Psychology, Ludwig Maximilian University Munich, Germany*
[¶]*Graduate School of Systemic Neurosciences, Ludwig Maximilian University of Munich, Munich, Germany*
[||]*Social Neuroscience Lab, Max-Planck Society, Berlin, Germany*
[**]*National Institute of Mental Health, Klecany, Czech Republic*
[††]*Department of Psychiatry and Medical Psychology, Third Faculty of Medicine, Charles University Prague, Prague, Czech Republic*

‡‡*Beijing Key Laboratory of Behavior and Mental Health, Peking University, Beijing, China*
§§*Department of Radiology, University Hospital, Ludwig Maximilian University of Munich, Germany*

Abstract

Faces and their aesthetic appreciation are a core element of social interaction. Although studies have been made on facial processing when looking at faces with different perspectives, a direct comparison of faces from the left to the right perspective is missing. Portraits in classical Western art indicate a preference for the left compared to the right perspective, but the neural underpinnings of such an asymmetry still have to be clarified. Using functional magnetic resonance imaging, this study focuses on the processing of three-quarter faces seen from different perspectives. Seventeen participants were asked to passively look at photographs of six male and six female faces with neutral expression; the photographs were taken from the left, right, and frontal perspectives while keeping their focus on the eyes. The results showed that specific brain areas were involved in processing the three-quarter faces in either symmetric or asymmetric ways. Viewing left and right three-quarter faces resulted in two mirror-like activations in the *striate* cortex corresponding to the symmetric layout of the left and right perspectives. Viewing the left face resulted additionally in an enhanced activation also in the left *extrastriate* cortex. The right perspective of male faces elicited a lower activation compared to other perspectives in face-selective areas of the brain. Our findings suggest that the preference of the left three-quarter face emerges already in the early visual pathway presumably prior to facial identification, emotional processing, and aesthetic

*This chapter was originally published in the special collection on neuroaesthetics in *PsyCh Journal*. Zeng, Y., Wang, X., Silveira, S., von Trott zu Solz, J., Simmank, F., Zaytseva, Y., Bao, Y., & Paolini, M. (2022). Symmetric in the striate but asymmetric in the extrastriate cortex when processing three-quarter faces: Neural underpinnings for aesthetic appreciations. *PsyCh Journal*, 11(5), 720–728. https://doi.org/10.1002/pchj.539.

appreciation. Our observations may have general importance in disentangling different neural components and processing stages in the spatiotemporal characteristics of artistic expressions.

***Keywords*:** aesthetic; face perception; functional magnetic resonance imaging; hemifaces; lateralization

Introduction

The biological basis of aesthetics and neural aspects of the arts have become an important topic in neuroscience, psychology, and ethology during the last decades (e.g., Rentschler *et al.*, 1988; Zeki, 1999). Using behavioral paradigms, imaging technologies, or ethological approaches, empirical studies have been focused on visual artworks with their spatial characteristics (e.g., Bao *et al.*, 2017b; Lutz *et al.*, 2013; Yang *et al.*, 2019), music with its emotional impact (e.g., Park *et al.*, 2015) or its access to personal identity (Zaytseva *et al.*, 2014), and poetry with its link to language processing in general and its temporal regularity, which is important for aesthetic appreciations (e.g., Yu & Bao, 2020; Zhao *et al.*, 2018). Interestingly, the temporal characteristics apply also to movies (Chen *et al.*, 2020).

Behavioral studies in "neuroaesthetics" and the identification of underlying neural mechanisms have also led to theoretical reflections about anthropological universals and cultural specifics (Bao *et al.*, 2016), about the general concept of complementarity as a generative principle in aesthetic appreciations (Bao *et al.*, 2017a), or even about aesthetics as a common denominator in different kinds of judgment (Wang *et al.*, 2020). Thus, research in this interdisciplinary field of neuroaesthetics with its different empirical approaches and theoretical reflections contributes to better understanding of cognitive mechanisms and their neural basis with respect to spatial characteristics and temporal dynamics. However, open questions still remain, and body representations — in particular those of faces, which have attracted major interest for ages (Sütterlin & Yu, 2021) — may provide unique access to *Cognitive Processing* and its neural correlates.

As frequent visual stimuli, faces have been considered the most specialized stimuli preserved in evolution (Leopold & Rhodes, 2010). Such specificity may give rise also to their unique status with respect to aesthetic appreciation. In particular, empirical studies have found a preference for depicting three-quarter portraits from the *left* perspective where the left hemiface composes most of the portrait (Blackburn & Schirillo, 2012; Grüsser *et al.*, 1988; Powell & Schirillo, 2009; Stoerig *et al.*, 1983), especially for European paintings during the 14th to 20th centuries (McManus & Humphrey, 1973; e.g., the *Mona Lisa* of Leonardo da Vinci, *Girl with a Pearl Earring* by Johannes Vermeer, the self-portraits of Rembrandt van Rijn [painted from a mirror], and Andy Warhol's *Marilyn Monroe* in the later American pop art period). Psychophysiological and psychophysical studies have provided supporting evidence revealing greater aesthetic ratings with increased pupil sizes when viewing left-side faces (Blackburn & Schirillo, 2012), and implicit detection of the original perspective of portraits particularly in the left view (Wang & Bao, 2020).

As aesthetic judgment might have a surprisingly fundamental role in various judgments (Wang *et al.*, 2020), it is conceivable to relate it to a relatively low level in the hierarchy of cognitive functions. More specifically, the aesthetic preference for the left perspective might be deeply rooted in the context of a basic perceptual bias; facial perception was found to be viewpoint-dependent (Ewbank *et al.*, 2008; Fang *et al.*, 2007), with three-quarter faces yielding better recognition over other perspectives (Krouse, 1981; Laeng & Rouw, 2001; Troje & Bülthoff, 1996), and the left view having a higher recognition rate than the right view (Arnold *et al.*, 2013; Kowatari *et al.*, 2004; Siéroff, 2001; Yamamoto *et al.*, 2005).

Neuroscientific investigations have pointed out right hemispheric lateralization on the basis of pathological observation (Barton *et al.*, 2002) and brain-imaging techniques (Weibert & Andrews, 2015; Zhang *et al.*, 2015). It is further suggested that the right hemispheric dominance of facial processing might give rise to the left-view superiority. For left three-quarter faces, most of the facial features fall into the left visual field and are, thus, projected to the right hemisphere resulting in better feature extraction (Arnold *et al.*, 2013). In line with this, Kowatari *et al.* (2004) found stronger brain activation in the right middle occipital and inferior frontal gyrus when recognizing familiar faces in the left than in the right view. However, the asymmetric layout of facial features would also lead to another type of leftward process enhancement, that is, pseudoneglect, which suggests a natural tendency to pay more attention to the left hemifield (Bowers & Heilman, 1980). Possibly, this perceptual bias of three-quarter faces is determined already in the "early visual pathway" (Zhou *et al.*, 2016), before facial information is holistically processed and faces are recognized.

While the superiority of left three-quarter faces has been examined with respect to identification (e.g., Arnold *et al.*, 2013), here we investigated whether its dominance precedes recognition by asking participants to passively view photos of faces taken from different perspectives (left, right, and front). By directly comparing blood-oxygen-level-dependent (BOLD) activity patterns among these conditions, we found that while the low-level activation in the striate cortex remained symmetric, a left superiority derived already in the face-related extrastriate cortex, suggesting a view-specific processing enhancement.

Materials and Methods

Participants

Seventeen right-handed (according to the Edinburgh Handedness Inventory; Oldfield, 1971) subjects participated (nine females; mean age 37 years, SD = 6.11 years). The participants had normal or corrected to normal vision, were informed about the study, and signed a consent form. They were financially rewarded for their participation. This study was approved by the ethics committee of the Medical Faculty of the Ludwig Maximilian University of Munich, Germany, and conducted in accordance with the Declaration of Helsinki.

Stimuli

During fMRI scanning, subjects were presented with photographs (see Figure 1, for examples) of faces taken with front-facing, or with a left and right perspective (shifted laterally by 38.5 degrees, assuring a three-quarter face presentation still depicting both eyes). Six male and six female volunteers (age range 25–45 years, mean age 37 years) were photographed with their eyes fixed on the camera, and without glasses or

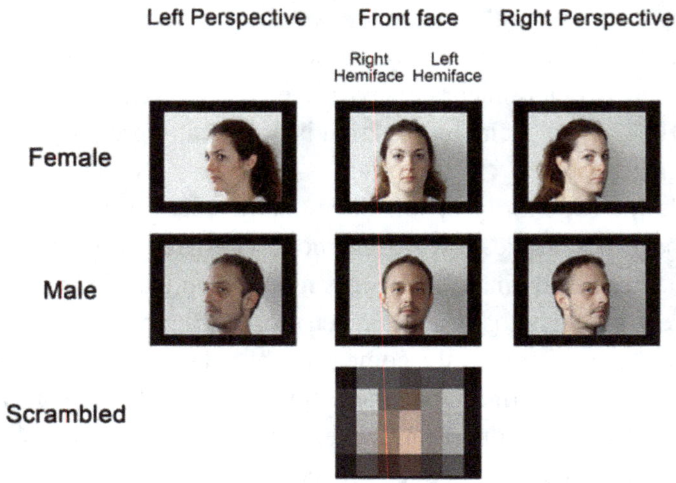

Figure 1. Three-quarter faces in different perspectives. Note that the "right" and "left" here are from the view of the model rather than the observer.

other face-covering devices or accessories, like earrings or piercings. Portraits were taken in front of a white background at a distance of 2 m. Instructions and feedback were used to maintain an emotionally neutral face with a closed mouth. The height of the camera was adjusted according to the height of the volunteer being photographed, with the tip of the nose as an anchor point. All portraits were taken under identical light conditions. Additionally, the luminance was adjusted using picture processing software (Adobe Photoshop CS 3/4). All pictures were cropped and adjusted to ensure comparability (Schäfer, Passbild Generator, 3.6a). Head sizes were resized to 4 degrees. The midpoint of the eyes in the frontal view, as well as the closer eye of three-quarter faces, were centered on the screen, leading to an asymmetric visual layout for the laterally shifted conditions. Scrambled versions of six portraits (three male and three female) were used as control stimuli (Photoshop CS 3/4).

Procedure

The experimental conditions included the genders of the photographed subjects (male or female) and the perspective of the faces (left, right, or front). The study was conducted as a block design with pseudo-randomized order, arranged by a stimulus delivery software (Presentation 15.1, Neurobehavioral Systems, Berkeley, CA, USA). The whole scanning session included 42 sequential blocks (six blocks per condition), each block containing four stimuli randomly selected from the same experimental condition; each stimulus was presented for 3 s, with a 1-s interstimulus interval and a 6-s inter-block interval to allow the hemodynamic responses to return to baseline. During the inter-block interval, a fixation cross was presented in order to maintain the participant's attention.

Participants were asked to passively look at the stimuli while focusing on the eyes of the portrait (for scrambled faces, they focused on potential areas where the eye could be located). For ensuring a high level of attention during the experiment, participants were required to press a button with their right hands on an MRI-compatible device (LUMItouch, Photon Control Inc, Burnaby, BC, Canada) whenever a new image appeared (the mean reaction time: 939.54 ± 259.28 ms). A 3.0 T whole-body system (ACHIEVA, Philips Healthcare, Best, Netherlands) equipped with a standard A head coil was used for fMRI scanning. Foam cushions were placed around the participants' heads to minimize movement. A T2-weighted echo-planar imaging (EPI) sequence was used for functional imaging (TR = 2,500 ms, TE = 30 ms, FA = 80°, number of slices = 38, slice thickness = 3 mm, no inter-slice gap, ascending acquisition, matrix = 64 × 64, in-plane resolution = 3 × 3 mm).

Data processing and analysis

Statistical analysis was done with the SPM8 fMRI toolbox (Wellcome Department of Cognitive Neurology, London, UK; http://www.fil.ion.ucl.ac.uk/spm) and the MarsBar toolbox (Brett *et al.*, 2002; http://marsbar.sourceforge.net/) implemented in MATLAB (MathWorks Inc.); analysis of variance (ANOVA) was performed with SPSS Statistics 19.0 (IBM). The anatomical description and visualization of the results were based on the xjView toolbox (http://www.alivelearn.net/xjview) referring to the Automatic Anatomic Labeling (Tzourio-Mazoyer *et al.*, 2002) atlas from the Wake Forest University Pickatlas (Advanced NeuroScience Imaging Research Laboratory, Winston-Salem, North Carolina, USA).

The first five functional volumes were discarded to avoid variable effects of blood oxygen saturation on the T1 relaxation due to the instabilities of the magnetic field. All functional images were head-motion corrected; realignment and spatial normalization to the EPI template were performed. Spatial smoothing was executed to minimize noise and residual inter-subject differences using a Gaussian kernel of 8 mm full width at half maximum (FWHM). Two analyses were then performed: contrasts between perspectives to reveal potential differences in neural activation patterns, and a two-way repeated-measures ANOVA on hemodynamic responses in face-selective regions of interest (ROIs) to reveal potential local differences in the strength of activation.

On the first level, each trial of photo presentation was specified in the model as 3-s events, and a hemodynamic response function was convolved with the design matrix for parameter estimation. T-contrasts were calculated on the individual level by subtracting the scramble-face control condition. After that, a within-subject ANOVA was calculated using thein dividual contrast images generated on the first level, resulting in a statistical model comprising one factor with three levels (left, right, and front-facing). The three levels of perspective were then contrasted against each other separately ($p < 0.05$, family-wise error [FWE]-corrected). Main effects and interaction of gender and perspective were tested using F-contrasts; voxels with significant interaction of these two factors indicate that results cannot be explained by a single factor, and thus were excluded from the interpretation of *t*-contrasts between perspective conditions.

For the second part of the analysis, the goal was to test whether different perspectives of faces induce different BOLD signal strength in face-related areas. Face-related ROIs were defined on the group level by comparing face with scrambled-face conditions, located at the left middle occipital gyrus (LMO), the right fusiform face area (RFFA), and the right middle occipital gyrus (RMO). Average signal changes caused by each condition were then estimated and compared in a two-way repeated-measures ANOVA (gender of photos * perspective of faces) for each ROI.

Results

For the first part of the analysis, contrasts between the left (L) and right perspective (R) revealed a higher activation for viewing the left three-quarter faces than the right (L > R) in the left primary visual cortex (V1, including the calcarine sulcus and the lingual gyrus) and extrastriate areas (the middle and the superior occipital gyrus). Similarly, higher activations in the right V1 were observed when comparing right perspective faces to the left perspective (R > L, see Figure 2). The left perspective resulted in higher activation in the left V1 compared to the front view (L > F), while the right perspective caused higher activation in the right V1 (R > F) and lower activation in the left V1 and extrastriate visual areas (F > R, see Table 1). A gender effect was tested with female compared to male faces (F > M) and male to female (M > F) faces; no significant difference was

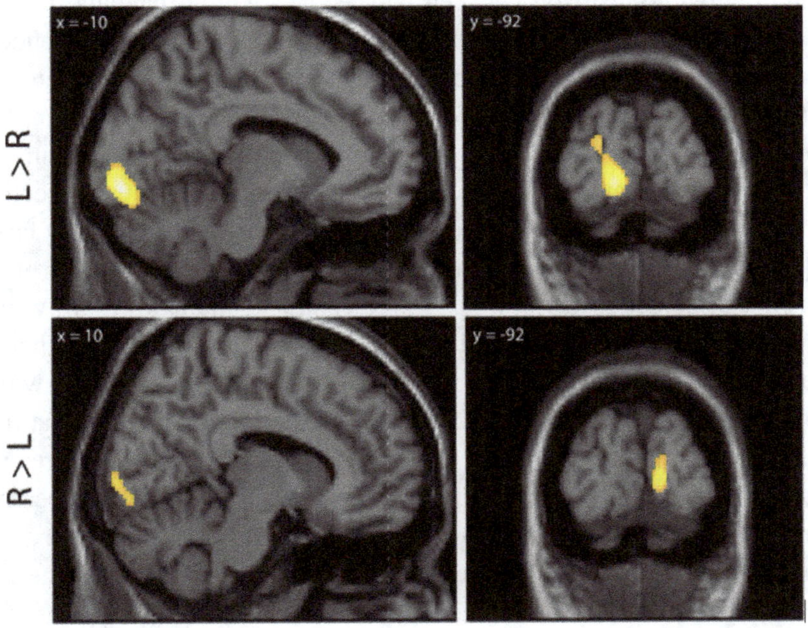

Figure 2. The results of group-level contrasts between the left view, the right, and the reverse. In both contrasts, the primary visual cortex showed similar higher activation in mirror positions. The contrast of L > R resulted in an additional activation in the extrastriate cortex. L, left three-quarter view; R, right three-quarter view.

Table 1. Significant brain regions in group-level *t*-contrasts between different perspective conditions.

Contrast	Region of clusters	Voxels	Hemisphere	Coordinates *x*	*y*	*z*	Max. *T*-statics
L > R	Lingual, Calcarine, Mid. Occipital, Sup. Occipital	510	L	−12	−88	−8	10.05
R > L	Calcarine, Lingual, Cuneus	183	R	14	−92	2	7.44
L > F	Lingual, Calcarine	94	L	−10	−88	−10	6.49
F > L	(n. s.)						
R > F	Calcarine, Lingual	34	R	12	−90	2	5.67
F > R	Mid. Occipital, Calcarine, Sup. Occipital, Lingual	253	L	−12	−92	−6	7.79

found. Paired comparison between perspectives within male and female face conditions showed similar results even when the two face-genders were collapsed, that is, neither a face-gender main effect nor an interaction between gender and perspective could be found. In conclusion, the lateralized view of faces caused a mirrored processing pattern in V1 between the left and the right view, that is, the left/right perspective caused higher activations in the left primary visual areas. Front-view faces resulted in a medial activation in both hemispheres.

As can be deduced from the simple contrasts, the extrastriate visual cortex, including the middle and superior occipital gyrus in the left hemisphere, showed a preference for the left perspective over the right, while the opposite effect was not found for R > L (see Figure 3). Note that this difference cannot be explained by low-level visual feature differences between hemifaces, as the stimuli used for the left and right perspective conditions were roughly mirror-symmetric. This difference in the extrastriate areas thus can be presumably attributed to the specialty of the left view based on a relatively higher-level face-related processing (Rossion, 2013; Young & Burton, 2018).

To further explore the effect in a detailed manner, three ROIs were chosen from the group-level comparison between face and scrambled face condition ($p < 0.05$, FWE-corrected, see Table 2 and Figure 4). The

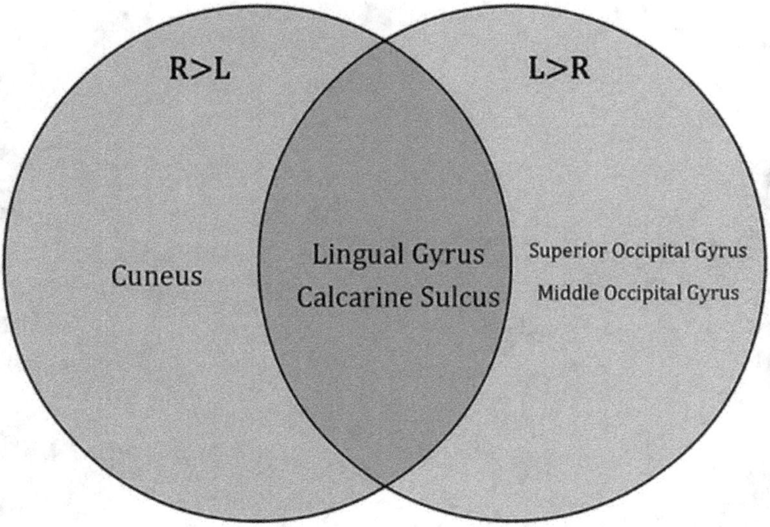

Figure 3. A Venn diagram of the results of the t–contrasts between the left three–quarter face and the right. The common activation in the V1, including the lingual gyrus and the calcarine sulcus, suggests a similar visual searching pattern of the left and right view. The process is no longer symmetric at the higher stage, indicated by an additional activation in the superior and middle occipital gyrus. L, left three–quarter view; R, right three–quarter view.

Table 2. Regions of interest defined by comparing face versus scrambled face condition.

		Coordinates			
Region of interest	**Voxels**	*x*	*y*	*z*	***T*-statics**
Left middle occipital gyrus (LMO)	36	−32	−94	−2	5.43
Right fusiform face area (RFFA)	18	40	−52	−18	5.17
Right middle occipital gyrus (RMO)	17	32	−86	−14	5.28

perspective and gender effect varied between the ROIs. RMO showed a significant perspective main effect ($F[2, 30] = 3.304$, $p = 0.050$, $\eta^2 = 0.181$) though with an interaction between perspective and gender of the photos that reached marginal significance ($F[2, 30] = 3.174$, $p = 0.056$, $\eta^2 = 0.175$). Left three-quarter faces induced larger signal changes compared

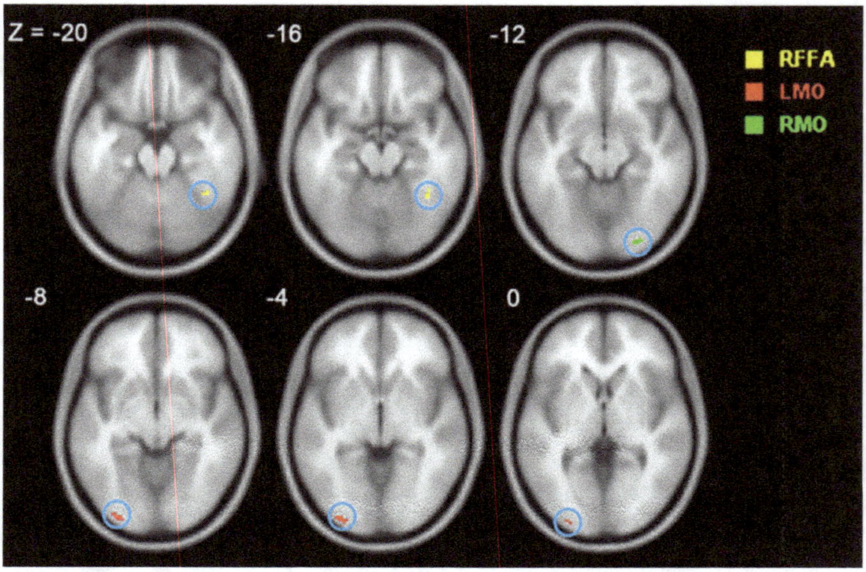

Figure 4. The locations of regions of interest are emphasized by blue circles. The three regions are colored in yellow (right fusiform face area [RFFA]), red (left middle occipital gyrus [LMO]), and green (right middle occipital gyrus [RMO]), respectively.

to the right, and this effect was particularly stronger for male faces (R < L, $p = 0.019 < 0.05$; R < F, $p = 0.005 < 0.01$, see Figure 5(a)). In the RFFA, a significant main effect for gender ($F[1, 15] = 5.851$, $p = .029 < .05$, $\eta^2 = .281$, see Figure 5(b)) was found as female faces generally induced stronger RFFA responses than male faces. A significant interaction between gender and perspective was found for LMO ($F[2, 30] = 7.083$, $p = 0.003 < 0.05$, $\eta^2 = 0.321$, see Figure 5(c)). For male faces, the pairwise comparison showed that the front-view male faces caused higher hemodynamic responses compared to the right ($p < 0.05$). For female faces, the left perspective activated significantly less than the right perspective ($p < 0.05$). One point worth noting is that in all three ROIs, the right three-quarter faces of male stimuli consistently induced smaller signal changes (see Figure 5), indicating a special kind of facial stimuli.

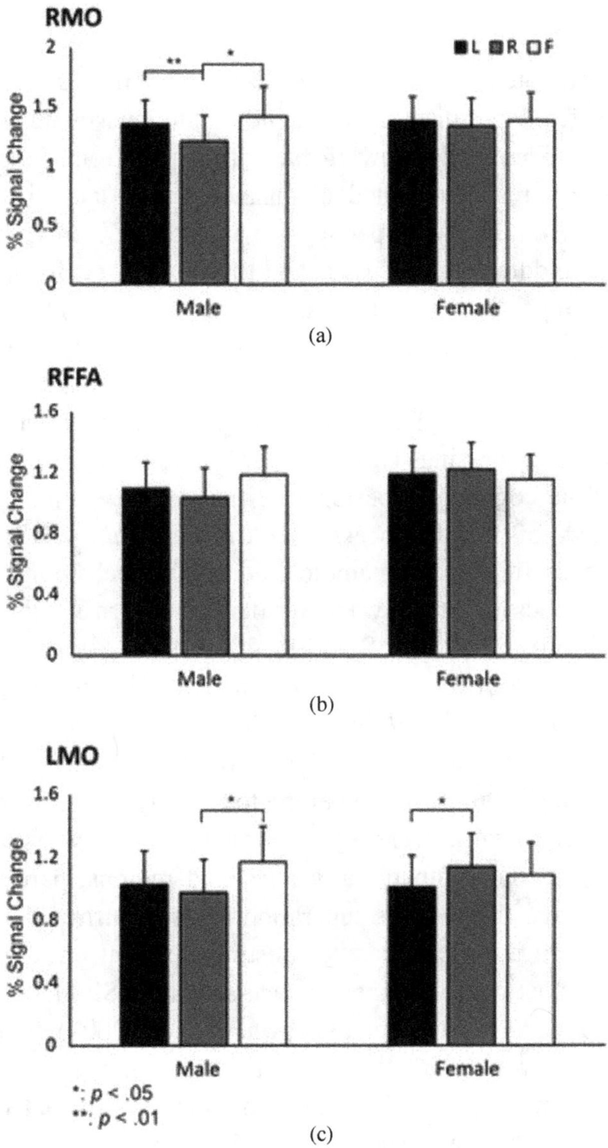

Figure 5. Average signal changes in the three regions of interest — (a) the right middle occipital gyrus (RMO), (b) the right fusiform face area (RFFA), and (c) the left middle occipital gyrus (LMO) — when viewing male and female faces from different views. Signal changes show a consistent decrease when viewing the male right three–quarter faces, compared with other perspectives.

Discussion

The neural correlates for a preference of the left three-quarter face have remained elusive. We utilized MRI technology to investigate and to compare neural patterns while viewing faces from the front, left, and right perspectives. The results revealed an enhancement of lateralized processing for the left three-quarter perspective in the left *extrastriate* cortex, including the middle occipital gyrus and the superior occipital gyrus. By contrast, we did not find similar enhanced lateralization in the *striate* cortex; the activity pattern in these areas was symmetric. While the extrastriate cortex is believed to carry out high-level, content-specific processes, and particularly the occipital gyri are reported to be related to face-selective feature processing (Duvernoy, 2005), our study provides further evidence for the cognitive superiority of the left viewpoint observed in several aspects of facial processing (Arnold *et al.*, 2013; Kowatari *et al.*, 2004; Siéroff, 2001; Yamamoto *et al.*, 2005); behavioral studies on empirical aesthetics report equivalent results (Blackburn & Schirillo, 2012; Grüsser *et al.*, 1988; Powell & Schirillo, 2009; Stoerig *et al.*, 1983).

The pairwise comparisons of L > R and R > L suggest a common activity roughly *symmetrical* in the primary visual cortex and additional *asymmetric* activation of middle and superior occipital areas for the L > R (Figure 3), all of which belong to an early stage in the anatomical hierarchy of visual processing. Importantly, the view-induced effect does not include more functionally advanced regions, particularly the fusiform face area (FFA). This separation is also confirmed by the ROI analysis, as there is no evidence suggesting a neural preference in the areas that are functionally defined as face-selective. Similar results have been obtained in an investigation by Kowatari *et al.* (2004). In their study, subjects were instructed to tell whether they recognized faces presented from different views; the BOLD signal in their study revealed a similar pattern as ours with an absence of the FFA, but activation in the middle occipital gyrus. Kowatari *et al.* (2004) concluded that the preference for the left view emerges during facial identification but not in the preceding facial processes in FFA. Later, it was found that the FFA might not only encode the individual face on a holistic level but also ignite the entire facial processing network (Rossion, 2008), including the inferior occipital gyrus where a fine-grained facial feature analysis is performed before facial identification (Arnold *et al.*, 2013; Kowatari *et al.*, 2004; Yamamoto *et al.*, 2005). Based on the anatomical adjacency and the result from our functional localization of face-selective regions, parts of other occipital gyri (middle/superior gyrus) might as well participate in these analytical feature processes. Considering our task did not require any explicit face identification yet still resulted in functional asymmetry in the occipital gyri, the preference of the left view may already exist during those fine-grained analyses independent of holistic facial identification and retrieving of personal semantic information revealed by Kowatari *et al.* (2004).

This result can be reconciled by the left spatial attentional bias reported in a variety of studies, including patients with hemineglect (Heilman *et al.*, 2008), line bisection (Bowers & Heilman, 1980; Werth & Pöppel, 1988), and visual search (Nicholls *et al.*, 2017). This general tendency to over-attend to the left space is proposed to result from a suspected hemispheric asymmetry in neural processing derived from observations when eye movement remains focused (Nicholls *et al.*, 2017). While people tend to fixate the eye that is closer to the observer when performing free-viewing on three-quarter faces (Bindemann *et al.*, 2009), key facial features of the left view (mouths, noses, etc.) would fall into the left hemifield and thus be better extracted and encoded with the extra attentional resource in the occipital gyri. By contrast, when it comes to a holistic facial processing/identification in the FFA, there is no such superiority for either the right or left view. (Nevertheless, our result does not exclude the fact that faces are represented in a viewpoint-specific manner in these regions, as shown by the neural adaptation effect; Harvey & Burgund, 2012; Pourtois *et al.*, 2009). Such dissociation between holistic and analytic processing is also found in other cognitive processes, such as time perception (Bao *et al.*, 2013).

One of the prevailing explanations of left three-quarter face preference in empirical arts is facial asymmetry (Arnold *et al.*, 2013). The left hemiface is believed to be involved more in emotional expressions with finer details, as it presumably has more anatomical projection to the emotional-dominant right hemisphere (Borod *et al.*, 1997). The left view mostly taken up by the left hemiface may consequently convey more information. Though faces tend to be less asymmetric with a neutral expression (Bennett *et al.*, 1987), and no emotional-related brain region was found activated when comparing faces with scrambled stimuli, the left hemiface may contribute to the additional activation in the occipital gyri while viewing the left view. Since it is nearly impossible to study three-quarter faces without emotional components, a better way to solve this dispute might be found in future experiments, for example using a study design based on three-quarter faces with strong emotional expressions. Respective trials investigating neural activation patterns at different emotional intensity levels may even fill up the gap between viewpoint and emotion studies, as most of the latter used either frontal or chimera faces (Borod *et al.*, 1997; Dimberg & Petterson, 2000; Heller & Levy, 1981).

Interestingly, in the ROIs that cover bilateral middle occipital areas, the right perspective of male faces was found uniformly with a lower activation compared to the other two perspectives and females. The effect was very robust, even when ROIs were defined in other ways (with sphere ROIs centered at local maximum voxels or defined on a subject level). Studies on portrait painting found more females than males being painted with their left cheek exposed (Powell & Schirillo, 2009), but studies on selfie photos and photo appreciation found gender does not interact with the left perspective preference (Blackburn & Schirillo, 2012; Bruno *et al.*, 2017; Lindell, 2017). The divergence may be explained by the different task requirements in this study. Nevertheless, our result suggests that the right perspective of male faces is a special type of stimuli related to lower activation within all face-related areas beyond the primary visual cortex.

Taken together, the perceptual mechanisms underlying such preference might be interpreted in a frame of hierarchical processing of visual information in the early visual pathway (Grill-Spector *et al.*, 2017; Zhou *et al.*, 2016). First, the perspective difference was found in the primary visual cortex, but in a mirror-symmetric manner, which suggests that exploration strategies for the left and right view of faces are similar. Asymmetric processing of different perspectives was found only in the extrastriate cortex, suggesting the processing superiority of the left view on a feature-based analytical level. The right perspective of male faces elicited a lower activation compared to other perspectives and face gender, suggesting that the representation of male faces is different on a neural level and that furthermore gender differences are already recognized in the early visual pathway. In general, the research presented here revealed cue exposure with human faces as a potential experimental tool to identify processing stages in perceptual tasks underlying aesthetic appreciations.

References

Arnold, G., Bécue, J. C., & Siéroff, E. (2013). Left-view sequences of faces are superior to right-view ones for view generalization. *Visual Cognition*, 21(6), 752–765. https://doi.org/10.1080/13506285.2013.823137.

Bao, Y., Szymaszek, A., Wang, X., Oron, A., Pöppel, E., & Szelag, E. (2013). Temporal order perception of auditory stimuli is selectively modified by tonal and non-tonal language environments. *Cognition*, 129(3), 579–585. https://doi.org/10.1016/j.cognition.2013.08.019.

Bao, Y., von Stosch, A., Park, M., & Pöppel, E. (2017a). Complementarity as generative principle: A thought pattern for aesthetic appreciations and cognitive appraisals in general. *Frontiers in Psychology*, 8, 727. https://doi.org/10.3389/fpsyg.2017.00727.

Bao, Y., Yang, T., Lin, X., Fang, Y., Wang, Y., Pöppel, E., & Lei, Q. (2016). Aesthetic preferences for eastern and Western traditional visual art: Identity matters. *Frontiers in Psychology*, 7(OCT), 1596. https://doi.org/10.3389/FPSYG.2016.01596/BIBTEX.

Bao, Y., Yang, T., Zhang, J. J., Zhang, J. J., Lin, X., Paolini, M., Pöppel, E., & Silveira, S. (2017b). The "third abstraction" of the Chinese artist LaoZhu: Neural and behavioral indicators of aesthetic appreciation. *PsyCh Journal*, 6(2), 110–119. https://doi.org/10.1002/pchj.167.

Barton, J. J. S., Press, D. Z., Keenan, J. P., & O'Connor, M. (2002). Lesions of the fusiform face area impair perception of facial configuration in prosopagnosia. *Neurology*, 58(1), 71–78. https://doi.org/10.1212/WNL.58.1.71.

Bennett, H. L., Delmonico, R. L., & Bond, C. F. (1987). Expressive and perceptual asymmetries of the resting face. *Neuropsychologia*, 25(4), 681–687. https://doi.org/10.1016/0028-3932(87)90059-5.

Bindemann, M., Scheepers, C., & Burton, A. M. (2009). Viewpoint and center of gravity affect eye movements to human faces. *Journal of Vision*, 9(2), 7–7, 716. https://doi.org/10.1167/9.2.7.

Blackburn, K. & Schirillo, J. (2012). Emotive hemispheric differences measured in real-life portraits using pupil diameter and subjective aesthetic preferences. *Experimental Brain Research*, 219(4), 447–455. https://doi.org/10.1007/s00221-012-3091-y.

Borod, J. C., Haywood, C. S., & Koff, E. (1997). Neuropsychological aspects of facial asymmetry during emotional expression: A review of the normal adult literature. *Neuropsychology Review*, 7(1), 41–60. https://doi.org/10.1007/BF02876972.

Bowers, D. & Heilman, K. M. (1980). Pseudoneglect: Effects of hemispace on a tactile line bisection task. *Neuropsychologia*, 18(4–5), 491–498. https://doi.org/10.1016/0028-3932(80)90151-7.

Brett, M., Jean-Luc, A., Romain, V., & Jean-Baptiste, P. (2002). Region of interest analysis using an SPM toolbox. The *8th International Conference on Functional Mapping of the Human Brain*, Sendai, Japan, 16(2), 497.

Bruno, N., Bode, C., & Bertamini, M. (2017). Composition in portraits: Selfies and wefies reveal similar biases in untrained modern youths and ancient masters*. *Laterality*, 22(3), 279–293. https://doi.org/10.1080/1357650X.2016.1185108.

Chen, S., Cai, M., & Bao, Y. (2020). The two- to three-second time window of shot durations in movies. *PsyCh Journal*, 9(4), 516–518. https://doi.org/10.1002/PCHJ.384.

Dimberg, U. & Petterson, M. (2000). Facial reactions to happy and angry facial expressions: Evidence for right hemisphere dominance. *Psychophysiology*, 37(5), 693–696. https://doi.org/10.1111/1469-8986.3750693.

Duvernoy, H. (2005). Brain anatomy. In R. I. Kuzniecky, & G. D. Jackson (eds.), *Magnetic Resonance in Epilepsy* (pp.29–97). Elsevier Inc. https://doi.org/10.1016/B978-012431152-7/50007-0.

Ewbank, M. P., Smith, W. A. P., Hancock, E. R., & Andrews, T. J. (2008). The M170 reflects a viewpoint-dependent representation for both familiar and unfamiliar faces. *Cerebral Cortex*, 18(2), 364–370. https://doi.org/10.1093/cercor/bhm060.

Fang, F., Murray, S. O., & He, S. (2007). Duration-dependent fMRI adaptation and distributed viewer-centered face representation in human visual cortex. *Cerebral Cortex*, 17(6), 1402–1411. https://doi.org/10.1093/cercor/bhl053.

Grill-Spector, K., Weiner, K. S., Kay, K., & Gomez, J. (2017). The functional neuroanatomy of human face perception. *Annual review of vision science*, 3, 167–196. https://doi.org/10.1146/annurev-vision-102016-061214.

Grüsser O. J., Selke, T., & Zynda, B. (1988). Cerebral Lateralization and Some Implications for Art, Aesthetic Perception, and Artistic Creativity. In I. Rentschler, B. Herzberger, & D. Epstein (Eds.), *Beauty and the brain*. Birkhäuser: Basel. https://doi.org/10.1007/978-3-0348-6350-6_12.

Harvey, D. Y. & Burgund, E. D. (2012). Neural adaptation across viewpoint and exemplar in fusiform cortex. *Brain and Cognition*, 80(1), 33–44. https://doi.org/10.1016/J.BANDC.2012.04.009.

Heilman, K. M., Valenstein, E., & Watson, R. T. (2008). Neglect and related disorders. *Seminars in Neurology*, 4(02), 209–219. https://doi.org/10.1055/S-2008-1041551.

Heller, W. & Levy, J. (1981). Perception and expression of emotion in right-handers and left-handers. *Neuropsychologia*, 19(2), 263–272. https://doi.org/10.1016/0028-3932(81)90110-X.

Kowatari, Y., Yamamoto, M., Takahashi, T., Kansaku, K., Kitazawa, S., Ueno, S., & Yamane, S. (2004). Dominance of the left oblique view in activating the cortical network for face recognition. *Neuroscience Research*, 50(4), 475–480. https://doi.org/10.1016/j.neures.2004.08.014.

Krouse, F. L. (1981). Effects of pose, pose change, and delay on face recognition performance. *Journal of Applied Psychology*, 66(5), 651–654. https://doi.org/10.1037/0021-9010.66.5.651.

Laeng, B. & Rouw, R. (2001). Canonical views of faces and the cerebral hemispheres. Laterality: Asymmetries of Body, *Brain and Cognition*, 6(3), 193–224. https://doi.org/10.1080/713754410.

Leopold, D. A. & Rhodes, G. (2010). A comparative view of face perception. *Journal of Comparative Psychology*, 124(3), 233–251. https://doi.org/10.1037/a0019460.

Lindell, A. K. (2017). Consistently showing your best side? Intra-individual consistency in #selfie pose orientation. *Frontiers in Psychology*, 8, 246. https://doi.org/10.3389/fpsyg.2017.00246.

Lutz, A., Nassehi, A., Bao, Y., Pöppel, E., Sztrókay, A., Reiser, M., … Gutyrchik, E. (2013). NeuroCognitive Processing of body representations in artistic and photographic images. *NeuroImage*, 66, 288–292. https://doi.org/10.1016/J.*NEUROIMAGE*.2012.10.067.

McManus, I. C. & Humphrey, N. K. (1973). Turning the left cheek. *Nature*, 243(5405), 271–272. https://doi.org/10.1038/243271a0.

Nicholls, M. E. R., Hobson, A., Petty, J., Churches, O., & Thomas, N. A. (2017). The effect of cerebral asymmetries and eye scanning on pseudoneglect for a visual search task. *Brain and Cognition*, 111, 134–143. https://doi.org/10.1016/J.BANDC.2016.11.006.

Oldfield, R. C. (1971). The assessment and analysis of handedness: The Edinburgh inventory. *Neuropsychologia*, 9(1), 97–113. https://doi.org/10.1016/0028-3932(71)90067-4.

Park, M., Gutyrchik, E., Welker, L., Carl, P., Pöppel, E., Zaytseva, Y., … Bao, Y. (2015). Sadness is unique: Neural processing of emotions in speech prosody in musicians and non-musicians. *Frontiers in Human Neuroscience*, 8(JAN), 1049. https://doi.org/10.3389/FNHUM.2014.01049/BIBTEX.

Pourtois, G., Schwartz, S., Spiridon, M., Martuzzi, R., & Vuilleumier, P. (2009). Object representations for multiple visual categories overlap in lateral occipital and medial fusiform cortex. *Cerebral Cortex*, 19(8), 1806–1819. https://doi.org/10.1093/CERCOR/BHN210.

Powell, W. R. & Schirillo, J. A. (2009). Asymmetrical facial expressions in portraits and hemispheric laterality: A literature review. *Laterality*, 14, 545–572. https://doi.org/10.1080/13576500802680336.

Rentschler, I., Herzberger, B., & Epstein, D. (1988). *Beauty and the Brain: Biological Aspects of Aesthetics*. Birkhäuser Verlag.

Rossion, B. (2008). Constraining the cortical face network by neuroimaging studies of acquired prosopagnosia. *NeuroImage*, 40(2), 423–426. https://doi.org/10.1016/J.*NEUROIMAGE*.2007.10.047.

Rossion, B. (2013). The composite face illusion: A whole window into our understanding of holistic face perception. *Visual Cognition*, 21(2), 139–253. https://doi.org/10.1080/13506285.2013.772929.

Siéroff, E. (2001). Feature processing and superiority of three-quarter views in face recognition. *Brain and Cognition*, 46(1–2), 272–276.

Stoerig, P., Sütterlin, C., & Pöppel, E. (1983). Rechts und links in Bildwerken: Ein neuropsychologischer Beitrag zum Kunstverständnis.(left and right in artwork. A neuropsychological contribution to the understanding of art). *Umschau in Wissenschaft und Technik*, 83, 427–428.

Sütterlin, C. & Yu, X. (2021). Aristotle's dream: Evolutionary and neural aspects of aesthetic communication in the arts. *PsyCh Journal*, 10(2), 224–243. https://doi.org/10.1002/PCHJ.416.

Troje, N. F. & Bülthoff, H. H. (1996). Face recognition under varying poses: The role of texture and shape. *Vision Research*, 36(12), 1761–1771. https://doi.org/10.1016/0042-6989(95)00230-8.

Tzourio-Mazoyer, N., Landeau, B., Papathanassiou, D., Crivello, F., Etard, O., Delcroix, N., Mazoyer, B., & Joliot, M. (2002). Automated anatomical labeling of activations in SPM using a macroscopic anatomical parcellation of the MNI MRI single-subject brain. *NeuroImage*, 15(1), 273–289. https://doi.org/10.1006/nimg.2001.0978.

Wang, X. & Bao, Y. (2020). Rembrandt portraits: Implicitly detecting the original perspective. *PsyCh Journal*, 9(4), 513–515. https://doi.org/10.1002/PCHJ.376.

Wang, X., Simmank, F., Zhang, D., Zeng, Y., Silveira, S., Sander, T., Bao, Y., & Paolini, M. (2020). Aesthetics as a common denominator for moral and commercial judgments. *PsyCh Journal.*, 10, 187–189. https://doi.org/10.1002/PCHJ.407.

Weibert, K. & Andrews, T. J. (2015). Activity in the right fusiform face area predicts the behavioural advantage for the perception of familiar faces. *Neuropsychologia*, 75, 588–596. https://doi.org/10.1016/J.*NEUROPSYCHOLOGIA*.2015.07.015.

Werth, R. & Pöppel, E. (1988). Compression and lateral shift of mental coordinate systems in a line bisection task. *Neuropsychologia*, 26(5), 741–745. https://doi.org/10.1016/0028-3932(88)90010-3.

Yamamoto, M., Kowatari, Y., Ueno, S., Yamane, S., & Kitazawa, S. (2005). Accelerated recognition of left oblique views of faces. *Experimental Brain Research*, 161(1), 27–33. https://doi.org/10.1007/s00221-004-2041-8.

Yang, T., Silveira, S., Formuli, A., Paolini, M., Pöppel, E., Sander, T., & Bao, Y. (2019). Aesthetic experiences across cultures: Neural correlates when viewing traditional eastern or western landscape paintings. *Frontiers in Psychology*, 10(MAR), 798. https://doi.org/10.3389/FPSYG.2019.00798/BIBTEX.

Young, A. W. & Burton, A. M. (2018). Are we face experts? *Trends in Cognitive Sciences*, 22(2), 100–110. https://doi.org/10.1016/j.tics.2017.11.007.

Yu, X. & Bao, Y. (2020). The three second time window in poems and language processing in general: Complementarity of discrete timing and temporal continuity. *PsyCh Journal*, 9(4), 429–443. https://doi.org/10.1002/PCHJ.390.

Zaytseva, Y., Gutyrchik, E., Bao, Y., Pöppel, E., Han, S., Northoff, G., Welker, L., Meindl, T., & Blautzik, J. (2014). Self processing in the brain: A paradigmatic fMRI case study with a professional singer. *Brain and Cognition*, 87(1), 104–108. https://doi.org/10.1016/J.BANDC.2014.03.012.

Zeki, S. (1999). Art and the brain. *Journal of Consciousness Studies*, 6(6–7), 76–96.

Zhang, J., Liu, J., & Xu, Y. (2015). Neural decoding reveals impaired face Configural processing in the right fusiform face area of individuals with developmental prosopagnosia. *Journal of Neuroscience*, 35(4), 1539–1548. https://doi.org/10.1523/JNEUROSCI.2646-14.2015.

Zhao, C., Zhang, D., & Bao, Y. (2018). A time window of 3 s in the aesthetic appreciation of poems. *PsyCh Journal*, 7(1), 51–52. https://doi.org/10.1002/PCHJ.194.

Zhou, B., Pöppel, E., Wang, L., Yang, T., Zaytseva, Y., & Bao, Y. (2016). Seeing without knowing: Operational principles along the early visual pathway. *PsyCh Journal*, 5(3), 145–160. https://doi.org/10.1002/pchj.141.

https://doi.org/10.1142/9789811298639_0022

Chapter 22

Beauty in Life: An Eye-Tracking Study on Young Adults' Aesthetic Evaluation and Vitality Judgment of Pictorial Representations of Sleeping and Dead Subjects*

Cinzia Di Dio[†,‡], Davide Massaro[†,‡], Federica A. Savazzi[§], Vittorio Gallese[¶,||], Tiziana Garau[†], Gabriella Gilli[†,‡] and Antonella Marchetti[†,‡]

[†]*Department of Psychology, Università Cattolica del Sacro Cuore, Milan, Italy*

[‡]*Research Unit on Theory of Mind, Università Cattolica del Sacro Cuore, Milan, Italy*

[§]*IRCCS Fondazione don Carlo Gnocchi ONLUS, Milan, Italy*

[¶]*Department of Neuroscience, University of Parma, Parma, Italy*

[||]*Department of Art History and Archaeology, Columbia University, New York, USA*

---

*This chapter was originally published in the special collection on neuroaesthetics in *PsyCh Journal*. Di Dio, C., Massaro, D., Savazzi, F.A., Gallese, V., Garau, T., Gilli, G., & Marchetti, A. (2020). Beauty in life: An eye-tracking study on young adults' aesthetic evaluation and vitality judgment of pictorial representations of sleeping and dead subjects. *PsyCh Journal*, 9(4), 458–471. https://doi.org/10.1002/pchj.285.

Abstract

Preferring life to death is deeply rooted in our biology. With this study, we explored two questions: (1) Can this inclination be transposed to aesthetics so that a living being is valued as more beautiful than a non-living being? and (2) Are there any differences in the visual exploration of portrayals of a living compared to a dead human? In particular, are there any specific facial features representing the vitality status of a living or dead subject? By answering both questions, young adults' eye gazing was analyzed while they observed, aesthetically judged, and judged the vitality status of faces extracted from paintings representing a sleeping or dead subject. The aesthetic preference for the stimuli as a function of vitality (living and dead) was assessed both during the eye-tracking study and during a follow-up priming behavioral experiment. The analysis of the responses given during the aesthetic judgment (AJ) task in the eye-tracking study revealed preference for the sleeping compared to the dead subjects, supporting proclivity to attribute greater aesthetic value to living beings. This evidence was substantially confirmed by the follow-up priming behavioral study, which further showed a significant effect of explicit vitality labeling on the aesthetic evaluation of the portrayals of sleeping subjects. As far as the visual exploration of the stimuli is concerned, the main eye-tracking results revealed great attention to the eye region of both sleeping and dead subjects, which was particularly enhanced for the sleeping compared to the dead subjects. For the latter stimuli, focused attention was also found in the mouth region. These results are discussed in light of different theoretical proposals, including the "embodied" theory of aesthetic perception based on the existence of mirror systems.

***Keywords*:** aesthetics; embodiment; eye-tracking; uncanny; vitality features

When interacting, people's attention is strongly attracted by each other's face to immediately detect visible elements that are vital to initiate and maintain proper communication and behavior. In fact, faces convey information about a person's identity, health state, emotional state, and attitude, which represent factors that play a fundamental role when socially relating to others (e.g., Adams & Kleck, 2003; Aharon *et al.*, 2001; O'Doherty *et al.*, 2003).

Attraction to faces occurs very early in life and, even within minutes of birth, infants exhibit great sensitivity to face-like stimuli (Hoehl & Peykarjou, 2012; Wilkinson, Paikan, Gredebäck, Rea, & Metta, 2014). As a matter of fact, a central non-verbal aspect of the mother–baby relationship is represented by eye contact and the mother's facial expression (Condon & Sander, 1974; Flom & Bahrick, 2007; Nadel & Butterworth, 2011; Rochat, 2009). Recent robotic and computational models (Wilkinson *et al.*, 2014) propose a system-level view grounded on an "innate, embodied" knowledge of faces, which would explain newborns' preferences for faces. In particular, there is evidence for a sensitive period in infancy for building a holistic face representation and for perceptual narrowing for faces of one's own species and race (for a review, see Hoehl & Peykarjou, 2012). Within Western culture, fixations during face identification typically follow a triangular sequence, sampling particularly the eyes, partially the mouth, and very little the nose (e.g., Althoff & Cohen, 1999; Groner *et al.*, 1984; Henderson, Williams, & Falk, 2005; Yarbus, 1965; see also, Caldara *et al.*, 2006; Davies, Ellis, & Shepherd, 1977; Fraser, Craig, & Parker, 1990; Gosselin & Schyns, 2001; Haig, 1986; Schyns, Bonnar, & Gosselin, 2002). Major attention to the eye region is of socio-cognitive importance. In his review, Emery (2000) discusses the evolution and neurobiology of gaze processing, providing ample evidence from studies with non-human primates and humans supporting the view that gaze is a crucial component of social interaction, the eye region being the first inspected. Additionally, some psychopathological traits associated with impaired social abilities have also been related to attention deficits in the eye region. Autism, just to cite a most commonly studied disorder associated with impaired social abilities, has been suggested to involve attention impairments to the eyes (e.g., Baron-Cohen, Campbell, Karmiloff-Smith, Grant, & Walker, 1995; Baron-Cohen, Wheelwright, & Jolliffe, 1997; Leekam, Baron-Cohen, Perrett, Milders, & Brown, 1997). Furthermore, deficits in fear recognition in children with psychopathic traits are in part owing to visual neglect of the eye region of other people's faces (Dadds *et al.*, 2006; see also, Adolphs *et al.*, 2005; Richell *et al.*, 2003).

Besides the eyes playing a pivotal role in all primate facial expressions (Andrew, 1963; Wada, 1961; as quoted in Emery, 2000), another important focus of visual attention is the mouth (e.g., McKelvie, 1976). Using implanted eye coils to record the eye movements of head-restrained monkeys when viewing neutral and expressive faces, Nahm, Perret, Amaral, and Albright (1997) showed that the subjects displayed a predominant interest in the eyes and the mouth. Gosselin & Schyns (2001) used the bubbles method to investigate different attentional strategies in humans. In the bubble paradigm, participants are shown a sequence of faces occluded by a grey field — except for portions seen through Gaussian-shaped regions (the bubbles) — and they have to categorize them by some specific properties (e.g., age and emotion). The authors found that attention varies as a function of the categorization task and of the spatial frequency of the observed images. When judging emotion, for example, observers pay attention primarily to the mouth at the high spatial frequencies but not at the lower spatial frequencies. Additionally, comparing cross-cultural differences in face exploration strategies, when asked to judge emotions, Westerners showed a tendency to focus on the mouth (Jack, Blais, Scheepers, Schyns, & Caldara, 2009).

In social cognition, it is now commonly accepted that a basic mechanism that allows us to relate to others, within cultures and across species, is the mirror neuron mechanism (Gallese, 2003, 2005; Grosbras, Beaton, & Eickoff, 2012; Molenberghs, Cunnington, & Mattingley, 2012; Rizzolatti, Fadiga, Gallese, & Fogassi, 1996; Rizzolatti & Sinigaglia, 2010). This mechanism lays at the basis of an embodied simulation process, according to which the same neural structures are activated by the actual execution of actions or experience of emotions and observation of others acting or expressing the same emotions and sensations (for review, see Gallese & Sinigaglia, 2011; see also, Di Cesare *et al.*, 2013, 2015, 2016). Through this mechanism, the visible — motor — aspects of behavior (actions and facial expressions) are rendered into corresponding emotions in the beholder, allowing individuals to understand and relate to one another. This has been shown in a number of neuroimaging studies where observation of faces and the embodied simulation it induces turn into an empathic encoding of others' basic motor intentions and emotions (e.g., Carr, Iacoboni, Dubeau, Mazziotta, & Lenzi, 2003; Enticott, Johnston, Herring, Hoy, & Fitzgerald, 2008; Wicker *et al.*, 2003; see also, Gallese & Di Dio, 2012).

Rooted or embodied attraction to faces is visible also when attending to human subjects portrayed in artworks. In artistic representations of human subjects, the face is the first part of the body to be scanned (Abbas & Duchaine, 2008; Graham, Friedenberg, McCandless, & Rockmore, 2010; Wallraven *et al.*, 2009). The eye-tracking data in Massaro *et al.* (2012; see also, Savazzi *et al.*, 2014) showed that the representation of a human figure within a painting primarily attracts the viewer's attention to the limbs when the portrayed subject is represented as performing an action (dynamic stimuli) and the face area when the subject is represented in a still posture (static stimuli). For the static stimuli, in fact, the face was shown to be the most salient region of inspection. Furthermore, focused attention on the limbs (action) of the human figures judged as dynamic as opposed to the face region (emotions) of those judged as static paintings correlated with increasing aesthetic rating (see also, Di Dio *et al.*, 2016, 2018). These results were interpreted in light of the embodied theory of aesthetic perception based on the above-described mirror mechanism and that, as suggested by Freedberg and Gallese (2007), might empirically ground the fundamental role of empathy in the aesthetic experience (see also, Ardizzi *et al.*, 2018; Chatterjee & Vartanian, 2014; Gernot, Pelowski, & Leder, 2017). The sociocognitive component underlying our inclination to grasp others' emotions and behavior thus appears to play a crucial role in the exploration pattern characterizing visual attention to elements conveying behavioral meanings, also when attending to works of art. If this were true, we would then expect a difference between the aesthetic value ascribed to visual stimuli presenting elements that carry information about the vitality status of the subject, with which one may potentially relate or "resonate," against the representation of dead subjects with no interaction potential.

To examine this idea, in this exploratory study, we presented participants with images of faces cropped from original representational paintings depicting a dead or a sleeping person and, while recording their eye movements, we asked them to view the artworks (free viewing [FV] task), to rate their beauty (aesthetic judgment [AJ] task), and to judge whether the observed face depicted a living or a dead person (vitality-status judgment [VSJ] task). Based on the literature reported above, we advanced the hypothesis that attention would particularly focus on the eye region of the living subjects, even though — in this study — all subjects (sleeping and dead) were represented with their eyes closed. Additionally, we predicted greater aesthetic preference for the stimuli depicting sleeping subjects than for those depicting the dead ones, in line with the embodied theory of aesthetic perception, according to which "resonance" potential is viable only with alive individuals.

The entire study was carried out in three phases. In the first phase (preliminary study), we administered an initial pool of stimuli to participants untrained in the arts, and we asked them to rate the vitality status of the portrayed subjects. The aim was to select stimuli that clearly represented living or dead subjects to be used in the second phase (i.e., the eye-tracking study). In the eye-tracking study, a different group of participants was required, in three independent tasks, to freely view, give an AJ, and judge the vitality status (living or dead) of the subjects represented in the selected stimuli while their gazing behavior was recorded. In the last phase, a follow-up priming experiment was carried out to behaviorally evaluate the effect of vitality status (living and dead) on the aesthetic evaluation of living and dead subjects. The participants of the priming experiment were divided into two experimental conditions: a congruent condition, in which — in half of the trials — the stimuli were primed by a congruent label of *sleeping* or *dead*, and, an incongruent condition, in which — in half of the trials — the stimuli were primed by an incongruent label of *sleeping* or *dead*. The general hypothesis was that recognition of the subjects' vitality status — independent of whether portrayed in the painting (implicit knowledge) or knowledge-driven (e.g., through explicit labeling) — would lead the participants to greater aesthetic evaluation of representations of life rather than death. In particular, we expected to observe differences in the aesthetic evaluation of the portrayals in the baseline condition (not primed condition) possibly driven by the visible features of the stimuli, and an effect of priming, according to which the explicit labeling of the stimuli as *sleeping* or *dead* would significantly affect their aesthetic value.

Methods

Participants

Preliminary study

Twenty-six Italian undergraduate students (14 males and 12 females, mean age = 25.38 years, *SD* = 5.32 years, range = 18–40 years) were recruited for the initial phase of this study to rate the vitality status of the portrayed subjects aimed at stimulus selection.

Eye-tracking

Twenty Italian undergraduate students (10 females; mean age = 24.33 years, *SD* = 4.40 years, age range = 19–34 years) took part in the eye-tracking study. The participants reported having no vision disorders that could interfere with the eye-tracking technique.

Priming experiment

Sixty participants took part in the behavioral study aimed at evaluating the relationship between AJ and VSJ of the stimuli. The participants were divided into two groups: (1) congruent condition group (*N* = 28, 16 females, 12 males; mean age = 26.82 years, *SD* = 6.94 years, age range = 18–49 years and (2) incongruent condition group (*N* = 32, 18 females, 14 males; mean age = 31.25 years, *SD* = 7.38 years, age range = 21–46 years).

For the preliminary and eye-tracking studies, the participants were recruited by means of fliers distributed at the university campus in Milan, and their participation was rewarded with a shopping voucher worth 10 euros. For the priming experiment, the participants were found from a work-group shared database, and they were recruited by e-mail. They agreed to participate without payment. All participants gave their written informed consent to the experimental procedure, and they declared to have no formal knowledge of the arts. With respect to the ethical issues related to the conduction of this study, not all academic institutions in Italy have ethical review boards and, because the authors' institution lacked such an arrangement at the time data were acquired, no formal ethical review of the method used in this study was conducted. The authors declare, however, that the procedure met all American Psychological Association (2002) ethical principles for the use of human subjects.

Experimental procedure and tasks

Preliminary study: Stimulus selection

In the preliminary study, eight representational paintings were identified depicting either a dead (D) or a sleeping (S) human subject. The images consisted of high-resolution digital versions of art paintings retrieved from the *Web Gallery of Art* searchable fine arts image database (www.wga.hu). From these images, the experimental stimuli were created by cropping the face area from the context using the Gimp program (vers. 2.8.14). In all images, both eyes were closed, in all instances visible, though in some cases one partially hidden by the nose (see Table 1). The faces were presented on a black foreground (800×600 px).

From the initial pool of images, four were selected for the eye-tracking exploratory study on the basis of the ratings recorded from 26 independent evaluators (see participants above, preliminary study), who judged, on a 7-point Likert scale ranging from 1 to 7, the vitality status of the represented subject by answering two questions for each image: "How much do you think he/she is alive?" and "How much do you think he/she is dead?" In total, five images survived the exclusion criteria, scoring a maximum of 2 and a minimum of 6 on the relative scales (e.g., for a sleeping subject, the mean scores had to be lower than or equal to 2 on the "dead" question and higher than or equal to 6 on the "alive" question; the opposite scores were applied to a dead subject). All stimuli falling in between these extremes were considered ambiguous and, thus, excluded from the final sample. One stimulus of the dead category representing a male subject, which met the inclusion criteria, was removed from the final selection to match the stimuli of the sleeping category by sex (both females; see Table 1 for a description of the selected paintings).

Table 1. Description of the paintings from which the selected stimuli were extracted: Author, title, year, repository.

Status	Author	Title	Year	Museum	Cropped stimuli
Sleeping	Giorgione	*Sleeping Venus*	1510	Gemäldegalerie, Dresden, Germany	
Sleeping	Caravaggio	*Riposo Durante la Fuga in Egitto*	1595–1596	Doria Pamphilj Gallery, Rome, Italy	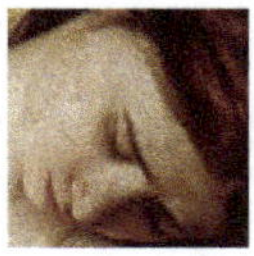
Dead	Guercino	*La Morte di Cleopatra*	1648	Musei di Strada Nuova, Genoa, Italy	
Dead	Caravaggio	*Death of the Virgin*	1605–1606	Louvre Museum, Paris, France	

Eye-tracking study

The participants were blind to the experimental aim. They were told that they would have seen images of faces extracted from representational paintings and that task-related instructions would have been provided during the experimental session. The stimuli were presented in three experimental tasks: FV, AJ, and VSJ. While FV was always presented as the first task with the aim to capture a most spontaneous visual response to the stimuli that was task-unbiased, the order of the two judgment tasks (AJ, VSJ) was counterbalanced across participants.

The stimuli were presented on a computer screen in a randomized order within each task. The presentation of the eye-tracking stimuli was created using the Tobii Studio 1.3 software (Tobii Technology AB). The participants were seated at a desk in a quiet room, at a distance of approximately 70 cm from the computer screen. Each trial began with the presentation of a central white cross on a black background for 1 s, followed by the presentation of the stimulus that lasted 3 s. Stimulus presentation time was based on previous eye-tracking works (Massaro et al., 2012; Savazzi *et al.*, 2014). Then, for the two judgment tasks, a task-related question appeared. Participants were instructed to answer the question on a scale ranging from –3 to 3 using the PC mouse, where –3 indicated *not at all* and +3 *very much*. The question was presented both at the beginning of each task and after each image presentation. When the answer was given, a new trial started. Each eye-tracker registration session lasted approximately 10 min. An initial calibration pattern was displayed to participants before running each task.

Eye-tracking data acquisition and ROI definition

The eye position was recorded using a Tobii Eye Tracker X120 set located on the desk in front of the subject between the subject and the monitor. The X120 Eye Tracker is a stand-alone eye-tracking unit that uses an infrared-based system for capturing reflections of the pupil and cornea in order to sample eye position every 1/120th of a second. The system is accurate to less than 0.5°.

The eye-movement indicator processed by the software (Tobii Studio 1.3) is fixations. Fixations occur when a target feature of interest is positioned on the fovea for a variable period of time (averaging about 300 ms per fixation). The data were collected in terms of total fixation duration (TFD), number of fixations (fixation count [FC]), and time to first fixation (TTFF).

For each image, regions of interest (ROIs) were manually created around the eye regions (one ROI for each eye), the nose, and the mouth (an example of ROI is in Figure 1). The ROI sizes were used for data normalization, as described in the following.

Eye-tracking statistical analysis

The eye-tracking data analysis was carried out by comparing attendance to each facial feature (mouth, nose, and eyes) for each stimulus category status (S, D) in the three tasks (FV, AJ, and VSJ). The dependent variables were TFD, FC (i.e., the total number of fixations), and TTFF (i.e., how long the subject took to enter into the first fixated area). The latter measure

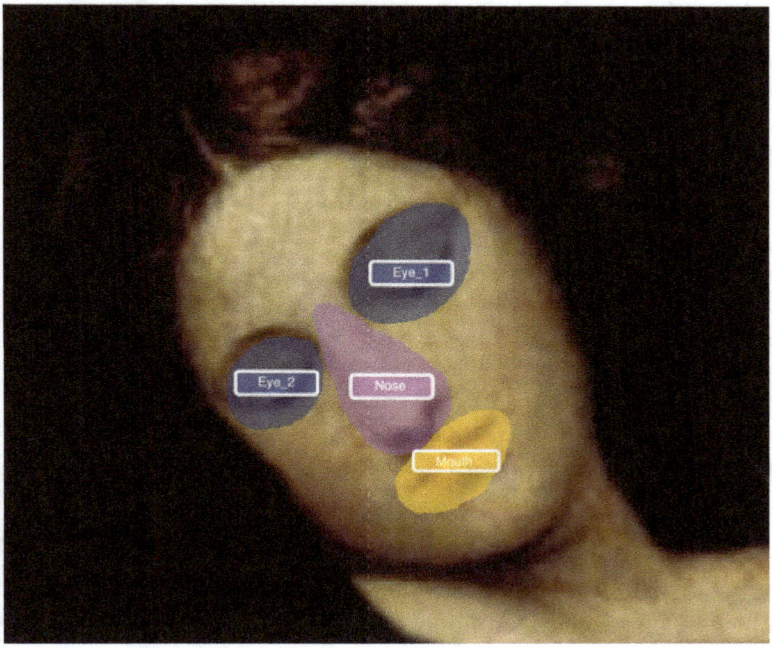

Figure 1. Example of regions of interest (ROI) on the eye, nose, and mouth. For data analysis, the normalized scores for the two eyes were summed to create a single ROI. The stimulus is a sleeping subject (Giorgione, 1510). [Color figure can be viewed at wileyonlinelibrary.com].

cues the first attended areas. TFD and FC data underwent a normalization that took into account the different ROI sizes. The normalization procedure followed the formula used by Massaro *et al.* (2012) and Savazzi *et al.* (2014), namely, the smallest ROI of both stimulus categories (S, D) was used as the reference size to which all values were adjusted: [time (TFD, FC)/(ROI size/smallest ROI size)]. As a result, DVs for two areas differing in size and fixated for an even amount/number of time/s would be recoded in favor of the smaller area.

For all variables, a 2×3×3×2 repeated-measures general linear model (GLM) analysis was carried out with two levels of vitality status (S and D), three levels of feature (eye, mouth, and nose), and three levels of task (FV, AJ, and VSJ) as the within-subject factors, and sex as the between-subject factor. While for TFD and FC the eye region was created by summing — after normalization — the values for the two eyes, for

TTFF, the eye region included only the values of one eye, and, namely, of the eye of each image that received attention. In fact, in most instances, TTFF was missing for one eye, which was thus excluded from the analysis. The Greenhouse–Geisser correction was applied when the sphericity assumption was violated ($p > 0.05$), whereas the Bonferroni correction was used for *post hoc* multiple comparisons.

Priming experiment

To assess the relation between aesthetic evaluation and the vitality status of the portrayed subjects, we carried out a behavioral study where the participants had to evaluate the aesthetic value of the stimuli used in the eye-tracking study, plus two images (one depicting a sleeping and the other a dead subject) that were considered ambiguous during the stimulus selection phase (preliminary study), totaling three S stimuli and three D stimuli. The within-subject dependent variable was the AJ of the images presented in the baseline condition, and, namely, in the condition in which the participants observed the images without any priming. The between-subject dependent variable was the AJ of the images, this time primed by the label *sleeping* or *dead* presented prior to each image (priming condition). For 28 participants, the label matched the presented image (congruent priming group), and for 32 participants, the label did not match the presented image (incongruent priming group).

The participants received an invitation e-mail to take part in the study (see Procedure above), and they were assigned to one or the other priming group taking into account sex and age, so as to acceptably counterbalance these variables across groups (see Participants section). The stimuli were run on a University Platform Qualtrics, which allows for the creation of experimental procedures that can be submitted to the participants online, as well as the recording of the participants' responses.

Sixty-three participants initially agreed to take part in the study. Three were excluded from data analysis because they declared having formal/academic background in the arts and were familiar with the stimuli. All the other participants declared having no formal knowledge of the arts and that they were not — or very poorly — familiar with the presented images, as assessed during the study. After agreeing to participate in the study, the participants were directed to a link where they took part in the study online. A description of the study (without making its goal explicit) was initially presented, and it was followed by a consent form. The participants could have access to the study only after giving their consent to take part in the study by flagging the appropriate box. The first part of the study aimed at recording demographic data and the level of expertise in the arts. The last part of the study assessed the level of familiarity with the presented images (*none*, *very poor*, *some*, and *very much*). The experimental procedure always started with the baseline condition, where the instruction specified to give an AJ to the observed images on a scale appearing after each stimulus presentation and ranging from –3 (*not at all beautiful*) to +3 (*very much beautiful*). The baseline condition was directly followed by the priming condition (congruent or incongruent depending on the experimental group), in which each stimulus was primed by the label *sleeping* or *dead* presented for 0.5 s.

Within each condition (baseline and priming), the stimuli were presented for a fixed time of 3 s in a random order with a repetition of three, totaling nine stimulus presentations for each S and D category. Evaluating response consistency across stimulus repetitions, we found a very high consistency in the participants' responses to the baseline-condition stimuli (Cronbach's alpha S = 0.94; Cronbach's alpha D = 0.96), indirectly providing some support to the internal validity of the psychophysical data recorded during the ET procedure (eye-tracking study), in which the stimuli were presented only once within each task.

Results

Preliminary study: Stimulus selection

A repeated-measures GLM analysis on the selected stimuli in the preliminary study confirmed the significant interaction between vitality status (S and D) and vitality judgment (alive and dead), $F(1,25) = 449.38$, $p<0.001$, $\eta^2 = 0.95$, $\delta = 1$, according to which S scored significantly higher on the alive scale and lower on the dead scale compared to D, and *vice versa.*

Eye-tracking study

Behavioral data

A repeated-measures GLM analysis was carried out with two levels of vitality status (S and D) for each judgment task (AJ and VSJ), independently. The results revealed that paintings depicting a sleeping face were rated as more beautiful than images representing dead faces, $F(1,19) = 8.79$, $p<0.01$, $\eta^2 = 0.39$, $\delta = 0.79$, with sleeping subjects receiving higher scores than portrayals of a dead person (sleeping M = 1.39, SD = 0.67; dead M = 0.50, SD = 1.03). Additionally, the participants correctly judged the faces as belonging to a sleeping or dead subject (VSJ task; sleeping $M = 1.41$, $SD = 0.81$; dead $M = -1.47$, $SD = 1.11$), $F(1, 19) = 230.09$; $p<0.001$, $\eta^2 = .94$, $\delta = 1$.

Eye-tracking data

Total fixation duration

The GLM analysis revealed a main effect of task, $F(2, 36) = 10.89$, $p<0.001$, $\eta^2 = 0.38$, $\delta = 0.99$, with greater mean TFD during VSJ than FV, $Mdiff = 0.02$, $SE = 0.004$, $p<0.001$; a main effect of status, S>D, $F(1, 18) = 15.06$, $p<0.001$, $\eta^2 = 0.46$, $\delta = 0.96$; a main effect of feature, $F(2, 36) = 12.60$, $p<0.001$, $\eta^2 = 0.41$, $\delta = 0.99$, with a greater mean TFD on the eye region than both the mouth, $Mdiff = 0.06$, $SE = 0.02$, $p<0.01$, and the nose, $Mdiff = 0.07$, $SE = 0.02$, $p<0.01$; and a significant Interaction Status × Feature, $F(2, 36) = 15.35$, $p<0.001$, $\eta^2 = 0.46$, $\delta = 1$. The *post hoc* analysis testing the interaction effect revealed that the main effect of feature just described particularly applied to S stimuli, for which mean TFD on the eye region was greater than both on the mouth, $Mdiff = 0.11$, $SE = 0.03$, $p<0.01$, and the nose, $Mdiff = 0.11$, $SE = 0.03$, $p<0.01$; whereas, for D stimuli, mean TFD on the eye region was greater only with respect to the nose, $Mdiff = 0.03$, $SE = 0.01$, $p<0.01$, indicating that, for D stimuli, the eye and the mouth were looked at for an equivalent amount of time, $p>0.05$. Additionally, significant TFD differences were found between S and D on both the eye region (S>D), $Mdiff = 0.08$,

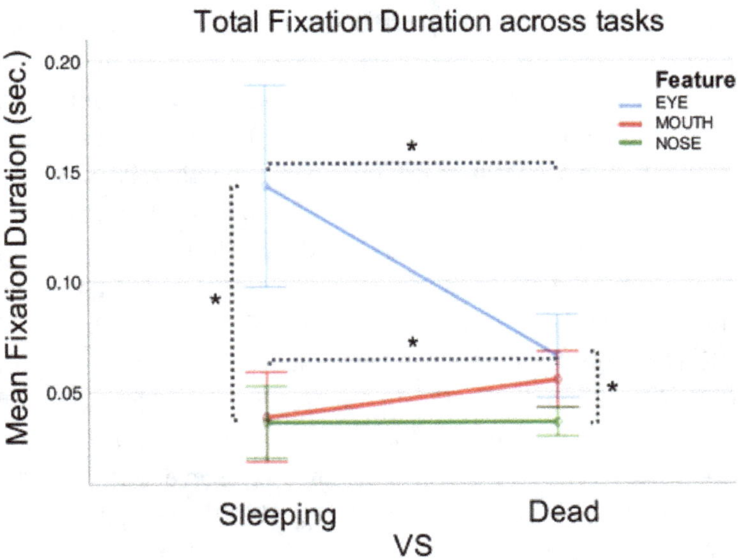

Figure 2. Mean total fixation duration (s) for each feature (eye, mouth, and nose) associated with the sleeping (S) and dead (D) subjects across tasks. The bars represent the standard error of the mean. *Significant differences. [Color figure can be viewed at wileyonlinelibrary.com].

SE= 0.02, $p<0.001$, and the mouth (D>S), *Mdiff* = 0.02, *SE*= 0.01, $p<0.05$ (see Figure 2). No differences were observed for TFD as a function of sex ($p>0.05$).

Fixation count

The results for FC were very similar to those found for TFD described above. In particular, the GLM analysis revealed a main effect of task, $F(2, 36) = 16.68$, $p<0.001$, $\eta^2 = .48$, $\delta = 1$, showing fewer mean FC during FV compared to both AJ, *Mdiff* = 0.05, *SE* = 0.01, $p<0.001$, and VSJ, *Mdiff* = 0.07, *SE* = 0.01, $p<0.001$; a main effect of status, S>D, $F(1, 18) = 25.09$, $p<0.001$, $\eta^2 = 0.58$, $\delta = 1$; and a main effect of feature, $F(2, 36) = 13.70$, $p<0.001$, $\eta^2 = 0.43$, $\delta = 1$, indicating greater mean FC on the eye compared to the mouth, *Mdiff* = 0.12, *SE* = 0.04, $p<0.05$, and the nose, *Mdiff* = 0.17, *SE* = 0.04, $p<0.001$. A significant Interaction

Feature × Task, $F(2,36) = 4.05$, $p < 0.05$, $\eta^2 = 0.18$, $\delta = 0.68$, indicated that the main effect of feature, with the eye fixated more times than the mouth and the nose, was true for FV and VSJ tasks but not for AJ, for which FC on the eye and mouth equaled, significantly differentiating from the nose, *Mdiff* eye = 0.14, *SE* = 0.04, $p < 0.001$, *Mdiff* mouth = 0.08, *SE* = 0.03, $p < 0.05$. Additionally, the eye was fixated more times during VSJ than in FV, *Mdiff* = 0.1, *SE* = 0.03, $p < 0.05$, and AJ, *Mdiff* = 0.06, *SE* = 0.02, $p < 0.05$; whereas the mouth was fixated less times during FV compared to AJ, *Mdiff* = 0.09, *SE* = 0.03, $p < 0.01$, and VSJ, *Mdiff* = 0.08, *SE* = 0.02, $p < 0.01$. The nose received a similar amount of inspections in all three tasks (this interaction is plotted in Figure 3(a)). Finally, a significant Interaction Status × Feature was found, $F(2, 36) = 19.20$, $p < 0.001$, $\eta^2 = 0.52$, $\delta = 1$, showing that, independent of task, S eye was fixated more times than D eye, *Mdiff* = 0.19, *SE* = 0.03, $p < 0.001$; the opposite fixation pattern was observed for the mouth, which was fixated more times for D than for S, *Mdiff* = 0.06, *SE* = 0.03, $p < 0.05$ (this interaction is plotted in Figure 3(b)). No differences were observed for FC as a function of sex, $p > 0.05$.

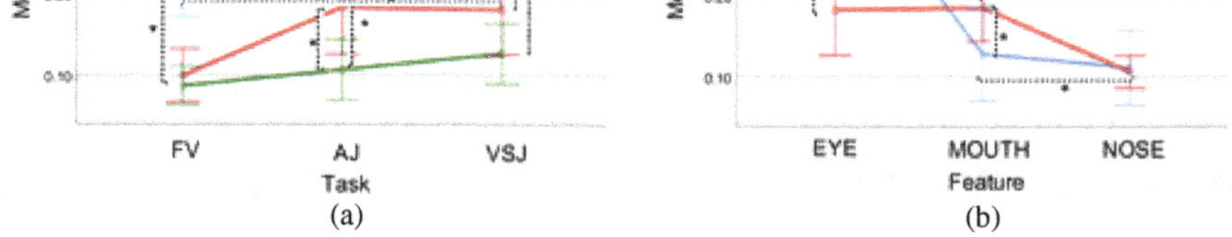

Figure 3. (a) Mean fixation count (s) for each feature (eye, mouth, and nose) associated with each task FV; AJ; VSJ across sleeping and dead stimuli. (b) Mean fixation count (s) for each feature (eye, mouth, and nose) associated with sleeping (S) and dead (D) stimuli across tasks. The bars represent the standard error of the mean. *Significant differences. [Color figure can be viewed at wileyonlinelibrary.com].

Time to first fixation

The analysis revealed a main effect of feature, $F(2, 38) = 4.26$, $p < 0.05$, $\eta^2 = 0.18$, $\delta = 0.7$, and an interaction effect of Feature × Status, $F(2, 38) = 15.99$, $p < 0.0001$, $\eta^2 = 0.46$, $\delta = 1$, showing that the eye was attended to significantly faster and the mouth significantly slower for the sleeping than the dead subjects, eye *Mdiff* = 0.31, *SE* = 0.12, $p < 0.05$; mouth *Mdiff* = 0.60, *SE* = 0.09, $p < 0.001$. Additionally, TTFF on the mouth of the sleeping subjects was slower than that for both the nose, *Mdiff* = 0.45, *SE* = 0.08, $p < 0.001$, and the eye, *Mdiff* = 0.69, *SE* = 0.12, $p < 0.001$, whereas for the dead subjects, no significant differences were observed between features, $p > 0.05$.

Priming experiment

To evaluate the relation between AJ and VSJ, data acquired in the priming experiment were entered in a GLM analysis with two levels of vitality status (VS: S and D) and two levels of condition (baseline and priming) as the within-subject factors, and sex and group (congruent and incongruent) as the between-subject factors. The results showed a main effect of vitality status (S > D), $F(1, 57) = 20.54$, $p<0.001$, $\eta^2 = 0.27$, $\delta = 0.99$, and a three-way interaction VS, condition, group, $F(1, 57) = 7.44$, $p < 0.001$, $\eta^2 = 0.12$, $\delta = 0.76$. No significant differences were found as a function of sex, $p>0.05$.

The *post hoc* analyses testing the interaction effect showed that S stimuli, but not D stimuli, were significantly affected by priming in both priming groups (congruent and incongruent) but with opposite effects. More specifically, after being primed congruently (i.e., S stimuli primed by the *sleeping* label), the S stimuli received higher aesthetic scores compared to the baseline condition, *Mdiff* = 0.21, *SE* = 0.09, $p<0.05$; whereas, after being primed incongruently (i.e., S stimuli primed by the *dead* label), the S stimuli received significantly lower aesthetic scores compared to the baseline condition, *Mdiff* = 0.22, *SE* = 0.09, $p<0.01$. Additionally, after incongruent priming, the aesthetic value of S stimuli did not significantly

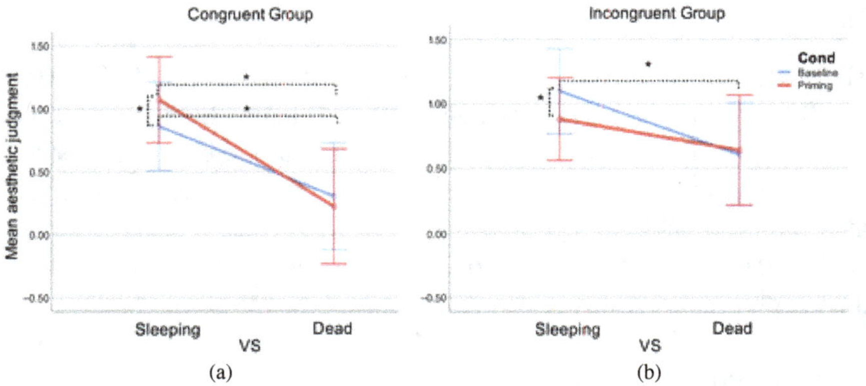

Figure 4. esthetic judgment for the sleeping and dead stimuli during the priming study for (a) the congruent group (i.e., correctly matching vitality label) and (b) the incongruent group (i.e., incorrectly matching vitality label). The bars represent the standard error of the mean. *Significant differences. [Color figure can be viewed at wileyonlinelibrary.com].

differ from the aesthetic value of D stimuli, $p > 0.05$. The effect of priming on S stimuli is shown in Figure 4.

Discussion

This exploratory study was aimed to investigate, by means of eye tracking, gazing differences between art images representing the faces of sleeping and dead subjects extracted from original representational paintings. The aesthetic value ascribed to the two stimulus categories was also evaluated in association with the vitality status (S and D) of the portrayed subjects.

In general, the eye-tracking data showed that fixations on the single features were more numerous and longer during the two judgment tasks compared to the free-viewing task, indicating a sparser observation pattern during FV and, as it can be plausibly accepted, more focused attention during the judgment tasks. Additionally, independent of task, the participants paid greater focused attention to the eye region, followed by the mouth and the nose, this latter being always fixated less, in line with the typical scan path reported — for Westerners — during a variety of categorization tasks (for a review, see Emery, 2000; see also, Kita *et al.*,

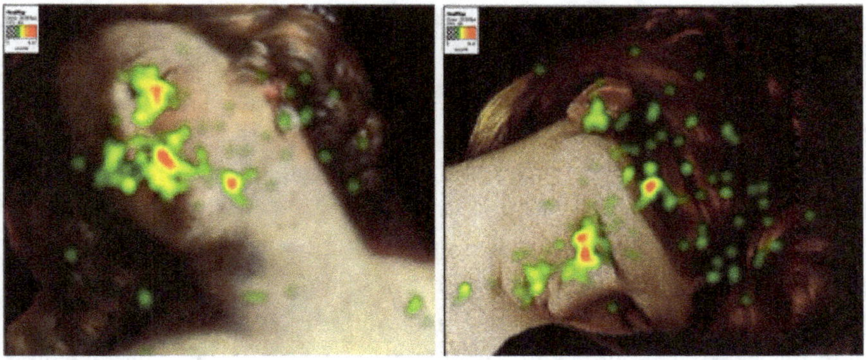

Figure 5. Example of the stimuli used with a superimposed heat map of the gaze pattern during AJ task. The red gradient reflects mean FCs. On the left, the face of a dead subject (Guercino, 1648); on the right, the face of a sleeping subject (Caravaggio, 1595–1596). [Color figure can be viewed at wileyonlinelibrary.com].

2010; Rodger, Kelly, Blais, & Caldara, 2010; Rowley, Baluja, & Kanade, 1998; Viola & Jones, 2004). An example of gazing pattern is in Figure 5.

In more detail, while for S stimuli, fixations on the eye region considerably outweighed fixations on the other ROIs, for D stimuli, fixations on the eye and the mouth regions outbalanced more, although the eye was still the region that was fixated more times (FC). It is not surprising then to find a substantial difference in fixation duration and count on the eye region in favor of S stimuli, and in fixation duration on the mouth in favor of D stimuli.

With respect to FC, the interaction between task and feature showed that, during FV, participants visited the eye region more often compared to the other ROIs. However, when the participants had to provide a judgment (AJ or VSJ), they focused also on the mouth region, although still privileging the eye during VSJ. These results indicate that evaluation of face portrayals involves the typical exploration pattern of the face as described above, according to which the eyes and the mouth are the most salient areas of inspection. However, focused attention to the eyes during VSJ reinforces the idea that this feature may be regarded as a benchmark representing vitality, especially of the sleeping subjects, as just discussed. In this light, and given the representational content of the stimuli, attention to the eyes also during FV — where the eyes were the prevalent area of attraction — may underpin a spontaneous tendency to wonder whether the depicted subject was sleeping or dead. This tentative — though plausible — interpretation further marks the eye as a vitality region (remember that during FV, the participants were still unaware of the following judgment tasks and, importantly, of vitality assessment).

The other region that received great attention was the mouth area, in particular when appraising D stimuli, whose mouth and eye regions were fixated for a comparable duration, and whose mouth was fixated longer with respect to S mouth. This result is consistent with FC data revealing that D mouth was generally fixated more often compared to S mouth. The prolonged fixations also on the mouth specifically for depictions of dead subjects is peculiar with respect to the popular saying "the sigh of life" to indicate the last breath before death, which was likely and not surprisingly well represented by the artists of these masterpieces. From this perspective, the two major vitality references — the eyes and the mouth — appear to best represent life and death, respectively.

Finally, TTFF further showed that, congruently with TFD and FC results, the eye was the quickest inspected region in S stimuli followed by the nose and, significantly later, by the mouth area, as opposed to D stimuli where time taken to inspect the first region was comparable among all features. These findings foster the view that, after a first global impression of the painting (i.e., "the painting gist"; Locher, 2015), the eyes of the sleeping subjects represent strong attractors of the viewer's attention, conceivably reflecting the element that best embodies vitality in living humans. In contrast, the first quick inspection of the nose may reflect the fact that, for some stimuli, the nose was the closest region to the fixation cross anticipating the stimulus appearance on the screen. First attention to the nose may have then represented a bias that can be overcome in future studies using a balanced lateralized cross instead of a central fixation cross. The idea of a bias toward the nose for TTFF is supported by TDF and FC data (corrected) suggesting a substantially less relevant role of the nose compared to both the eye and mouth regions.

Turning now our attention to the aesthetic ratings of the stimuli, we asked the question of whether AJ is somehow related to the portrayed subjects' vitality status. The behavioral data collected during the AJ task showed that S stimuli were aesthetically preferred to D stimuli, providing the first support to the idea of greater aesthetic preference for living beings. To further support the relation between AJ and the stimuli vitality status, in the follow-up priming experiment, S and D stimuli were anticipated by the label *sleeping* or *dead* either congruently or incongruently. The results confirmed a general preference for S stimuli. Additionally, and crucially, they further showed that vitality labeling significantly affected the aesthetic evaluation of the S stimuli, increasing it when the stimuli were labeled as *sleeping*, and decreasing it when the stimuli were labeled as *dead*. This is in line with the hypothesis that recognition of life is associated with a greater aesthetic value and that, on the other hand, recognition of death is associated with lower aesthetic appeal. The observed effects may conceivably lay on different reasons.

According to the embodied theory of aesthetic perception (Freedberg & Gallese, 2007), a possible interpretation of these effects comes from the discovery of a biological "resonance" system, mediated by the activity of neurons with "mirror" properties (for review, see Rizzolatti & Sinigaglia, 2010). These neurons fire both when performing an act or experiencing an emotion in the first person and when observing somebody else acting or experiencing an emotion. Therefore, in principle, since our body can potentially relate or "resonate" only with living beings — which in our study were represented by S subjects — and since, according to the embodied theory of aesthetic perception, this bodily involvement would please our senses, it would also favor the aesthetic value of portrayals of life rather than death.

Furthermore, we may prefer a living being because — by contrast — the idea of death is refused, at least within some Western cultural frames. In this respect, the priming effects that were found are quite enlightening. Priming had in fact a substantial effect only on S stimuli in both congruent and incongruent conditions: we found enhanced aesthetic preference for sleeping subjects compared to baseline in the congruent condition, and decreased aesthetic preference compared to baseline when the S stimuli were primed incongruently. This result provides us with a twofold, not mutually exclusive, interpretation. According to the first, perceptual, interpretation, S stimuli were possibly more ambiguous in terms of vitality than representations of death, and therefore they were more susceptible to variations induced by top-down effects (labeling). On the other hand, and most importantly, the lack of effect for D stimuli suggests that death is a *very powerful* construct, to the extent (1) to significantly negatively affect the aesthetic experience (i.e., D labeling decreasing AJ for S images) and (2) not to sensibly suffer from high-order influences (i.e., S labeling unaffecting AJ for D images). That is, once our perceptual system has captured a sign of death, not even the explicit indication of life (top-down modulation) can overcome the initial trigger of threat evoked by death.

As a matter of fact, fear of death appears to be deeply rooted in ourselves as living beings (see, Becker, 1973; Haidt, McCauley, & Rozin, 1994), and it is related to the maturation of consciousness during childhood (Melaer, 1973; Natterson & Knudson, 1960; Piaget, 1960; Slaughter, 2005) when awareness of ourselves as individuals in time is reached. A term that appears to capture the psychological attitude toward death is *uncanny*. At the beginning of the 20th century, Ernst Jentsch (1997) first introduced in psychology the concept of "uncanny" (*Das Unheimliche*), defined as the intellectual uncertainty of whether an evidently animated object is actually alive or not; or, opposite, whether a unanimated object could be somehow endowed with an autonomous life (see, Marchetti, Manzi, Itakura, & Massaro, 2018). Freud (1995) used the term *uncanny* to express within aesthetics a particular attitude in the more general feeling of fear that develops when a thing (or a person, an impression, a fact, or a situation) is felt as familiar and unfamiliar at the same time, generating general anguish along with an unpleasant sensation of confusion and unfamiliarity. This conflict may then result in a feeling of unease that is evoked right when a threatening content, such as death, is beautifully represented.

In sum, the results of this study provide evidence prompting an assessment of the relationship between life, death, and their representation in art, offering a range of possible experimental and hypothetical ideas that can be approached by different disciplinary lines. Deepening this topic would shed light on our inclination to life, our attitude to death, and aesthetics as a mediating vehicle between these two extremes of the same continuum.

Concluding Remarks

This preliminary study suggests that there is a proclivity to aesthetically prefer images of living beings, even if portrayed in an artwork. Our data also show a different gazing pattern between the faces of sleeping and dead individuals, with representations of a sleeping subject receiving more focused attention to the eye, as opposed to representations of dead subjects receiving more evenly distributed fixations between the eye and mouth regions.

Even if our results show a tendency to aesthetically prefer living beings, it has to be emphasized that the images shown in this study were faces cropped from a whole pictorial context to stress the importance of vitality. However, embedded in the pictorial content, where we can admire — not only the context — but also the pictorial elements that define the quality of the artwork, the representation of death could be less effective on aesthetic depreciation. In fact, although it is true that there is a common tendency that life is much preferred than death in everyday life, when it comes to art, death is one of the essential subjects in many forms and it can be highly appreciated as evidenced, for example, in many tragedy plays. This means that aesthetic values in artworks are more than just "life-is-good and death-is-bad." In support of this idea, we appreciate that catastrophic contents or even depiction of death can be valued in art. Why is it so? One hypothesis is that there is a sort of short circuit between true death and its representation. In art, death becomes plausible and, not only can it be valued given the artistic quality of its depiction, but, in its negative valence, through art we can exert the illusion of control over death.

References

Abbas, Z.-A. & Duchaine, B. (2008). The role of holistic processing in judgments of facial attractiveness. *Perception*, 37, 1187–1196. https://doi.org/10.1068/p5984.

Adams, R. B. & Kleck, R. E. (2003). Perceived gaze direction and the processing of facial displays of emotion. *Psychological Science*, 14, 644–647. https://doi.org/10.1046/j.0956-7976.2003.psci_1479.x.

Adolphs, R., Gosselin, F., Buchanan, T. W., Tranel, D., Schyns, P., & Damasio, A. R. (2005). A mechanism for impaired fear recognition after amygdala damage. *Nature*, 433, 68–72. https://doi.org/10.1038/nature03086.

Aharon, I., Etcoff, N., Ariely, D., Chabris, C. F., O'Connor, E., & Breiter, H. C. (2001). Beautiful faces have variable reward value. *Neuron*, 32, 537–551. https://doi.org/10.1016/s0896-6273(01)00491-3.

Althoff, R. R. & Cohen, N. J. (1999). Eye-movement-based memory effect: A reprocessing effect in face perception. *Journal of Experimental Psychology: Learning, Memory, and Cognition,* 25, 997–1010. https://doi.org/10.1037//0278-7393.25.4.997.

American Psychological Association (2002). Ethical principles of psychologists and code of conduct. *American Psychologist*, 57, 1060–1073. https://doi.org/10.1037/0003-066X.57.12.1060.

Andrew, R. J. (1963). Evolution of facial expression. *Science*, 142, 1034–1041. https://doi.org/10.1126/science.142.3595.1034.

Ardizzi, M., Ferroni, F., Siri, F., Umiltà, M. A., Cotti, A., Calbi, M., Fadda, E., Freedberg, D., & Gallese, V. (2018). Beholders' sensorimotor engagement enhances aesthetic rating of pictorial facial expressions of pain. *Psychological Research*, 1–10. https://doi.org/10.1007/s00426-018-1067-7.

Baron-Cohen, S., Campbell, R., Karmiloff-Smith, A., Grant, J., & Walker, J. (1995). Are children with autism blind to the mentalistic significance of the eyes. *British Journal of Developmental Psychology*, 13, 379–398. https://doi.org/10.1111/j.2044-835x.1995.tb00687.x.

Baron-Cohen, S., Wheelwright, S., & Jolliffe, A. T. (1997). Is there a "language of the eyes"? Evidence from normal adults, and adults with autism or Asperger syndrome. *Visual Cognition*, 4, 311–331. https://doi.org/10.1080/713756761.

Becker, E. (1973). *The Denial of Death*. New York, NY: Free Press.

Caldara, R., Seghier, M. L., Rossion, B., Lazeyras, F., Michel, C., & Hauert, C. A. (2006). The fusiform face area is tuned for curvilinear patterns with more high-contrasted elements in the upper part. *NeuroImage*, 31, 313–319. https://doi.org/10.1016/j.*NeuroImage*.2005.12.011.

Caravaggio (1595–1596). *Riposo durante la fuga in Egitto* [Painting]. Rome, Italy: Doria Pamphilj Gallery.

Caravaggio (1605–1606). *Death of the Virgin* [Painting]. Paris, France: Louvre Museum.

Carr, L., Iacoboni, M., Dubeau, M. C., Mazziotta, J. C., & Lenzi, G. L. (2003). Neural mechanisms of empathy in humans: A relay from neural systems for imitation to limbic areas. *Proceedings of the National Academy of Sciences of the United States of America*, 100, 5497–5502. https://doi.org/10.1073/pnas.0935845100.

Chatterjee, A. & Vartanian, O. (2014). Neuroaesthetics. *Trends in Cognitive Sciences*, 18, 370–375. https://doi.org/10.1016/j.tics.2014.03.003.

Condon, W. S. & Sander, L. W. (1974). Neonate movement is synchronized with adult speech: Interactional participation and language acquisition. *Science*, 183, 99–101. https://doi.org/10.1126/science.183.4120.99.

Dadds, M. R., Perry, Y., Hawes, D. J., Merz, S., Riddell, A. C., Haines, D. J., & Abeygunawardane, A. I. (2006). Attention to the eyes and fear-recognition deficits in child psychopathy. *British Journal of Psychiatry*, 189, 280–281. https://doi.org/10.1192/bjp.bp.105.018150.

Davies, G., Ellis, H., & Shepherd, J. (1977). Cue saliency in faces as assessed by the "Photofit" technique. *Perception*, 6, 263–269. https://doi.org/10.1068/p060263.

Di Cesare, G., Di Dio, C., Marchi, M., & Rizzolatti, G. (2015). Expressing our internal states and understanding those of others. *Proceedings of the National Academy of Sciences of the United States of America*, 112, 10331–10335. https://doi.org/10.1073/pnas.1512133112.

Di Cesare, G., Di Dio, C., Rochat, M. J., Sinigaglia, C., Bruschweiler-Stern, N., Stern, D. N., & Rizzolatti, G. (2013). The neural correlates of 'vitality form' recognition: An fMRI study: This work is dedicated to Daniel Stern, whose immeasurable contribution to science has inspired our research. *Social Cognitive and Affective Neuroscience*, 9, 951–960. https://doi.org/10.1093/scan/nst068.

Di Cesare, G., Valente, G., Di Dio, C., Ruffaldi, E., Bergamasco, M., Goebel, R., & Rizzolatti, G. (2016). Vitality forms processing in the insula during action observation: A multivoxel pattern analysis. *Frontiers in Human Neuroscience*, 10, 267. https://doi.org/10.3389/fnhum.2016.00267.

Di Dio, C., Ardizzi, M., Massaro, D., Di Cesare, G., Gilli, G., Marchetti, A., & Gallese, V. (2016). Human, nature, dynamism: The effects of content and movement perception on brain activations during the aesthetic judgment of representational paintings. *Frontiers in Human Neuroscience*, 9, 705. https://doi.org/10.3389/fnhum.2015.00705.

Di Dio, C., Berchio, C., Massaro, D., Lombardi, E., Gilli, G., & Marchetti, A. (2018). Body esthetic preference in preschoolers and attraction to canons violation: An exploratory study. *Psychological Reports*, 121, 1053–1071. https://doi.org/10.1177/0033294117744560.

Emery, N. J. (2000). The eyes have it: The neuroethology, function and evolution of social gaze. *Neuroscience and Biobehavioral Reviews*, 24, 581–604. https://doi.org/10.1016/s0149-7634(00)00025-7.

Enticott, P. G., Johnston, P. J., Herring, S. E., Hoy, K. E., & Fitzgerald, P. B. (2008). Mirror neuron activation is associated with facial emotion processing. *Neuropsychologia*, 46, 2851–2854. https://doi.org/10.1016/j.*Neuropsychologia*.2008.04.022.

Flom, R. & Bahrick, L. E. (2007). The development of infant discrimination of affect in multimodal and unimodal stimulation: The role of intersensory redundancy. *Developmental Psychology*, 43, 238–252. https://doi.org/10.1037/0012-1649.43.1.238.

Fraser, I. H., Craig, G. L., & Parker, D. M. (1990). Reaction time measures of feature saliency in schematic faces. Perception, 19, 661–673. https://doi.org/10.1068/p190661.

Freedberg, D. & Gallese, V. (2007). Motion, emotion and empathy in esthetic experience. *Trends in Cognitive Sciences*, 11, 197–203. https://doi.org/10.1016/j.tics.2007.02.003.

Freud, S. (1995). The "uncanny." In James Strachey (ed.), *The Standard Edition of the Complete Psychological Works of Sigmund Freud*, Vol. 17, pp. 217–256. London, England: Hogarth Press (Original Work Published 1919).

Gallese, V. (2003). The manifold nature of interpersonal relations: The quest for a common mechanism. *Philosophical Transactions of the Royal Society B: Biological Sciences*, 358, 517–528. https://doi.org/10.1098/rstb.2002.1234.

Gallese, V. (2005). Embodied simulation: From neurons to phenomenal experience. *Phenomenology and the Cognitive Sciences*, 4, 23–48. https://doi.org/10.1007/s11097-005-4737-z.

Gallese, V. & Di Dio, C. (2012). Neuroesthetics: The body in esthetic experience. In: V. S. Ramachandran (ed.). *The Encyclopedia of Human Behavior*, Vol. 2, pp. 687–693. New York, NY: Academic Press.

Gallese, V. & Sinigaglia, C. (2011). What is so special about embodied simulation. *Trends in Cognitive Sciences*, 15, 512–519. https://doi.org/10.1016/j.tics.2011.09.003.

Gernot, G., Pelowski, M., & Leder, H. (2017). Empathy, Einfühlung, and aesthetic experience: The effect of emotion contagion on appreciation of representational and abstract art using fEMG and SCR. *Cognitive Processing*, 19, 147–165. https://doi.org/10.1007/s10339-017-0800-2.

Giorgione (1510). *Sleeping Venus* [Painting]. Dresden: Gemäldegalerie.

Gosselin, F. & Schyns, P. G. (2001). Bubbles: A technique to reveal the use of information in recognition tasks. *Vision Research*, 41, 2261–2271. https://doi.org/10.1016/s0042-6989(01)00097-9.

Graham, D. J., Friedenberg, J. D., McCandless, C. H., & Rockmore, D. N. (2010). Preference for art: Similarity, statistics, and selling price. In B. E. Rogowitz & T. N. Pappas (eds.), *Proceedings of the International Society for Optics and Photonics: Vol. 7527. Human Vision and Electronic Imaging XV*, p. 75271A. San Jose, CA: SPIE Electronic Imaging. https://doi.org/10.1117/12.842398.

Groner, R., Walder, F., & Groner, M. (1984). Looking at faces: Local and global aspects of scanpaths. *Advances in Psychology*, 22, 523–533. https://doi.org/10.1016/S0166-4115(08)61874-9.

Grosbras, M. H., Beaton, S., & Eickoff, S. B. (2012). Brain regions involved in human movement perception: A quantitative voxel-based meta-analysis. *Human Brain Mapping*, 33, 431–454. https://doi.org/10.1002/hbm.21222.

Guercino (1648). *La Morte di Cleopatra* [Painting]. Genoa, Italy: Musei di Strada Nuova.

Haidt, J., McCauley, C., & Rozin, P. (1994). Individual differences in sensitivity to disgust: A scale sampling seven domains of disgust elicitors. *Personality and Individual Differences*, 16, 701–713. https://doi.org/10.1016/0191-8869(94)90212-7.

Haig, N. D. (1986). Exploring recognition with interchanged facial features. *Perception*, 15, 235–247. https://doi.org/10.1068/p150235.

Henderson, J. M., Williams, C. C., & Falk, R. J. (2005). Eye movements are functional during face learning. *Memory and Cognition*, 33, 98–106. https://doi.org/10.3758/bf03195300.

Hoehl, S. & Peykarjou, S. (2012). The early development of face processing — What makes faces special. *Neuroscience Bulletin*, 28, 765–788. https://doi.org/10.1007/s12264-012-1280-0.

Jack, R. E., Blais, C., Scheepers, C., Schyns, P. G., & Caldara, R. (2009). Cultural confusions show that facial expressions are not universal. *Current Biology*, 19, 1543–1548. https://doi.org/10.1016/j.cub.2009.07.051.

Jentsch, E. A. (1997). On the psychology of the uncanny. *Angelaki: Journal of the Theoretical Humanities*, 2, 7–16 (Original Work Published 1906).

Kita, Y., Gunji, A., Sakihara, K., Inagaki, M., Kaga, M., Nakagawa, E., & Hosokawa, T. (2010). Scanning strategies do not modulate face identification: Eye-tracking and near-infrared spectroscopy study. *PLoS One*, 5, e11050. https://doi.org/10.1371/journal.pone.001105.

Leekam, S., Baron-Cohen, S., Perrett, D., Milders, M., & Brown, S. (1997). Eye-direction detection: A dissociation between geometric and joint attention skills in autism. *British Journal of Developmental Psychology*, Vol. 15, pp. 77–95. Basel, Switzerland: International Publishing AG. https://doi.org/10.1111/j.2044-835x.1997.tb00726.x.

Locher, P. J. (2015). The aesthetic experience with visual art "at first glance". In P. Bundgaard & F. Stjernfelt (eds.), *Investigations into the Phenomenology and the Ontology of the Work of Art: Contributions to Phenomenology*, pp. 75–88. https://doi.org/10.1007/978-3-319-14090-2_5.

Marchetti, A., Manzi, F., Itakura, S., & Massaro, D. (2018). Theory of mind and humanoid robots from a lifespan perspective. *Zeitschrift für Psychologie*, 226, 98–109. https://doi.org/10.1027/2151-2604/a000326.

Massaro, D., Savazzi, F., Di Dio, C., Freedberg, D., Gallese, V., Gilli, G., & Marchetti, A. (2012). When art moves the eyes: A behavioral and eye-tracking study. *PLoS One*, 7, e37285. https://doi.org/10.1371/journal.pone.0037285.

McKelvie, S. J. (1976). The role of eyes and mouth in the memory of a face. *American Journal of Psychology*, 89, 311. https://doi.org/10.2307/1421414.

Melaer, J. D. (1973). Children's conception of death. *Journal of Genetic Psychology*, 360, 122–123.

Molenberghs, P., Cunnington, R., & Mattingley, J. B. (2012). Brain regions with mirror properties: A meta-analysis of 125 human fMRI studies. *Neuroscience & Biobehavioral Reviews*, 36, 341–349. https://doi.org/10.1016/j.neubiorev.2011.07.004.

Nadel, J. & Butterworth, G. (2011). *Imitation in Infancy*. Cambridge, England: Cambridge University Press.

Nahm, F. K., Perret, A., Amaral, D. G., & Albright, T. D. (1997). How do monkeys look at faces? *Journal of Cognitive Neuroscience*, 9, 611–623. https://doi.org/10.1162/jocn.1997.9.5.611.

Natterson, J. M., & Knudson, A. G., Jr. (1960). Observations concerning fear of death in fatally ill children and their mothers. *Psychosomatic Medicine*, 22, 456–465. https://doi.org/10.1097/00006842-196011000-00006.

O'Doherty, J., Winston, J., Critchley, H., Perrett, D., Burt, D. M., & Dolan, R. J. (2003). Beauty in a smile: The role of medial orbitofrontal cortex in facial attractiveness. *Neuropsychologia*, 41, 147–155. https://doi.org/10.1016/S0028-3932(02)00145-8.

Piaget, J. (1960). *The Child's Conception of Physical Causality*. London, England: Routledge.

Richell, R. A., Mitchell, D. G. V., Newman, C., Leonard, A., Baron-Cohen, S., & Blair, R. J. R. (2003). Theory of mind and psychopathy: Can psychopathic individuals read the 'language of the eyes'. *Neuropsychologia*, 41, 523–526. https://doi.org/10.1016/S0028-3932(02)00175-6.

Rizzolatti, G., Fadiga, L., Gallese, V., & Fogassi, L. (1996). Premotor cortex and the recognition of motor actions. *Cognitive Brain Research*, 3, 131–141. https://doi.org/10.1016/0926-6410(95)00038-0.

Rizzolatti, G. & Sinigaglia, C. (2010). The functional role of the parieto-frontal mirror circuit: Interpretations and misinterpretations. *Nature Reviews Neuroscience*, 11, 264–274. https://doi.org/10.1038/nrn2805.

Rochat, P. (2009). *Others in Mind: Social Origins of Self-Consciousness*. Cambridge, England: Cambridge University Press.

Rodger, H., Kelly, D. J., Blais, C., & Caldara, R. (2010). Inverting faces does not abolish cultural diversity in eye movements. *Perception*, 39, 1491–1503. https://doi.org/10.1068/p6750.

Rowley, H. A., Baluja, S., & Kanade, T. (1998). Neural network-based face detection. *IEEE Transactions on Pattern Analysis and Machine Intelligence*, 20, 23–38. https://doi.org/10.1109/34.655647.

Savazzi, F. M., Massaro, D., Di Dio, C., Gallese, V., Gilli, G., & Marchetti, A. (2014). Exploring responses to art in adolescence: A behavioral and eye-tracking study. *PLoS One*, 9, e102888. https://doi.org/10.1371/journal.pone.0102888.

Schyns, P. G., Bonnar, L., & Gosselin, F. (2002). Show me the features! Understanding recognition from the use of visual information. *Psychological Science*, 13, 402–409. https://doi.org/10.1111/1467-9280.00472.

Slaughter, V. (2005). Young children's understanding of death. *Australian Psychologist*, 40, 179–186. https://doi.org/10.1080/00050060500243426.

Viola, P. & Jones, M. J. (2004). Robust real-time face detection. *International Journal of Computer Vision*, 57, 137–154. https://doi.org/10.1023/B:VISI.0000013087.49260.fb.

Wada, J. A. (1961). Modification of cortically induced responses in brain stem by shift of attention in monkeys. *Science*, 133, 40–42. https://doi.org/10.1126/science.133.3445.40-a.

Wallraven, C., Fleming, R., Cunningham, D., Rigau, J., Feixas, M., & Sbert, M. (2009). Categorizing art: Comparing humans and computers. *Computers and Graphics*, 33, 484–495. https://doi.org/10.1016/j.cag.2009.04.003.

Wicker, B., Keysers, C., Plailly, J., Royet, J. P., Gallese, V., & Rizzolatti, G. (2003). Both of us disgusted in my insula: The common neural basis of seeing and feeling disgust. *Neuron*, 40, 655–664. https://doi.org/10.1016/S0896-6273(03)00679-2.

Wilkinson, N., Paikan, A., Gredebäck, G., Rea, F., & Metta, G. (2014). Staring us in the face? An embodied theory of innate face preference. *Developmental Science*, 17, 809–825. https://doi.org/10.1111/desc.12159.

Yarbus, A. L. (1965). *Role of Eye Movements in the Visual Process*. Moscow, Russia: Nauka.

https://doi.org/10.1142/9789811298639_0023

Chapter 23

Rembrandt Portraits: Implicitly Detecting the Original Perspective*

Xuanyu Wang[†] and Yan Bao[†,‡]

[†]*School of Psychological and Cognitive Sciences, Peking University, Beijing, China*

[‡]*Institute of Medical Psychology, Ludwig Maximilian University, Munich, Germany*

Abstract

The original left-front perspective of portraits by Rembrandt was detected by Chinese students with a higher recognition rate as compared to the right-front perspective or the mirror reversals of both perspectives. Oculomotor patterns indicated that the eye regions provided essential information for such implicit detection.

***Keywords*: eye movements; face perception; implicit knowledge; Rembrandt portraits; visual perspective**

---

*This chapter was originally published in the special collection on neuroaesthetics in *PsyCh Journal*. Wang, X. & Bao, Y. (2020), Rembrandt portraits: Implicitly detecting the original perspective. *PsyCh Journal*, 9(4), 513–515. https://doi.org/10.1002/pchj.376.

Human faces are asymmetric: The left hemiface is more expressive of emotions, especially negative ones (Nicholls *et al.*, 2004). Asymmetry of faces is also observed in visual art; artists prefer to paint a portrait with the left hemiface in front (McManus & Humphrey, 1973). Typical examples of such left-front perspective portraits are *The Mona Lisa* by Leonardo da Vinci and *Girl with a Pearl Earring* by Johannes Vermeer. The left-front perspective leads also to a higher aesthetic appreciation (Blackburn & Schirillo, 2012). On the basis of these observations, we asked whether the left-front faces are perceptually different from the right-front faces, taking portraits by Rembrandt as an experimental sample.

To address this question, we performed a judgment task with 24 Chinese students (18–24 years old, M = 21.19, SD = 1.88, gender-balanced) without knowledge of Rembrandt portraits. Informed consent was obtained from all participants prior to their participation. The study was approved by the ethical committee of the School of Psychological and Cognitive Sciences, Peking University, in agreement with the Declaration of Helsinki. Forty-eight gender-balanced portraits were selected from Rembrandt's works of different periods (1632–1668) with equal numbers of left-front and right-front perspectives. A mirror-reversed version of each original portrait was generated by Adobe Photoshop CS6, resulting in a total of 96 portrait stimuli. The portraits were downloaded in high resolution from open-source collections and resized into 800 × 1,064 pixels with the face in a central frame of 400 × 532 pixels (7.6° × 10.1°). Adjustments were made to balance luminance and contrast. The stimulus was displayed on a 27-in. LCD screen with a viewing distance of 70 cm.

Participants were informed that following a fixation point, either the original portrait or its mirror reversal was shown for 12 s (the center of the portrait was 15° above the fixation), and they had to judge whether the figure was original or mirror-reversed by pressing one of the two preassigned keys. Eye movements were measured with Eyelink 1,000 Plus (SR Research Ltd., Ontario, Canada) during the viewing time within which they were asked to imagine themselves to be in a museum. A self-evaluation on confidence with a 7-point Likert scale of 1 (*very unconfident*) to 7 (*very confident*) was required after each judgment.

Each participant performed 48 trials in the main experiment with 24 original portraits and 24 mirror reversals of the other portraits not presented in their original perspectives (both with left-front and right-front perspectives balanced). Thus, each portrait was presented only once to a single subject, either in its original or mirror-reversed form, and was balanced across subjects for an equal probability of presenting as original or mirror reversal. Following four practice trials with four additional portraits, the 48 experimental trials were presented in an individually randomized order with a short break in the middle.

Analyses of variance (ANOVA) statistics with the source of portraits (original/mirror) and the original perspective (left-front/right-front) as two within-subjects variables were performed over the percentage of correct judgment. The results showed a significant two-way interaction, $F(1, 23) = 14.79, p<0.001, \eta^2 = 0.391$; further analysis revealed a significantly higher recognition rate for the original left-front perspective portraits as compared to the other three conditions (i.e., original right-front perspective and the two mirror reversals of both perspectives, $p<0.001$ for all pairs), which were all close to chance level (see Figure 1). Similar ANOVA statistics were performed on confidence rating, and none of the main effects as well as the two-way interaction were found (all $p_s > 0.05$), suggesting that the higher recognition rate for the original left-front perspective portraits was presumably due to implicit visual processing.

For eye movement recordings, we analyzed separately the percentage of gaze-dwelling time and the number of saccades toward different *areas of interest* (*AOI*); namely, the whole face, the left/right hemiface, the left/right eye, the nose, and the mouth. By performing three-way ANOVA statistics with correct/incorrect judgment, left-front/right-front perspectives, and original/mirror portraits as three within-subjects variables, we found that the eye regions of the portraits seemed to be particularly important for the correct judgment. When participants made correct responses (recognizing both the originals and the mirror reversals) relative to incorrect responses, they spent a longer time viewing in eye regions only ($p<0.05$). Furthermore, independent of correct/incorrect judgment, the viewing time in the eye regions was longer for the left-front originals than for the mirror reversals of the right-front originals ($p<0.05$), although

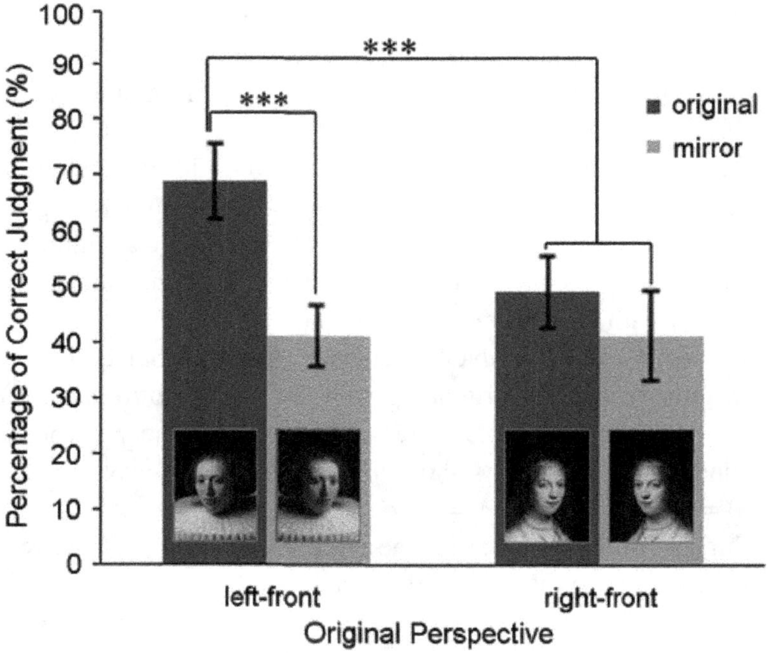

Figure 1. Percentage of correct judgment for different experimental conditions. Original portraits marked in dark gray, mirror reversals in light gray (***p≤0.001, error bars indicated).

both of them appeared in a similar left-hemiface-front manner. Moreover, when viewing left-front originals, more saccades were made toward the eye regions than when viewing the mirror reversals of the right originals ($p < 0.05$).

It can be concluded that in addition to the higher recognition rate, the oculomotor pattern with more fixations and longer dwelling time at the eye regions revealed an implicit perceptual saliency of the left-front perspective portraits but only when the portrait was originally painted with such a perspective. Since viewers were not confident about their judgments, the successful recognition of the original left-front perspective portraits was presumably based on *implicit* instead of *explicit semantic knowledge*. The attentional bias toward the eyes also revealed *visual knowledge* as a third complementary mode of knowledge underlying visual aesthetics (Bao, von Stosch, Park, & Pöppel, 2017; Pöppel & Bao,

2011). This perceptual saliency in the eye regions of the portraits is possibly due to a particular spatial outlay originally created by the artist, who was especially successful with the left-front perspective portraits.

It has to be emphasized that this surprising effect in extracting the correct perspective of portraits is observed some 400 years later with participants having different cultural backgrounds. On a more general level, this result indicates that our visual system on an implicit processing level creates perceptual trust in individual and cultural identity (Bao *et al.*, 2016). Furthermore, our discovery supports the concept of anthropological universals in visual processing. Artists without explicit knowledge about principles of visual processing can intuitively make use of these principles (i.e., "Artists are in some sense neurologists;" Zeki, 2002, p. 10). Moreover, this study also proves that researchers can benefit from artwork to get new insight into basic cognitive mechanisms and even make new discoveries which might be shadowed by simple experimental stimuli like Gabor patches or line drawings.

Acknowledgment

This work was supported by the National Natural Science Foundation of China (31771213 and 31371018) to Yan Bao.

References

Bao, Y., von Stosch, A., Park, M., & Pöppel, E. (2017). Complementarity as generative principle: A thought pattern for aesthetic appreciations and cognitive appraisals in general. *Frontiers in Psychology*, 8, 727. https://doi.org/10.3389/fpsyg.2017.00727.

Bao, Y., Yang, T., Lin, X., Fang, Y., Wang, Y., Pöppel, E., & Lei, Q. (2016). Aesthetic preferences for eastern and Western traditional visual art: Identity matters. *Frontiers in Psychology*, 7, 1596. https://doi.org/10.3389/fpsyg.2016.01596.

Blackburn, K. & Schirillo, J. (2012). Emotive hemispheric differences measured in real-life portraits using pupil diameter and subjective aesthetic preferences. *Experimental Brain Research*, 219(4), 447–455. https://doi.org/10.1007/s00221-012-3091-y.

McManus, I. C. & Humphrey, N. K. (1973). Turning the left cheek. *Nature*, 243(5405), 271–272. https://doi.org/10.1038/243271a0.

Nicholls, M. E. R., Ellis, B. E., Clement, J. G., & Yoshino, M. (2004). Detecting hemifacial asymmetries in emotional expression with three-dimensional computerized image analysis. *Proceedings of the Royal Society B: Biological Sciences*, 271(1540), 663–668. https://doi.org/10.1098/rspb.2003.2660.

Pöppel, E. & Bao, Y. (2011). Three modes of knowledge as basis for intercultural cognition and communication: A theoretical perspective. In S. Han & E. Pöppel (eds.), *Culture and Neural Frames of Cognition and Communication*, pp. 215–231. Berlin, Heidelberg: Springer. https://doi.org/10.1007/978-3-642-15423-2_14.

Zeki, S. (2002). *Inner Vision: An Exploration of Art and the Brain*, p. 10. Oxford: Oxford University Press.

https://doi.org/10.1142/9789811298639_0024

Chapter 24

The Power of External Influences to Modify Judgments of Facial and Moral Beauty*

Marcus J. Glennon and Semir Zeki

Laboratory of Neurobiology, Division of Cell & Developmental Biology, University College London, London, UK

Abstract

Empirical evidence shows that the often-made positive correlation between human physical and moral beauty is tenuous. In this study, we aimed to learn whether facial and moral beauty can be psychophysically separated. Participants (n = 95) provided beauty and goodness (i.e., trustworthiness) ratings for pictures of faces, after which they were presented with a fictitious peer rating for the same face and asked to re-rate the face. We used the difference between the initial and final ratings to quantify the degree of resistance to external influence. We found that judgments of facial beauty were more resistant to external influence than judgments of facial "goodness"; in addition, there was significantly

*This chapter was originally published in the special collection on neuroaesthetics in *PsyCh Journal*. Glennon, M. J. & Zeki, S. (2022). The power of external influences to modify judgments of facial and moral beauty. *PsyCh Journal*, 11(5), 707–714. https://doi.org/10.1002/pchj.492.

higher agreement within beauty ratings than within goodness ratings. These findings are discussed in light of our Bayesian–Laplacian classification of priors, from which we conclude that moral beauty relies more upon acquired "artifactual" priors and facial beauty more on inherited biological priors.

***Keywords*:** beauty judgments; facial beauty; judgmental conformity; moral beauty; neuroaesthetics

Introduction

Individuals commonly infer personality traits of strangers based on physical appearance, with many studies indicating links between high facial attractiveness and various positive attributes (Langlois *et al.*, 2000). Chief among these links is the often stated one between beauty and moral "goodness," encapsulated by a single word, *Kalon*, in ancient Greek culture and now more commonly referred to as the "beauty is good stereotype" (Dion *et al.*, 1972; Jenkins & Turner, 2009). Even in spite of its ancient history, how strongly the two are linked is uncertain since the strength of the linkage varies substantially between studies (Eagly *et al.*, 1991). In this study, we addressed this question in what seemed to us a more manageable way, by restricting ourselves to facial beauty, as representative of physical beauty on the one hand, and to moral beauty (or trustworthiness) on the other.

Past studies have linked facial attractiveness to perceived moral beauty (i.e., trustworthiness; D. S. Ma *et al.*, 2015, table 3; Shinners, 2009; Xu *et al.*, 2012). This linkage has implications in many walks of life, particularly in the courtroom, where more attractive individuals benefit from having a higher probability of winning (Zebrowitz & McDonald, 1991). Neuroimaging evidence shows that the medial orbitofrontal cortex, which is engaged during the experience of beauty derived from different sources (Ishizu & Zeki, 2011; Kawabata & Zeki, 2004; Zeki *et al.*, 2014), is also engaged when subjects experience facial and moral beauty (Kranz & Ishai, 2006; O'Doherty *et al.*, 2003; Tsukiura & Cabeza, 2011), while judgments of whether a face is beautiful or neutral or whether a scene is morally beautiful or neutral engage the lateral orbitofrontal cortex

(Wang *et al.*, 2015), which is engaged in both aesthetic and perceptual judgments (Ishizu & Zeki, 2013). The issue of a link between facial and moral beauty is possibly confounded further by the common conflation between facial attractiveness and facial beauty, the former often being used as indicative of the latter; these two terms in fact describe separate concepts, which are not always linked to each other, since one can perceive each in a face without necessarily perceiving the other (Yang *et al.*, 2021). In this study, we are concerned with beauty, not attractiveness.

Our general hypothesis here is that the perception of physical beauty, even if linked to that of moral beauty, can nevertheless be dissociated from it psychophysically and thus can be shown to be a separate entity, one that is not hostage to external opinion to the same degree as moral beauty. This prediction stems from our Bayesian–Laplacian classification of priors into biological (β) and artifactual (α) categories (Zeki & Chén, 2020) where facial beauty is considered to belong more in the biological category. This proposition is due, *inter alia*, to the finding that infants of varying age, on average, orient preferentially toward face-like stimuli, as well as toward faces judged to be attractive by adult humans, thus indicating the presence of an innate, or rapidly acquired, perceptual mechanism that determines the beauty of a face (Goren *et al.*, 1975; Langlois *et al.*, 1991; Slater *et al.*, 1998). On the other hand, history teaches us that moral values are far more flexible and significantly influenced by prevailing opinions and conditions. Indeed, although some believe that many moral values are universally inherited (see Krebs, 2008, for a summary), others suppose that they are established throughout life, with the internalization of a set of standards allowing for self-regulation of behavior and the development of moral values (Bandura, 1991).

The purpose of this study was to identify the extent to which judgments of facial and moral beauty can be separated from one another psychophysically. Ma *et al.* (2016) drew a dissociation between facial attractiveness and moral beauty by finding significantly higher agreement amongst subjects' attractiveness ratings compared to their ratings of the trustworthiness of faces. Concentrating on beauty rather than attractiveness, we wanted to take this a stage further and learn whether two can be dissociated psychophysically, which would be a guide to the tenuous linkage between them. As well as comparing agreement in ratings of facial and moral beauty, we compared the resistance of each to external influence (defined here as peer opinion). Previous studies have addressed conformity solely in the context of physical beauty (e.g., Bignardi *et al.*, 2020; Klucharev *et al.*, 2009; Zaki *et al.*, 2011). Both Klucharev *et al.* (2009) and Zaki *et al.* (2011) found that, on average, participants change their ratings to conform to those of their peers when assessing the attractiveness of faces while Bignardi *et al.* (2020) found that ratings of facial beauty also conformed to peer opinion, but to a significantly lesser extent than ratings of the beauty of abstract art.

In this study, participants were presented with pictures of faces and asked to provide a beauty or goodness (i.e., trustworthiness) rating both before and after exposure to the average rating of peers for the same faces (similar to the design used in Bignardi *et al.*, 2020). We use the terms "goodness" and "trustworthiness" interchangeably throughout and participants were so informed. Our hypothesis was that we would be able to separate beauty and goodness ratings psychophysically through two different means: First, we predicted that beauty ratings would be more resistant to external information than goodness ratings, and second, that there would be significantly greater agreement amongst individuals in their beauty ratings compared to their goodness ratings.

Method

Participants

One hundred and seven subjects were recruited through an online participant recruitment platform (https://www.callforparticipants.com/) and by word of mouth; of these, 95 (49 female, 5 undisclosed; age range 18 to 61 years, with a mean of 26.29 years [SD = 7.18 years]) were taken through to analysis (see Results section for exclusion justifications). We recruited subjects who spoke fluent English and were between 18 and 65 years of age. Prior to consent, participants were informed that the study involved rating a selection of faces on several different traits.

Participants gave their informed consent and provided, optionally, details of their gender, sexual orientation, age, ethnicity, and primary country of residence (i.e., the country they have lived longest in), all of which could possibly be factors in biasing beauty ratings for faces.

Participants were nationals of 14 different countries, with ethnicity also varying substantially (see Supplementary Materials 1 for details); they were given the option of declaring their sexual orientation using an adapted Kinsey scale (as used in Zeki & Romaya, 2010) which scales sexuality from 1 (*completely heterosexual*) to 7 (*completely homosexual*). Only integer responses were permitted.

Stimuli

The stimuli used in this study were pictures of faces obtained from several face databases (the Face Research Lab London Set [FRLLS]; DeBruine & Jones, 2017; Chicago Face Database [CFD]; Ma *et al.*, 2015) and online sources (Adobe Stock, Pinterest, and Google Images). Based on beauty ratings given in a preliminary study, we selected 32 beautiful faces, 16 average ones, and 16 faces in the "not beautiful" category; more beautiful faces were selected because the study aimed to learn about the relationship between beauty and morality. Half were female and half male, and they were of varying ethnicity (see Supplementary Materials 2); each was set to 500 × 500 pixels and presented against a plain white background.

Procedure

The experimental procedure was designed in PsychoPy (Peirce *et al.*, 2019) and implemented online via Pavlovia (https://pavlovia.org/).

The experiment consisted of two blocks of 64 trials each (during which subjects rated the beauty and "goodness" of faces, separately). Both blocks were preceded by three-trial practice blocks to familiarize participants with the task (see Figure 1). Each trial had four components: an initial beauty/goodness rating, a certainty rating for the initial rating, a presentation of external information (see the following for details), and a

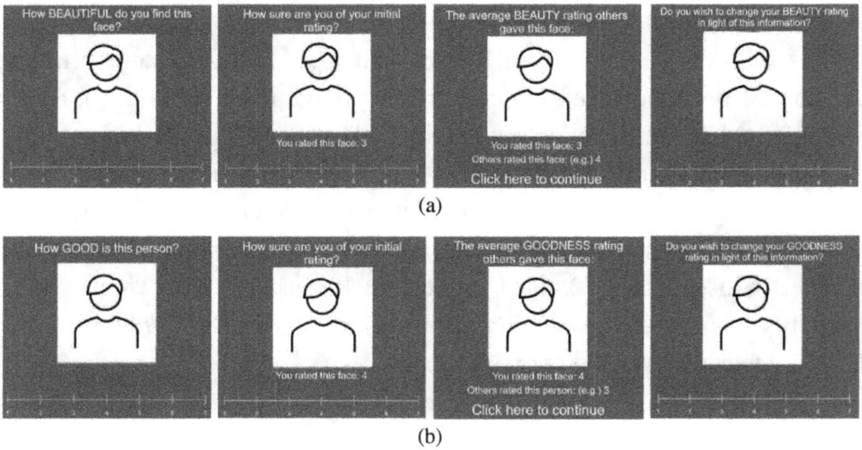

Figure 1. Practice experimental paradigm. There were four steps to each trial. Participants gave an initial beauty or goodness rating on a discrete scale of 1 to 7. Then they were asked to express their certainty regarding their initial rating on a continuous scale of 1 to 7. Following this, they were presented with external information (i.e., the fictitious peer opinion). Finally, once they had clicked to continue, they made a final beauty or goodness rating in light of the peer opinion. (a) A practice beauty trial and (b) a practice goodness trial.

final beauty/goodness rating in light of the external information (see Supplementary Materials 3 for instructions provided to participants). Ratings were made on a 7-point scale, with one corresponding to *not beautiful/good at all*, and seven corresponding to *extremely beautiful/good*. Originally, participants had a 4-s response window for the initial, certainty, and final ratings. This was increased to an 8-s window after recruiting the first 32.6% of participants due to a high number of mistrials in which participants ran out of time to respond. Increasing the response window led to fewer mistrials and did not significantly affect average rating change (RC) ($p = 0.691$ for beauty and $p = 0.871$ for goodness).

The external information consisted of a peer opinion which, participants were informed, was the average rating of other participants in the study. However, we used fictitious, pseudo-randomized peer opinions instead of real ones. Facial beauty is primarily reliant upon stable, biological priors, meaning that ratings of facial beauty are likely to have high agreement among different groups, whereas ratings of facial goodness may be more variable due to their suspected reliance on artifactual priors. Therefore, if we were to use real peer opinions, we may observe an unequal distribution of discrepant peer opinions between beauty and goodness, with peer ratings of beauty being more similar to the ratings of participants than peer ratings of goodness. This would lead to a higher number of opportunities for conformity amongst goodness ratings compared to beauty ratings, thereby making it difficult to compare the data obtained for the two attributes. For this reason, we used fictitious peer opinions, with both beauty and goodness peer ratings having a 50% chance of being different to, and a 50% of being the same as, the initial ratings of participants.

As in Zaki *et al.* (2011), the fictitious peer opinions were generated using the following procedure: When participants rated a face lower than 3 (i.e., 1 or 2), peer opinion was just as likely to be the same or 2 to 3 points higher than the initial rating (i.e., 3, 4, or 5). Likewise, when participants gave a rating higher than 5 (i.e., 6 or 7), peer opinion was presented as being either the same or 2 to 3 points lower than the initial rating (i.e., 3, 4, or 5), with a 50% chance of being the same and a 50% chance of being different. Finally, for initial ratings between 3 and 5, peer opinion had a 50% chance of being equal to the initial rating, a 25% chance of being 2 to 3 points lower, and a 25% chance of being 2 to 3 points higher.

The order in which trials appeared in each block was randomized, and the blocks were counterbalanced to minimize any order effects.

Analysis

Firstly, we undertook a two-way analysis of covariance (ANCOVA) to determine whether beauty ratings given to faces were affected by the gender and sexual orientation of participants. The gender of the image was a within-subjects factor, the gender of the participant was a between-subjects factor, and sexual orientation (as measured on a 7-point scale) was listed as a covariate.

We next wanted to learn whether beauty and goodness ratings correlate with one another in our sample. We took the average beauty and goodness ratings per face and used a Spearman rank-order correlation coefficient to determine the strength and direction of the association between them. Three faces were removed from this analysis: two of which had exceptionally low average goodness ratings (<2.4) and another of which had an exceptionally high goodness rating (>5.1). Removing these outliers from the analysis did not alter the significance of the results.

To address our hypothesis that subjects would be more resistant to changing their beauty ratings compared to their goodness ratings in light of external opinion, we undertook five within-subject comparisons (paired-sample sign tests; 2 for beauty, 2 for goodness, and 1 addressing both). A sign test is a non-parametric test that identifies whether there are consistent differences between the averages of two dependent samples. We began by determining whether external information that differed from the initial rating given by a participant led to a final rating that was different from the initial rating; next, we investigated whether such external influence was higher when the initial rating of the participant was medium or extreme, that is to say, ratings between 3 and 5, on the one hand, and ratings of 1, 2, 6, or 7, on the other; a final sign test was done to determine whether external information influenced beauty ratings and goodness ratings to the same or to different degrees. As there was a total of five sign tests included in this analysis, Holm–Bonferroni adjustments were made to correct for the false discovery rate. Rating change (final rating minus initial rating) was the key metric used in the within-subjects comparisons, as it represents the strength of the influence of external information. We calculated absolute values (i.e., removed the negative sign from negative values) to ensure comparability across trials.

After this, we used the certainty ratings given by participants to assess whether they affected susceptibility to external influence and whether participants were any more certain of their beauty ratings compared to their goodness ratings. A Spearman's rank-order correlation was used to determine whether there was a link between the average certainty and RC exhibited by participants. In addition, another sign test was used to examine whether participants, on average, were more certain of their beauty and goodness ratings.

We then used the Means Minus One test (MM1; Vessel *et al.*, 2018) to calculate the average agreement within beauty ratings and within goodness ratings, separately. The MM1 was calculated by comparing the ratings of each participant to the average ratings of all other participants for the same faces. This process is repeated for each participant, and it results in individual *r* values that represent the extent to which a participant's ratings correlate with group mean ratings. Subsequently, the *r* values were converted to *z* values (using the Fisher *r* to *z* transformation), as this has been shown to reduce bias in estimates when determining agreement between participants (Bronstad & Russell, 2007). The average *z* value was calculated and then converted back to an *r* value. A paired-sample sign test was used to compare participants' *z* values for individual beauty and goodness agreement scores.

All analyses were undertaken in either IBM SPSS (Statistics version 27) or R-Studio using R version 4.0.3 (R-Studio Team, 2020).

Results

Sexual orientation of participants

Of the 90 out of 95 participants who provided their sexual orientation, 57% identified as either completely homosexual or heterosexual, with the remainder of participants distributed between the two (see Supplementary Materials 1). This corresponds well with previous data on sexual orientation among the younger British population by a 2015 YouGov poll for 18–24-year-olds in the United Kingdom (https://yougov.co.uk/topics/lifestyle/articles-reports/2015/08/16/half-young-not-heterosexual).

Mean beauty and goodness ratings

Twelve participants were excluded from the analysis for several reasons: Eight had too many mistrials in their data (>30% of trials), failing to make analysis feasible; two exhibited extremely low average response times to faces in goodness trials (<1.8 s) when compared to the average of others (mean = 3.74, *SD* = 0.66), implying lack of attention; and, finally, two exhibited extremely high average RCs for all (feedback and control) beauty trials (>2.5) when compared with the mean of others (mean = 0.24, *SD* = 0.20) with no discernible pattern in their responses. Therefore, 95 participants were taken forward for analysis.

As found previously, average beauty ratings for faces from female participants were highly correlated with average beauty ratings from male participants ($r = 0.99$, $p<0.001$; Bignardi *et al.*, 2020).

To check whether beauty ratings given to faces were affected by the gender or sexual orientation of participants, we undertook a two-way ANCOVA. Data from participants who did not wish to disclose their gender or sexual orientation were excluded from this analysis (therefore, $n = 85$). It revealed that, on average, male participants did not give significantly different ratings for images of female faces, and vice versa ($F[1, 82] = 0.138$, $p = 0.711$). However, there was a significant effect of the gender of the faces being rated, with female faces averaging a higher beauty rating than male faces (mean = 4.20, $SD = 0.63$ and mean = 3.64, $SD = 0.73$, respectively; $F[1, 82] = 51.021$, $p<0.001$; as found by Bignardi *et al.*, 2020). Sexual orientation did not significantly affect beauty ratings ($F[1, 82] = 0.579$, $p = 0.449$). Effect sizes are given in Supplementary Materials 4.

The Spearman's rank correlation coefficient revealed that there was a strong, positive correlation between beauty and goodness ratings of the same faces, which was statistically significant ($r_s[60] = 0.786$, $p<0.001$; see Figure 2). However, visual inspection indicates that it is not without considerable variation.

Main analysis: Effect of external influence on beauty and goodness ratings

Rating change, quantified as the difference between the initial and final ratings, was the value used to represent the effect of the external influence. The higher the RC, the more susceptible the initial rating to modification through feedback provided by external influence. Given that the magnitude, as opposed to the direction, of the effect was the focus of this study, negative RCs were converted to positive values. This allowed all values to be compared.

Half of the trials in the study were feedback trials, in which there was a mismatch between the participants' initial rating and the pseudo-randomized peer rating they were presented with. The remaining trials

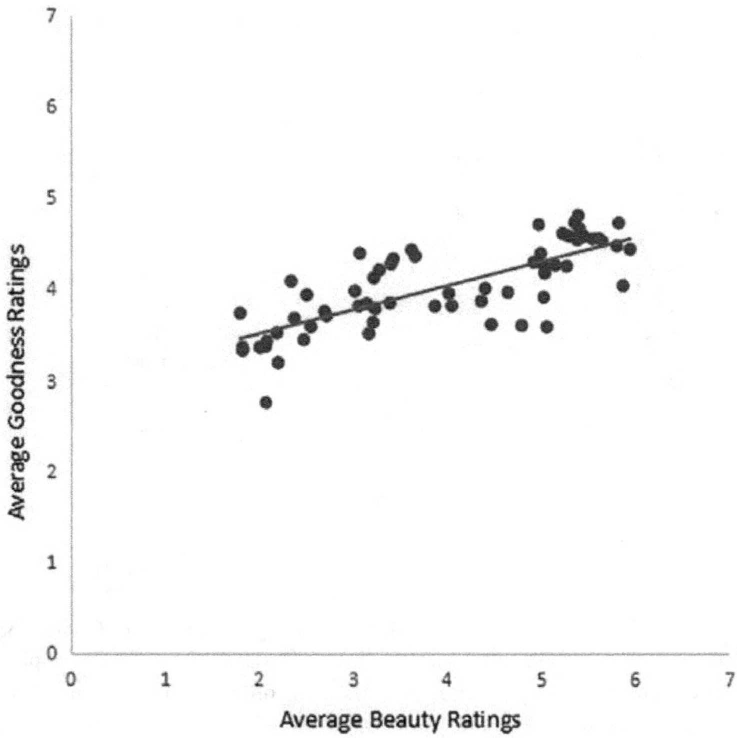

Figure 2. Scatter plot showing the positive correlation between average beauty and goodness ratings per face. Beauty and goodness ratings were taken on a 7-point scale and averaged for each face stimulus in preparation for the Spearman rank-order correlation analysis.

were control ones, in which the initial and peer ratings were equal. To learn whether external information affected RC, paired-sample sign tests were used; these compared the average RC produced by feedback trials to that of control trials (see Figure 3). For beauty trials, there was a significant difference between the average RC of feedback versus control trials (mean RC = 0.30, *SD* = 0.24 and mean RC = 0.17, *SD* = 0.20, respectively; $p<0.001$, one-tailed, Holm–Bonferroni adjusted), with 71 of the 95 participants exhibiting a higher average RC for feedback trials. A significant difference was also found for goodness trials when comparing feedback and control trials (mean RC = 0.34, *SD* = 0.26 and mean RC = 0.14,

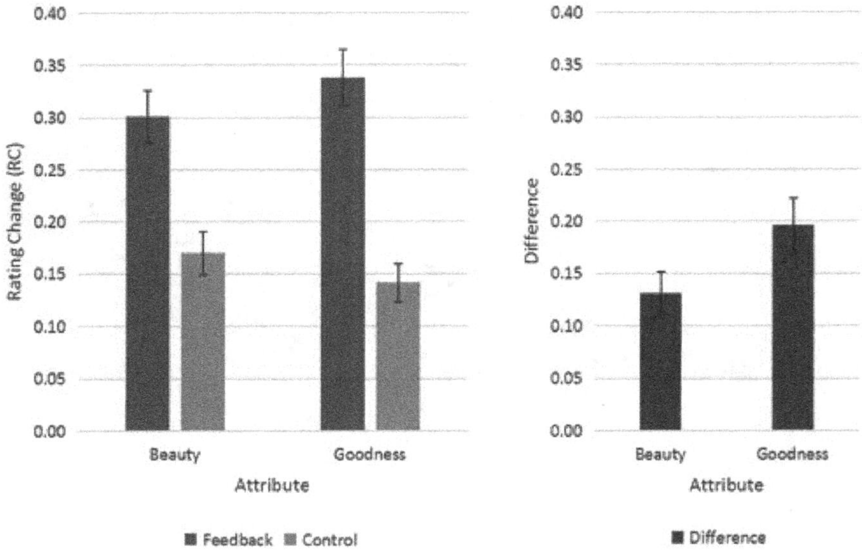

(a) (b)

Figure 3. Bar chart showing RC in the different study conditions. (a) Average RC was calculated for feedback and control trials for both beauty and goodness and is plotted on the *y*-axis. (b) The bars labeled "difference" show the difference in average RC between feedback and control trials. For both panels, attribute (beauty or goodness) is plotted on the *x*-axis. Error bars represent standard error.

SD = 0.18, respectively; $p < 0.001$, one-tailed, Holm–Bonferroni adjusted), with 72 of the 95 participants exhibiting a higher average RC for feedback trials. Although these results show that external information influenced both beauty and goodness ratings, the results below show that there was a difference in the magnitude of the effect.

The next sign tests aimed to establish whether external influence affects medium versus extreme ratings differentially. For beauty trials, no significant difference was detected between extreme and medium ratings (mean RC = 0.30, *SD* = 0.29 and mean RC = 0.30, *SD* = 0.28, respectively; $p = 0.460$, one-tailed, Holm–Bonferroni adjusted). The same was true for goodness trials (mean RC = .40, *SD* = 0.33 and mean RC = 0.34, *SD* = 0.26, respectively; $p = 0.861$, one-tailed, Holm–Bonferroni adjusted). This shows that external information influences extreme and medium ratings in a similar manner.

To address our hypothesis that beauty ratings would be more resistant to external influence than goodness ratings, we undertook another sign test. To ensure we were using values solely pertaining to the effect of external information, we subtracted participants' average control trial RC from their average feedback trial RC (i.e., the bars labeled "Difference" in Figure 3). The values obtained were used for the sign test, which showed a significant difference between the average RC for beauty compared to goodness trials (mean RC = 0.13, $SD = 0.20$ and mean RC = 0.20, $SD = 0.26$, respectively; $p = 0.045$, one-tailed, Holm–Bonferroni adjusted). Of the 94 out of 95 participants who exhibited a difference, 58 had a higher average RC for goodness trials than beauty trials, and the remaining 36 exhibited the reverse. This shows that, on average, participants were less resistant to external information when re-evaluating goodness ratings than beauty ratings.

Analysis of certainty and consensus

In each trial, participants gave certainty ratings regarding their initial judgments of beauty and goodness. Despite a Spearman's correlation analysis, undertaken within each category, showing no significant correlation between the average certainty and RC ($r_s[93] = -0.170$, $p = 0.099$ and $r_s[93] = 0.135$, $p = 0.193$, for beauty and goodness judgments, respectively), a sign test showed that the average certainty rating for beauty judgments was significantly higher than that for goodness judgments (mean = 4.73, $SD = 1.05$ and mean = 4.27, $SD = 0.93$, respectively; $p = 0.004$, one-tailed). This is in line with the finding of a higher resistance to external influence for beauty compared to goodness judgments.

We next examined agreement between participants for both beauty and goodness ratings by way of a MM1 analysis (see Method section for details). The analysis showed that the average of all the correlations between each participant's ratings and the average ratings of all other participants was higher for beauty ratings than for goodness ratings (MM1 = 0.81, 95% confidence interval [CI] = [0.74, 0.88] and MM1 = 0.46, 95% CI = [0.41, 0.51], respectively; see Figure 4). A paired-sample sign test carried out on the Fisher Z-transformed values revealed that agreement within beauty ratings was significantly greater than agreement

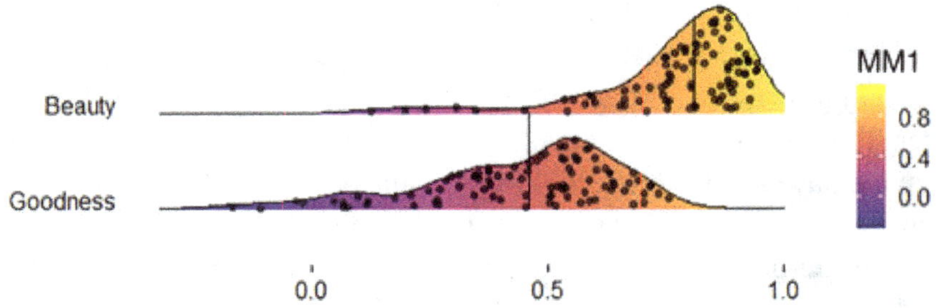

Figure 4. Individual and average agreement scores amongst beauty and goodness ratings. The ratings of each participant were correlated with the average ratings of all the other participants combined. The values obtained are presented in the graph, with each dot representing an individual agreement score. The vertical black lines represent the overall MM1 value for each attribute.

within goodness ratings (mean z-score = 1.13, SD = 0.36 and mean z-score = 0.50, SD = 0.25, respectively; $p<0.001$, one-tailed). The results of the MM1 analysis were replicated in an intraclass correlation coefficient analysis, a similar method used to examine inter-rater agreement (see Supplementary Materials 5). Both the certainty and agreement analyses fit with our hypothesis by demonstrating that beauty and goodness judgments can be psychophysically separated.

Discussion

We aimed to identify the extent to which judgments of facial and moral beauty can be separated from one another psychophysically. Our results show that, regardless of the correlation between the two, goodness ratings were more hospitable to external influence than beauty ratings. There was also significantly less agreement amongst subjects in their ratings for moral beauty, as well as significantly less certainty for them. Each of these findings is in line with our hypothesis, which was that we could separate beauty and goodness ratings using measures of conformity, agreement, and certainty (see also Vessel *et al.*, 2018). As well as providing evidence for their differentiation, these findings shed light on how we can classify facial and moral beauty in relation to our theory of the distinction between biological and artifactual categories of priors (Zeki & Chén, 2020).

Biological and artifactual categories refer to the extreme ends of a spectrum of priors (Zeki & Chén, 2020). In color vision, an inherited concept — that of ratio-taking (Land, 1986) — is applied to incoming chromatic visual signals. This ratio-taking mechanism generates stable biological priors (namely, the constant color categories and the hues attached to them; Zeki *et al.*, 2020). These priors result in highly consistent color categorization that is independent of learning, memory, and judgment and remains unchanged when confronted with changes in the wavelength composition of the light reflected from surfaces (Zeki & Chén, 2020).

The recognition of a stimulus as a face is another example of the interfacing of incoming visual signals with an inherited or rapidly acquired brain concept which identifies a certain significant configuration as constituting a face. It is difficult to override this even with prolonged daily exposure (for a month) to mutilated faces; the brain's perceptive system will continue to identify the latter as departures from the norm (Chen & Zeki, 2011).

Facial beauty is another example of a process that is dictated by biological priors (see Introduction), but it is not quite as resistant to external influence as the categorization of color or the determination of a significant configuration as indicating a face (Klucharev *et al.*, 2009; Zaki *et al.*, 2011); on the other hand, it is more resistant to external influence than artistic judgments (which constitute an example of reliance on artifactual priors; Bignardi *et al.*, 2020).

Where moral beauty fits on this scale of priors is unclear; there has been much discussion on whether a sense of morality is inherited or acquired. Some accounts emphasize the role of evolution in moral acquisition (e.g., Darwin, 1874; Krebs, 2008), whereas others focus on the role of social learning (e.g., Bandura, 1991). There are some moral norms that are apparent across the vast majority of cultures, such as reciprocity (Gouldner, 1960) and obligation to fulfill social duties (Berkowitz, 1972). However, individuals from different cultures, and even individuals within the same culture, define and prioritize these moral norms differently (Vasquez *et al.*, 2001), as is clear from the history of conflicts in the 20th century, indicating the importance of social context on the development of a sense of morality. We hypothesize that the capacity to develop moral values is inherited, but that the moral values themselves are acquired and subject to influence. The results of this study, alongside the fact that the acquisition of moral values is heavily dependent upon social experience, support the proposition that moral beauty is reliant primarily upon artifactual priors as opposed to biological ones, regardless of the evolved mechanisms providing the capacity for moral development.

In conclusion, we have demonstrated that facial and moral beauty can be psychophysically separated. Given this, and previous accounts that have emphasised the importance of social learning and environment in the development of moral values, we conclude that moral beauty relies more upon acquired artifactual priors than biological ones.

Acknowledgments

This work was supported by the Leverhulme Trust, London.

References

Bandura, A. (1991). Social cognitive theory of moral thought and action. In W. M. Kurtines & J. L. Gewirtz (eds.), *Handbook of Moral Behavior and Development*, Vol. 1, pp. 45–103. Hillsdale, NJ: Erlbaum.

Berkowitz, L. (1972). Social norms, feelings, and other factors affecting helping and altruism. In *Advances in Experimental Social Psychology*, Vol. 6, pp. 63–108. San Diego, CA: Academic Press.

Bignardi, G., Ishizu, T., & Zeki, S. (2020). The differential power of extraneous influences to modify aesthetic judgments of biological and artifactual stimuli. *PsyCh Journal*, 10, 190–199. https://doi.org/10.1002/pchj.415.

Bronstad, P. M. & Russell, R. (2007). Beauty is in the 'we' of the beholder: Greater agreement on facial attractiveness among close relations. *Perception*, 36(11), 1674–1681. https://doi.org/10.1068/p5793.

Chen, C. H. & Zeki, S. (2011). Frontoparietal activation distinguishes face and space from artifact concepts. *Journal of Cognitive Neuroscience*, 23(9), 2558–2568. https://doi.org/10.1162/jocn.2011.21617.

Darwin, C. (1874). The descent of man and selection in relation to sex. UK: Rand, McNally & Company.

DeBruine, L. & Jones, B. (2017). Face research lab London set. *Figshare*. https://doi.org/10.6084/m9.figshare.5047666.v3.

Dion, K., Berscheid, E., & Walster, E. (1972). What is beautiful is good. *Journal of Personality and Social Psychology*, 24(3), 285–290. https://doi.org/10.1037/h0033731.

Eagly, A. H., Ashmore, R. D., Makhijani, M. G., & Longo, L. C. (1991). What is beautiful is good, but…: A meta-analytic review of research on the physical attractiveness stereotype. *Psychological Bulletin*, 110(1), 109–128. https://doi.org/10.1037/0033-2909.110.1.109.

Goren, C. C., Sarty, M., & Wu, P. Y. (1975). Visual following and pattern discrimination of face-like stimuli by newborn infants. *Pediatrics*, 56(4), 544–549.

Gouldner, A. W. (1960). The norm of reciprocity: A preliminary statement. *American Sociological Review*, 25, 161–178.

Ishizu, T. & Zeki, S. (2011). Toward a brain-based theory of beauty. *PLoS One*, 6(7), e21852. https://doi.org/10.1371/journal.pone.0021852.

Ishizu, T. & Zeki, S. (2013). The brain's specialized systems for aesthetic and perceptual judgments. *European Journal of Neuroscience*, 37(9), 1413–1420. https://doi.org/10.1111/ejn.12135.

Jenkins, I. D. & Turner, V. (2009). *The Greek Body*. Los Angeles, CA: Getty Publications.

Kawabata, H. & Zeki, S. (2004). Neural correlates of beauty. *Journal of Neurophysiology*, 91(4), 1699–1705. https://doi.org/10.1152/jn.00696.2003.

Klucharev, V., Hytönen, K., Rijpkema, M., Smidts, A., & Fernández, G. (2009). Reinforcement learning signal predicts social conformity. *Neuron*, 61(1), 140–151. https://doi.org/10.1016/j.neuron.2008.11.027.

Kranz, F. & Ishai, A. (2006). Face perception is modulated by sexual preference. *Current Biology*, 16(1), 63–68. https://doi.org/10.1016/j.cub.2005.10.070.

Krebs, D. L. (2008). Morality: An evolutionary account. *Perspectives on Psychological Science*, 3(3), 149–172. https://doi.org/10.1111/j.1745-6924.2008.00072.x.

Land, E. H. (1986). An alternative technique for the computation of the designator in the retinex theory of color vision. *Proceedings of the National Academy of Sciences*, 83(10), 3078–3080. https://doi.org/10.1073/pnas.83.10.3078.

Langlois, J. H., Kalakanis, L., Rubenstein, A. J., Larson, A., Hallam, M., & Smoot, M. (2000). Maxims or myths of beauty? A meta-analytic and theoretical review. *Psychological Bulletin*, 126(3), 390–423. https://doi.org/10.1037/0033-2909.126.3.390

Langlois, J. H., Ritter, J. M., Roggman, L. A., & Vaughn, L. S. (1991). Facial diversity and infant preferences for attractive faces. *Developmental Psychology*, 27(1), 79–84. https://doi.org/10.1037/0012-1649.27.1.79

Ma, D. S., Correll, J., & Wittenbrink, B. (2015). The Chicago face database: A free stimulus set of faces and norming data. *Behavior Research Methods*, 47(4), 1122–1135. https://doi.org/10.3758/s13428-014-0532-5.

Ma, F., Xu, F., & Luo, X. (2016). Children's facial trustworthiness judgments: Agreement and relationship with facial attractiveness. *Frontiers in Psychology*, 7, 499. https://doi.org/10.3389/fpsyg.2016.00499.

O'Doherty, J., Winston, J., Critchley, H., Perrett, D., Burt, D. M., & Dolan, R. J. (2003). Beauty in a smile: The role of medial orbitofrontal cortex in facial attractiveness. *Neuropsychologia*, 41(2), 147–155. https://doi.org/10.1016/S0028-3932(02)00145-8.

Peirce, J., Gray, J. R., Simpson, S., MacAskill, M., Höchenberger, R., Sogo, H., Kastman, E., & Lindeløv, J. K. (2019). PsychoPy2: Experiments in behavior made easy. *Behavior Research Methods*, 51(1), 195–203. https://doi.org/10.3758/s13428-018-01193-y.

Shinners, E. (2009). Effects of the "what is beautiful is good" stereotype on perceived trustworthiness. *UW-L Journal of Undergraduate Research*, 12, 1–5.

Slater, A., Von der Schulenburg, C., Brown, E., Badenoch, M., Butterworth, G., Parsons, S., & Samuels, C. (1998). Newborn infants prefer attractive faces. *Infant Behavior and Development*, 21(2), 345–354. https://doi.org/10.1016/S0163-6383(98)90011-X.

R-Studio Team (2020). *RStudio: Integrated Development for R*. Boston, MA: RStudio, PBC.

Tsukiura, T. & Cabeza, R. (2011). Shared brain activity for aesthetic and moral judgments: Implications for the beauty-is-good stereotype. *Social Cognitive and Affective Neuroscience*, 6(1), 138–148. https://doi.org/10.1093/scan/nsq025.

Vasquez, K., Keltner, D., Ebenbach, D. H., & Banaszynski, T. L. (2001). Cultural variation and similarity in moral rhetorics: Voices from The Philippines and the United States. *Journal of Cross-Cultural Psychology*, 32(1), 93–120. https://doi.org/10.1177/0022022101032001010.

Vessel, E. A., Maurer, N., Denker, A. H., & Starr, G. G. (2018). Stronger shared taste for natural aesthetic domains than for artifacts of human culture. *Cognition*, 179, 121–131. https://doi.org/10.1016/j.cognition.2018.06.009.

Wang, T., Mo, L., Mo, C., Tan, L. H., Cant, J. S., Zhong, L., & Cupchik, G. (2015). Is moral beauty different from facial beauty? Evidence from an fMRI study. *Social Cognitive and Affective Neuroscience*, 10(6), 814–823. https://doi.org/10.1093/scan/nsu123.

Xu, F., Wu, D., Toriyama, R., Ma, F., Itakura, S., & Lee, K. (2012). Similarities and differences in Chinese and Caucasian adults' use of facial cues for trustworthiness judgments. *PLoS One*, 7(4), e34859. https://doi.org/10.1371/journal.pone.0034859.

Yang, T., Formuli, A., Paolini, M., & Zeki, S. (2021). The neural determinants of beauty. *bioRxiv*. https://doi.org/10.1101/2021.05.21.444999.

Zaki, J., Schirmer, J., & Mitchell, J. P. (2011). Social influence modulates the neural computation of value. *Psychological Science*, 22(7), 894–900. https://doi.org/10.1177/0956797611411057.

Zebrowitz, L. A. & McDonald, S. M. (1991). The impact of litigants' baby-facedness and attractiveness on adjudications in small claims courts. *Law and Human Behavior*, 15(6), 603–623. https://doi.org/10.1007/BF01065855.

Zeki, S. & Chén, O. Y. (2020). The Bayesian-Laplacian brain. *European Journal of Neuroscience*, 51(6), 1441–1462. https://doi.org/10.1111/ejn.14540.

Zeki, S., Javier, A., & Mylonas, D. (2020). The biological basis of the experience and categorization of colour. *European Journal of Neuroscience*, 51(2), 670–680. https://doi.org/10.1111/ejn.14557.

Zeki, S. & Romaya, J. P. (2010). The brain reaction to viewing faces of opposite- and same- sex romantic partners. *PLoS One*, 5(12), e15802. https://doi.org/10.1371/journal.pone.0015802.

Zeki, S., Romaya, J. P., Benincasa, D. M., & Atiyah, M. F. (2014). The experience of mathematical beauty and its neural correlates. *Frontiers in Human Neuroscience*, 8, 68. https://doi.org/10.3389/fnhum.2014.00068.

https://doi.org/10.1142/9789811298639_0025

Chapter 25

The Differential Power of Extraneous Influences to Modify Aesthetic Judgments of Biological and Artifactual Stimuli*

Giacomo Bignardi[†,‡], Tomohiro Ishizu[†,§] and Semir Zeki[†]

[†]*Laboratory of Neurobiology, Division of Cell and Developmental Biology, University College London, London, UK*
[‡]*Max Planck School of Cognition, Leipzig, Germany*
[§]*Department of Psychology, Kansai University, Osaka, Japan*

Abstract

We addressed the question of the extent to which external information is capable of modifying aesthetic ratings given to two different categories of stimuli — images of faces (which belong to the biological category) and those of abstract paintings with no recognizable objects (which sit in the artifactual category). A total of 51 participants of different national origins rated the beauty of both sets of stimuli, indicating the

*This chapter was originally published in the special collection on neuroaesthetics in *PsyCh Journal*. Bignardi, G., Ishizu, T., & Zeki, S. (2021). The differential power of extraneous influences to modify aesthetic judgments of biological and artifactual stimuli. *PsyCh Journal*, 10(2), 190–199. https://doi.org/10.1002/pchj.415.

certainty of their rating; they then re-rated them after being exposed to the opinions of others on their aesthetic status. Of these 51 participants, 42 who met our criteria were selected to complete the experiment. The results showed that individuals were less prone to modifying their ratings of stimuli belonging to the biological category compared to those falling into the artifactual category. We discuss this finding in light of our theoretical Bayesian–Laplacian model and on the evidence given by previous empirical research.

***Keywords*:** artifactual beauty; biological beauty; neuroesthetics

It is common knowledge that our judgments in general may be affected by several factors; these include our mental state, previous experience, and psychological predisposition when making a judgment. Our opinions and judgments may be especially hostage to external influences, which may include the financial or aesthetic value attributed by others to the object or situation being judged, its supposed authenticity, or the context in which it is judged (Gartus & Leder, 2014; Huang, Bridge, Kemp, & Parker, 2011; Kirk, Skov, Hulme, Christensen, & Zeki, 2009). These general extraneous influences have been variously referred to as contextual effects, conformity effects, and schema changes, *inter alia*. We prefer in this work to use the term "external information". The effects of external influences in modifying judgments have been studied in a variety of situations, including consumer behavior and political and social decision-making (Cialdini & Goldstein, 2004; Wagner, Menninghaus, Hanich, & Jacobsen, 2014; Zaki, Schirmer, & Mitchell, 2011). Here, we address the question of the extent to which external influences, consisting of the opinions of others, modify our aesthetic judgments.

While accepting that aesthetic judgments may be hostage to external opinions (Izuma & Adolphs, 2013), we set out to learn something about the distribution and potency of such external influences. In particular, we hypothesized that aesthetic judgments of images that belong to the biological categories, such as human faces, would be much more resistant to external opinion than aesthetic judgments of objects that may be considered to fall better under the category of artifactual, man-made objects; we consider abstract, non-representational paintings to fall into the latter category. This can be traced back to our general Bayesian–Laplacian classification of experiences into biological (β) and non-biological or artifactual (α) priors (Zeki & Chén, 2020). At the root of this classification is our belief (Zeki, 2011), supported by recent empirical evidence (Vessel, Maurer, Denker, & Starr, 2018), that there is broader agreement between humans in general in judgments regarding biological categories than in those regarding artifactual categories. This is not to say that all biological categories are totally immune from external influences, as previous studies on facial attractiveness suggest, but only that they are significantly more so than artifactual categories (Klucharev, Hytönen, Rijpkema, Smidts, & Fernández, 2009; Zaki *et al.*, 2011). But to what extent is the judgment of how beautiful a face is (an aesthetic judgment) subject to external influence? Faces judged to be beautiful by individuals belonging to one ethnic or cultural grouping are, broadly, also regarded to be beautiful by those belonging to a different culture. Moreover, infants orient more toward beautiful faces, regardless of ethnicity, implying that there is some significant cross-cultural agreement about facial aesthetics (Cunningham, Roberts, Barbee, Druen, & Wu, 1995; Fink & Neave, 2005; Langlois *et al.*, 2000; Langlois, Ritter, Roggman, & Vaughn, 1991). Given this, it was interesting to enquire further into the extent to which judgments of facial beauty are resistant to external influence.

There is, of course, a wide variety of ways of addressing this problem. In this initial approach, we chose to begin by comparing two sets of stimuli in terms of how external information, defined as the aesthetic judgment given by others, affects the aesthetic rating given by participants to faces and to abstract paintings with no biologically recognizable elements (thus falling into the artifactual category). We did so by asking participants to rate the two sets of stimuli aesthetically before and after exposing them to external influences. Our hypotheses were threefold: first, that there will be greater unanimity among individuals in aesthetic ratings of images of faces than those of abstract artworks; next, that aesthetic ratings given to faces will be more resistant to external influences than those given to abstract artworks; and finally, that extreme aesthetic ratings will be less susceptible to external influences. Our results supported the first and second of these hypotheses but not the third.

Method

Participants

One hundred and eleven participants were recruited through Prolific Academic, social media, and by word of mouth. Of these, only 51 completed the entire experiment (28 females and 23 males; age range: 18 to 65 years, mean age: 28.8 years [*SD* = 11.27 years]). Eight participants reported having had more than 6 years of general training in either fine art or art history, while the remaining 43 reported only an average of 1.56 years (*SD* = 1.59). Participants were nationals of 19 different countries (see Supplementary 1). Of these, 48 were native English speakers or spoke English fluently, while for the remaining three English was a second language; however, they declared themselves to have not had any trouble in understanding and executing the instructions. Forty-five percent of the participants described themselves as being not completely heterosexual, which was in line with previous reports on sexual orientation in the younger British population, as reported by an adapted Kinsey Scale (Zeki & Romaya, 2010) and by the recent YouGov poll among 18–24-year-olds in the United Kingdom (https://yougov.co.uk/topics/lifestyle/articles-reports/2015/08/16/half-young-not-heterosexual). All participants were informed that the study was designed to study beauty judgments. All aspects of our study conformed to the guidelines of the 2013 World Medical Association Declaration of Helsinki and were approved by the University College London Ethics Committee.

Stimuli

We used two types of stimuli. The first consisted of photographs of faces derived from different databases (the Face Research Lab London Set; DeBruine & Jones, 2017; and the Chicago Face Database; Ma, Correll, & Wittenbrink, 2015) and from a variety of other online sources (Pexels, Adobe Stock, Pinterest, and Google Images). Based on the attractiveness ratings, scored from 1 to 7, given in those databases, in addition to our preliminary study in which we assessed the average attractiveness of the images available from the other online sources, we selected the following:

20 images with the lowest attractiveness ratings (10 male and 10 female), 20 images with the highest attractive ratings (10 male and 10 female), and 20 below and 20 above the median attractiveness ratings (20 male and 20 female; see Supplementary 2 for details).

We next collected photographs of abstract paintings obtained from an online source containing averaged beauty ratings scored from 1 to 9 (Sidhu, McDougall, Jalava, & Bodner, 2018). From these, we selected the following: 20 images with the lowest beauty ratings, 20 with the highest beauty ratings, and 20 below and 20 above the median ratings. Three abstract artworks that resembled bodily-like shapes were removed and replaced with artworks with similar beauty ratings prior to the experiment. The size of all images was converted into 600 × 600 pixels in Photoshop, with the background set to white.

Procedure

The experimental procedure was designed in PsychoPy 3.0 (Peirce *et al.*, 2019) and executed using the online platform Pavlovia (https://pavlovia.org/). After giving informed consent, participants were asked to give their sex and sexual orientation (to ascertain whether there was a sexual-orientation-biased aesthetic rating for faces), and their years of training in either fine arts or art history. Sexual orientation was obtained by using an adapted Kinsey Scale (Kinsey, Pomeroy & Martin, 2003; Zeki & Romaya, 2010), which scales the sexuality of the participants from 1 (*completely heterosexual*) to 7 (*completely homosexual*). Years of training in the arts were similarly collected using a 1 to 7 scale (from 0 *years* to 6+ *years* of training).

Participants were then instructed on the behavioral task and, to familiarize them with it, one practice session was administered before beginning the experiment.

The experiment consisted of two blocks (faces/abstract artworks) of 80 trials each. Each trial had four components (questions): Participants were first asked to give (a) their beauty rating and (b) how certain they were of their rating. They were then exposed to (c) external information (of which the details are given in the following). They were, finally, asked

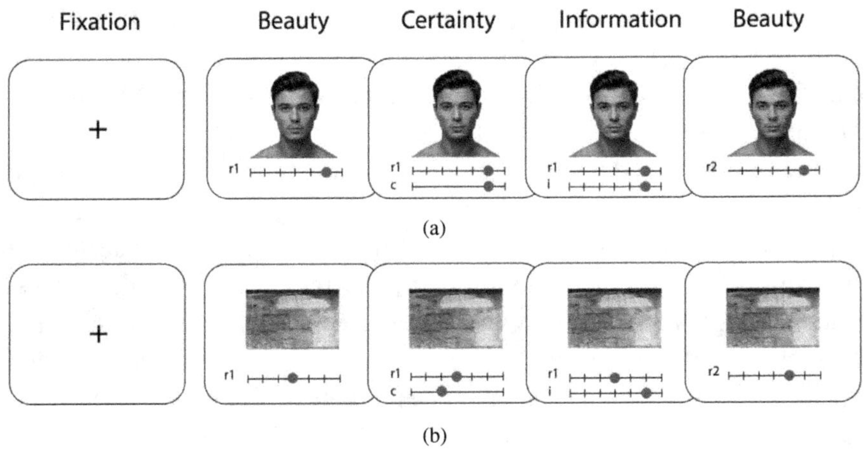

Figure 1. Online experimental paradigm. Participants were asked to complete 80 trials per category (total of 160 images). After a brief presentation of a fixation cross (500 ms), each trial consisted of four steps. A first beauty rating: Participants were asked to give their beauty rating (r1. "how beautiful?" from 1 to 7). Their first rating (r1) was displayed for the entire duration of the trial. Certainty: Subjects were then asked to indicate how certain they were of their rating, on a continuous 1–7 scale, from not at all to totally. Immediately following this, participants were told to compare their rating with the fictitious average rating (i). A final beauty rating: After being exposed to the fictitious average rating, participants were asked to give their final beauty rating (r2). (a) Example of the timeline for one trial of beauty rating for faces. (b) Example of the timeline for one trial of beauty rating for artworks.

once more (d) to give their final beauty rating in light of the external input (see Figure 1). During the external information phase, participants were shown fictitious average ratings which, they had been told, represented the average opinion of other participants (see Supplementary 3 for the instruction given to the participants). Fictitious rather than real ratings were used because we expected the ratings of biological stimuli to be more consistent across individuals than those of artifactual ones; hence using real ratings to sway opinion would have been less effective for biological than artifactual stimuli. Similar to Zaki *et al.* (2011), ratings were manipulated as follows: When participants rated any image less than three, the fictitious average rating was presented as being either equal to the initial participant rating or to two or three points higher than the original rating, the two being given with equal probability. Similarly, when participants rated any image more than five, the fictitious average rating was presented as being either equal to the initial participant rating or to two or three points lower than that rating, the two being given with equal probability. Finally, for ratings between three and five, the average rating was presented with a 50% chance of being equal, 25% of being two or three points higher, and 25% of being two or three points lower.

The order of presentation of each trial in each block was randomized and the experiment was counterbalanced to avoid order-effect.

Analysis

We started with a mixed (within–between) analysis of covariance to test whether beauty ratings given to faces were affected by the sex of the participants (i.e., whether female participants rated images of males more beautiful than those of females on average and vice versa). Next, we computed the mean minus one (MM1; Vessel *et al.*, 2018) to inspect the average agreement between beauty ratings for images of faces and abstract artworks. The MM1 is calculated by computing the correlation between all the ratings given by one participant with the average ratings of every other participant per image, a procedure repeated for every participant. This resulted in individual preference scores (r values), which reflect the extent to which individual ratings are in agreement with the mean rating (Germine *et al.*, 2015; Sutherland *et al.*, 2020). We then applied a Fisher transformation on the r values (z values — Fisher r to z transformation); such a transformation has been shown to result in less biased estimates when assessing agreement between participants (Vessel *et al.*, 2018). After this, we took the mean of the z value and, following convention, transformed it back to r (Fisher z-to-r transformation; see Vessel *et al.*, 2018, for details).

To test our main hypothesis, that external information influences differentially aesthetic judgments of faces (biological) and abstract artwork (artifactual), we undertook three within-subject comparisons: First, we tested whether external information influences the experience of beauty in the sense of leading to a re-rating that departs from the original rating for both categories. Next, we tested whether such influences were higher when the initial beauty rating of the participant was medium (i.e., when a rating between 3 and 5 was given) than when it was extreme (i.e., when a rating between 1 and 2 or between 6 and 7 was given). Finally, we tested whether this external information effect acts differently on the judgment of beauty of faces versus that of abstract artworks. We then explored, using the average certainty expressed by the participants, individual differences in resistance to external information. All analyses were conducted in R Studio Version 1.2.5033 using R version 3.6.2 (R Core Team, 2013).

Results

Mean beauty ratings for faces and abstract artworks

Mean beauty ratings for faces and abstract images were 3.45 and 3.72, with standard deviations of 1.41 and 0.61, respectively. Consistent with previous studies, average beauty ratings for faces from female participants were highly correlated with average beauty ratings from male participants, $r(78) = 0.97$, $p<0.001$ (Germine *et al.*, 2015). Also, a two-way mixed analysis of covariance indicated that the sex of the participants did not significantly interact with beauty ratings for faces of the opposite sex, $F(1, 49) = 2.24$, $p = 0.141$; that is, both males and females rated the faces of the opposite sex in a similar way. Regardless of the sexual orientation of the participants, $F(1, 48) = 0.32$, $p = 0.575$, images of female faces were rated significantly more beautiful than those of males, with mean (*SD*) female beauty ratings of 3.67 (0.68), and mean (*SD*) male beauty ratings of 3.24 (0.81), $F(1, 49) = 20.93$, $p<0.001$ (see Supplementary 4, for effect sizes).

Average ratings of beauty for faces correlated highly with their average attractiveness rating, as reported in previous studies ($r_s = 0.89$, $p<0.001$; DeBruine & Jones, 2017; Ma *et al.*, 2015; see Supplementary 5 for scatter plot). This suggests that images of faces that are rated as attractive are, on average, also rated as beautiful. For the ratings of abstract artworks, however, there was only a weak correlation between our participants' ratings and ratings given in the previous study by Sidhu *et al.* (2018; $r_s = 0.27$, $p = 0.015$). Moreover, an analysis of the agreement in beauty ratings between our participants showed that the average of all the correlations (Fisher Z-transformed) between the set of ratings given by one participant, with the averages of the ratings of the same images given by every other participant per category, was high for faces, while it was low for abstract images (with MM1 = 0.79, 95% confidence interval [CI] = [0.69, 0.87] and MM1 = 0.35, 95% CI = [0.13, 0.54], respectively). In sum, the agreement in the ratings of beauty for faces was significantly higher than that for abstract artworks, as shown by a paired t-test between the z values of the individual preference scores, $t(50) = 13.96$, $p<0.001$ (see Figure 2 and Supplementary 4 for effect sizes).

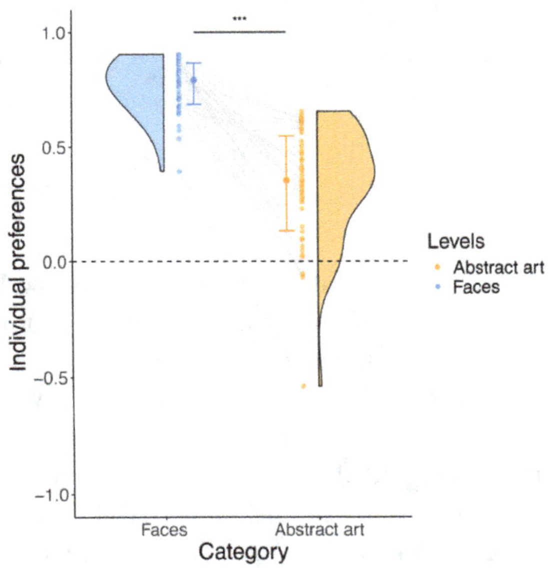

Figure 2. Mean minus one (MM1) beauty rating agreement across participants. Half violin plots of the distribution of agreement for beauty ratings for faces and abstract art. Dots represent individual preferences (i.e., the correlation between one's own beauty ratings and the average ratings given to each image by others). Higher scores indicate less idiosyncratic judgments. Gray lines connect individual preferences for the two categories (faces and abstract art) and central dots represent the MM1. Error bars represent 95% confidence interval (CI); the dashed horizontal line represents a total absence of agreement between participants' ratings. For further details, see text. Image computed adopting van Langen's (2020) tutorial. *** $p<0.001$.

Main analysis: Effect of external influences on beauty

Eight participants who reported having had more than 6 years of art or art history training were excluded from our analysis because previous reports suggest that art expertise confers a resistance to reassessment of ratings when valuing artworks (Kirk, Harvey, & Montague, 2011). Another participant was excluded because she was extremely uncertain of her ratings, giving an average certainty of 2.17, compared to the average certainty ratings of the others, which was 6.18. (SD = 0.77; total sample = 42). Excluding the outlier did not influence the results.

For each trial, the strength of the external influence was quantified as the difference between the second and the first beauty rating (rating change [RC]). To test the magnitude of the external influences on beauty ratings, we inverted the distribution of RC when the mismatching information was negative. That is:

$$RC(n) = \begin{cases} -(r2 - r1) & \text{if } i < r1 \\ r2 - r1 & \text{if } i > r1 \end{cases},$$

where n corresponds to the number of the trial, $r1$ to the beauty rating before exposure to external information, and $r2$ to the beauty rating after; i refers to the value of the external information given to the participant.

The first paired-sample sign test showed that there was a significant difference in RC between conditions in which there was mismatch between the first rating and the external opinion, and the conditions in which there was no such mismatch ($p < 0.001$, one-tailed, Holm–Bonferroni adjusted; see Figure 3(a) legend a for descriptive). That is, the average RC was higher for 33 participants, compared to the condition where there was no mismatch. This indicates that external information can influence the rating of beauty.

A second paired-sample sign test showed that there was no significant difference in the effect of external information on ratings that were extreme or medium; that is, the effect was more or less similar when the first rating was higher than 5 or lower than 3 versus when the rating was between 3 and 5 ($p = 0.37$, one-tailed, Holm–Bonferroni adjusted; see the Figure 3* legend for descriptive). This indicates that external information impacts extreme and average beauty ratings similarly.

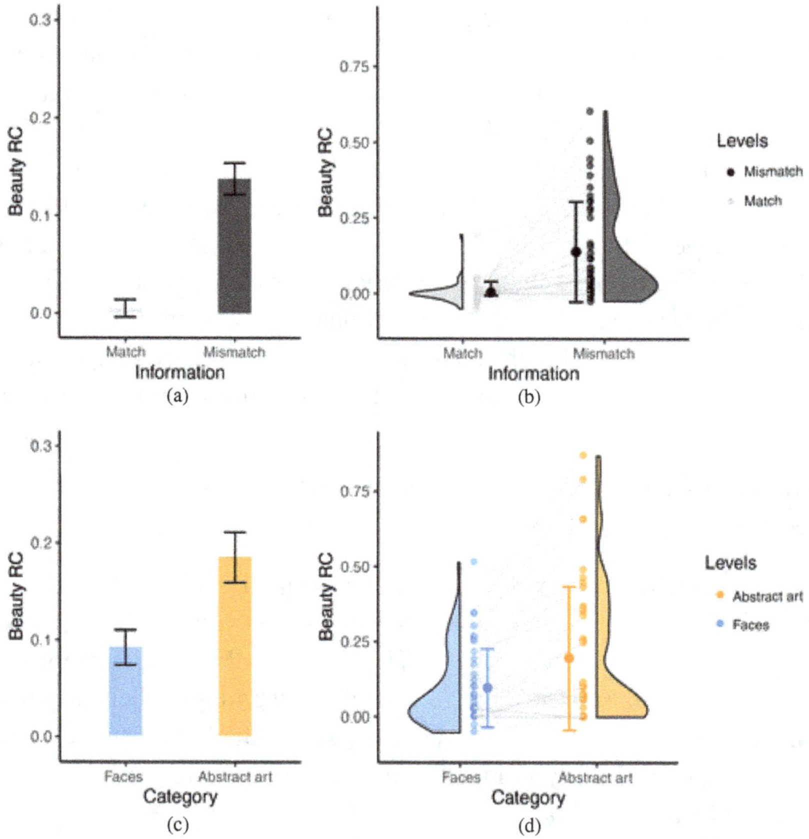

Figure 3. Effect of external information on beauty ratings. The y–axis represents the average beauty rating change (RC) between the first and the second ratings, while the x–axis represents the different conditions (external information and category). (a) Effect of external information (fictitious ratings) when the external information matched the first rating given (i.e., where the two ratings were equal) and when the two did not match, mean average RC = 0.01 (SD = 0.04), median = 0, mean average RC = 0.14 (SD = 0.17), median = 0.05, respectively. (b) Same as panel (a) but with individual differences shown. (c) Average effect of external information (fictitious ratings) on beauty ratings for images of faces versus images of abstract artworks; mean average RC = 0.09 (SD = 0.13), median = 0.03, mean average RC = 0.19 (SD = 0.24), median = 0.07, respectively. (d) Same as panel (c) but with individual differences shown. The error bars in (a) and (c) represent the 95% confidence interval (CI). (b) and (d) error bars represent the SD and the gray lines connect rating changes between the conditions (match, mismatch) or categories (faces, abstract art). Images computed adopting van Langen's (2020) tutorial. *The effect of external information on ratings that were extreme and medium; mean average RC = 0.13 (SD = 0.19), median = 0.06, mean average RC = 0.14 (SD = 0.20), median = 0.05, respectively.

Turning next to learn whether there was a difference in the effect of external information on beauty ratings for faces (biological) compared to ratings for abstract artwork (artifactual), we used a third paired sign test. This showed that the effect of external information on beauty ratings for images of faces was significantly lower than that for abstract artworks ($p = 0.017$, one-tailed, Holm–Bonferroni adjusted; see Figure 3(c) legend for descriptive). Thus, participants were on average more likely to change their beauty ratings in light of external information when evaluating images of abstract artwork than when evaluating faces; more specifically, of those participants who changed their ratings (35 out of 42), 25 did so more for abstract artworks than for faces, while only 10 did the reverse.

Analysis of individual differences in resistance to external influences

While on average, the effect of external information impacted beauty ratings significantly, the degree to which different individuals showed resistance to external influences varied. The slopes in Figure 3(b) represent participants' tendencies to change their beauty ratings when presented with external information; the steeper the slope, the higher the tendency of changing one's beauty ratings toward the average rating in light of the external information obtained.

We found that the more certain participants were of their ratings on average, the less they were prone to changing their initial ratings, and hence the more resistant they were to external information. This is reflected by the moderate correlation between the steepness of the slopes and the amount of certainty expressed by the individual, $r_s = -0.54$, $p<0.001$.

Moreover, subjects who showed the greatest tendency to modify their ratings in response to external opinion were those who modified their ratings more for abstract artworks (artifactual stimuli) than for faces (biological stimuli). This was reflected by the strong correlation between the individual slopes in Figure 3(b) and those in Figure 3(d), $r_s = 0.80$, $p<0.001$ (see also Supplementary 6 for details).

Discussion

In this work, we set out to explore the extent to which received opinion, which we refer to as external information, modifies an individual's aesthetic ratings for images of faces and abstract artworks. Past studies have shown that external information can influence judgments in general, including aesthetic ones, a result with which our present findings are consistent (Berns *et al.*, 2005; Cialdini & Goldstein, 2004; Kirk *et al.*, 2009; Zaki *et al.*, 2011). What remains unknown is whether this effect works uniformly for all kinds of aesthetic experiences. We have shown in this study that the capacity of such extraneous influences to modify the aesthetic judgments of images that fall under the category of biological stimuli is much less pronounced than that for those falling under the artifactual category. This was consistent with our theoretical formulation of the distinction between biological and artifactual experiences and the resistance of the former to be updated in light of experience compared to the latter (Zeki & Chén, 2020).

Studying the neural determinants of the experience of beauty brings with it difficulties quite unique and unlike anything that is encountered in studies of other sensory experiences. Individuals normally judge faces as beautiful even in spite of the fact that the faces they so rate differ significantly in many aspects — for example, more or differently colored hair, different shaped lips, varying extents of forehead, and so on; it is in fact the totality of the face that is being judged as being beautiful and so rated. Equally with artifactual beauty, the judgment is immediate even in spite of the fact that objects being judged may vary significantly; for example, both a building and a painting may be judged as beautiful or not, even in spite of the fact that the two belong to different categories. We tried to minimize the latter by restricting our category of artifactual objects to modern abstract paintings and ask subjects to give their rating on the basis of the whole painting. It stands to reason to expect that, had we asked subjects to rate paintings according to, say, the colors in them, we may have had different ratings. Equally, if we had asked subjects to rate faces according to specific features, such as the eyes, we might well expect to have different ratings. Instead, for both categories, we asked merely for the beauty of the entire stimulus, which we believe to be a perfectly justifiable way of proceeding. Future studies may well want to break these holistic stimuli into their components, which would be an interesting thing to do. One may ask subjects, for example, to rate the eyes only according to beauty or to rate only the colors in a painting; such future studies may yield interesting results.

Underlying our approach are some beliefs that are worth a brief mention. Along with Immanuel Kant, we believe that all experiences must be interfaced through concepts. However, we depart from Kant by partitioning aesthetic experiences into a spectrum ranging between two extremes, biological and artifactual (Zeki & Chén, 2020). Kant (1790/1987) supposed that aesthetic experiences, as opposed to "utilitarian" experiences, are interfaced through "indeterminate" concepts and, as a consequence, the experiencing individual could assume that his or her aesthetic judgment has universal assent. We, on the other hand, suppose that there should be some considerable agreement between individuals belonging to different ethnic and cultural groups in the aesthetic judgment of faces because they are interfaced through inherited brain concepts that are shared among all humans (Zeki, 2011; Zeki & Chén, 2020), which appears to be the case (see also Langlois *et al.*,1991, 2000). In line with our supposition, other studies indicate that assent on the perceived attractiveness of faces between individuals is substantially higher than assent for stimuli that fall under the artifactual category (Vessel *et al.*, 2018; Zeki & Chén, 2020); in other words, an individual's aesthetic judgment of faces corresponds better to his or her imagined aesthetic judgment that other individuals might make of the same face compared to the judgments that other individuals might make of abstract artworks (Leder, Goller, Rigotti, & Forster, 2016).

In this study, we use our hypothesis and results from previously published work to enquire into whether external influences have equal sway with aesthetic rating of faces (as representing biological stimuli) compared to abstract paintings (representing artifactual stimuli). In our study, too, agreement on the aesthetic ratings of faces was also higher than that for abstract artworks, and the aesthetic ratings of faces given by our participants were highly similar to those reported in previous studies. In contrast, the average aesthetic ratings of abstract artworks were only weakly related to the average aesthetic ratings of such works reported previously (Sidhu *et al.*, 2018).

An interesting corollary to the supposition that there are inherited brain concepts through which aesthetic judgments are interfaced is the extent to which individuals can update such concepts, or beliefs, in light of new experience, as is generally assumed in Bayesian inferential models. Some of our previous results showed that experiences such as color, interfaced through biologically inherited concepts, are more resistant to updating than those interfaced through acquired concepts (Zeki, 2011; Zeki & Chén, 2020). We emphasize that, even in the updating of beliefs based on biological concepts in light of experience, there are gradations; color categorizations represent perhaps an extreme in not allowing any significant updating, as do stimuli that have certain "significant configurations" that mark them as constituting a human face (Zeki, 2013; Zeki, Javier, & Mylonas, 2020). Humans are unlikely to modify their experience of what a normal face looks like even if exposed repetitively to disfigured faces (see also Chen & Zeki, 2011; Zeki & Ishizu, 2013, for a brain-based perspective of this issue). Equally, humans are unlikely to categorize a leaf as red even when it reflects more long-wave (red) light, as it does at dawn and dusk, because of brain operations resulting from inherited brain concepts that, through a comparison process, result in constant color categorization, as discussed at length elsewhere (Land, 1974, 1983; Zeki & Chén, 2020; Zeki *et al.*, 2020). In summary, the Bayesian system of updating beliefs is more potent the more stimuli depart from the biological toward the artifactual category because the acquired concepts through which they are interfaced are less rigid and therefore more modifiable.

The general view expressed above is reflected in the results of this study, which show that the capacity of extraneous influences to modify aesthetic judgment of faces, although present, is much less pronounced than that for abstract artworks and more pronounced than that for other biological categories, such as color. This implies that there is a gradation in the degree to which external influences can modify judgments within the biological category. Indeed, while aesthetic ratings of faces are substantially shared across individuals, there may in addition be individual differences (Germine *et al.*, 2015; Hönekopp, 2006; Vessel *et al.*, 2018) and such differences may mostly be accounted for by environmental rather than genetic factors (Germine *et al.*, 2015; Sutherland *et al.*, 2020). Hence, we believe that what is shared is not shared because of environmental influence but because of a common inheritance that leads to common brain concepts. Our belief is strengthened not only by the results of this study but also by studies that have shown that infants tend to look more at faces that are reliably classified as attractive by adult individuals (Slater *et al.*, 1998). We note that although our participants were drawn from 19 countries, most were nevertheless from Western countries. Hence, there was less ethnic than cultural diversity. Future studies will no doubt correct this imbalance.

We also found that there was no significant difference in the external influence on stimuli that had been highly rated and those that were not. This came as a surprise to us, since we had previously hypothesized that the more beautiful an image is perceived to be, the less likely the experiencing individual is to change their opinion of how beautiful it is. One possible interpretation of this null result is that the effect was too small to be detected by our sample size. Future work may shed more light on this counter-intuitive result.

Finally, we found that the capacity of external influences to modify ratings was related to the certainty with which individuals made their first aesthetic rating: The more certain they were of that, the less effective was an external influence in modifying their initial rating. This was true for both the biological and the artifactual categories used in this study.

We conclude by noting simply that the results presented here are consistent with our hypothesis of the presence of an inherited brain concept that can influence the experience of beauty of biological stimuli, such as faces.

Acknowledgment

This work was undertaken with a grant from the Leverhulme Trust, London.

References

Berns, G. S., Chappelow, J., Zink, C. F., Pagnoni, G., Martin-Skurski, M. E., & Richards, J. (2005). Neurobiological correlates of social conformity and independence during mental rotation. *Biological Psychiatry*, 58(3), 245–253. https://doi.org/10.1016/j.biopsych.2005.04.012.

Chen, C.-H. & Zeki, S. (2011). Frontoparietal activation distinguishes face and space from artifact concepts. *Journal of Cognitive Neuroscience*, 23(9), 2558–2568. https://doi.org/10.1162/jocn.2011.21617.

Cialdini, R. B. & Goldstein, N. J. (2004). Social influence: Compliance and conformity. *Annual Review of Psychology*, 55, 591–621. https://doi.org/10.1146/annurev.psych.55.090902.142015.

Cunningham, M. R., Roberts, A. R., Barbee, A. P., Druen, P. B., & Wu, C.-H. (1995). 'Their ideas of beauty are, on the whole, the same as ours': Consistency and variability in the cross-cultural perception of female physical attractiveness. *Journal of Personality and Social Psychology*, 68(2), 261–279. https://doi.org/10.1037/0022-3514.68.2.261.

DeBruine, L. & Jones, B. (2017). Face research lab London set. *Figshare*. https://doi.org/10.6084/m9.figshare.5047666.v3.

Fink, B. & Neave, N. (2005). The biology of facial beauty. *International Journal of Cosmetic Science*, 27(6), 317–325. https://doi.org/10.1111/j.1467-2494.2005.00286.x.

Gartus, A. & Leder, H. (2014). The white cube of the museum versus the gray cube of the street: The role of context in aesthetic evaluations. *Psychology of Aesthetics, Creativity, and the Arts*, 8(3), 311–320. https://doi.org/10.1037/a0036847.

Germine, L., Russell, R., Bronstad, P. M., Blokland, G. A. M., Smoller, J. W., Kwok, H., Anthony, S. E., Nakayama, K., Rhodes, G., & Wilmer, J. B. (2015). Individual aesthetic preferences for faces are shaped mostly by environments, not genes. *Current Biology*, 25(20), 2684–2689. https://doi.org/10.1016/j.cub.2015.08.048.

Huang, M., Bridge, H., Kemp, M. J., & Parker, A. J. (2011). Human cortical activity evoked by the assignment of authenticity when viewing works of art. *Frontiers in Human Neuroscience*, 5, 134. https://doi.org/10.3389/fnhum.2011.00134.

Hönekopp, J. (2006). Once more: Is beauty in the eye of the beholder? Relative contributions of private and shared taste to judgments of facial attractiveness.

Journal of Experimental Psychology: Human Perception and Performance, 32(2), 199–209. https://doi.org/10.1037/0096-1523.32.2.199.

Izuma, K. & Adolphs, R. (2013). Social manipulation of preference in the human brain. *Neuron*, 78(3), 563–573. https://doi.org/10.1016/j.neuron.2013.03.023.

Kant I. (1790/1987). *Critique of Judgment*, p. 576. Indianapolis, IN and Cambridge: Hackett Publishing Company.

Kinsey A., C., Pomeroy W., R., & Martin, C. E. (2003). Sexual behavior in the human male. *American Journal of Public Health*, 93(6), 894–898. http://dx.doi.org/10.2105/ajph.93.6.894.

Kirk, U., Harvey, A., & Montague, P. R. (2011). Domain expertise insulates against judgment bias by monetary favors through a modulation of ventromedial prefrontal cortex. *Proceedings of the National Academy of Sciences*, 108(25), 10332–10336. https://doi.org/10.1073/pnas.1019332108.

Kirk, U., Skov, M., Hulme, O., Christensen, M. S., & Zeki, S. (2009). Modulation of aesthetic value by semantic context: An fMRI study. *NeuroImage*, 44(3), 1125–1132. https://doi.org/10.1016/j.NeuroImage.2008.10.009.

Klucharev, V., Hytönen, K., Rijpkema, M., Smidts, A., & Fernández, G. (2009). Reinforcement learning signal predicts social conformity. *Neuron*, 61(1), 140–151. https://doi.org/10.1016/j.neuron.2008.11.027.

Land, E. H. (1974). The retinex theory of color vision. *Proceedings of the Royal Institution of Great Britain*, 47, 23–58.

Land, E. H. (1983). Recent advances in retinex theory and some implications for cortical computations: Color vision and the natural image. *Proceedings of the National Academy of Science USA*, 80, 5163–5169. https://doi.org/10.1073/pnas.80.16.5163.

Langlois, J. H., Kalakanis, L., Rubenstein, A. J., Larson, A., Hallam, M., & Smoot, M. (2000). Maxims or myths of beauty? A meta-analytic and theoretical review. *Psychological Bulletin*, 126(3), 390–423. https://doi.org/10.1037/0033-2909.126.3.390.

Langlois, J. H., Ritter, J. M., Roggman, L. A., & Vaughn, L. S. (1991). Facial diversity and infant preferences for attractive faces. *Developmental Psychology*, 27(1), 79–84. https://doi.org/10.1037/0012-1649.27.1.79.

Leder, H., Goller, J., Rigotti, T., & Forster, M. (2016). Private and shared taste in art and face appreciation. *Frontiers in Human Neuroscience*, 10, 155. https://doi.org/10.3389/fnhum.2016.00155.

Lindeløv, J. K. (2019). PsychoPy2: Experiments in behavior made easy. *Behavior Research Methods*, 51(1), 195–203. https://doi.org/10.3758/s13428-018-01193-y.

Ma, D. S., Correll, J., & Wittenbrink, B. (2015). The Chicago Face Database: A free stimulus set of faces and norming data. *Behavior Research Methods*, 47(4), 1122–1135. https://doi.org/10.3758/s13428-014-0532-5.

R Core Team (2013). *R: A Language and Environment for Statistical Computing*. Vienna, Austria: R Foundation for Statistical Computing. http://www.R-project.org/.

Sidhu, D. M., McDougall, K. H., Jalava, S. T., & Bodner, G. E. (2018). Prediction of beauty and liking ratings for abstract and representational paintings using subjective and objective measures. *PLoS One*, 13(7), e0200431. https://doi.org/10.1371/journal.pone.0200431.

Slater, A., Von der Schulenburg, C., Brown, E., Badenoch, M., Butterworth, G., Parsons, S., & Samuels, C. (1998). Newborn infants prefer attractive faces. *Infant Behavior and Development*, 21(2), 345–354. https://doi.org/10.1016/S0163-6383(98)90011-X.

Sutherland, C. A. M., Burton, N. S., Wilmer, J. B., Blokland, G. A. M., Germine, L., Palermo, R., Collova, J. R., & Rhodes, G. (2020). Individual differences in trust evaluations are shaped mostly by environments, not genes. *Proceedings of the National Academy of Sciences*, 117(19), 10218–10224. https://doi.org/10.1073/pnas.1920131117.

van Langen, J. (2020). *Open-Visualizations in R and Python* [Jupyter Notebook]. https://github.com/jorvlan/open-visualizations.

Vessel, E. A., Maurer, N., Denker, A. H., & Starr, G. G. (2018). Stronger shared taste for natural aesthetic domains than for artifacts of human culture. *Cognition*, 179, 121–131. https://doi.org/10.1016/j.cognition.2018.06.009.

Wagner, V., Menninghaus, W., Hanich, J., & Jacobsen, T. (2014). Art schema effects on affective experience: The case of disgusting images. *Psychology of Aesthetics, Creativity, and the Arts*, 8(2), 120–129. https://doi.org/10.1037/a0036126.

Zaki, J., Schirmer, J., & Mitchell, J. P. (2011). Social influence modulates the neural computation of value. *Psychological Science*, 22(7), 894–900. https://doi.org/10.1177/0956797611411057.

Zeki, S. & Chén, O. Y. (2020). The Bayesian-Laplacian brain. *European Journal of Neuroscience*, 51(6), 1441–1462. https://doi.org/10.1111/ejn.14540.

Zeki, S. & Ishizu, T. (2013). The "visual shock" of Francis Bacon: An essay in neuroesthetics. *Frontiers in Human Neuroscience*, 7, 850. https://doi.org/10.3389/fnhum.2013.00850.

Zeki, S. & Romaya, J. P. (2010). The brain reaction to viewing faces of opposite- and same-sex romantic partners. *PLoS One*, 5(12), e15802. https://doi.org/10.1371/journal.pone.0015802.

Zeki, S. (2011). *Splendors and Miseries of the Brain: Love, Creativity, and The Quest for Human Happiness*. Chichester, England: John Wiley & Sons.

Zeki, S. (2013). Clive Bell's "significant form" and the neurobiology of aesthetics. *Frontiers in Human Neuroscience*, 7, 730. https://doi.org/10.3389/fnhum.2013.00730

Zeki, S., Javier, A., & Mylonas, D. (2020). The biological basis of the experience and categorization of colour. *European Journal of Neuroscience*, 51(2), 670–680. https://doi.org/10.1111/ejn.14557.

https://doi.org/10.1142/9789811298639_0026

Chapter 26

Human Faces and Face-Like Stimuli are More Memorable*

Marianna Kapsetaki and Semir Zeki

Laboratory of Neurobiology, Department of Cell & Developmental Biology, University College London, London, UK

Abstract

We have previously suggested a distinction in the brain processes governing biological and artifactual stimuli. One of the best examples of the biological category consists of human faces, the perception of which appears to be determined by inherited mechanisms or ones rapidly acquired after birth. In extending this work, we inquire here whether there is a higher memorability for images of human faces and whether memorability declines with increasing departure from human faces; if so, the implication would add to the growing evidence of differences in the processing of biological versus artifactual stimuli. To do so, we used images and memorability scores from a large dataset of 58,741 images to compare the relative memorability of the following image categories: real human faces versus buildings, and extending this to a comparison of real human faces with five image categories that differ in their grade of

*This chapter was originally published in the special collection on neuroaesthetics in *PsyCh Journal*. Kapsetaki, M. E. & Zeki, S. (2022). Human faces and face-like stimuli are more memorable. *PsyCh Journal*, 11(5), 715–719. https://doi.org/10.1002/pchj.564.

resemblance to a real human face. Our findings show that, in general, when we compare the biological category of faces to the artifactual category of buildings, the former is more memorable. Furthermore, there is a gradient in which the more an image resembles a real human face, the more memorable it is. Thus, the previously identified differences in biological and artifactual images extend to the field of memory.

Keywords: artifactual images; biological images; faces; illusory faces; memorability

Introduction

Our ability to remember stimuli depends both on our internal states and on external factors which are intrinsic properties of the stimuli. An example of the former is the impoverishment of memorability in general during depressive states (Williams *et al.*, 2000), whereas an example of the latter is the dependence of the memorability of images containing animals on the number of animals in the image (Dubey *et al.*, 2015). Although many other stimulus properties have been examined with regard to their effect on memorability (e.g., Isola *et al.*, 2011), stimuli have not been previously studied for their memorability as a function of the two broad categories to which they belong, namely, the biological and artifactual ones.

We use the term "biological" to refer to non-man-made stimuli (e.g., human faces and bodies, trees, flowers, and animals), and the term "artifactual" to refer to man-made stimuli (e.g., motor vehicles, buildings, and technological devices). Our previous work has shown that there is a difference in the extent to which esthetic judgments for different categories of stimuli are resistant to revision in light of external opinion, with biological stimuli being more resistant than artifactual ones (Bignardi *et al.*, 2021; Chen & Zeki, 2011; Glennon & Zeki, 2021; Zeki & Chén, 2020; Zhang & Zeki, 2022). This distinction has been addressed mainly in the field of visual perception and has not yet been extended to the field of memory.

In this study, we examined memorability for stimuli by concentrating on faces as representative of biological stimuli compared to buildings as representative of artifactual stimuli. We chose our stimuli to represent different categories, beginning with real human faces compared to buildings and progressing through a comparison of real human faces to five categories that differed increasingly from human faces. Previous studies have examined whether there are differences in recognition memory and in neural activity patterns across object categories that differ in their animacy and size (Blumenthal *et al.*, 2018) and differences in neural activity patterns for faces, chairs, and buildings (Martin *et al.*, 2016). We wanted to address the question of how memorability varies, if at all, with stimuli that depart increasingly from resemblance to a real biological stimulus, namely, human faces.

We used a very large number of images derived from a large dataset. This has both advantages and disadvantages; the former revolves around the fact that the large number of stimuli includes both the biological and the artifactual categories as well as the subcategories within each, thus making it possible to make a broad comparison. The disadvantage is that the stimuli were not strictly controlled for a number of characteristics such as color, size, angle of view, and so on. We nevertheless hypothesized that if the division between the biological category of faces and artifactual categories is real, a difference in the memorability of human faces versus other categories would emerge even in the absence of strict controls for the factors mentioned.

Methods

In all three of our experiments, we used images and memorability scores from the LaMem dataset (http://memorability.csail.mit.edu/index.html; Khosla *et al.*, 2015). This dataset consists of 58,741 images with memorability scores for each of the images. Each image contains data from 80 Amazon Mechanical Turk workers on average. In Khosla *et al.*'s (2015) study, each image was presented for 600 ms, and participants were asked whether they had previously seen it during the experiment. Memorability scores can range from 0 (*lowest memorability*) to 1 (*highest memorability*). The memorability score of an image was computed as the hit rate of that image minus the false alarm rate of that image. For each image, we used the mean of the five memorability scores provided in the dataset.

Five scores were available because Khosla *et al.* split their data into five parts, each containing the training, validation, and testing sets. To determine the training and validation scores, they used a random half of their subjects and used the other half to find the testing scores.

We inspected carefully the 58,741 images and placed them into different categories depending on our question of interest. We excluded images that showed a word or words. The experimenters in this study were blind to the previously provided memorability scores. To examine differences in memorability scores between the different image categories, we used Welch's *t*-test in the first experiment and Welch's analysis of variance and the Games-Howell *post hoc* test in the second experiment. These tests were used because there was no homogeneity of variances, $p < 0.001$, in Levene's test.

In our first experiment, we examined the memorability of stimuli depicting only buildings or real human faces (Figure 1) because previous studies examining biological and artifactual stimuli have most often included these types of stimuli. We found 363 images that contained a building or buildings; we excluded images of famous buildings (e.g., the Twin Towers in New York or the Eiffel Tower in Paris). We found 129 images that contained a real human face; here, we excluded images that showed more than one face or showed only parts of a face (e.g., only one eye or only the mouth). No representation of faces or buildings in paintings was used.

Figure 1. Categories of images analyzed in Experiment 1. (a) Real human faces and (b) buildings. The yellow bar illustrates the increase in memorability from (b) to (a).

We followed this by comparing the memorability scores for stimuli falling into six different image categories, each successive category departing increasingly from real human faces: real human faces (Figure 2(a)), face-like stimuli (Figure 2(b)), animal faces resembling

Figure 2. Categories of images analyzed in Experiment 2. (a) Real human faces, (b) face-like stimuli, (c) animal faces resembling human ones, (d) animal faces with lesser resemblance to human faces, (e) imagined-face-like images, and (f) non-face-like images. The yellow bar illustrates the increase in memorability from (f) to (a).

human ones (Figure 2(c)), animal faces with lesser resemblance to human faces (Figure 2(d)), imagined-face-like stimuli (Figure 2(e)), and non-face-like images (Figure 2(f)). For the real human faces category, we used the 129 images from our first experiment. The face-like category (1,317 images) included replicas of real (animal or human) faces such as those shown in the form of a cartoon image, paintings, sculptures, skulls, dolls, or other toys. It also included examples of face pareidolia (i.e., the perception of illusory faces in random stimuli) in which it was easy to see the face-like configuration. In the "animal faces resembling human ones" category (2,022 images), we did not include the following: (a) images that apart from the animal face also contained artifactual stimuli or (b) animals whose head did not resemble a human face (e.g., butterfly, centipede, and jellyfish). We had a separate category named "animal faces with lesser resemblance to human faces" (508 images) in which we included images of animals whose head did not much resemble a human face (e.g., butterfly, centipede, jellyfish, snail, spider, and bee) and animals whose face could not be clearly seen (e.g., birds shown from a long distance). In the imagined-face-like category (9,379 images), we included examples of face pareidolia in which it was difficult to see the face-like configuration, that is, images that could look like a face if one searched for a long time and/or had some imagination and/or excellent ability to detect faces among many distracting stimuli. In the non-face-like category (13,750 images), we included images that did not contain a face and did not look face-like if seen for a short or long duration.

In our third experiment, we selected the 50 most memorable images and the 50 least memorable ones in the LaMem dataset. In that test, the memorability scores for the first category ranged from 0.9894 to 1 and in the latter category ranged from 0.2658 to 0.3644.

Results

The single most memorable category in our results belongs to real human faces, a result obtained by comparing real human faces against all other categories, $p<0.001$, $\eta^2 = 0.055$ (Figure 2 and Table 1) as well as real human faces against buildings (M = 0.859, SEM = 0.0050 for real

Table 1. Memorability scores for different image categories.

Image categories	M	SEM	Games-Howell test
Real human faces	0.859	0.0050	
All face-like stimuli (including face-like artifactual stimuli)	0.835	0.0021	$p < 0.001$
Animal faces resembling human ones	0.773	0.0021	$p < 0.001$
Animal faces with lesser resemblance to human faces	0.767	0.0044	$p < 0.894$
Imagined-face-like	0.732	0.0012	$p < 0.001$
Non-face-like	0.718	0.0011	$p < 0.001$

human faces; $M = 0.697$, $SEM = 0.0051$ for buildings; $p<0.001$, $\eta^2 = 0.5084$, Figure 1). We have noticed that of the 50 most memorable images in the LaMem dataset, 32 contained faces or face-like stimuli; of the 50 least memorable images, only 3 contained a face or face-like stimuli.

Discussion

In this study, we set out to explore whether the previously identified differences in the extent to which esthetic judgments of biological and artifactual images are resistant to updating in light of external opinion (Bignardi *et al.*, 2021; Zeki & Chén, 2020; Zhang & Zeki, 2022) extend to the field of memory. As expected, the largest differences in memorability were found when comparing biological stimuli that were those of real human faces and artifactual ones that were those of buildings. In fact, our results show that there is a gradient in which the more an image looks like a real human face, the more memorable it is (see Figure 2).

Three previous studies (Brady *et al.*, 2019; Dubey *et al.*, 2015; Maguire *et al.*, 2001) have examined memorability for somewhat similar types of stimuli to ours with a different approach and methodology. Maguire *et al.* (2001) examined memory for buildings, human faces, and animal faces, and Brady *et al.* (2019) compared memorability for stimuli that had different grades of resemblance to faces (from non-faces to unambiguous faces). These previous studies, in addition to using only black and white stimuli (unlike our stimuli, most of which were colored) and analyzing a smaller number of images, also presented them for durations that differ from ours, significantly so in the work of Maguire *et al.* (at 3,000 ms) and less so in the work of Brady *et al.* (at 500 ms). Dubey *et al.*'s (2015) comparison was between persons and animals, but their stimuli, whether those of humans or animals, were not restricted to their faces. Despite these differences, our results are similar to theirs in finding that memorability scores were higher for persons than for animals. Consistent with the results of Brady *et al.*, we also found that the more the images looked like a face, the more memorable they were.

A clear gradient in memorability, depending on how much images look like a real human face, may depend on several factors: It may be because subjects are more familiar (i.e., have more expertise) with real human faces than with non-face stimuli, or stimuli that have more self-related information (e.g., stimuli more like ourselves) are more important than ones containing non-self-related information and therefore more demanding of our attention. There seems to be a preference for face-like stimuli across vertebrate species that is present even at birth and is independent of experience (Di Giorgio *et al.*, 2017; Sugita, 2008). Another possibility is that faces may register more easily in our memory because they possess characteristics such as symmetry. Our eyes tend to look for symmetric structures in the world (Kootstra *et al.*, 2011). This increase in gaze would likely lead to more attention and to greater ability to remember that stimulus (Aly & Turk-Browne, 2016). Lastly, the judgment of the beauty of faces seems to be more resistant to revision in light of external influences (Bignardi *et al.*, 2021); therefore, when they are encoded, it may be that they are not affected so much by proactive and retroactive interference. It has also been argued that the brain has an inherited, or very rapidly acquired, template for faces and facial configuration, making their recognition easier (Yang *et al.*, 2022; Zeki, 2009).

Functional neuroimaging studies have shown that the fusiform face area and occipital face area play a role in processing both faces and face-like stimuli (e.g., Liu *et al.*, 2014; Wardle *et al.*, 2020). Importantly, these areas can distinguish human faces from face-like stimuli and also face-like stimuli from non-face-like stimuli (Wardle *et al.*, 2020). However, there is variability across participants and across illusory face images in the ability to perceive a face-like pattern and how face-like this pattern is perceived to be (Liu *et al.*, 2014; Wardle *et al.*, 2022). This variability may have been the reason why there was quite large variability in memorability scores in the imagined-face-like category of our second experiment.

We believe that complexity did not affect our results. Gilbert & Schleuder (1990) and Saraee *et al.* (2020) found that memorability was higher for more complex images. Therefore, according to their results, we would have expected that scenes would be more memorable than single stimuli such as faces, which is the opposite of what we found.

The limitations of using the LaMem dataset are that participant demographics were unavailable, the number of images per category was unbalanced, remember/know responses and confidence ratings were not collected, and it is unknown whether participants were previously familiar with the images shown (i.e., before seeing them in the experiment). Another limitation of our study is that we only examined one modality (vision) and only restricted ourselves to two-dimensional stimuli within that modality. It has been previously shown that there is a differentiation in memory for real three-dimensional stimuli versus photographs of real three-dimensional stimuli (Kapsetaki *et al.*, 2022; Snow *et al.*, 2014). Thus, future studies could examine whether our findings translate to actual stimuli in real life (i.e., real three-dimensional stimuli rather than images of them) and to stimuli in other modalities (e.g., audition and touch).

Acknowledgments

We would like to thank Sophia Sia for her assistance in categorizing some of the images. This work was supported by a grant from The Leverhulme Trust (RPG-2017-341), London, to S. Zeki.

References

Aly, M. & Turk-Browne, N. B. (2016). Attention promotes episodic encoding by stabilizing hippocampal representations. *Proceedings of the National Academy of Sciences*, 113(4), E420–E429. https://doi.org/10.1073/pnas.1518931113.

Bignardi, G., Ishizu, T., & Zeki, S. (2021). The differential power of extraneous influences to modify aesthetic judgments of biological and artifactual stimuli. *PsyCh Journal*, 10, 190–199. https://doi.org/10.1002/pchj.415.

Blumenthal, A., Stojanoski, B., Martin, C. B., Cusack, R., & Köhler, S. (2018). Animacy and real-world size shape object representations in the human medial temporal lobes. *Human Brain Mapping*, 39(9), 3779–3792. https://doi.org/10.1002/hbm.24212.

Brady, T. F., Alvarez, G. A., & Störmer, V. S. (2019). The role of meaning in visual memory: Face-selective brain activity predicts memory for ambiguous face stimuli. The *Journal of Neuroscience*, 39(6), 1100–1108. https://doi.org/10.1523/JNEUROSCI.1693-18.2018.

Chen, C. H. & Zeki, S. (2011). Frontoparietal activation distinguishes face and space from artifact concepts. *Journal of Cognitive Neuroscience*, 23(9), 2558–2568. https://doi.org/10.1162/jocn.2011.21617.

Di Giorgio, E., Loveland, J. L., Mayer, U., Rosa-Salva, O., Versace, E., & Vallortigara, G. (2017). Filial responses as predisposed and learned preferences: Early attachment in chicks and babies. *Behavioural Brain Research*, 325, 90–104. https://doi.org/10.1016/j.bbr.2016.09.018.

Dubey, R., Peterson, J., Khosla, A., Yang, M.-H., & Ghanem, B. (2015). What makes an object memorable? In *Proceedings of the IEEE International Conference on Computer Vision*, pp. 1089–1097. Institute of Electrical and Electronics Engineers (IEEE).

Gilbert, K. & Schleuder, J. (1990). Effects of color and complexity in still photographs on mental effort and memory. *Journalism Quarterly*, 67(4), 749–756. https://doi.org/10.1177/107769909006700429.

Isola, P., Parikh, D., Torralba, A., & Oliva, A. (2011). Understanding the intrinsic memorability of images. In J. Shawe-Taylor, R. Zemel, P. Bartlett, F. Pereira, & K. Weinberger (eds.), *Advances in Neural Information Processing Systems (NIPS)*, pp. 2429–2437. Neural Information Processing Systems Foundation Inc. (NIPS).

Kapsetaki, M. E., Militaru, I. E., Sanguino, I., Boccanera, M., Zaara, N., Zaman, A., Loreto, F., Malhotra, P. A., & Russell, C. (2022). Type of encoded material and age modulate the relationship between episodic recall of visual perspective and autobiographical memory. *Journal of Cognitive Psychology*, 34(1), 142–159. https://doi.org/10.1080/20445911.2021.1922417.

Khosla, A., Raju, A. S., Torralba, A., & Oliva, A. (2015). Understanding and predicting image memorability at a large scale. In *Proceedings of the IEEE International Conference on Computer Vision*, pp. 2390–2398. Institute of Electrical and Electronics Engineers (IEEE).

Kootstra, G., de Boer, B., & Schomaker, L. R. B. (2011). Predicting eye fixations on complex visual stimuli using local symmetry. *Cognitive Computation*, 3(1), 223–240. https://doi.org/10.1007/s12559-010-9089-5.

Liu, J., Li, J., Feng, L., Li, L., Tian, J., & Lee, K. (2014). Seeing Jesus in toast: Neural and behavioral correlates of face pareidolia. *Cortex*, 53, 60–77. https://doi.org/10.1016/j.cortex.2014.01.013.

Maguire, E. A., Frith, C. D., & Cipolotti, L. (2001). Distinct neural systems for the encoding and recognition of topography and faces. *NeuroImage*, 13(4), 743–750. https://doi.org/10.1006/nimg.2000.0712.

Martin, C. B., Cowell, R. A., Gribble, P. L., Wright, J., & Köhler, S. (2016). Distributed category-specific recognition-memory signals in human perirhinal cortex. *Hippocampus*, 26(4), 423–436. https://doi.org/10.1002/hipo.22531.

Saraee, E., Jalal, M., & Betke, M. (2020). Visual complexity analysis using deep intermediate-layer features. *Computer Vision and Image Understanding*, 195, 102949. https://doi.org/10.1016/j.cviu.2020.102949.

Snow, J. C., Skiba, R. M., Coleman, T. L., & Berryhill, M. E. (2014). Real-world objects are more memorable than photographs of objects. *Frontiers in Human Neuroscience*, 8, 837. https://doi.org/10.3389/fnhum.2014.00837.

Sugita, Y. (2008). Face perception in monkeys reared with no exposure to faces. *Proceedings of the National Academy of Sciences*, 105(1), 394–398. https://doi.org/10.1073/pnas.0706079105.

Wardle, S. G., Paranjape, S., Taubert, J., & Baker, C. I. (2022). Illusory faces are more likely to be perceived as male than female. *Proceedings of the National Academy of Sciences*, 119(5), 1–12. https://doi.org/10.1073/pnas.2117413119.

Wardle, S. G., Taubert, J., Teichmann, L., & Baker, C. I. (2020). Rapid and dynamic processing of face pareidolia in the human brain. *Nature Communications*, 11(1), 1–14. https://doi.org/10.1038/s41467-020-18325-8.

Williams, R. A., Hagerty, B. M., Cimprich, B., Therrien, B., Bay, E., & Oe, H. (2000). Changes in directed attention and short-term memory in depression. *Journal of Psychiatric Research*, 34(3), 227–238. https://doi.org/10.1016/S0022-3956(00)00012-1.

Yang, T., Formuli, A., Paolini, M., & Zeki, S. (2022). The neural determinants of beauty. *European Journal of Neuroscience*, 55(1), 91–106. https://doi.org/10.1111/ejn.15543.

Zeki, S. (2009). *Splendors and Miseries of the Brain: Love, Creativity and The Quest for Human Happiness*, 1st Edition. Chichester, UK and Malden, MA: John Wiley & Sons.

Zeki, S. & Chén, O. Y. (2020). The Bayesian-Laplacian brain. *European Journal of Neuroscience*, 51(6), 1441–1462. https://doi.org/10.1111/ejn.14540.

Zhang, H. & Zeki, S. (2022). Judgments of mathematical beauty are resistant to revision through external opinion. *PsyCh Journal*, 1–7. https://doi.org/10.1002/pchj.556.

Part IV

Art in Time

https://doi.org/10.1142/9789811298639_0027

Chapter 27

The Three-Second Time Window in Poems and Language Processing in General: Complementarity of Discrete Timing and Temporal Continuity*

Xinchi Yu[†,‡] and Yan Bao[†,§,¶]

[†]*School of Psychological and Cognitive Sciences, Peking University, Beijing, China*
[‡]*Department of Chinese Language and Literature, Peking University, Beijing, China*
[§]*Institute of Medical Psychology, Ludwig Maximillian University, Munich, Germany*
[¶]*Beijing Key Laboratory of Behavior and Mental Health, Peking University, Beijing, China*

*This chapter was originally published in the special collection on neuroaesthetics in *PsyCh Journal*. Yu, X. & Bao, Y. (2020). The three second time window in poems and language processing in general: Complementarity of discrete timing and temporal continuity. *PsyCh Journal*, 9(4), 429–443. https://doi.org/10.1002/pchj.390.

Abstract

The existence of discrete time windows has triggered the search for permanence and continuity for artists (including poets) in multiple cultures throughout history. In this chapter, we argue that there exists a 3-s window in the temporal structure of poems as well as in the aesthetic appreciation of poetry by reviewing previous literature on the temporal aspects of poems. This 3-s window can also be considered to be a general temporal machinery underlying human behavior, including language production and perception in general. The reafference principle has provided us with a unique frame for understanding cognitive processes. However, "time" was absent in the original two-stage reafference principle. Therefore, we propose a three-stage cycling model of language perception, taking into account time and time windows. We also inspect the possible neural implementations of the three stages: the generation, maintenance, and comparison of predictions (as well as the integration of predictions into the representational context). These three stages are embedded in a temporal window of ~3 s and are repeated in a cycling mode, resulting in the representational context being continuously updated. Thus, it is possible that "semantics" could be carried forward across different time windows, being a "glue" linking the discrete time windows and thus achieving the subjective feeling of temporal continuity. Candidates of such "semantic glue" could include semantic and syntactic structures as well as identity and emotion.

Keywords: language perception; poem appreciation; reafference principle; temporal perception; time window

Observations of discrete time windows have triggered the search for permanence and continuity for artists and cognitive scientists alike. Reconciling discrete temporal windows and perceptual continuity has always been a challenge in the field of cognitive science. In this chapter, we first review current evidence for a specific time window, the 3-s window, in the recitation and appreciation of poems as well as in the production and perception of language; then, we propose a three-stage model as a possible basis for perceptual continuity despite the discreteness of time windows, which not only accounts for perceptual continuity in language processing but also potentially in *Cognitive Processing* in general.

The Pursuit of Permanence

East and West, poets in all cultures and at all ages refer to time in their poems. Many poets have emphasized the discreteness of human life; every human owns only one time window of hopefully some tens of years. These time windows come to an end, and all humans are subject to a finite life span, as expressed, for instance, by the German poet Friedrich Schiller (1759–1805) in "Nänie:" "Auch das Schöne muß sterben! Das Menschen und Götter bezwinget" ("Even the beautiful must perish! It conquers human and gods alike"), or by the Spanish poet Federico García Lorca (1898–1936) in "Alma Ausente:" "No te conoce el niño ni la tarde, porque te has muerto para siempre" ("The child and the afternoon do not know you, because you have died for ever"). The Chinese poet Zhang Ruoxu (660–720) was a pessimist: "人生代代无穷已/江月年年只相似" ("Generations and generations, there is no end/Years and years, the moon on the river is just the same"); the only glue of time seems to be the merciless moon.

However, many artists have also tried to overcome such discreteness to create continuity or permanence and even infinity. This pursuit for permanence is a common sentiment across cultures, and it is reflected in different forms of art. Paintings, photographs, poems, and essays capture a scene and make it permanent as long as the artwork exists. Epic paintings make a moment of an event permanent; scenery paintings make beautiful scenes permanent, and travel notes and diaries keep the episodes alive whenever and wherever someone reads them. The existence of artwork itself marks the permanence of the work and the artist who created it. The artists, although living within the time window of their existence only, create permanence through their works, passing them down through generations. This has been long expressed by poets: They know that they will not perish and will glue the time windows of the human life spans together to achieve continuity and permanence through their works and thoughts. The Roman poet Horace (Quintus Horatius Flaccus, 65 B.C.–8 B.C.) claimed that "Annorum series et fuga temporum./Non omnis moriar multaque pars mei/Vitabit Libitinam: isque ego postera...." (I would not entirely die and a large part of me will avoid Libitina [the goddess of funerals and burial]; fresh, I continually would grow with future praise...).

Similarly, the English poet William Shakespeare (1564–1616) expressed in Sonnet 18 that "So long as men can breathe, or eyes can see,/So long lives this, and this gives life to thee;" "this" being the poem. Poems, stories, and theories (in the world of science) will continue to exist after the time window of one's own existence ends; later, other people can still pick them up to read, to praise, and to elaborate. The discrete time windows of life can be glued not only by a merciless moon (as for Zhang Ruoxu in 春江花月夜 *A Moonlit Night On The Spring River*) but also by the cycle of generating and receiving information and knowledge (Pöppel & Bao, 2011). These works of art collectively indicate that the seemingly impermanence and discreteness of human life can become permanent and continuous through information processed and knowledge created that connect generations across ages in history.

A Temporal Constant of ~3 s in Poetry

In addition to the pursuit of permanence, one can identify another temporal characteristic in poetry when poems are recited or read aloud. Poems are composed of verses or *lines* (for a definition of lines, see Turner & Pöppel, 1988). Each line in a poem takes up a certain duration of time. Is there a universal time constant for poetic lines when people read them aloud or recite them by heart? Turner & Pöppel (1988) collected over 20 types of poetry, varying from East to West and from modern societies to indigenous cultures, and found a constant of ~3 s for the duration of poetic lines.

One may argue that this observation is a consequence of the number of syllables, as there is the possibility that every culture may have a similar syllable number per poetic line. Given the temporal constancy of syllable articulation, the observation of ~3 s might be the result of mere item-based (i.e., syllable-number-based) characteristics. We reexamined the data presented by Turner & Pöppel (1988) where the poems have a constant syllabic length; the result is plotted in Figure 1. The mean of the temporal length of lines is 3.09, with an *SD* of 0.64 s, as plotted in this figure (Lines indicate $M \pm SD$.) We did not get a significant Pearson correlation, $p = 0.348$, $r = 0.261$, nor a Spearman correlation, $p = 0.314$, $\rho = 0.279$, of these two variables based on the poetry types with constant

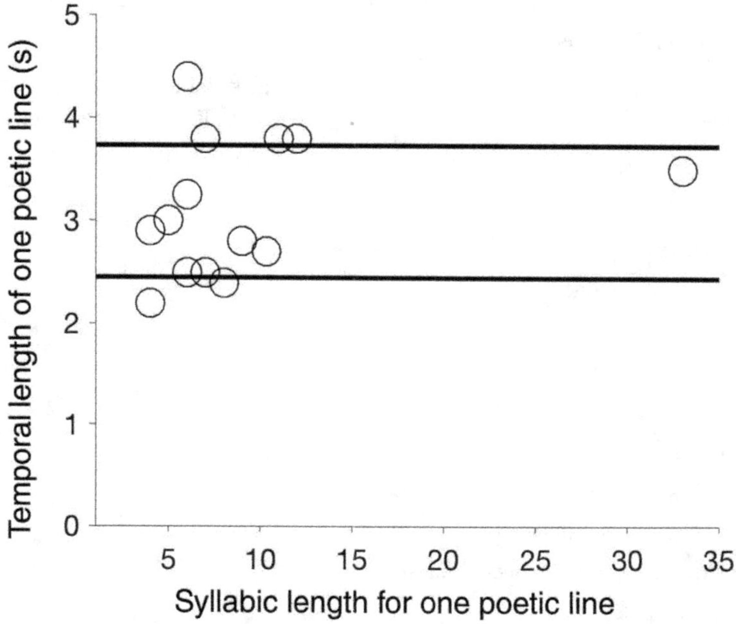

Figure 1. Visualizing syllabic length and temporal length of one line in different types of poems from Turner & Pöppel (1988).

syllables (which are Japanese epic meter and tanka; Chinese four, five, and seven-syllable line; English seven-syllable trochaic line and octosyllabic ballad meter, ancient Greek dactylic hexameter half-line, trochaic tetrameter half-line, iambic trimester, anapestic tetrameter half-line, Latin elegiac couplet, dactylic hexameter half-line, hendecasyllabic, and French 12-syllabic alexandrine; see Table 1). Note that a significant correlation is still not present after removing the data point of "Latin elegiac couplet". Bearing the risk of interpreting null results in statistics, this result suggests that the temporal constant of poetic lines is not a result of constancy in syllabic length but presumably a temporal constant.

Our conclusion fits with our experience that we automatically speak slower with less syllables in one poetic line and relatively faster if there are more syllables; if there are too many syllables in one "sentence," we will enter a pause in the middle and separate them into multiple lines in recitation (Pöppel, 1988). One typical example of slowing down comes

Table 1. Syllabic length and temporal length of poem types with constant syllable number from Turner & Pöppel (1988).

Poem type	Average syllabic length per line	Average temporal length per line(s)
Japanese epic meter	6	3.25
Japanese Tanka	10.33	2.70
Chinese four-syllable line	4	2.20
Chinese five-syllable line	5	3.00
Chinese seven-syllable line	7	3.80
English seven-syllable trochaic line	7	2.50
English ballad meter (octosyllabic)	8	2.40
Ancient Greek dactylic hexameter (half-line)	9	2.80
Ancient Greek trochaic tetrameter (half-line)	4	2.90
Ancient Greek iambic trimeter	6	4.40
Ancient Greek anapestic tetrameter (half-line)	6	2.50
Latin elegiac couplet	33	3.50
Latin dactylic hexameter (half-line)	9	2.80
Latin hendecasyllabic	11	3.80
French alexandrine (12-syllable)	12	3.80

from the poem "Il pleure dans mon cœur" by the French poet Paul Verlaine (1844–1896): "C'est bien la pire peine/De ne savoir pourquoi/Sans amour et sans haine,/Mon cœur a tant de peine." (And the worst pain of all/Must be not to know why/Without love and without hate/My heart feels such pain (Richard Stokes, Trans.). In this example, one automatically speaks slower when there are fewer syllables in one line. Less words are employed in one line to express depression in this case; the pace of citing is slowed down to fit into the 3-s time window.

There was, however, severe criticism of Turner & Pöppel (1988) by Fabb (2013), who claimed that the 3 s constant in poetic lines does not exist. Fabb made this conclusion mainly based on a recording corpus of English poems that he collected and argued that there are plenty of poetic lines exceeding 3 s. There may have been, however, a misunderstanding by Fabb concerning the 3-s time window; it is not a window with a ceiling of exactly 3 s as one would expect from a physical constant but is an operational platform based on statistical analyses reflecting an endogenous neural mechanism with some variance being typical for biological processes. This means that many poetic lines have a duration of ~3 s, with some variance to give rise to a median/mean value of ~3 s. This mean value can be shown also in Fabb's own corpus. Fabb (2013) recorded 55 English poems, and the data of the median/mean line duration for *each poem* are listed in the article. In Figure 2, we plot the distribution of the mean and the median of line duration in Fabb's article. The mean of median duration (for each poem) is 3.28 s, $SD = 0.853$ s, and the mean of mean duration is 3.38 s, $SD = 0.893$ s. As two slightly skewed distributions (skewness of median duration = 0.240, skewness of mean duration is 0.146, note that the two distributions all pass a Shapiro–Wilk normality test, $p > 0.05$, whereas the distribution of median duration exhibits a $p = 0.049$ in the Kolmogorov–Smirnov test with Lilliefors correction), the median of median duration is 3.10 s, and the median of mean duration is 3.22 s. This result, in contrast to what Fabb argued, actually shows the robustness of the 3-s time window, as it applies to a new set of poems collected by another independent researcher. Exceeding the exact value of ~3 s does not invalidate the existence of this time window in poems; it is not unusual that human cognitive processes are subject to variation (e.g., Bao, Yang, Lin, & Pöppel, 2016).

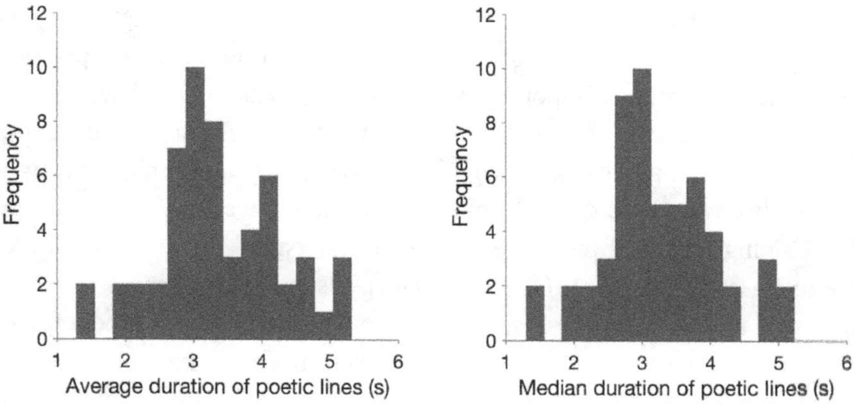

Figure 2. The distribution of the temporal length of poetic lines, based on data from Fabb (2013).

One might still argue that this notion of a 3-s time window is trivial, as it matches approximately the normal human respiratory cycle of 12 to 15 breaths per minute (Barrett, Barman, Boitano, & Brooks, 2012). One reason for this explanation to be unlikely is that poems as works of art do not have any intrinsic characteristics to match the human respiratory cycles. Furthermore, humans are not forced to obey the respiratory cycle in speech (Parkes, 2006). In this context, it is useful to take a more general biological perspective comparing the 3-s window in humans with a potential equivalent in other species. It has been observed that the 3-s window is also a frame for (repetitive) behavioral segments in humans and other higher mammals (Gerstner & Cianfarani, 1998; Gerstner & Fazio, 1995; Gerstner & Goldberg, 1994; Kien, Schleidt, & Schöttner, 1991; Schleidt, Eibl-Eibesfeldt, & Pöppel, 1987). This time window being the same in speech and other motor activities in different species indicates that they share a common logistic temporal mechanism (Pöppel, 1997a). However, the breathing rate of higher mammals does not necessarily match with 3 s. For example, the median duration of motion segments for a giraffe is 2.61 s and the mean is 3.61 s (Gerstner & Fazio, 1995), whereas the mean respiratory rate is 6.57 s (Langman, Bamford, & Maloiy, 1982). Therefore, the pattern of 3 s in speech and other motor behaviors is not a passive reflection of breathing cycles, although we do not dismiss the possibility that respiratory cycles can modulate the duration of such time windows.

Another reason that the 3-s window does not reflect a trivial phenomenon is that poem appreciation is not only productional but also perceptual, thus providing an important synchrony of production and perception on the aesthetic level. By simply altering the temporal length of a poetic line (without changing the syllable number), Zhao, Zhang, and Bao (2018) discovered a cross-culturally robust preference for poetic lines in the 3-s range. Even if the subjects cannot understand the meaning of the poetic lines (in this case, German subjects listening to Chinese poems), they still prefer lines of ~3 s. This observation indicates a production–perception synchrony of ~3 s, which means that we not only tend to recite poems (and speech in a more general sense; discussed later) with a 3-s pattern but also appreciate poems aesthetically within the same temporal frame. This temporal preference for a 3-s pattern and not a 1-s or a 10-s pattern, which linguistically would be possible, indicates presumably a profound evolutionary basis. The temporal modulation effect of the 3-s window on aesthetic appreciation may also motivate us to look for other concepts and phenomena of the cognitive and neural basis of aesthetic perception in general and in detail, as has been partly already done for decision processes, the visual arts and music (Avram *et al.*, 2013; Bao *et al.*, 2016, 2017; Park *et al.*, 2014, 2015; Pöppel, 1989a).

A Necessary Distinction Between Content and Logistic Functions

Speech is of a transient nature, which obviously means that what has been said no longer physically exists after it has been said; this implies that one has to utilize a segmentation process to store and process information (Christiansen & Chater, 2016) and that this segmentation process most likely has its limits. Some have addressed the nature of such limits as item-based, like the magical number of 7 ± 2 by Miller (1956) or 4 ± 1 by Cowan (2001). Others have addressed the nature of such segmentation as time-based, as suggested by the 3-s time window by Pöppel (1997a), which is the main focus of this chapter. However, no matter if such segmentation is item-based or time-based, one has to stress that this segmentation process is logistic. This means that segmentation itself is relatively independent of the content that is segmented. No matter what one is segmenting, the segmentation limit is always 7 ± 2 or 4 ± 1 if item-based or ~3 s if time-based, indicating that this limit has a logistic characteristic or is "pre-semantic" (Pöppel, 1997a, 2009). We avoid using the ambiguous term *chunk*, as chunking can happen on multiple levels (Christiansen & Chater, 2016); one has to chunk at least one whole sentence as one to understand it.

The logistic-content distinction appears to be necessary for a deeper understanding of cognitive processes. It is not controversial that percepts, memories, emotions, or volitions represent the content of our cognitive systems. However, temporal organization, attentional control, and general activation or arousal are not on the same categorical level as the content functions. The temporal organization of different content functions cannot be classified at the same logical level as content functions themselves. Furthermore, attentional systems operate *on* content functions rather than *with* them in parallel (Bao *et al.*, 2011; Bao & Pöppel, 2007; B. Zhou, Bao, Sander, Trahms, & Pöppel, 2010). Thus, attentional processes can also not be classified at the same logical level as content functions. In addition, arousal modulated by the activation system represents another neural process that is logically different from content functions; it is functioning like a "power supply" of the central nervous system. Therefore, it is necessary to separate the logistic functions (temporal control, attentional mechanisms, and activation of functions) from the content functions. The logistic functions are the "how" functions providing the necessary neuronal infrastructure for "the state of being conscious" (Pöppel, 1997b), whereas the content functions are the "what" functions that are related to conscious representation (Pöppel, 1989b; B. Zhou *et al.*, 2016).

Such logistic-content distinction can be seen in physiological and imaging studies of auditory speech processing. For example, the acoustic analysis of the auditory temporal structure is observed in the superior temporal sulcus whereas speech-related activations appear in more brain areas (Ding, Melloni, Zhang, Tian, & Poeppel, 2016; Overath, McDermott, Zarate, & Poeppel, 2015). Following these findings, we can conjecture that the logistic "containers" root partly in the functionally or temporally earlier areas in the processing routes. For example, bistable perception of the Necker cube can be decoded even in early visual areas in the posterior visual cortex (M. Wang, Arteaga, & He, 2013). The concept of central timing mechanisms has been discussed for quite some time, based, for instance, on the similar temporal order threshold of tens of milliseconds or the multimodal response distributions of reaction time and eye movement latencies, and the low-frequency time window of several seconds across modalities (Hirsh & Sherrick, 1961; Pöppel, 1970, 1994, 1997a; Pöppel & Logothetis, 1986; Wittmann, 1999). Taken together, the distinction between content and logistic functions appears to be necessary, reflecting fundamental biological mechanisms, and temporal processing serves as one of these logistic functions.

A Logistic Time Window of ~3 s in Language

The 3-s synchrony in the recitation and appreciation of poetry is similarly observed in language in general: Linguistic utterances have the tendency to be expressed with temporal segments of ~3 s (Kien & Kemp, 1994; Kowal, O'Connell, & Sabin, 1975; Schleidt & Kien, 1997; Vollrath, Kazenwadel, & Krüger, 1992; note that the results of Kowal *et al.*, 1975, were re-analyzed for more clarity; see Table 2). These findings indicate that there is a periodic nature of ~3 s in human verbal behavior, meaning that a conspicuous pause often appears approximately every 3 s. We argue that besides being a logistic segmentation window for language production, the logistic 3-s time window is also underlying language perception. Note that in this context, we do not distinguish systematically between language perception and language processing (as in Hickok & Poeppel, 2007; note that in the citations of this chapter, "Poeppel" refers to Dr. David Poeppel and "Pöppel" refers to Dr. Ernst Pöppel).

Both behavioral and electrophysiological evidence for other cognitive functions supports the concept that the interval of 3 s is something special in temporal processing (Bao *et al.*, 2013; Elbert, Ulrich, Rockstroh, & Lutzenberger, 1991; Fraisse, 1984; Kagerer, Wittmann,

Table 2. Average duration per speech segment (re-analysis based on Kowal *et al.*, 1975).

Age	Syllables per second (Kowal *et al.*, 1975)	Syllables per speech segment (Kowal *et al.*, 1975)	Average duration per speech segment (s)
Kindergarten	2.15 (0.75)	5.5 (1.9)	2.56
Grade 2	2.86 (0.53)	7.4 (2.2)	2.59
Grade 4	3.24 (0.51)	7.6 (2.8)	2.35
Grade 6	3.26 (0.66)	7.3 (2.5)	2.24
Grade 8	3.83 (0.50)	9.4 (2.4)	2.45
Sophomores	4.00 (0.51)	9.9 (3.2)	2.48
Seniors	3.84 (0.52)	10.0 (3.2)	2.60

Note: Numbers in parentheses represent *SD*s. The average duration per speech segment is calculated from syllables per speech segment divided by syllables per second.

Szelag, & Steinbüchel, 2002; Pöppel, 1972, 1997a). A 3-s time window has, for instance, been demonstrated in experiments on auditory bistable perception, indicating perceptual segmentation. When listening to ambiguous (i.e., bistable) auditory stimuli such as "KU-BA-KU," people switch their perception spontaneously approximately every 3 s between "KUBA" and "BAKU" (Radilová, Pöppel, & Ilmberger, 1990; von Steinbüchel, 1998). This phenomenon of perceptual alteration applies also to bistable visual figures (Gómez, Argandoña, Solier, Angulo, & Vázquez, 1995; Polgári, Causin, Weiner, Bertschy, & Giersch, 2020; von Steinbüchel, 1998). A 3-s time window in auditory temporal segmentation is also indicated electrophysiologically by the event-related potential (ERP) component of mismatch negativity (MMN; L. Wang *et al.*, 2016; L. Wang, Lin, Zhou, Pöppel, & Bao, 2015) and magnetoencephalography experiments (Sams, Hari, Rif, & Knuutila, 1993). A larger MMN for deviant stimuli is elicited if the interval between successive auditory stimuli is around 3 s; the MMN amplitude decreases after 3 s but restores again after around 6 s, indicating an oscillatory neural mechanism with a period of ~3 s. The fact that the human cognitive system is more open to new stimuli approximately every 3 s is sometimes misunderstood by White (2017); he claimed that the MMN being observed also after an interval of 9 to 12 s would contradict the hypothesis of temporal integration on a time scale of ~3 s; this is certainly a misunderstanding. Note, however, that the concept of a 3-s time window or a high-frequency time window of ~30 to 40 ms does not exclude time windows between these periods; for example, the window of 150 to 300 ms (e.g., Ding *et al.*, 2017; Hickok & Poeppel, 2007; Poeppel & Assaneo, 2020), in which suprasegmental information is carried. This inclusion of other temporal windows has actually already been stated by Szelag & Pöppel (2000).

Abundant behavioral and neurophysiological evidence has proven the existence of a 3-s time window also in language perception. Early behavioral evidence came from Sachs (1974), where subjects were asked to report if a sentence was identical to a previous one. The two sentences were distanced by sentences with different numbers of syllables. There is a significant drop and a plateau after the 3-s window both for auditory stimuli (between 3 s and 7.5 s) and visual stimuli (between 1 s and 4 s). Such results suggest that subjects in these experiments were more sensitive to changes in a sentence if they were in the same temporal segment of ~3 s. Wagers & Phillips (2009) discovered that the plausibility effect of a sentence is weaker for longer dependencies than are shorter ones in a self-paced reading task. More significant differences in reading time after the critical verb onset were observed for short dependency conditions than for long dependency ones. In short dependency conditions, the distance, that is, the interstimulus interval (ISI), between the sentence-initial noun and the critical verb, is 1 (~450 ms, as estimated from the figure as the original data were not provided), whereas the distance is 6 (~3 s) in long dependency conditions. Thus, a difference within and beyond the 3-s window is observed in this experiment. One may argue that more interval conditions are needed to confirm that this effect is truly a staircase effect (indicating a window) rather than a simple linear decreasing effect. Nonetheless, Wagers & Phillips (2009) provided some positive implications about the existence of this time window. Also note that such findings are unlikely to be a trivial result of the human memory limits of item (i.e., 7 ± 2 or 4 ± 1), as one needs to chunk the whole sentence holistically in language perception, and there was no significant difference in answering the follow-up questions, indicating that the subjects' memory can equally hold up short and long sentences in this experimental condition. Schremm, Horne, and Rolls (2015) result has indicated that the reaction time to grammatical and semantic incongruency can be modulated by a 3-s window. Subjects reacted slower if the critical words had a distance longer than ~3 s when the number of words between them was the same.

Electrophysiological evidence also has supported the existence of a 3-s time window in language perception. Strong evidence has come from Roll, Lindgren, Alter, and Horne (2012), who examined the ERP component of the closure positive shift (CPS), a centroparietal positivity related to perceptual boundaries in language and music perception. A more salient CPS effect was observed only when the clause boundaries coincided with the temporal boundary of ~2.7 s. This indicates that there is a logistic wrap-up effect enhancing the wrap-up effect of content. Roll, Gosselke, Lindgren, and Horne (2013) studied the component of the left anterior negativity (LAN) elicited by the anomaly of grammatical agreement. They found that the LAN effect holds for an (SOA) of 1.75 s and 2.25 s, but is not significant for an SOA of 2.75 s. For an SOA beyond the 3-s window (3.25 s), an anterior negativity with different scalp topography was observed, indicating different neural mechanisms inside and outside the 3-s window. Phillips, Kazanina, and Abada (2005) discovered that longer filler-gap dependency (ISI = 4,000 ms) will result in a delayed P600 effect, as compared to a shorter dependency (ISI = 2,000 ms). A similar experiment with ISIs of 700 ms and 2,600 ms did not report such a result on the verb position (Fiebach, Schlesewsky, & Friederici, 2002). This can be explained by the 3-s time window, as an ISI of 700 ms, 2,000 ms, and 2,600 ms falls into the 3-s window whereas an ISI of 4,000 ms does not, thus resulting in a qualitative difference for the P600 latency.

Synchronization in Verbal Communication

Based on these many different observations, one can conclude that the 3-s window acts as a segmentation frame for both language production and perception. This production–perception synchrony can be considered as the basis for synchronization in verbal communications because it allows the creation of shared temporal windows of communicators. They can synchronize their mutual "nowness" or "subjective present" and share a temporal frame of reference in social interactions (Bao *et al.*, 2015; Miyake, 2012; Nagy, 2011; Pöppel, 1988, 1994, 1997b, 2009).

Synchronization on the basis of a 3-s time frame can account for many aspects of human linguistic behaviors (e.g., in conversations). Sometimes one does not finish while speaking a sentence or omits parts of a sentence; despite this, one can still successfully transmit a verbal message. The fact that human speech segments follow a 3-s time frame (Kowal *et al.*, 1975; Vollrath *et al.*, 1992) also explains that one makes deletions to fit utterances into this 3-s time frame. The implicit consensus that speakers and listeners all operate within a 3-s basis also saves energy, as one does not have to shift between different temporal frames from moment to moment. Thus, synchronization is also the operative basis for anticipation; one implicitly expects a speech segment to stop at around 3 s, and another topic in conversation can be initiated after this time.

This 3-s temporal production–perception synchrony may be based on similar neural mechanisms for speech production and the perception systems as a result of an evolutionary co-adaptation of the time windows. The basic idea of a social co-adaptation is that humans synchronize with each other during numerous social interactions, adjusting individual temporal windows to a common length (here, of ~3 s). Such synchronization in social interactions might involve synchronization on both behavioral and physiological levels (e.g., Dikker *et al.*, 2017; Hale, Ward, Buccheri, Oliver, & Hamilton, 2020). The neural and functional relationship between speech production and perception is not a new topic; it has been discussed with respect to the perception–production functions of Wernicke's area (Binder, 2015) and the motor theory of speech perception (Liberman & Mattingly, 1985). Such low-frequency auditory–motor interaction might also involve low-frequency neural oscillatory coupling (e.g., with delta waves); a similar scenario has been observed for the syllabic level (Assaneo & Poeppel, 2018).

The continuity of 3-s windows can be modeled with oscillations. One has, however, to distinguish between two different classes: pendulum oscillations and relaxation oscillations. The continuity of 3-s windows can technically be described as a pendulum oscillation, and physiologically it can perhaps be related to the delta band wave in the electroencephalogram. The frequency of the delta band (0.5–4 Hz) perhaps matches the cycle of such a window, and there is evidence that the delta band is related to the processing of intonational phrase boundaries (e.g., Bourguignon *et al.*, 2013; Molinaro, Lizarazu, Lallier, Bourguignon, & Carreiras, 2016; for a review, see Meyer, 2018), which often fall into the window of ~3 s. However, the 3-s time window can also be modeled by relaxation oscillations (Pöppel, 1970, 1994) because the ongoing pendulum oscillation can be interrupted at any phase, and a new oscillatory process with the same or similar period is entrained with very short latency, a characteristic of relaxation oscillations (for a recent discussion on exogenous and endogenous cortical rhythms in language processing, see Meyer, Sun, & Martin, 2019). Neural and behavioral phenomena below 1 Hz are often difficult to identify (H. Zhou, Melloni, Poeppel, & Ding, 2016), but they certainly account for a great part of human behavior (e.g., Donhauser & Baillet, 2019; Teng *et al.*, 2020). Individual variance of the period of an oscillatory process can possibly mask an oscillatory phenomenon below 1 Hz, as seen in studies by L. Wang *et al.* (2016; L. Wang *et al.*, 2015). Therefore, single-case studies can be extremely useful in elucidating such low-frequency oscillatory phenomena (Bao *et al.*, 2015; Bao *et al.*, 2017c; Pöppel & Bao, 2017; Zaytseva & Bao, 2015).

A Three-Stage Cycling Model in Language Processing

The concept of the reafference principle (von Holst & Mittelstaedt, 1950, 1971) roots from the exploration of the stability of vision by von Helmholtz (1896), and it was later generalized using the term *corollary discharge* (Teuber, 1960). The later notion of "predictive coding" could also be seen as inspired by the reafference principle (e.g., Clark, 2013; Rao & Ballard, 1999; Stefanics, Kremláček, & Czigler, 2014). In the original version of the reafference principle, an efference copy is generated in parallel to an action command for whatever one is doing. This copy is then compared with the sensory information after the execution of the action (i.e., the reafference). Thus, there are two successive operations in the original reafference principle: (a) the generation of an efference copy and (b) the process of reafference allowing the comparison between the executed movement and the efference copy. What is missing in the original concept and the derived theories is the factor "time" (i.e., how long the duration of an efference copy has to be maintained). Thus, the reafference principle should actually comprise three components: (a) generation of an efference copy (which serves as the basis for prediction) in parallel to the motor command, (b) maintenance of the efference copy or the prediction in a temporal buffer, and (c) comparison of the prediction with the sensory input after a specific temporal interval. Another cycle with all its components then can be initiated, be it either periodically under endogenous temporal control or in a voluntarily determined active mode.

A new cycle will not result in a simple repetition of what is represented because there is always an update in each cycle, which will result each time in modified predictions (Pöppel & Bao, 2011). As has been suggested, many cycles could operate simultaneously in a hierarchical system, such as the higher frequency cycles being embedded in the lower frequency cycles (Tanida & Pöppel, 2006). This theoretical concept provides an explanation for planning and confirming what has been planned as a basis for the continuity of action and perception. This can also be considered as a framework for realizing the original attempt of von Helmholtz (1896) in discussing the nature of visual continuity despite saccadic eye movements (for a review, see Wurtz, 2018) and visual stability across different areas of the visual field which are characterized by different mechanisms of spatial attention (Lei, Bao, Wang, & Gutyrchik, 2012; Pöppel & Bao, 2012).

The reafference principle has indeed been suggested as a theoretical frame for perceptual processes in general and has been suggested to explain "pre-emptive perception" (Bodis-Wollner, 2008), "proactive perception" (Rimmele, Morillon, Poeppel, & Arnal, 2018), the discontinuity of thinking in schizophrenia (Feinberg, 1978), and speech production (Tian & Poeppel, 2010, 2013; Tian, Zarate, & Poeppel, 2016). Although the reafference principle has been invited into the field of language perception (Gambi & Pickering, 2017; Pickering & Clark, 2014; Pickering & Gambi, 2018; Pickering & Garrod, 2013), the reasoning is basically on the computational and algorithmic level (Marr, 1982). Neurobiological features of the principle refer mainly to perception–production connections (Dick & Andric, 2013; Pickering & Garrod, 2013) rather than to the basis of the two (or three, as we propose) operations in the reafference principle. Pickering & Gambi (2018) introduced a cognitive model of reafference in language perception but did not consider its neural implementation. We suggest possible neural mechanisms (i.e., the hardware implementations) of reafference in language processing, in poetry, in particular, and in cognitive processes, in general, while being compatible with current linguistic concepts.

When processing a sentence, predictions can be generated before the "critical" word actually appears. For example, when we hear/see the first part of a sentence (e.g., You never forget how to ride. . . .), the following part can be predicted (e.g., a bicycle or an elephant). Predictions can have different weights. Here, "a bicycle" is more likely to appear, but if "an elephant" appears, we know the sentence is still congruent, but more integration efforts should be implemented because "an elephant" is less likely to appear here. Such a notion of prediction and integration, however, is abstract, and it is difficult to be mapped directly onto neural activities in the brain.

We propose that our three-stage model (as outlined earlier) is compatible with language perception. These three stages are not just a theoretical construct but are validated by neural evidence and can be mapped onto certain neural underpinnings. In our framework, the prediction-integration process consists of three consecutive stages (see Figure 3): (a) the generation of predictions, (b) the maintenance of predictions, and (c) the comparison between prediction and the current input (as well as the integration of the current input into the representational context). Note that comparison is not the end of the story: After comparison, the input word is integrated into the current semantic context ("narrative") and syntactic

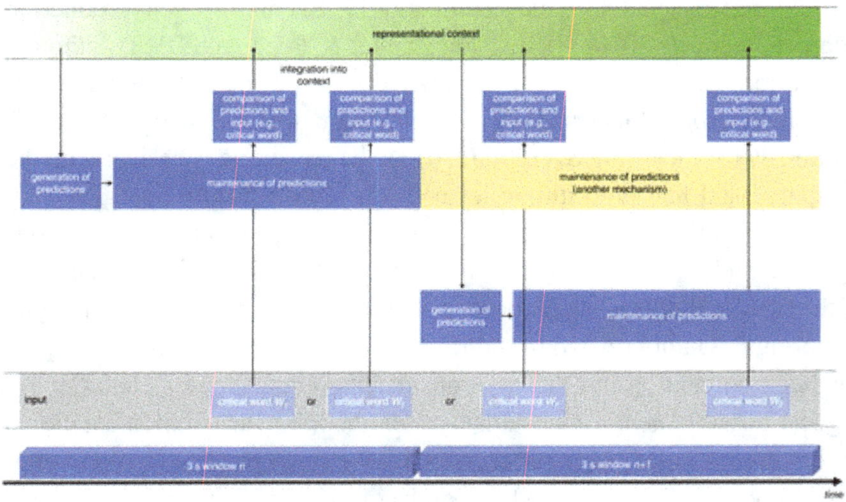

Figure 3. An illustration of the three-stage cycling model in language processing. Note that the change from blue to yellow in the "maintenance of predictions" stage indicates a shift of processing machinery within and outside the 3-s window. Critical word W1 is compared and integrated with predictions maintained by different mechanisms, based on its temporal onset (within/beyond the same 3-s window). If W1 is presented after the previous 3-s window and related to the predictions of the former 3-s window, then it is compared and integrated into a different machinery, as the brain is currently in another 3-s window, generating new predictions, and looking for new information. The cycling three-stage process as well as the continuously updated representational context ("semantics") allow for a temporal continuity across discrete time windows.

context (i.e., hierarchical structure), which, along with the predictions, has also been persisting through time. To be more precise, we would refer to the "context" as a "gestalt." Therefore, the whole process should also include the generation and maintenance of context and the integration of the current information into context; this process apparently falls into a larger temporal window. Note that the "integration" we use here can be different from the way it is used in some psycholinguistic research. One prominent feature of our model is that this three-stage cycle operates in a 3-s time window. After one cycle, the brain actively "looks for" something new, with the launching of a new cycle, which has a range of ~3 s (L. Wang *et al.*, 2015, 2016). However, this does not mean that the information maintained in the previous temporal window is lost. Rather, it is taken over by a different cognitive machinery (discussed earlier, in the "A logistic time window of ~3 seconds in language" section), indicated by behavioral and electrophysiological studies (Fiebach *et al.*, 2002; Phillips *et al.*, 2005; Roll *et al.*, 2013; Sachs, 1974; Schremm *et al.*, 2015; Wagers & Phillips, 2009).

Predictions are generated before the "critical word" actually appears, as the brain activities predicting semantically similar words are similar (L. Wang, Kuperberg, & Jensen, 2018). L. Wang *et al.* (2018) also demonstrated that such predictions are likely to be generated from the left inferior/medial temporal lobe. It often takes some time after predictions for a certain word to be generated before it actually appears; therefore, the prediction needs to be held for a certain time. The cognitive demand of maintaining predictions cannot be done without the continuity of specific neural activities. A possible candidate for such neural persistence is sustained anterior negativity (e.g., Kluender & Kutas, 1993; Lau, 2018; Lau & Liao, 2017; Phillips *et al.*, 2005). During this sustaining process, knowledge from long-term memory could be retrieved to narrow down and re-weight the saliency of predictions, known as the notion of "prediction as memory retrieval" (Chow, Momma, Smith, Lau, & Phillips, 2016). The third stage is comparing the current input with the predictions; such comparison relies on the decodability of neural activities (e.g., de Lange, Heilbron, & Kok, 2018), making it possible to compare "new" percepts with the "old" predictions which are maintained. The new percepts are then integrated into the current gestalt (context). The well-known evoked potential components in some psycholinguistics studies (e.g., N400 and P600) are likely to reflect the third stage due to its temporal characteristics; it appears after the critical word onset (e.g., Kutas & Hillyard, 1980; Zhang, Yu, & Boland, 2010). Every time each three-stage cycle ends, offline knowledge (i.e., long-term memory or experience) is updated and can be utilized in future predictions. In this way, "online" information becomes "offline" and gets stored for future use.

In terms of oscillation, we conjecture that the critical word for prediction (e.g., "ride" given in the example) re-entrains a relaxation oscillation and initiates a new time window; the oscillatory process continues, "waiting" for an appropriate predictive operation (e.g., coupling) until the target word (e.g., "a bicycle") appears. Thus, in this model, the 3-s time window being implemented by successive periods of oscillation plays the decisive role for understanding.

Continuity Across Discrete Time Windows to Form Discourse Gestalts

One might ask an obvious question: If information processing is framed by successive discrete 3-s time windows, why does one not sense the "bumps" or interruptions when entering a new time window? This seemingly killer question can be resolved after our reasonings on the 3-s window and the reafference principle. Predictions do not vanish to nowhere at the end of the 3-s windows. The discrete time window does not stop the predictions from existing; information from the previous time window continues to exist as predictions for the following time windows, and through comparison and integration, the new inputs in the recent time window are connected with information of the previous time window(s). Thus, continuity can be achieved across time windows. Through this process, utterance segments (with a duration of ~3 s) can be "glued" temporally through prediction, comparison, and integration to form a coherent discourse. The reafference principle can function as a framework in understanding how discrete perceptual units through time can be perceived as a holistic gestalt. Contents in every single temporal window are integrated into a gestalt. Such persisting gestalts form the prediction of future input. The predictions can be compared and integrated with future inputs or future gestalts, resulting in a "full" gestalt or a new gestalt leading to future predictions. The three-stage process is used to different extents across individuals, which can be validated by the fact that different people have different abilities in discourse comprehension and by experiments on holistic and analytical processing strategies (Bao *et al.*, 2013).

The discrete windows are "glued" together because the content ("semantics" in a broader sense, including semantic and syntactic structures) is carried over from previous time windows. Thus, semantics is the glue for the sequence of time windows. Beyond semantic and syntactic structures, such glues are functional in our model, given two presuppositions: the concept of identity "of something as something" and the time constants in the representation of emotions. Identity is the defining characteristic by which a percept, a thought, or a mental image is recognized as a persisting entity over time (Pöppel, 2010; B. Zhou, Pöppel, & Bao, 2014). Identity can be the sameness of an object (or objects); for example, the computer screen (or the paper) one is staring at in the previous time window is still the same in the current time window. Identity also can refer to mental images created, for instance, by sequential lines in a poem (see the example of the poem by Verlaine). Thus, the maintenance of the identity of a percept or a mental image is the necessary (but not sufficient) condition for the creation of a meaningful permanence. The second presupposition in our model to glue time windows together is related to emotional or evaluative neural representations. Emotional states normally last longer and can have a duration of many seconds, minutes, hours, or even days (Verduyn & Lavrijsen, 2015). On that basis, the neural representation of emotional evaluations could act as a functional glue that bridges the sequence of 3-s time windows; in that case, the process of connecting could stay at an implicit level (Pöppel & Bao, 2011). The long-term emotional tonality would provide the frame for continuity. Failure in using semantics as a temporal glue might account for subjective discontinuities, as in schizophrenic patients, for example. Such patients often exhibit discontinuities in their mental activity, and it has been suggested that the disturbance of the semantic connection of successive units could be the reason for this problem (Bleuler, 1969; Martin *et al.*, 2014; Pöppel, 1989b).

Conclusion and Summary

As has been stated, behavioral as well as neural evidence support the concept that cognitive functions operate with multiple time windows (Hickok & Poeppel, 2007; Martin, 1972; Szelag & Pöppel, 2000; L. Wang *et al.*, 2015, 2016; for an earlier review, see Pöppel, 1997a). We have argued further on the basis of experimental evidence that there is indeed a 3-s time window in poetry and other linguistic behaviors and that these discrete windows do not disprove perceptual and conceptual continuity. To illustrate the latter point, we introduced the reafference principle as a concept and argued that the three-stage process creates the basis for continuity across temporal windows. This reasoning brings us back to the initial remarks; although one owns only one time window in history, stories, poems or theories will be passed down through generations. The continuity across discrete windows is achieved through complementarity as a generative principle of passing and receiving. Thus, our model on continuity of windows can also be seen as being inspired by human anthropological universals (Bao & Pöppel, 2012) from different poets in different windows and different cultures. Not only can the principle of reafference bring about continuity but it also can bring about a holistic gestalt in perception, as has been discussed in discourse comprehension.

We also provided the possible neural and cognitive bases for the successful operation of reafference in language (and possibly other cognitive processes). To further elucidate the process of reafference in language, we also need to settle the underpinnings for reafference on the neurobiological level in future research. It is also worth researching whether different levels of linguistic stimuli (e.g., syllable level and word level; word category level, semantic level, and syntax level) operate on the same or different neurobiological routes. The reafference principle is also a demonstration of complementarity. Therefore, our reasoning also demonstrates the power of complementarity as a generative principle (Bao, von Stosch, Park, & Pöppel, 2017). Finally, note that poems from different cultures being characterized by anthropological universals provide an ideal data source to reflect on discrete time and temporal permanence.

Acknowledgments

This work was supported by the National Natural Science Foundation of China Projects 31771213 and 31371018 to Yan Bao.

References

Assaneo, M. F. & Poeppel, D. (2018). The coupling between auditory and motor cortices is rate-restricted: Evidence for an intrinsic speech-motor rhythm. *Science Advances*, 4(2), eaao3842. https://doi.org/10.1126/sciadv.aao3842.

Avram, M., Gutyrchik, E., Bao, Y., Pöppel, E., Reiser, M., & Blautzik, J. (2013). Neurofunctional correlates of esthetic and moral judgments. *Neuroscience Letters*, 534, 128–132. https://doi.org/10.1016/j.neulet.2012.11.053.

Bao, Y. & Pöppel, E. (2007). Two spatially separated attention systems in the visual field: Evidence from inhibition of return. *Cognitive Processing*, 8(1), 37–44. https://doi.org/10.1007/s10339-006-0151-x.

Bao, Y. & Pöppel, E. (2012). Anthropological universals and cultural specifics: Conceptual and methodological challenges in cultural neuroscience. *Neuroscience and Biobehavioral Reviews*, 36(9), 2143–2146. https://doi.org/10.1016/j.neubiorev.2012.06.008.

Bao, Y., Pöppel, E., & Zaytseva, Y. (2017). Single case studies as a prime example for exploratory research. *PsyCh Journal*, 6(2), 107–109. https://doi.org/10.1002/pchj.176.

Bao, Y., Pöppel, E., Wang, L., Lin, X., Yang, T., Avram, M., Blautzik, J., Paolini, M., Silveira, S., Vedder, A., Zaytseva, Y., & Zhou, B. (2015). Synchronization as a biological, psychological and social mechanism to create common time: A theoretical frame and a single case study. *PsyCh Journal*, 4(4), 243–254. https://doi.org/10.1002/pchj.119.

Bao, Y., Sander, T., Trahms, L., Pöppel, E., Lei, Q., & Zhou, B. (2011). The eccentricity effect of inhibition of return is resistant to practice. *Neuroscience Letters*, 500(1), 47–51. https://doi.org/10.1016/j.neulet.2011.06.003.

Bao, Y., Szymaszek, A., Wang, X., Oron, A., Pöppel, E., & Szelag, E. (2013). Temporal order perception of auditory stimuli is selectively modified by tonal and non-tonal language environments. *Cognition*, 129(3), 579–585. https://doi.org/10.1016/j.cognition.2013.08.019.

Bao, Y., von Stosch, A., Park, M., & Pöppel, E. (2017). Complementarity as generative principle: A thought pattern for aesthetic appreciations and cognitive appraisals in general. *Frontiers in Psychology*, 8(727). https://doi.org/10.3389/fpsyg.2017.00727.

Bao, Y., Wang, Z., Liang, W., Wang, Y., Pöppel, E., & Li, H. (2013). Inhibition of return at different eccentricities in the visual field share the same temporal window. *Neuroscience Letters*, 534, 7–11. https://doi.org/10.1016/j.neulet.2012.11.046.

Bao, Y., Yang, T., Lin, X., & Pöppel, E. (2016). Donders revisited: Discrete or continuous temporal processing underlying reaction time distributions? *PsyCh Journal*, 5(3), 177–179. https://doi.org/10.1002/pchj.142.

Bao, Y., Yang, T., Lin, X., Fang, Y., Wang, Y., Pöppel, E., & Lei, Q. (2016). Aesthetic preferences for eastern and western traditional visual art: Identity matters. *Frontiers in Psychology*, 7(1596). https://doi.org/10.3389/fpsyg.2016.01596.

Bao, Y., Yang, T., Zhang, J., Zhang, J., Lin, X., Paolini, M., Pöppel, E., & Silveira, S. (2017). The "third abstraction" of the Chinese artist LaoZhu: Neural and behavioral indicators of aesthetic appreciation. *PsyCh Journal*, 6(2), 110–119. https://doi.org/10.1002/pchj.167.

Barrett, K. E., Barman, S. M., Boitano, S., & Brooks, H. (2012). *Ganong's Review of Medical Physiology*, Vol. 24. New York, NY: McGraw-Hill Medical.

Binder, J. R. (2015). The Wernicke area: Modern evidence and a reinterpretation. *Neurology*, 85(24), 2170–2175. https://doi.org/10.1212/WNL.0000000000002219.

Bleuler, M. (1969). *Lehrbuch der Psychiatrie* (*Textbook of Psychiatry*, 11th Edition. Berlin, Germany: Springer. https://doi.org/10.1007/978-3-662-00125-7.

Bodis-Wollner, I. (2008). Pre-emptive perception. *Perception*, 37(3), 462–478. https://doi.org/10.1068/p5880.

Bourguignon, M., De Tiege, X., de Beeck, M. O., Ligot, N., Paquier, P., Van Bogaert, P., Goldman, S., Hari, R., & Jousmäki, V. (2013). The pace of prosodic phrasing couples the listener's cortex to the reader's voice. *Human Brain Mapping*, 34(2), 314–326. https://doi.org/10.1002/hbm.21442.

Chow, W. Y., Momma, S., Smith, C., Lau, E., & Phillips, C. (2016). Prediction as memory retrieval: Timing and mechanisms. *Language, Cognition and Neuroscience*, 31(5), 617–627. https://doi.org/10.1080/23273798.2016.1160135.

Christiansen, M. H. & Chater, N. (2016). The now-or-never bottleneck: A fundamental constraint on language. *Behavioral and Brain Sciences*, 39, e62. https://doi.org/10.1017/s0140525x1500031x.

Clark, A. (2013). Whatever next? Predictive brains, situated agents, and the future of cognitive science. *Behavioral and Brain Sciences*, 36(3), 181–204. https://doi.org/10.1017/S0140525X12000477.

Cowan, N. (2001). The magical number 4 in short-term memory: A reconsideration of mental storage capacity. *Behavioral and Brain Sciences*, 24(1), 87–114. https://doi.org/10.1017/s0140525x01003922.

de Lange, F. P., Heilbron, M., & Kok, P. (2018). How do expectations shape perception? *Trends in Cognitive Sciences*, 22(9), 764–779. https://doi.org/10.1016/j.tics.2018.06.002.

Dick, A. S. & Andric, M. (2013). The neurobiology of receptive-expressive language interdependence. *Behavioral and Brain Sciences*, 36(4), 352–353. https://doi.org/10.1017/s0140525x12002543.

Dikker, S., Wan, L., Davidesco, I., Kaggen, L., Oostrik, M., McClintock, J., *et al.* (2017). Brain-to-brain synchrony tracks real-world dynamic group interactions in the classroom. *Current Biology*, 27(9), 1375–1380. https://doi.org/10.3389/fnhum.2017.00554.

Ding, N., Melloni, L., Zhang, H., Tian, X., & Poeppel, D. (2016). Cortical tracking of hierarchical linguistic structures in connected speech. *Nature Neuroscience*, 19(1), 158–164. https://doi.org/10.1038/nn.4186.

Ding, N., Patel, A. D., Chen, L., Butler, H., Luo, C., & Poeppel, D. (2017). Temporal modulations in speech and music. *Neuroscience & Biobehavioral Reviews*, 81, 181–187. https://doi.org/10.1016/j.neubiorev.2017.02.011.

Donhauser, P. W. & Baillet, S. (2019). Two distinct neural timescales for predictive speech processing. *Neuron*, 105(2), 385–393. https://doi.org/10.1016/j.neuron.2019.10.019.

Elbert, T., Ulrich, R., Rockstroh, B., & Lutzenberger, W. (1991). The processing of temporal intervals reflected by CNV-like brain potentials. *Psychophysiology*, 28(6), 648–655. https://doi.org/10.1111/j.1469-8986.1991.tb01009.x.

Fabb, N. (2013). There is no psychological limit on the duration of metrical lines in performance: Against Turner and Pöppel. *International Journal of Literary Linguistics*, 2(1). https://doi.org/10.15462/ijll.v2i1.14.

Feinberg, I. (1978). Efference copy and corollary discharge: Implications for thinking and its disorders. *Schizophrenia Bulletin*, 4(4), 636–640. https://doi.org/10.1093/schbul/4.4.636.

Fiebach, C. J., Schlesewsky, M., & Friederici, A. D. (2002). Separating syntactic memory costs and syntactic integration costs during parsing: The processing of German WH-questions. *Journal of Memory and Language*, 47(2), 250–272. https://doi.org/10.1016/s0749-596x(02)00004-9.

Fraisse, P. (1984). Perception and estimation of time. *Annual Review of Psychology*, 35(1), 1–37. https://doi.org/10.1146/annurev.psych.35.1.1.

Gambi, C. & Pickering, M. J. (2017). Models linking production and comprehension. In M. J. Traxler & M. Ann Gernsbacher (eds.), *Handbook of Psycholinguistics*, pp. 157–182. https://doi.org/10.1002/9781118829516.ch7.

Gerstner, G. E. & Cianfarani, T. (1998). Temporal dynamics of human masticatory sequences. *Physiology & Behavior*, 64(4), 457–461. https://doi.org/10.1016/s0031-9384(98)00107-3.

Gerstner, G. E. & Fazio, V. A. (1995). Evidence of a universal perceptual unit in mammals. *Ethology*, 101(2), 89–100. https://doi.org/10.1111/j.1439-0310.1995.tb00348.x.

Gerstner, G. E. & Goldberg, L. J. (1994). Evidence of a time constant associated with movement patterns in six mammalian species. *Ethology and Sociobiology*, 15(4), 181–205. https://doi.org/10.1016/0162-3095(94)90013-2.

Gómez, C., Argandoña, E. D., Solier, R. G., Angulo, J. C., & Vázquez, M. (1995). Timing and competition in networks representing ambiguous figures. *Brain and Cognition*, 29(2), 103–114. https://doi.org/10.1006/brcg.1995.1270.

Hale, J., Ward, J. A., Buccheri, F., Oliver, D., & Hamilton, A. F. D. C. (2020). Are you on my wavelength? Interpersonal coordination in dyadic conversations. *Journal of Nonverbal Behavior*, 44(1), 63–83. https://doi.org/10.1007/s10919-019-00320-3.

Hickok, G. & Poeppel, D. (2007). The cortical organization of speech processing. *Nature Reviews Neuroscience*, 8(5), 393–402. https://doi.org/10.1038/nrn2113.

Hirsh, I. J. & Sherrick, C. E. S., Jr. (1961). Perceived order in different sense modalities. *Journal of Experimental Psychology*, 62(5), 423–432. https://doi.org/10.1037/e572322012-046.

Kagerer, F. A., Wittmann, M., Szelag, E., & Steinbüchel, N. V. (2002). Cortical involvement in temporal reproduction: Evidence for differential roles of the hemispheres. *Neuropsychologia*, 40(3), 357–366. https://doi.org/10.1016/s0028-3932(01)00111-7.

Kien, J. & Kemp, A. (1994). Is speech temporally segmented? Comparison with temporal segmentation in behavior. *Brain and Language*, 46(4), 662–682. https://doi.org/10.1006/brln.1994.1036.

Kien, J., Schleidt, M., & Schöttner, B. (1991). Temporal segmentation in hand movements of chimpanzees (Pan troglodytes) and comparisons with humans. *Ethology*, 89(4), 297–304. https://doi.org/10.1111/j.1439-0310.1991.tb00375.x.

Kluender, R. & Kutas, M. (1993). Subjacency as a processing phenomenon. *Language and Cognitive Processes*, 8(4), 573–633. https://doi.org/10.1080/01690969308407588.

Kowal, S., O'Connell, D. C., & Sabin, E. J. (1975). Development of temporal patterning and vocal hesitations in spontaneous narratives. *Journal of Psycholinguistic Research*, 4(3), 195–207. https://doi.org/10.1007/bf01066926.

Kutas, M. & Hillyard, S. A. (1980). Reading senseless sentences: Brain potentials reflect semantic incongruity. *Science*, 207(4427), 203–205. https://doi.org/10.1126/science.7350657.

Langman, V. A., Bamford, O. S., & Maloiy, G. M. O. (1982). Respiration and metabolism in the giraffe. *Respiration Physiology*, 50(2), 141–152. https://doi.org/10.1016/0034-5687(82)90013-5.

Lau, E. (2018). Neural indices of structured sentence representation: State of the art. In K. Federmeier & E. Schotter (eds.), *Psychology of Learning and Motivation*, Vol. 68, pp. 117–142. Elsevier. https://doi.org/10.1016/bs.plm.2018.08.004.

Lau, E. & Liao, C. H. (2017). Linguistic structure across time: ERP responses to coordinated and uncoordinated noun phrases. *Language, Cognition and Neuroscience*, 33(5), 633–647. https://doi.org/10.1080/23273798.2017.1400081.

Lei, Q., Bao, Y., Wang, B., & Gutyrchik, E. (2012). fMRI correlates of inhibition of return in Perifoveal and peripheral visual field. *Cognitive Processing*, 13, 223–227. https://doi.org/10.1007/s10339-012-0487-3.

Liberman, A. M. & Mattingly, I. G. (1985). The motor theory of speech perception revised. *Cognition*, 21(1), 1–36. https://doi.org/10.1093/acrefore/9780199384655.013.404.

Marr, D. (1982). *Vision: A Computational Approach.* San Francisco, CA: Freeman & Co.

Martin, B., Wittmann, M., Franck, N., Cermolacce, M., Berna, F., & Giersch, A. (2014). Temporal structure of consciousness and minimal self in schizophrenia. *Frontiers in Psychology*, 5, 1175. https://doi.org/10.3389/fpsyg.2014.01175.

Martin, J. G. (1972). Rhythmic (hierarchical) versus serial structure in speech and other behavior. *Psychological Review*, 79(6), 487–509. https://doi.org/10.1037/h0033467.

Meyer, L. (2018). The neural oscillations of speech processing and language comprehension: State of the art and emerging mechanisms. *European Journal of Neuroscience*, 48(7), 2609–2621. https://doi.org/10.1111/ejn.13748.

Meyer, L., Sun, Y., & Martin, A. E. (2019). Synchronous, but not entrained: Exogenous and endogenous cortical rhythms of speech and language processing. *Language, Cognition and Neuroscience*, 1–11. https://doi.org/10.1080/23273798.2019.1693050.

Miller, G. A. (1956). The magical number seven, plus or minus two: Some limits on our capacity for processing information. *Psychological Review*, 63(2), 81–352. https://doi.org/10.1037/0033-295x.101.2.343.

Miyake, Y. (2012). kyōsōteki komyunikesyon to "ma" [coordinative communication and "Ma"]. *Baiomekanisumu Gakkaishi [Journal of the Society of Biomechanisms Japan]*, 36(2), 97–103. https://doi.org/10.3951/sobim.36.97.

Molinaro, N., Lizarazu, M., Lallier, M., Bourguignon, M., & Carreiras, M. (2016). Out-of-synchrony speech entrainment in developmental dyslexia. *Human Brain Mapping*, 37(8), 2767–2783. https://doi.org/10.1002/hbm.23206.

Nagy, E. (2011). Sharing the moment: The duration of embraces in humans. *Journal of Ethology*, 29(2), 389–393. https://doi.org/10.1007/s10164-010-0260-y.

Overath, T., McDermott, J. H., Zarate, J. M., & Poeppel, D. (2015). The cortical analysis of speech-specific temporal structure revealed by responses to sound quilts. *Nature Neuroscience*, 18(6), 903–911. https://doi.org/10.1038/nn.4021.

Park, M., Gutyrchik, E., Bao, Y., Zaytseva, Y., Carl, P., Welker, L., Pöppel, E., Reiser, M., Blautzik, J., & Meindl, T. (2014). Differences between musicians and non-musicians in neuro-affective processing of sadness and fear expressed in music. *Neuroscience Letters*, 566, 120–124. https://doi.org/10.1016/j.neulet.2014.02.041.

Park, M., Gutyrchik, E., Welker, L., Carl, P., Pöppel, E., Zaytseva, Y., Meindl, T., Blautzik, J., Reiser, M., & Bao, Y. (2015). Sadness is unique: Neural processing of emotions in speech prosody in musicians and non-musicians. *Frontiers in Human Neuroscience*, 8, 1049. https://doi.org/10.3389/fnhum.2014.01049.

Parkes, M. J. (2006). Breath-holding and its breakpoint. *Experimental Physiology*, 91(1), 1–15. https://doi.org/10.1113/expphysiol.2005.031625.

Phillips, C., Kazanina, N., & Abada, S. H. (2005). ERP effects of the processing of syntactic long-distance dependencies. *Cognitive Brain Research*, 22(3), 407–428. https://doi.org/10.1016/j.cogbrainres.2004.09.012.

Pickering, M. J. & Clark, A. (2014). Getting ahead: Forward models and their place in cognitive architecture. *Trends in Cognitive Sciences*, 18(9), 451–456. https://doi.org/10.1016/j.tics.2014.05.006.

Pickering, M. J. & Gambi, C. (2018). Predicting while comprehending language: A theory and review. *Psychological Bulletin*, 144(10), 1002–1044. https://doi.org/10.1037/bul0000158.

Pickering, M. J. & Garrod, S. (2013). An integrated theory of language production and comprehension. *Behavioral and Brain Sciences*, 36(4), 329–347. https://doi.org/10.1017/s0140525x12001495.

Poeppel, D. & Assaneo, M. F. (2020). Speech rhythms and their neural foundations. *Nature Reviews Neuroscience*, 21(6), 1–13. https://doi.org/10.1038/s41583-020-0304-4.

Polgári, P., Causin, J.-B., Weiner, L., Bertschy, G., & Giersch, A. (2020). Novel method to measure temporal windows based on eye movements during viewing of the Necker cube. *PLoS One*, 15(1), e0227506. https://doi.org/10.1371/journal.pone.0227506.

Pöppel, E. (1970). Excitability cycles in central intermittency. *Psychologische Forschung*, 34, 1–9. https://doi.org/10.1007/BF00422860.

Pöppel, E. (1972). Oscillations as possible basis for time perception. In *The Study of Time*, pp. 219–241. Berlin, Germany: Springer. https://doi.org/10.1007/978-3-642-65387-2_16.

Pöppel, E. (1988). *Mindworks: Time and Conscious Experience*. Harcourt Brace Jovanovich.

Pöppel, E. (1989a). The measurement of music and the cerebral clock: A new theory. *Leonardo*, 22, 83–89. https://doi.org/10.2307/1575145.

Pöppel, E. (1989b). Taxonomy of the subjective: An evolutionary perspective. In *Neuropsychology of Visual Perception*, pp. 219–232. Hillsdale, NJ: Lawrence Erlbaum Associates Publishers. https://doi.org/10.4324/9781315441849-11.

Pöppel, E. (1994). Temporal mechanisms in perception. *International Review of Neurobiology*, 37, 185–202. https://doi.org/10.1016/s0074-7742(08)60246-9.

Pöppel, E. (1997a). A hierarchical model of temporal perception. *Trends in Cognitive Sciences*, 1(2), 56–61.20https:/doi.org/10.1016/s1364-6613(97)01008-5.

Pöppel, E. (1997b). Consciousness versus states of being conscious. *Behavioral and Brain Sciences*, 20(1), 155–156. https://doi.org/10.1017/s0140525x97300053.

Pöppel, E. (2009). Pre-semantically defined temporal windows for cognitive processing. *Philosophical Transactions of the Royal Society B: Biological Sciences*, 364(1525), 1887–1896. https://doi.org/10.1098/rstb.2009.0015.

Pöppel, E. (2010). Perceptual identity and personal self: Neurobiological reflections. In T. Maruszewski, M. Fajkowska, & M. M. Eysenck (eds.), *Personality from Biological, Cognitive, and Social Perspectives*, pp. 75–82. Clinton Corners, NY: Eliot Werner Public.

Pöppel, E. & Bao, Y. (2011). Three modes of knowledge as basis for intercultural cognition and communication: A theoretical perspective. In *Culture and Neural Frames of Cognition and Communication*, pp. 215–231. Berlin, Germany: Springer. https://doi.org/10.1007/978-3-642-15423-2_14.

Pöppel, E. & Bao, Y. (2012). The visual field paradox: A theoretical account on the reafference principle providing a common frame for the homogeneity and inhomogeneity of visual representation. *Cognitive Processing*, 13(1), 285–287. https://doi.org/10.1007/s10339-012-0489-1.

Pöppel, E. & Bao, Y. (2017). The power of single cases: Examples from the early visual pathway and from visual art. In *Cultural Patterns and Neurocognitive Circuits II: East-West Connections*, pp. 33–62. Singapore: World Scientific Press.

Pöppel, E. & Logothetis, N. (1986). Neuronal oscillations in the human brain. *Naturwissenschaften*, 73(5), 267–268. https://doi.org/10.1007/bf00367781.

Radilová, J., Pöppel, E., & Ilmberger, J. (1990). Auditory reversal timing. *Activitas Nervosa Superior*, 32, 13.

Rao, R. P. N. & Ballard, D. H. (1999). Predictive coding in the visual cortex: A functional interpretation of some extra-classical receptive-field effects. *Nature Neuroscience*, 2(1), 79–87. https://doi.org/10.1038/4580.

Rimmele, J. M., Morillon, B., Poeppel, D., & Arnal, L. H. (2018). Proactive sensing of periodic and aperiodic auditory patterns. *Trends in Cognitive Sciences*, 22(10), 870–882. https://doi.org/10.1016/j.tics.2018.08.003.

Roll, M., Gosselke, S., Lindgren, M., & Horne, M. (2013). Time-driven effects on processing grammatical agreement. *Frontiers in Psychology*, 4, 01004. https://doi.org/10.3389/fpsyg.2013.01004.

Roll, M., Lindgren, M., Alter, K., & Horne, M. (2012). Time-driven effects on parsing during reading. *Brain and Language*, 121(3), 267–272. https://doi.org/10.1016/j.bandl.2012.03.002.

Sachs, J. S. (1974). Memory in reading and listening to discourse. *Memory & Cognition*, 2(1), 95–100. https://doi.org/10.3758/bf03197498.

Sams, M., Hari, R., Rif, J., & Knuutila, J. (1993). The human auditory sensory memory trace persists about 10 sec: Neuromagnetic evidence. *Journal of Cognitive Neuroscience*, 5(3), 363–370. https://doi.org/10.1162/jocn.1993.5.3.363.

Schleidt, M. & Kien, J. (1997). Segmentation in behavior and what it can tell us about brain function. *Human Nature*, 8(1), 77–111. https://doi.org/10.1007/s12110-997-1005-7.

Schleidt, M., Eibl-Eibesfeldt, I., & Pöppel, E. (1987). A universal constant in temporal segmentation of human short-term behavior. *Naturwissenschaften*, 74(6), 289–290. https://doi.org/10.1007/bf00366417.

Schremm, A., Horne, M., & Roll, M. (2015). Time-driven effects on processing relative clauses. *Journal of Psycholinguistic Research*, 45(5), 1033–1044. https://doi.org/10.1007/s10936-015-9391-1.

Stefanics, G., Kremláček, J., & Czigler, I. (2014). Visual mismatch negativity: A predictive coding view. *Frontiers in Human Neuroscience*, 8, 00666. https://doi.org/10.3389/fnhum.2014.00666.

Szelag, E. & Pöppel, E. (2000). Temporal perception: A key to understanding language. *Behavioral and Brain Sciences*, 23(1), 52–52. https://doi.org/10.1017/s0140525x0055239x.

Tanida, K. & Pöppel, E. (2006). A hierarchical model of operational anticipation windows in driving an automobile. *Cognitive Processing*, 7(4), 275–287. https://doi.org/10.1007/s10339-006-0152-9.

Teng, X., Ma, M., Yang, J., Blohm, S., Cai, Q., & Tian, X. (2020). Constrained structure of ancient Chinese poetry facilitates speech content grouping. *Current Biology*, 30, 1299–1305.e7. https://doi.org/10.1016/j.cub.2020.01.059.

Teuber, H. L. (1960). Perception. In *Handbook Physiology Neurophysiol*, Vol. III, pp. 1595–1668.

Tian, X. & Poeppel, D. (2010). Mental imagery of speech and movement implicates the dynamics of internal forward models. *Frontiers in Psychology*, 1, 00166. https://doi.org/10.3389/fpsyg.2010.00166

Tian, X. & Poeppel, D. (2013). The effect of imagination on stimulation: The functional specificity of efference copies in speech processing. *Journal of Cognitive Neuroscience*, 25(7), 1020–1036. https://doi.org/10.1162/jocn_a_00381.

Tian, X., Zarate, J. M., & Poeppel, D. (2016). Mental imagery of speech implicates two mechanisms of perceptual reactivation. *Cortex*, 77, 1–12. https://doi.org/10.1016/j.cortex.2016.01.002.

Turner, F. & Pöppel, E. (1988). Metered poetry, the brain, and time. In *Beauty and the Brain*, pp. 71–90. Basel, Switzerland: Birkhäuser. https://doi.org/10.1007/978-3-0348-6350-6_4.

Verduyn, P. & Lavrijsen, S. (2015). Which emotions last longest and why: The role of event importance and rumination. *Motivation and Emotion*, 39(1), 119–127. https://doi.org/10.1007/s11031-014-9445-y.

Vollrath, M., Kazenwadel, J., & Krüger, H. P. (1992). A universal constant in temporal segmentation of human speech. *Naturwissenschaften*, 79(10), 479–480. https://doi.org/10.1007/bf01139205.

von Helmholtz, H. (1896). *Handbuch der physiologischen Optik* (*Handbook of Physiological Optics*). Hamburg und Leipzig: Verlag von Leopold Voss, 2 Auflage.

von Holst, E. & Mittelstaedt, H. (1950). Das Reafferenzprinzip. Wechselwirkungen zwischen Zentralnervensystern und Peripherie (the Reafference principle. Interactions between the central nervous system and the periphery). *Naturwissenschaften*, 37, 464–476. https://doi.org/10.1007/bf00622503.

von Holst, E. & Mittelstaedt, H. (1971). The principle of reafference: Interactions between the central nervous system and the peripheral organs. In P. Dodwell (ed.), *Perceptual Processing: Stimulus Equivalence and Pattern Recognition*, pp. 41–72.

von Steinbüchel, N. (1998). Temporal ranges of central nervous processing: Clinical evidence. *Experimental Brain Research*, 123(1–2), 220–233. https://doi.org/10.1007/s002210050564.

Wagers, M. W. & Phillips, C. (2009). Multiple dependencies and the role of the grammar in real-time comprehension. *Journal of Linguistics*, 45(2), 395–433. https://doi.org/10.1017/s0022226709005726.

Wang, L., Bao, Y., Zhang, J., Lin, X., Yang, L., Pöppel, E., & Zhou, B. (2016). Scanning the world in three seconds: Mismatch negativity as an indicator of temporal segmentation. *PsyCh Journal*, 5(3), 170–176. https://doi.org/10.1002/pchj.144.

Wang, L., Kuperberg, G., & Jensen, O. (2018). Specific lexico-semantic predictions are associated with unique spatial and temporal patterns of neural activity. *eLife*, 7, e39061. https://doi.org/10.1080/01690969308407588.

Wang, L., Lin, X., Zhou, B., Pöppel, E., & Bao, Y. (2015). Subjective present: A window of temporal integration indexed by mismatch negativity. *Cognitive Processing*, 16(1), 131–135. https://doi.org/10.1007/s10339-015-0687-8.

Wang, M., Arteaga, D., & He, B. J. (2013). Brain mechanisms for simple perception and bistable perception. *Proceedings of the National Academy of Sciences*, 110(35), E3350–E3359. https://doi.org/10.1073/pnas.1221945110.

White, P. A. (2017). The three-second "subjective present": A critical review and a new proposal. *Psychological Bulletin*, 143(7), 735–756. https://doi.org/10.1037/bul0000104.

Wittmann, M. (1999). Time perception and temporal processing levels of the brain. *Chronobiology International*, 16(1), 17–32. https://doi.org/10.3109/07420529908998709.

Wurtz, R. H. (2018). Corollary discharge contributions to perceptual continuity across saccades. *Annual Review of Vision Science*, 4, 215–237. https://doi.org/10.1146/annurev-vision-102016-061207.

Zaytseva, Y. & Bao, Y. (2015). Special collection on single case studies. *PsyCh Journal*, 4(4), 177–177. https://doi.org/10.1002/pchj.122.

Zhang, Y., Yu, J., & Boland, J. E. (2010). Semantics does not need a processing license from syntax in reading Chinese. *Journal of Experimental Psychology: Learning, Memory, and Cognition*, 36(3), 765–781. https://doi.org/10.1037/a0019254.

Zhao, C., Zhang, D., & Bao, Y. (2018). A time window of 3 s in the aesthetic appreciation of poems. *PsyCh Journal*, 7(1), 51–52. https://doi.org/10.1002/pchj.194.

Zhou, B., Bao, Y., Sander, T., Trahms, L., & Pöppel, E. (2010). Dissociation of summation and peak latencies in visual processing: An MEG study on stimulus eccentricity. *Neuroscience Letters*, 483(2), 101–104. https://doi.org/10.1016/j.neulet.2010.07.070.

Zhou, B., Pöppel, E., & Bao, Y. (2014). In the jungle of time: The concept of identity as a way out. *Frontiers in Psychology*, 5(844). https://doi.org/10.3389/fpsyg.2014.00844.

Zhou, B., Pöppel, E., Wang, L., Yang, T., Zaytseva, Y., & Bao, Y. (2016). Seeing without knowing: Operational principles along the early visual pathway. *PsyCh Journal*, 5(3), 145–160. https://doi.org/10.1002/pchj.141.

Zhou, H., Melloni, L., Poeppel, D., & Ding, N. (2016). Interpretations of frequency domain analyses of neural entrainment: Periodicity, fundamental frequency, and harmonics. *Frontiers in Human Neuroscience*, 10(274). https://doi.org/10.3389/fnhum.2016.00274.

https://doi.org/10.1142/9789811298639_0028

Chapter 28

A Time Window of 3 s in the Aesthetic Appreciation of Poems*

Chen Zhao[†], Dongxue Zhang[†] and Yan Bao[†,‡]

[†]*School of Psychological and Cognitive Sciences, Peking University, Beijing, China*

‡*Institute of Medical Psychology, Ludwig Maximilian University, Munich, Germany*

Abstract

We studied the effect of time windows of a few seconds on the aesthetic appreciation of poems. Both Chinese and German subjects rated traditional Chinese poetic verses more beautiful in a time window of approximately 3 s, irrespective of understanding the poetic content. This observation suggests a common temporal preference for poetry appreciation across a different language background.

Keywords: culture; emotion; poetry appreciation; time window

*This chapter was originally published in *PsyCh Journal*. Zhao, C., Zhang, D., & Bao, Y. (2018), A time window of 3 s in the aesthetic appreciation of poems. *PsyCh Journal*, 7(1), 51–52. https://doi.org/10.1002/pchj.194.

Temporal processing in neural systems is characterized by a time window of approximately 3 s. This temporal segmentation is reflected in cognitive processes like sensorimotor synchronization, intentional movements, temporal perception, verbal communication, attentional modulation, and, on the neural level, in the temporal modulation of cortical sensitivity (Wang, Lin, Zhou, Pöppel, & Bao, 2015; for an overview and mechanisms involved, see Bao *et al.*, 2015; Pöppel & Bao, 2014). With respect to verbal communication, it is observed that spontaneous speech is also embedded in this time window, giving it a rhythmic structure (Vollrath, Kazenwadel, & Krüger, 1992). This temporal segmentation is also reflected in poetry observed in almost every culture. Specifically, the duration of a spoken verse typically lasts approximately 3 s (Turner & Pöppel, 1988). Apparently, poets of all cultures have implicit knowledge of how to organize their poetic thoughts in time, and the reciting of poems is, thus, naturally embedded in a 3-s frame accordingly. Possibly, this temporal frame plays an important role in the aesthetic appreciation of poems. Therefore, this study aims to investigate whether a temporal window of 3 s provides an optimal frame for the aesthetic appreciation of poems. Furthermore, whether this time window is modulated by emotional valence is also tested.

Thirteen Chinese and 13 German university students participated in this study, the latter without any knowledge of the Chinese language. They listened to recordings of unfamiliar traditional Chinese poetic sentences. Each sentence contained two verses with five Chinese characters per verse. The original recordings were pre-rated with respect to their poetic emotions (happy, sad, or neutral). Since poetry is rarely neutral, we selected for this study 30 sentences with two emotional types (half happy and half sad) as experimental material. All 15 sentences in each emotional type were modified using SONY Sound Forge into four high-quality versions of different poetic durations (i.e., each verse of the poetic sentence lasted for 1.5, 2, 3, or 4 s, respectively). The resulting 120 sentences were used twice each in random sequence, making a total of 240 trials arranged in eight blocks. Each trial started with a 300-ms pure tone followed by the recording of a poetic sentence. After a blank screen for 800 ms, the question "How beautiful does the sentence sound?" (in German or Chinese), and a bipolar 9-point Likert scale (1 = *very ugly* to 9 = *very beautiful*)

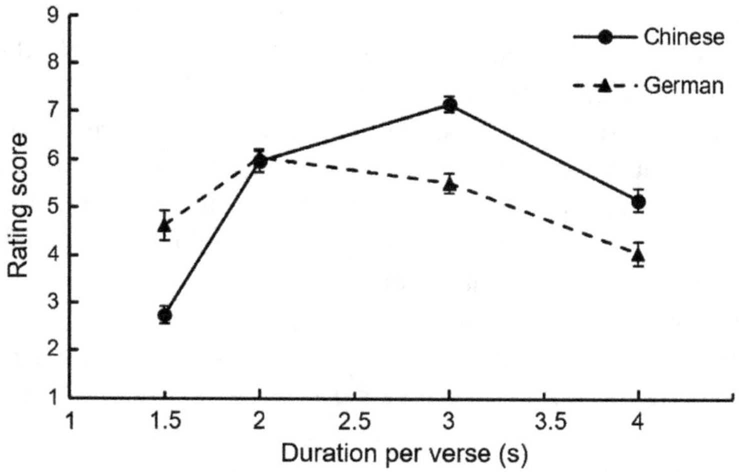

Figure 1. The aesthetic ratings of Chinese and German subjects on poetic sentences with different durations.

Notes: In the 9-point Likert scale, 1 = very ugly and 9 = very beautiful.

appeared simultaneously and remained on the screen until the subject responded. The intertrial interval varied between 1 and 2 s. A practice session with neutral sentences of the same four duration levels was arranged before the main experiment to help the subjects become familiar with the task and set up a rating reference.

The data were subjected to a three-way (duration vs. culture vs. emotion) analysis of variance, which revealed a significant main effect only for poetic duration, $F = 44.009$, $p < 0.001$, $\eta^2 = 0.857$. Interactions were non-significant except for the interaction between duration and culture, $F = 7.913$, $p < 0.001$, $\eta^2 = 0.519$, as shown in Figure 1. For German subjects, the adjacent difference between the 2- and 3-s duration levels was not significant, nor was the difference between the 1.5- and 4-s levels. However, the difference between the 1.5- and 2-s levels as well as the difference between the 3- and 4-s levels were both significant ($p < 0.001$ for both cases). For Chinese subjects, all rating differences were significant ($p < 0.05$ for all pairs) except the difference between the 2- and 4-s duration levels. The ratings of the two cultural groups were significantly different from each other at all duration levels ($p < 0.05$ for all pairs) except the 2-s duration.

The aesthetic ratings of the two groups were highest for the 2- and 3-s durations. These results confirm the effect of an optimal time window on poetry appreciation. For both the Chinese subjects, who could understand the language, and the Germans, who could not, only when the duration of each verse fitted into a 2–3-s time window did the poetry sound most beautiful. This aesthetic preference suggests an optimal time window that creates a temporal frame for processing sensory information. Within this frame, information is integrated into a coherent percept, and the rhythm of information flow is appreciated.

Despite sharing a common time window, the two cultural groups differed significantly in the patterns of their poetry appreciation. For the Chinese subjects, the shortest duration verses were experienced as ugly, while the longest ones were acceptable. No such difference between the shortest and longest durations was found for the German subjects. Accordingly, the German subjects had only two stages of rating (1.5 and 4 s in the lower stage and 2 and 3 s in the higher stage), while the Chinese had three (1.5 s in the lower stage, 2 and 4 s in the middle stage, and 3 s in the higher stage), plus a larger rating range. These patterns suggest that Chinese subjects seem to be more sensitive in discriminating the aesthetic differences of poetry, presumably resulting from their comprehension of the language. Although all poems selected for this study were unfamiliar to the Chinese subjects, the Chinese subjects still might have benefited from the semantic processing of the Chinese poems, while for the German subjects, this was not possible. The common ground and the differences in poetry appreciation between Chinese and German subjects suggest that although poetry appreciation is modulated by culture, its temporal frame for appreciating the aesthetic value is independent of the specific poetic content. The time window of 3 s is not determined by the linguistic content but is implemented by a pre-semantic neural process that provides an operational platform for information to be processed (Bao *et al.*, 2015; Pöppel & Bao, 2014).

Intuitively, we expected an emotional modulation on poetry appreciation, but no such effect was observed between the happy and sad poems. This might be related to the unfamiliar material we deliberately selected. Even for the Chinese, who understood the language, the poetic content could not be fully grasped, thus weakening the extraction of poetic emotions.

This study shows, on the one hand, that a psychological approach can help better understand the cognitive processes underlying aesthetic appreciations and, on the other hand, that aesthetic material, such as poems, can be used as experimental stimuli to study human cognition. Since only Chinese poems were used in this cross-cultural experiment, further studies are needed to check whether the observed poetic appreciation pattern can be replicated when Chinese and German participants are exposed to German poems.

Acknowledgments

This study was supported by the National Training Program of Innovation for Undergraduates of China and the National Natural Science Foundation of China (Projects 31371018 and 31771213).

References

Bao, Y., Pöppel, E., Wang, L., Lin, X., Yang, T., Avram, M., Blautzik, J., Paolini, M., Silveira, S., Vedder, A., Zaytseva, Y., & Zhou, B. (2015). Synchronization as a biological, psychological and social mechanism to create common time: A theoretical frame and a single case study. *PsyCh Journal*, 4, 243–254. https://doi.org/10.1002/pchj.119.

Pöppel, E. & Bao, Y. (2014). Temporal windows as a bridge from objective to subjective time. In D. Lloyd & V. Arstila (eds.), *Subjective Time: The Philosophy, Psychology, and Neuroscience of Temporality*, pp. 241–261. Cambridge, MA: MIT Press.

Turner, F. & Pöppel, E. (1988). Metered poetry, the brain, and time. In I. Rentschler, B. Herzberger, & D. Epstein (eds.), *Beauty and the Brain: Biological Aspects of Aesthetics*, pp. 71–90. Basel, Switzerland: Birkhäuser.

Vollrath, M., Kazenwadel, J., & Krüger, H.-P. (1992). A universal constant in temporal segmentation of human speech. *Naturwissenschaften*, 79, 479–480. https://doi.org/10.1007/BF01139205.

Wang, L., Lin, X., Zhou, B., Pöppel, E., & Bao, Y. (2015). Subjective present: A window of temporal integration indexed by mismatch negativity. *Cognitive Processing*, 16(Suppl. 1), 131–135. https://doi.org/10.1007/s10339-015-0687-8.

https://doi.org/10.1142/9789811298639_0029

Chapter 29

The Two- to Three-Second Time Window of Shot Durations in Movies*

Shuai Chen[†], Mengtong Cai[†] and Yan Bao[†,‡,§]

[†]*School of Psychological and Cognitive Sciences, and Beijing Key Laboratory of Behavior and Mental Health, Peking University, Beijing, China*
[‡]*Institute of Medical Psychology, Ludwig Maximilian University, Munich, Germany*
[§]*Parmenides Center for Art and Science, Pöcking, Germany*

Abstract

Movie shots of singular scenes have a preferential duration of 2 to 3 s regardless of producers, movie types, and cultural environments. This observation suggests that the temporal structure of movies matches a neural mechanism of information processing in the time domain.

Keywords: duration; movie; temporal integration; time window

*This chapter was originally published in the special collection on neuroaesthetics in *PsyCh Journal*. Chen, S., Cai, M., & Bao, Y. (2020), The two- to three-second time window of shot durations in movies. *PsyCh Journal*, 9(4), 516–518. https://doi.org/10.1002/pchj.384.

Given that one phenomenon can be observed with various tasks, it is reasonable to assume a global mechanism. The time window of 2 to 3 s has been observed with many different experimental paradigms such as the spontaneous alteration of ambiguous figures, temporal reproduction, sensorimotor synchronization, speech segmentation, intentional movements, or attentional control (Bao *et al.*, 2015) and also on the neural level such as the amplitude modulation of mismatch negativity indicating an oscillatory temporal regularity (Wang *et al.*, 2016). In addition, this temporal constancy extends to artistic artifacts such as motifs of classical music and verses of poems in different cultures (Zhao *et al.*, 2018). Does this time window also exist in other artistic forms such as movies? Does the virtual reality world on the screen also obey basic principles of information processing in the brain? To answer these questions, this study focused on the durations of movie shots.

As examples, we analyzed the duration of segments in two movies, *The Stranger*, directed by Orson Welles (1946), and *The Lifeboat*, directed by Alfred Hitchcock (1994) on the basis of the transcript written by Strobel (1981, 1983). The duration of such segments or shots in these movies was originally recorded with a temporal resolution of 0.5 and 1 s, respectively. We converted the data into a unified 1 s resolution using the moving average method to better compare the distribution between the two different movies. Both datasets showed a typical positively skewed curve with a mode at 2 s, as shown in Figures 1(a) and (c). The distribution is asymmetric; the positive skewness is due to a large number of longer shots. To overcome some fluctuations, we conducted a model fitting of the raw data. According to previous research, the distribution of movie shot durations follows the log-normal distribution (DeLong, 2015). The data we analyzed fit well with log-normal distribution with an R-square higher than 0.95 for all cases. Note that the distributions of the shot lengths do not follow an exponential decay, which would have indicated a passive and not an active temporal control.

Having observed a similar mode of shot length distribution in the two sample movies, we further asked whether various movies directed by different directors with diverse cultural backgrounds also show in their shot lengths a similar time window. Thus, we further applied the same method to analyze the shot durations of another two movies:

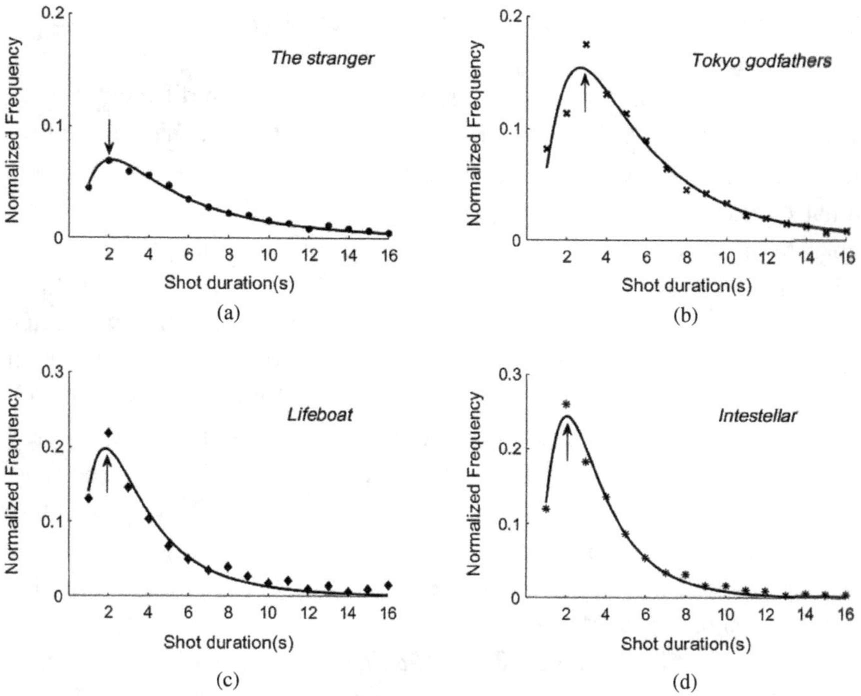

Figure 1. Normalized frequency distribution of shot durations in movies. Fitting curves are illustrated using log-normal distribution function. The mode of each distribution is indicated by an arrow.

Interstellar, directed by Christopher Nolan in 2014 (http://www.cinemetrics.lv), and the animation *Tokyo Godfathers*, directed by Satoshi Kon in 2003 (*Storyboard Collection Book: Tokyo Godfathers* by Satoshi Kon in 2018. Tokyo, Japan: Fukkan.com). The distribution and fitting curves of the shot durations in these two new examples are shown in Figures 1(b) and (d). Although the shooting time of the four analyzed movies varies from the early 20th century to the 21st century, and the directors come from Europe, North America, and Japan, all modes of these shot duration distributions lie within 2 to 3 s, regardless of producers, movie types, and cultural environments. It seems that there is an optimal time window in the cutting of movies, irrespective of the habits of the directors, the various forms of art, and the different shooting periods of the movies. Note that we emphasize the modes other than the average shot length because they are a more representative index for a skewed distribution.

The striking feature is that the same pattern exists in different movies from different regions and in different periods. The fact that movie producers favor a shot length of 2 to 3 s seems to be a universal phenomenon. What could be the reason? Typically, in a movie, the director creates a story taking into account both logic and aesthetics. Except for some experimental movies that use only one shot throughout the movie (e.g., *Rope* by Hitchcock in 1948), shot length is defined or implicitly selected by the content of the story. Long shots correspond to peaceful stories or magnificent scenes whereas short shots are used more in chasing, violence, horror, thrilling, and other high-tempo situations. When a new scene appears, attention is drawn to it and gradually decays over time. The purpose of film cutting is to replace a scene with a new scene at an appropriate rhythm such that the attention can be constantly captured and maintained, and the abruptness is reduced to the minimum. We hypothesize that a shot length of 2 to 3 s fits such a purpose. For example, when a ship appears on the screen for 3 s in *The Stranger*, the audience forms the concept that the main character is on board and heading to the United States for the hunt. The duration should not be too short to confuse the audience nor should it be too long to be boring. In a movie when one scene is followed by the next one, what is expected to happen has to guarantee the continuous flow of the story. It turns out that a time window of 2 to 3 s (Bao *et al.*, 2015) provides an optimal temporal frame for anticipations and, thus, allows for maintaining the identity of the story. If anticipations are violated too much due to too short or too long temporal frames, the flow or coherence of the story and its identity would break down, and the viewer would no longer be involved in the story and could not identify effortlessly with the sequence of events. The concept of maintaining identity is considered to be a general principle for aesthetic appreciation. Besides the identity of the artwork, the identity of the viewer also matters because it provides a basis for belongingness to a cultural environment in which the artwork is embedded (Bao *et al.*, 2016).

Taken together, this study shows that a time window of 2 to 3 s also applies to the shot durations in movies. Such temporal integration is suggested to be a universal mechanism of *Cognitive Processing* in the human brain. This time window of 2 to 3 s helps the producer to better tailor the movie, making the cut more dynamic and the story more coherent, thus more easily capturing the attention of the audience.

Acknowledgments

This work was supported by the National Natural Science Foundation of China (Projects 31771213 and 31371018).

References

Bao, Y., Pöppel, E., Wang, L., Lin, X., Yang, T., Avram, M., Blautzik, J., Paolini, M., Silveira, S., Vedder, A., Zaytseva, Y., & Zhou, B. (2015). Synchronization as a biological, psychological and social mechanism to create common time: A theoretical frame and a single case study. *PsyCh Journal*, 4(4), 243–254. https://doi.org/10.1002/pchj.119.

Bao, Y., Yang, T., Lin, X., Fang, Y., Wang, Y., Pöppel, E., & Lei, Q. (2016). Aesthetic preferences for eastern and Western traditional visual art: Identity matters. *Frontiers in Psychology*, 7, 1596. https://doi.org/10.3389/fpsyg.2016.01596.

DeLong, J. (2015). Horseshoes, handgrenades, and model fitting: The lognormal distribution is a pretty good model for shot-length distribution of Hollywood films. *Digital Scholarship in the Humanities*, 30(1), 129–136. https://doi.org/10.1093/llc/fqt030.

Strobel, R. (1981). *Orson Welles: "The stranger"/"Die Spur des Fremden" Ein Transcript*. Tübingen, Germany: Gunter Narr Verlag.

Strobel, R. (1983). *Alfred Hitchcock: "lifeboat"/"das Rettungsboot" Ein transcript*. Tübingen, Germany: Gunter Narr Verlag.

Wang, L., Bao, Y., Zhang, J., Lin, X., Yang, L., Pöppel, E., & Zhou, B. (2016). Scanning the world in three seconds: Mismatch negativity as an indicator of temporal segmentation. *PsyCh Journal*, 5, 170–176. https://doi.org/10.1002/pchj.144.

Zhao, C., Zhang, D., & Bao, Y. (2018). A time window of 3 s in the aesthetic appreciation of poems. *PsyCh Journal*, 7(1), 51–52. https://doi.org/10.1002/pchj.194.

https://doi.org/10.1142/9789811298639_0030

Chapter 30

Composing in the Scanner: A Functional Magnetic Resonance Imaging Single Case Study on Visual and Auditory Imagery*

Morteza Izadifar[†], Taoxi Yang[†], Marco Paolini[‡], Arusu Formuli[†] and Yan Bao[†,§,¶]

[†]*Institute of Medical Psychology, Ludwig, Maximilian University Munich, Germany*
[‡]*Department of Radiology, University Hospital, LMU Munich, Germany*
[§]*School of Psychological and Cognitive Sciences, Peking University, Beijing, China*
[¶]*Beijing Key Laboratory of Behavior and Mental, Health, Peking University, Beijing, China*

*This chapter was originally published in the special collection on neuroaesthetics in *PsyCh Journal*. Izadifar, M., Yang, T., Paolini, M., Formuli, A., & Bao, Y. (2022). Composing in the scanner: A functional magnetic resonance imaging single case study on visual and auditory imagery. *PsyCh Journal*, 11(5), 684–690. https://doi.org/10.1002/pchj.522.

Abstract

"Seeing with the mind's eye" and "hearing with the mind's ear" are two common indicators of musical imagery, and they can be referred to as "visual" and "auditory" musical imagery. However, a question remains open, that is, whether visual and auditory imagery of the same musical composition share the same neural mechanisms. Moreover, how can neural mechanisms guarantee the temporal flow of "musical imagery"? To answer these questions, we report here a preliminary single case study using functional magnetic resonance imaging with an eminent composer who imagined one of his compositions in two states of mind as compared to his resting-state activity. In the visual imagery condition, he imagined visually the score of his composition in a continuous way. In the auditory imagery condition, he imagined auditorily the same musical composition with pauses. In spite of the modality and temporal differences, the two types of mental imagery showed similar temporal durations for the same musical composition. However, different patterns of neural activation were observed for visual and auditory imagery with one important exception, that is, a common activation pattern was observed in the left medial temporal gyrus in both visual and auditory imagery. We speculate that the left medial temporal gyrus may play an important role in the creation of apparent temporal continuity in musical imagery and perhaps even in conscious information processing in general.

Keywords: auditory imagery; functional magnetic resonance imaging; music; temporal continuity; visual imagery

Introduction

For a better understanding of how cognitive contents like thoughts, feelings, memories, or images are neuronally represented, different behavioral and imaging paradigms have been used, including mental imagery procedures (e.g., von Trott Zu Solz *et al.*, 2017). One open question is whether "seeing with the mind's eye" or "hearing with the mind's ear" as an indicator of mental imagery may be equivalent (or at least similar) to the perceptual process itself. Independent of controversies about potential similarities in neural activation patterns of imagery and visual or auditory perception, the critical point has been raised that imagery suffers severe limitations and should not be used as an experimental paradigm due to its subjectivity (e.g., Pylyshyn, 1973; Watson, 1913). However, in spite of such methodological criticism, important insight has been gained about the neuro-cognitive machinery using imagery as a tool (e.g., Griffiths, 2000; von Trott Zu Solz *et al.*, 2017; Zatorre & Halpern, 1993). Although the term "imagery" might suggest that it refers only to the visual modality, auditory (musical) and tactile imagery are also recognized (e.g., Halpern *et al.*, 1995; Kraemer *et al.*, 2005; McGuire *et al.*, 1996; von Trott Zu Solz *et al.*, 2017). In fact, musical imagery (Bailes, 2019; Cotter, 2019; Halpern, 2001; Weber & Brown, 1986; Zatorre & Halpern, 1993) can be quite dominant like echoing a musical motif in our mind for hours. An impressive individual example of the power of auditory imagery is Ludwig van Beethoven who became deaf later in his life and in spite of his deafness still was a creative composer. This process of composing presumably happened on the basis of auditory imagery initiated and then followed by the visual equivalent in the musical scores. This duality of the two involved sensory modalities is another example of complementarity as a generative principle which can be observed in several domains of human cognition such as in top-down *and* bottom-up visual processing, in content of conscious representation *and* logistic functions (like musical motifs *and* their temporal dynamics), or aesthetic evaluations when linking aesthetic appreciation *and* cognitive appraisal (Bao & Pöppel, 2012; Bao, von Stosch, *et al.*, 2017b; Bao, Yang, Lin, Fang, *et al.*, 2016a; Zhang *et al.*, 2017; Zhou *et al.*, 2016).

Considering that both auditory and visual imageries for the same musical composition share identical musical content, one question is whether the underlying neural mechanisms of auditory and visual imageries are similar or different. Furthermore, another question is as follows: How can neural mechanisms guarantee the temporal flow of "musical imagery"? With respect to the latter point, we would like to cite a statement by the eminent conductor Bruno Walter (1957, p. 36): "Kontinuität des Tempos gehört zu den wichtigsten Grundsätzen des Musizierens" ("The continuity of tempo is one of the most important principles to make music"). Thus, temporal control is essential in music, and the open question is whether such temporal control is equivalent (or at least similar) in imagery compared to the perceptual process itself. It has been suggested that specific "time windows" of finite durations are essential for temporal control. In the low-frequency domain, a time window of 2–3 s has been experimentally identified for different kinds of cognitive processes like speech segmentation, intentional movement control, sensorimotor synchronization, bistable perception, or aesthetic evaluation (Bao *et al.*, 2015; Wang *et al.*, 2015; Yu & Bao, 2020; Zhao *et al.*, 2018). Furthermore, this time window may also be essential for the control of tempo in music as suggested previously (Pöppel, 1989).

Although auditory imagery has been studied experimentally using behavioral paradigms or imaging techniques like magnetoencephalography (Schürmann *et al.*, 2002), positron emission tomography (Halpern & Zatorre, 1999; Zatorre *et al.*, 1996), functional magnetic resonance imaging (fMRI; Halpern *et al.*, 2004; Kraemer *et al.*, 2005; Yoo *et al.*, 2001) with healthy subjects or brain-injured patients (Zatorre & Halpern, 1993), there has been a neglect in analyzing the temporal continuity of imagery. Thus, temporal processing is an issue with respect to the question of how information processing in the time domain is implemented on the neural level. Does the brain process information in a continuous form or in a sequence of discrete instants? Introspectively, subjective time has been considered as a continuous flow. William James (1890) has been one of the pioneers assuming consciousness as "sensibly continuous". Indeed, our intuition suggests that time is continuous rather than characterized by discrete temporal steps. However, experimental evidence shows that the intuition of a continuous time in perception and cognitive activities in general is misleading and that this intuition can perhaps be considered as an illusion (Bao *et al.*, 2015; Bao, Yang, Lin, & Pöppel, 2016b; Eriksen & Collins, 1968; Pöppel, 1970, 1971, 1997; Zhou *et al.*, 2014). For instance, if one takes a look at the receptor surface of information processing in the visual and auditory modality, then discrete time sampling with distinct time windows is actually a necessity. The different transduction time in these sensory modalities, the physical characteristics of stimuli with different intensities or distances to the observer, or the divergent neuroanatomical projection to central regions of the brain has as a consequence that physical time cannot be mapped directly onto central areas of the brain like the primary visual or auditory cortices. The resultant distortion can, however, be overcome by time windows of finite duration within which information being processed at different physical times is treated as co-temporal. Such a mechanism of overcoming ill-defined temporal information can technically be implemented by neural oscillations with different frequencies (Bao *et al.*, 2015; Pöppel, 1970, 1971, 1997). It is then a general question of how information represented in such sequential time windows with different frequencies are connected with each other to result in apparent continuity. What are the neural mechanisms behind these constructions of continuity, in particular with respect to the "functional glue" of the time window of 2–3 s that has been considered as the "subjective present" (Bao *et al.*, 2015; Pöppel, 1997; Yu & Bao, 2020)? Perhaps by focusing on mental imagery with and without interruptions to the same musical composition, one could scratch the surface of an answer.

Thus, in the experiment described, we aimed to obtain a better understanding of the underlying neural mechanisms of auditory and visual imagery with an identical musical composition, and furthermore, to get a hint of how apparent continuity might be created in musical imagery. Since there exists only a small number of exploratory results implemented on musicians who have received long-term musical training to do both auditory and visual musical imagery, and different musical compositions might result in diverse mental processes in different individuals due to the different degrees of familiarity and understanding of the compositions, etc., we decided to conduct an exploratory single case fMRI experiment trusting in the power of such studies (Flyvbjerg, 2006). Single case studies have been proven to be of heuristic value in many domains of cognitive processes such as in better understanding speech, referential or procedural memory, as well as visual, musical, or tactile imagery (Bao, Pöppel, *et al.*, 2017a, Bao, Yang, *et al.*, 2017c; Pöppel & Bao, 2018; von Trott Zu Solz *et al.*, 2017).

Materials and Methods

Participant(s)

The 71-year-old German composer P.M.H. took part in the experiment after giving informed written consent. He had corrected-to-normal visual acuity, and he had no history of neurological diseases. He was enthusiastic about being involved in the study, and he understood the demands of his time, effort, and experimental conditions. The study was approved by the ethical committee of Ludwig Maximilian University Munich, Germany, in agreement with the Declaration of Helsinki. Prior to the experiment the composer had composed a piece of some 6 min duration without prescribed limitations to genre or instrumentation.

Material

The composition was used as stimulus material for musical imagery in the fMRI experiment with a block design. The composer was first measured the resting-state activity for 7.5 min; during this condition, the composer was instructed to keep his eyes closed and was told not to think about anything in particular. After this initial phase, two experimental conditions with auditory and visual imagery were tested. The reason why we implemented the experiment into two auditory and visual conditions lies in two facts: (a) We did not see in the literature the different temporal aspects of auditory and visual imagery among musicians and (b) we wanted to know whether different brain patterns are observed during two diverse kinds of imagery.

Procedure

A pause of 5 min was inserted between adjacent conditions. In the auditory condition, the composer was asked to imagine auditorily his composition; in this experimental phase, the musical imagery was separated into four parts, each lasting 1.5 min. The reason behind separation of auditory imagery into four sections was determined according to the composer's idea. Between sequential parts, he was required to pause for some 30 s. No explicit feedback was given with respect to the amount of time that had passed. The composer indicated the start and the end of each part by pressing a button with his right index finger. In the visual condition, the composer used visual imagery going through the score of his composition. No pause was inserted into the continuous imagery. The composer simply indicated the beginning and the end of his visual imagery. Both visual and auditory imagery are routine tasks of professional musicians. In a report 2 months later after the experiment, the composer confirmed that he imagined his composition in the interrupted auditory phase as music "hearing it with the mind's ear", and in the uninterrupted visual phase, he went through the score "seeing it with the mind's eye".

Tools/apparatus

Brain imaging data were obtained with a 3T standard clinical MRI scanner (Ingenia; Philips Medical Systems Nederland) with a 32ch receiving head coil. For BOLD signals, T_2*-weighted EPI sequences were used (TR = 2,500 ms; TE = 30 ms; flip angle = 90°; acquisition matrix = 80 × 80; number of slices = 43; slice thickness = 3 mm, no gap between slices). In total, one run of 372 functional volumes was acquired for the subject. Structural data were acquired with a T_1-weighted scan of each participant's brain anatomy (1 mm × 1 mm × 1 mm; 240 × 240 matrix, field of view = 220 mm).

All neuroimaging data were preprocessed and analyzed using SPM12 (Statistical Parametric Mapping V12, http://www.fil.ion.ucl.ac.uk/spm). The functional scan volumes were subjected to spatial realignment to correct for head motion. In further preprocessing analysis, the mean functional image was co-registered to the anatomical image, normalized to the Montreal Neurological Institute (MNI) template provided in SPM12, and spatially smoothed to reduce noise using a Gaussian kernel of 8 mm full width at half maximum. The task was modeled as a block design. We applied a general linear model (GLM) using predictors convoluted with a typical hemodynamic response function. Two experimental conditions (auditory imagery and visual imagery) were compared to the resting-state condition respectively to generate statistical parametric maps of the t-tests (auditory condition vs. resting-state condition, and visual condition vs. resting-state condition). For these statistical tests, significant voxels initially passed a voxel-wise statistical threshold of $p \leq 0.001$, and a cluster-level threshold was obtained at the statistical significance level of $p < 0.01$.

Results

Before we focused on the underlying neural mechanisms of auditory and visual imagery of musical composition, we first checked the behavioral data with respect to the time of imagery. Compared to the consciously perceived time of the musical composition, that is, 6 min when the composition was made by the composer, does visually imagining the same musical piece without interruption share the same temporal duration? Moreover, when the imagery is interrupted with its four segments as in the auditory condition, does the mental imagery still hold the same temporal constant?

When the composer made his continuous visual imagery with his "inner eye" on the score of the composition, he showed a rather accurate temporal control; the visual imagery phase lasted 6 min and 9 s. The durations of the four sequential parts of the auditory imagery were 80.8, 80.9, 85.3, and 129.3 s, respectively; thus, the three first parts were shorter than the target duration of 90 s, and the fourth part was considerably longer compared to the preceding three parts and the target duration of 90 s. To put the four parts together, the whole auditory imagery lasted 6 min and 27 s.

With respect to the imaging data, the neural network for visual imagery versus resting state included the supramarginal gyrus and the postcentral gyrus, both in the left hemisphere. In the temporal cortex, the middle temporal gyrus and the superior temporal gyrus in the left hemisphere were also activated. In addition, an activation of the left caudate nucleus was observed, which is considered part of the basal ganglia (Table 1, Figure 1a). In the comparison of auditory imagery versus the resting state, an activation pattern was observed in the middle and the inferior temporal gyrus of the left hemisphere. In addition, there was a significant activation of the temporal pole in the left hemisphere (Table 1, Figure 1b). The Venn diagram illustrates the separation and the overlap of the two experimental conditions (Figure 2). Although the patterns of neural activation are different, there is one important exception, that is, the left middle temporal gyrus (LMTG) is shared both by visual and auditory imagery.

Table 1. Brain regions that respond to the comparisons of auditory imagery versus resting state, and the visual imagery versus resting state.

Brain regions	Montreal Neurological Institute coordinates *x*	*y*	*z*	*z* scores	Number of voxels
Auditory imagery versus baseline					
LMTG	−57	−10	−25	5.27	71
LITG	−48	−13	−37	4.37	26
LTP	−54	5	−31	3.79	70
Visual imagery versus baseline					
LMTG	−54	−10	−25	10.94	172
LSTG	−60	−31	11	8.98	276
LMTG	−60	−19	−7	7.87	48
LCN	−24	−1	23	7.78	37
LSG	−45	−40	35	7.78	258
LPG	−45	−22	35	7.64	26

Note: The resting-state activity serves as a baseline in the experiment. Abbreviations: LMTG = left middle temporal gyrus; LTP = left temporal pole; LITG = left inferior temporal gyrus; LCN = left caudate nucleus; LPG = left postcentral gyrus; LSG = left supramarginal gyrus; LSTG = left superior temporal gyrus.

Discussion

Our behavioral results of the total imagery timing showed rather similar temporal durations for both visual and auditory imagery, which were quite close to the consciously perceived time of the musical composition. However, the experimental segmentation of the composition may have had an influence on the temporal control of mental imagery. If one takes only the first three segments of auditory imagery and compares this duration with visual imagery, one may conclude that the visual imagery for the corresponding part had a longer duration; with this conclusion, one assumes that following the score of the composition visually did not change in "mental speed" throughout the entire composition of 6 min. However, as the sum of the four segments in auditory imagery is not too

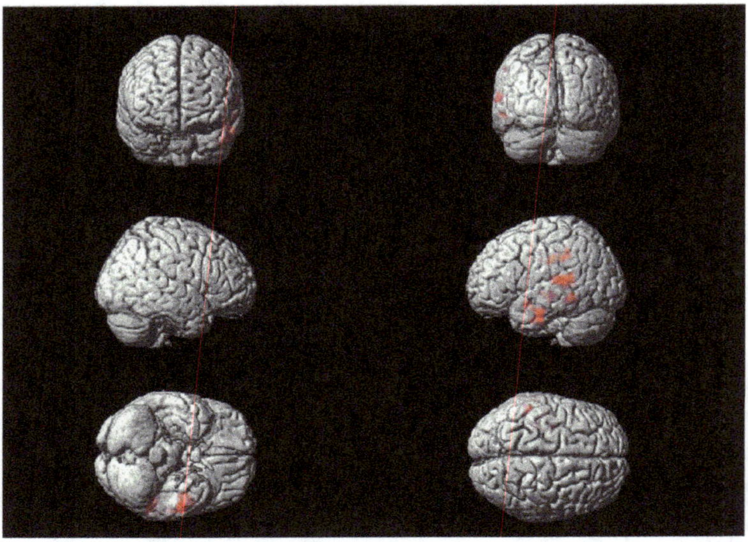

(a)

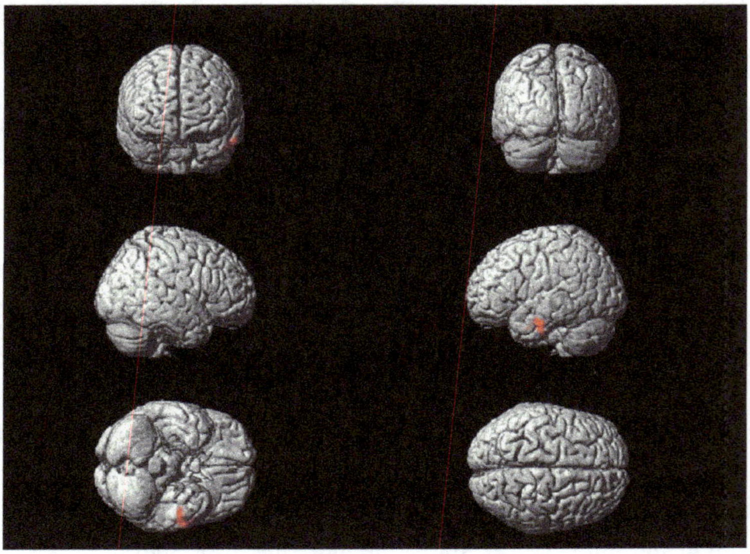

(b)

Figure 1. Differences in brain activation for visual imagery without a pause versus resting state (a) and for auditory imagery with pauses versus resting state (b). Activations are cluster-level corrected with p (familywise error; FWE) < 0.05. The x, y, and z coordinates are normalized using a standard brain from the Montreal Neurological Institute and displayed in radiological convention.

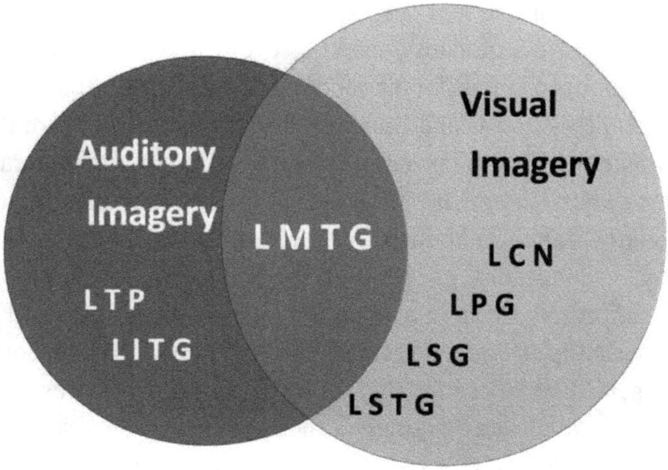

Figure 2. Venn diagram illustrating the overlap between auditory imagery and visual imagery. Circle size is proportional to the size of the networks identified. Abbreviations indicate the brain regions where activities were observed: LMTG = left middle temporal gyrus; LTP = left temporal pole; LITG = left inferior temporal gyrus; LCN = left caudate nucleus; LPG = left postcentral gyrus; LSG = left supramarginal gyrus; LSTG = left superior temporal gyrus.

much different from the target duration, one may also speculate that the different parts of the score had actually comparable durations during visual and auditory imagery. Another possibility could be that the composer reported that he got tired during visual imagery at the end and therefore he may also have slowed down during the last part of visual imagery, similar to the auditory imagery. An answer to this question on the behavioral level has to remain open. Whether an answer can be obtained with imaging technology is a major challenge as it requires identification of distinct patterns of neural activation both for visual and auditory imagery at corresponding phases of the composition. After having analyzed the fMRI data in this experiment it has to be concluded that it is not possible to get an answer as the patterns of neural activation are very different for visual and auditory imagery.

The imaging data in our experiment clearly showed very different neural patterns for auditory imagery and visual imagery. This was consistent with mental imagery as a unique mental phenomenon being separated from the auditory and visual sensory channels in general. Thus, mental imagery seems to be modality-specific, even though common musical composition is used for different modality imagery. On the other hand, since both auditory and visual imagery in our experiment share the same semantic content with similar temporal duration, a common neural mechanism, if it exists, should be related to such commonality. Indeed, the temporal gyrus in the left hemisphere was activated for both auditory and visual imagery conditions, as shown in the Venn diagram (Figure 2). This new observation may lead to a potential answer to the question of how apparent continuity is created at neural level. The fact that discontinuous mental imagery in another modality (auditory modality) shares a common neural underpinning with continuous mental imagery in visual modality has actually provided better evidence for understanding the temporal continuity of musical composition since otherwise it will be difficult to exclude the modality contribution if discontinuous and continuous mental imagery are compared in the same modality. Thus, we speculate that the LMTG with its common neural pattern in the two experimental conditions could be a signature of temporal continuity in musical imagery, which serves the function of filling the temporal gap of sequential musical representations.

Is there any other evidence available for supporting such a speculation mentioned above? It has for instance been shown that the LMTG is an important region for the semantic component in language (Acheson & Hagoort, 2013). As semantics is not limited in typical situations to separate time windows, it can be assumed that apparent continuity (or the "illusion of continuity") may be a consequence of the semantic component in language with its longer time constant. What an utterance means within one time window is carried over to the meaning in the next time window; thus, a long-term narrative can provide in a top manner a temporal glue. Supporting evidence comes from some lesion studies; injuries in this region may cause alexia and agraphia for *kanji* characters used in Japanese writing, thus destroying the capacity to extract the meaning (Sakurai *et al.*, 2008). The LMTG is also activated during poem composition (Liu *et al.*, 2012), and it has been shown to be part of a conceptual/semantic network (Wei *et al.*, 2012). Another study shows that theta-band oscillations in the LMTG imitate novel word consolidation (Bakker-Marshall *et al.*, 2018). In some schizophrenic patients it has been reported that auditory verbal hallucinations are related to cortical thinning in the LMTG (Cui *et al.*, 2018). Furthermore, an increased LMTG was found in both low- and high-function autism in another study (Jia *et al.*, 2018). Although these different results can be considered only as circumstantial evidence in favor of our speculation and not as a proof, they at least do not contradict it.

In conclusion, auditory and visual imagery of the same musical composition are characterized by different neural patterns, but meanwhile share a common activation of the LMTG. On the basis of previous studies (Cui *et al.*, 2018; Jia *et al.*, 2018; Liu *et al.*, 2012) together with the present experiment, the LMTG might be involved in creating temporal continuity of mental imagery. If one considers some general features shared by language and music from a cultural and evolutionary perspective (Jackendoff, 2009), our experiment also suggests similarities in composing music and linguistic artifacts. However, it has to be further investigated whether the LMTG and its network are indeed causally linked to neural mechanisms that create continuity of conscious information processing.

Acknowledgments

Financial support for this study has come from the Hanns Seidel Foundation Munich, Germany, to M.I., the Chinese Scholarship Council to T.Y., the Zonta-Club II, Munich, Germany, to A.F., and the National Natural Science Foundation of China (Projects 31771213 and 31371018) to Y.B.

References

Acheson, D. J. & Hagoort, P. (2013). Stimulating the brain's language network: Syntactic ambiguity resolution after TMS to the inferior frontal gyrus and middle temporal gyrus. *Journal of Cognitive Neuroscience*, 25(10), 1664–1677. https://doi.org/10.1162/jocn_a_00430.

Bakker-Marshall, I., Takashima, A., Schoffelen, J. M., Van Hell, J. G., Janzen, G., & McQueen, J. M. (2018). Theta-band oscillations in the middle temporal gyrus reflect novel word consolidation. *Journal of Cognitive Neuroscience*, 30(5), 621–633. https://doi.org/10.1162/jocn_a_01240.

Bailes, F. (2019). Empirical musical imagery beyond the "Mind's Ear". In M. M. Grimshaw-Aagaard, M. Walther-Hansen, & M. Knakkergaard (eds.) *The Oxford Handbook of Sound and Imagination*, pp. 445–466. Oxford University Press.

Bao, Y. & Pöppel, E. (2012). Anthropological universals and cultural specifics: Conceptual and methodological challenges in cultural neuroscience. *Neuroscience & Biobehavioral Reviews*, 36, 2143–2146. https://doi.org/10.1016/j.neubiorev.2012.06.008.

Bao, Y., Pöppel, E., Wang, L., Lin, X., Yang, T., Avram, M., Blautzik, J., Paolini, M., Silveira, S., Vedder, A., Zaytseva, Y., & Zhou, B. (2015). Synchronization as a biological, psychological and social mechanism to create common time: A theoretical frame and a single case study. *PsyCh Journal*, 4, 243–254. https://doi.org/10.1002/pchj.119.

Bao, Y., Pöppel, E., & Zaytseva, Y. (2017a). Editorial: Single case studies as a prime example for exploratory research. *PsyCh Journal*, 6, 107–109. https://doi.org/10.1002/pchj.176.

Bao, Y., von Stosch, A., Park, M., & Pöppel, E. (2017b). Complementarity as generative principle: A thought pattern for aesthetic appreciations and cognitive appraisals in general. *Frontiers in Psychology*, 8, 727. https://doi.org/10.3389/fpsyg.2017.00727.

Bao, Y., Yang, T., Lin, X., Fang, Y., Wang, Y., Pöppel, E., & Lei, Q. (2016a). Aesthetic preferences for eastern and western traditional visual art: Identify matters. *Frontiers in Psychology*, 7, 1596. https://doi.org/10.3389/fpsyg.2016.01596.

Bao, Y., Yang, T., Lin, X., & Pöppel, E. (2016b). Donders revisited: Discrete or continuous temporal processing underlying reaction time distributions? *PsyCh Journal*, 5, 177–179. https://doi.org/10.1002/pchj.142.

Bao, Y., Yang, T., Zhang, J., Zhang, J., Lin, X., Paolini, M., Pöppel, E., & Silveira, S. (2017c). The "third abstraction" of the Chinese artist LaoZhu: Neural and behavioral indicators of aesthetic appreciation. *PsyCh Journal*, 6, 110–119. https://doi.org/10.1002/pchj.167.

Cotter, K. N. (2019). Mental control in musical imagery: A dual component model. *Frontiers in Psychology*, 10, 1904. https://doi.org/10.3389/fpsyg.2019.01904.

Cui, Y., Liu, B., Song, M., Lipnicki, D. M., Li, J., Xie, S., …, Wang, H. (2018). Auditory verbal hallucinations are related to cortical thinning in the left middle temporal gyrus of patients with schizophrenia. *Psychological Medicine*, 48(1), 115–122. https://doi.org/10.1017/S0033291717001520.

Eriksen, C. W. & Collins, J. F. (1968). Sensory traces versus the psychological moment in the temporal organization of form. *Journal of Experimental Psychology,* 77(3), 376–382. https://doi.org/10.1037/h0025931.

Flyvbjerg, B. (2006). Five misunderstandings about case-study research. *Qualitative Inquiry*, 12(2), 219–245.

Griffiths, M. (2000). Internet addiction-time to be taken seriously? *Addiction Research*, 8(5), 413–418. https://doi.org/10.3109/16066350009005587.

Halpern, A. (2001). *Cerebral Substrates of Musical Imagery*. Lewisburg, PA: Bucknell University Press.

Halpern, A. R., Bartlett, J. C., & Dowling, W. J. (1995). Aging and experience in the recognition of musical transpositions. *Psychology and Aging,* 10(3), 325–342. https://doi.org/10.1037/0882-7974.10.3.325.

Halpern, A. R. & Zatorre, R. J. (1999). When that tune runs through your head: A PET investigation of auditory imagery for familiar melodies. *Cerebral Cortex*, 9(7), 697–704. https://doi.org/10.1093/cercor/9.7.697.

Halpern, A. R., Zatorre, R. J., Bouffard, M., & Johnson, J. A. (2004). Behavioral and neural correlates of perceived and imagined musical timbre. *Neuropsychologia*, 42(9), 1281–1292. https://doi.org/10.1016/j.Neuropsychologia.2003.12.017.

Jackendoff, R. (2009). Parallels and non-parallels between language and music. *Music Perception: An Interdisciplinary Journal*, 26(3), 195–204. https://doi.org/10.1525/mp.2009.26.3.195.

James, W. (1890/2013). *The Principles of Psychology*, Vol. 1. Read Books Ltd.

Jia, C., Hu, X., Guo, K., Yang, P., SItu, M., & Huang, Y. (2018). Increased left inferior temporal gyrus was found in both low function autism and high function autism. *Frontiers in Psychiatry*, 9, 542. https://doi.org/10.3389/fpsyt.2018.00542.

Kraemer, D. J., Macrae, C. N., Green, A. E., & Kelley, W. M. (2005). Musical imagery: Sound of silence activates auditory cortex. *Nature*, 434(7030), 158. https://doi.org/10.1038/434158a.

Liu, S., Chow, H. M., Xu, Y., Erkkinen, M. G., Swett, K. E., Eagle, M. W., Rizik-Baer, D. A., & Braun, A. R. (2012). Neural correlates of lyrical improvisation: An fMRI study of freestyle rap. *Scientific Reports*, 2, 834. https://doi.org/10.1038/srep00834.

McGuire, P. K., Silbersweig, D. A., Wright, I., Murray, R. M., Frackowiak, R. S. J., & Frith, C. D. (1996). The neural correlates of inner speech and auditory verbal imagery in schizophrenia: Relationship to auditory verbal hallucinations. *British Journal of Psychiatry*, 169(2), 148–159. https://doi.org/10.1192/bjp.169.2.148.

Pöppel, E. (1970). Excitability cycles in central intermittency. *Psychologische Forschung*, 34(1), 1–9. https://doi.org/10.1007/BF00422860.

Pöppel, E. (1971). Oscillations as possible basis for time perception. *Studium Generale*, 24, 85–107; J.T. Fraser *et al.* (1972). *The Study of Time*, pp. 219–241. Berlin: Springer.

Pöppel, E. (1989). The measurement of music and the cerebral clock: A new theory. *Leonardo*, 22, 83–89. https://doi.org/10.2307/1575145.

Pöppel, E. (1997). A hierarchical model of temporal perception. *Trends in Cognitive Sciences*, 1, 56–61. https://doi.org/10.1016/S1364-6613(97)01008-5.

Pöppel, E. & Bao, Y. (2018). The power of single cases: Examples from the early visual pathway and from visual art. In B. Gulyas & J. Vasbinder (eds.), *Cultural Patterns and Neurocognitive Circuits II — East-West Connections*, pp. 33–62. World Scientific Press.

Pylyshyn, Z. W. (1973). What the mind's eye tells the mind's brain: A critique of mental imagery. *Psychological Bulletin*, 80(1), 1–24. https://doi.org/10.1037/h0034650.

Sakurai, Y., Mimura, I., & Mannen, T. (2008). Agraphia for kanji resulting from a left posterior middle temporal gyrus lesion. *Behavioural Neurology*, 19(3), 93–106. https://doi.org/10.1155/2008/393912.

Schürmann, M., Raij, T., Fujiki, N., & Hari, R. (2002). Mind's ear in a musician: Where and when in the brain. *NeuroImage*, 16(2), 434–440. https://doi.org/10.1006/nimg.2002.1098.

von Trott Zu Solz, J., Paolini, M., & Silveira, S. (2017). Does touch inhibit visual imagery? A case study on acquired blindness. *PsyCh Journal*, 6, 139–140. https://doi.org/10.1002/pchj.162.

Walter, B. (1957). *Von der Musik und vom Musizieren* (About Music and To Make Music). Germany: S. Fischer Verlag.

Wang, L., Lin, X., Zhou, B., Pöppel, E., & Bao, Y. (2015). Subjective present: A window of temporal integration indexed by mismatch negativity. *Cognitive Processing*, 16, 131–135. https://doi.org/10.1007/s10339-015-0687-8.

Watson, J. B. (1913). Psychology as the behaviorist views it. *Psychological Review*, 20(2), 158–177.

Weber, R. J. & Brown, S. (1986). Musical imagery. *Music Perception*, 3(4), 411–426.

Wei, T., Liang, X., He, Y., Zang, Y., Han, Z., Caramazza, A., & Bi, Y. (2012). Predicting conceptual processing capacity from spontaneous neuronal activity of the left middle temporal gyrus. *Journal of Neuroscience*, 32(2), 481–489. https://doi.org/10.1097/00001756-200110080-00013.

Yoo, S. S., Lee, C. U., & Choi, B. G. (2001). *Human Brain Mapping* of auditory imagery: Event-related functional MRI study. *Neuroreport*, 12(14), 3045–3049. https://doi.org/10.1097/00001756-200110080-00013.

Yu, X. & Bao, Y. (2020). The three second time window in poems and language processing in general: Complementarity of discrete timing and temporal continuity. *PsyCh Journal*, 9, 429–443. https://doi.org/10.1002/pchj.390.

Zatorre, R. J. & Halpern, A. R. (1993). Effect of unilateral temporal-lobe excision on perception and imagery of songs. *Neuropsychologia*, 31(3), 221–232. https://doi.org/10.1016/0028-3932(93)90086-f.

Zatorre, R. J., Meyer, E., Gjedde, A., & Evans, A. C. (1996). PET studies of phonetic processing of speech: Review, replication, and reanalysis. *Cerebral Cortex*, 6(1), 21–30. https://doi.org/10.1093/cercor/6.1.21.

Zhang, J., Yang, T., Bao, Y., Li, H., Pöppel, E., & Silveira, S. (2017). Sadness and happiness are amplified in solitary listening to music. *Cognitive Processing*, 19(1), 133–139. https://doi.org/10.1007/s10339-017-0832-7.

Zhao, C., Zhang, D., & Bao, Y. (2018). A time window of 3 seconds in the aesthetic appreciation of poems. *PsyCh Journal*, 7, 51–52.

Zhou, B., Pöppel, E., & Bao, Y. (2014). In the jungle of time: The concept of identity as an escape. *Frontiers in Psychology*, 5, 844. https://doi.org/10.3389/fpsyg.2014.00844.

Zhou, B., Pöppel, E., Wang, L., Yang, T., Zaytseva, Y., & Bao, Y. (2016). Seeing without knowing: Operational principles along the early visual pathway. *PsyCh Journal*, 5(3), 145–160. https://doi.org/10.1002/pchj.141.

www.ingramcontent.com/pod-product-compliance
Lightning Source LLC
LaVergne TN
LVHW010323220926
848484LV00004BB/835